普通高等教育双语系列教材

Hydraulic and Pneumatic Transmission

(English-Chinese Bilingual)

液压与气压传动

（英汉双语）

第2版

陈淑梅 编著

机 械 工 业 出 版 社

本书分为两篇（Part One and Part Two），共12章。1~9章为液压传动，10~12章为气压传动。与同类型教材相比不同点在于：全书采用英汉双语形式编写与排版，可作为高等院校机械设计制造及自动化、机械电子工程（机电一体化）、车辆工程、冶金、水利工程、材料成形及控制工程、模具设计与制造、轻工机械等机械工程类专业的英汉双语教材。

同时配套有英汉双语多媒体电子教案（教师版）和各章节的练习题（英文版），可供教师在课堂授课和学生做课外练习题时与本教材配套使用。授课教师可根据书末的“信息反馈表”向主编索取。

本书还附有部分液压与气压传动专业词汇（英汉双语），供学生在自学本教材时更快地查阅并掌握专业术语。同时本教材也可以作为机械工程类研究生的专业英语辅助教材或阅读材料。

图书在版编目（CIP）数据

液压与气压传动：英汉双语/陈淑梅编著．—2版．—北京：机械工业出版社，2013.9（2024.8重印）

ISBN 978-7-111-43311-8

Ⅰ.①液…　Ⅱ.①陈…　Ⅲ.①液压传动—双语教学—高等学校—教材—英、汉②气压传动—双语教学—高等学校—教材—英、汉　Ⅳ.①TH137②TH138

中国版本图书馆CIP数据核字（2013）第158425号

机械工业出版社（北京市百万庄大街22号　邮政编码100037）
策划编辑：余　皞　责任编辑：余　皞　李　超
版式设计：霍永明　责任校对：丁丽丽
封面设计：张　静　责任印制：郜　敏
中煤（北京）印务有限公司印刷
2024年8月第2版第8次印刷
184mm×260mm·23.25印张·582千字
标准书号：ISBN 978-7-111-43311-8
定价：69.80元

电话服务
客服电话：010-88361066
010-88379833
010-68326294

网络服务
机　工　官　网：www.cmpbook.com
机　工　官　博：weibo.com/cmp1952
金　书　网：www.golden-book.com
机工教育服务网：www.cmpedu.com

Preface

Mechanical engineering and automation technology are advancing rapidly in our country. Therefore it is imperative to cultivate highly qualified application-oriented talents to meet the current needs, which requires new development in the higher education.

For filling in the widening gap between social needs and status quo of higher education, the Auditing and Editing Committee for "Planned Textbooks of Higher Education" is set up by the Machinery Electronics Committee, which subordinates to the China Machinery Education Association, and the China Machine Press, which aims to publish a series of practical textbooks based on the principle of "attaching more importance to basic theories and engineering application rather than the process of deduction". The main features of the series are:

(1) Scientific objective. The series is targeted for training all-round application-oriented talents and distinguished from other high-school teaching materials. It has high technicality emphasizing on engineering application and innovation.

(2) Selective categories. The series includes mechanics, hydraulic and pneumatic transmission, cartography, design, numerical control, practical training, materials, bilingual teaching, etc. tank are available for teachers to choose from.

(3) Multimedia teaching. Each volume of the series has related courseware and accessorial books and a discussion room on the Internet.

The China Machine Press, one of the earliest large-scale publishing houses and high-education textbooks bases, is a prestigious authority in publishing books in mechanical and electrical engineering. With concerted efforts from all circles this series of books will contribute greatly to bringing up high-qualified engineers for our country.

Peigen Li,
Academician
Director of the Machinery Electronics Teaching Committee
China Machinery Education Association
Huazhong University of Science and Technology

序

为了适应我国机械工程及自动化专业迅速发展的需要，培养大批素质高、应用能力与实践能力强的应用综合型人才已成为当务之急。这同时对高等教育的办学理念、体制、模式、机制和人才培养等方面提出了全新的要求。

为了打通新形势下高等教育和社会需求之间的瓶颈，本着“重基本理论、基本概念，淡化过程推导，突出工程应用”的原则，组织教材编写工作，并力求使本套教材突出以下特点：

（1）定位科学。本套教材主要面向应用的综合型人才的培养。既不同于培养研究型人才的教材，也不同于一般应用型本科的教材；在保持高学术水准的基础上，突出工程应用，强调创新思维。

（2）品种齐全。这套教材设有“力学”、“液压与气压传动”、“制图”、“设计”、“数控”、“实训”、“材料”、“双语”等模块，方便学校选用。

（3）立体化程度高。每本教材均要求配备CAI课件和相关的教辅材料，并在网站上为本套教材开设研讨专栏。

机械工业出版社是我国成立最早、规模最大的科技出版社之一，是国家级优秀出版社，是国家高等教育的教材出版基地之一，在机电类教材出版领域具有很高的地位。相信这套教材在中国机械工业教育协会机电类学科委员会和机械工业出版社的精心组织下，通过全国几十所学校老师的仔细、认真的编写，一定能够为我国高等教育应用综合型人才的培养提供更好用、更适用的教材。

教育部机械工程及自动化专业分教学指导委员会主任
中国机械工业教育协会高等学校机械工程及自动化学科教学委员会主任
李培根院士
于华中科技大学

Foreword to The Second Edition

Motivation of this book. The higher education in China is now entering a new era of reform which encourages bilingual teaching in foundation courses, major foundation courses and specialized courses. The main objective of the reform is to promote the internationalization of higher education and to cultivate application-oriented talents with creativity and quality education. Bilingual teaching is thus becoming one of the ways to link our higher education to the world-facing various challenges of the new century and assure high-quality and comprehensive elitists. The reform in bilingual teaching is a strategic choice for our higher education in the 21st century. This book thus intends to provide a solid base for bilingual teaching and to meet the immediate needs of our higher education reform for interfacing with the world.

Arrangement of the material. This book was written following the *Basic Requirements in Teaching Descriptive Hydraulic and Pneumatic Course Developed* by the Mechanical Engineering Teaching Supervisory Committee for Higher Engineering Education. It is introduced based on the current practice and demands of education reform in teaching power engineering. It targets at fostering application-oriented talents for the 21st century. The development of this book is guided by advanced mechanical engineering techniques. On one hand, It is focused on learning centered activities and aims at quality education. The contents of this book cover selected topics simplified and extracted for a wide range of readers. On the other hand, this book was written based on the characteristics of foundation courses. It inherits the essence of a traditional course content and integrates extensive teaching experience of the authors in the past years. It focuses on the present education needs in fostering talents with comprehensive quality education and creative thinking. This book exhibits following characteristics:

(1) Bilingual teaching The bilingual edition represents the achievements of higher education reform in teaching and curriculum development for the 21st century. It emphasizes on the training of application-oriented talents in the international context and quality education in basic engineering subjects.

(2) Multimedia teaching The accompanied CAI multimedia with case-based teaching materials, which animates and covers the complete contents of this book published simultaneously. Various powerful softwares have been used to make the videos which demonstrate the dynamic, static and continuous actions. For example, pressure characteristics of hydraulic oil display clearly the oil flowing through the inner configuration of elements and system circuits, which represent clearly the operating principle of elements and flowing dynamic experience for system circuits. Therefore the two main features of pressure and motion transmissions in hydraulic/pneumatic transmission system are represented visually. The large number of engineering models, animations, various function of multimedia, and intuitive presentation of live examples with visual virtual software effects are used to

stimulate students, interest in learning, which will achieve better teaching effect.

(3) English exercises This book has been prepared to provide students and users with well-connected English-Chinese bilingual exercises. It will enable students to think and solve problems in the English context. It covers various simple and complex questions that are suited for a wide range of students and enables students to grip professional vocabulary, ensuring students of international application on the subject of hydraulic and pneumatic transmission.

(4) Proving a re-edited CAI version The electric CAI multimedia version prepared using Powerpoint is an open course material with a large benefit that it could be easily modified (insert, cancel and revision) by teachers to meet the needs of their own teaching.

Editors for this book. The editor of this book is Professor Chen Shumei from Fuzhou University, Fujian Province of China. The participants are: Xu Xiaoqin, Chen Xiaomei, Yang Kunquan, Chen Yuangui and Chen Tianyan. This book is audited by Professor Qin Ning from University of Sheffield, Great Britain and by Professor Lin Xinzhang from College of Mechanical Engineering of Fuzhou University, China. The authors would sincerely appreciate and like specially to thank them for providing a thorough review and giving many invaluable suggestions. In addition, multimedia videos are also accomplished by Tian Lichao, Wang Lingwei, Tang Xiaoteng and Xu Qiaodan.

Readership of this book. Hydraulic and pneumatic transmission is one of the key technical foundation courses with wide application fields, such as mechanical design and manufacture, engineering machines, mining, metallurgy, light industry machines and so on. The authors thus feel that this book is suitable as a bilingual teaching material for mechanical engineering students such as mechatronics engineering, material fabricating molding and control engineering, mould design and manufacture, aerospace engineering, shipping machines and light industry. At the same time, it could be reading and supplementary materials for postgraduate students who are interested in the subject in the later years of their course of study. However a certain amount of background knowledge of mathematics and applied mechanics is required.

Acknowledgments. His book was written with the reference to some excellent books and teaching materials as shown in the section of references. The authors highly appreciate, especially Associate Professors Xu Fuling and Chen Yaoming of Huazhong University of Science and Technology and Professor Zuo Jianmin from Nanjing Institute of Technology. Thanks go to the China Machine Press, College of Mechanical Engineering and Educational Administration of Fuzhou University. Thanks also go to many others who helped in the course of the publication of this book. Special support from Fuzhou University through Teaching Material Development Funds and Fujian Province of China, Outstanding Courses Construction Funds are all gratefully acknowledged.

In addition, it is unavoidable that this book may still contain mistakes. All critics and corrections from experts and readers are welcome.

The authors
At Fuzhou University

第2版前言

当前，我国的高等教育进入了新一轮的改革阶段，即大力开展基础课、专业基础课和专业课的双语教学，加速推进我国高等教育的国际化和培养大批素质高、应用能力与实践能力强的综合型人才。因此，双语教学是我国高等教育与国际接轨的重要途径之一，是培养大批素质高、应用能力与实践能力强的综合型人才的有力保证，是迎接新世纪挑战和教育改革发展的必然趋势，也是中国高等教育在21世纪必须作出的战略选择。

本教材是依照高等学校工科“液压与气压传动”教学指导委员会制订的课程教学基本要求，结合近年来本课程的教学改革特点，围绕培养面向21世纪的素质高、应用能力与实践能力强的综合型人才进行编写的。一方面，本教材将面向现代机械工程技术，并紧紧围绕以“学”为中心、以“素质提高”为目的的指导思想，力求简明扼要、质量上乘、覆盖面广。另一方面，本教材将针对专业基础课程的特点，继承传统内容的精华，融入编者多年积累的教学经验，着眼于新时期对人才培养的要求，以加强对学生综合素质及创新能力的培养为出发点。本教材具有如下特点：

（1）突出英汉双语的特色　本教材采用英汉对照编排形式，体现高等教育面向21世纪教学内容和课程体系改革的成果，立足于与国际接轨的综合型人才的培养，重视基本工程素质教育。

（2）多媒体教学　与本教材配套的多媒体CAI电子教案光盘与本书同时出版，更加增强了本教材的感观效果。该CAI课件覆盖教材的全部内容，运用各种软件的功能，灵活地将视频的动态、间断和连续动作等表现手法进行有机整合，结合了液/气压传动系统回路压力的变化特性，将液压油（流体介质）在各种液压元件及各种控制回路中的流动形式可视化地描述，从而剖析了各种液压元件的结构特征和工作原理以及各种复杂液压回路的实现过程。这种采用大量反映实物模型的动画演示可激发学生的学习兴趣，将会取得较好的教学效果。

（3）英语练习题　练习题以英文形式出现，进一步培养学生在专业英语环境下进行本课程的思考与练习。练习题型博采众长，由浅入深，覆盖面宽，难度适宜，兼顾典型性和通用性，为培养学生掌握液压与气压传动专业英语词汇，进一步从事国际化的液压与气压传动设计的能力培养打下基础。

（4）配套的CAI课件可编辑性强　该课件采用较易掌握的Powerpoint工具软件编制，为开放式课件。其最大的好处是：可由任课教师根据课程需要及教学习惯，方便地自行增加、删减或重组有关内容，或按自己的风格和特色进行改编，以满足个性化教学的要求。授课教师可根据书末的“信息反馈表”向主编索取。

本书由福州大学机械工程及自动化学院陈淑梅教授编著，参编者有：许晓勤、陈小梅、杨坤全、陈元贵和陈天炎。全书由英国谢菲尔德（Sheffield）大学机械工程系覃宁教授和福州大学机械工程及自动化学院林信彰教授审核。编者对两位专家、教授所给予的大力支持和提出的宝贵建议深表感谢。另外参加本教材多媒体动画制作的还有田立超、王玲玮、唐晓腾，许乔丹。

液压与气压传动课程是机械工程的一门专业基础主干课程，具有很强的工程应用背景（如机械设计与制造、工程机械、矿山机械、冶金机械和轻工机械等）。本书可作为高等院校机械设计制造及自动化、机械电子工程（机电一体化）、车辆工程、材料成形及控制工程、模具设计与制造、轻工机械等机械工程类专业的英汉双语教材，同时也可以作为机械工程类研究生的专业英语辅助教材或阅读材料，具有广泛的适用性。学习者在学习本教材之前需要掌握一定的高等数学和应用流体力学的基础知识。

在本书的编写过程中我们参考和引用了国内外优秀的同类参考书籍，详见本教材的参考文献。在此特别对华中科技大学许福玲和陈尧明、左健民以及其他所有参考书籍的作者表示衷心的感谢。在本教材的编写过程中还得到了机械工业出版社、福州大学机械工程及自动化学院和教务处等有关部门的大力支持，并得到了福州大学教材建设基金和福建省精品课程建设基金的资助，在此表示由衷的感谢；同时对所有关心和帮助本书出版的各界人士表示真诚的谢意！

由于编者水平有限，书中难免有错误和疏漏，敬请各位专家及广大读者批评指正。

编者

于福州大学

CONTENTS

目　录

Part One

Hydraulic Transmission

第1篇

液压传动

Chapter 1 Introduction to Hydraulic and Pneumatic Transmission

1.1 Study on Hydraulic and Pneumatic Transmission

Hydraulic and pneumatic transmissions are a discipline that is based on fluid medium energy of compressive fluid (pressure oil or compressive air) to accomplish mechanical transmission and automatic control. Hydraulic and pneumatic transmissions are similar in operating principle and control means. Special sub-circuits built by various hydraulic (or pneumatic) components are used to build up particular hydraulic/pneumatic transmission systems to realize energy transfer and control. Having a sound grip of knowledge in physical performances and static/dynamic characters of fluid, operating principle and construct several hydraulic element and sub-circuit and hydraulic/pneumatic system designs are the best way to study power (hydraulic/pneumatic) transmission and control technology.

1.2 Operating Principles of Hydraulic Transmission

There is an important physics law, i. e. ,the Pascal's law in the hydraulic pressure. In the seventeenth century (1620) , Blaise Pascal, a French physicist formulated the fundamental law on which modern hydraulics is based. Pascal's law states "pressure exerted on a confined liquid is transmitted undiminished in all directions and acts with equal force on all equal areas".

Operating principle(such as hydraulic jack)is shown in the Fig. 1-1.

In Fig. 1-1, the jack contains a hand-operated pump, which draws oil from a reservoir. When one pulls up the lever 2, the small piston is moved up,the volume enlarges and becomes vacuum in the volume below in the small actuator 1 and non-return valve 7 is shut. Oil in the reservoir 6 is pumped past a non-return valve 8 into the bottom of the linear actuator under atmospheric pressure. When you apply force to the lever and press down the lever, it compresses the oil in the small actuator, which increases the oil force. Here, non-return valve 8 is shut;the oil in the bottom small actuator is pumped past a non-return valve 7 into the bottom of the big actuator 4;the oil flowing into the big actuator forces the big piston to move heavy load. By pulling up and pushing down the lever alternately the heavy load 3 will be moved up continuously. If lever is stopped working, the nonreturn valve 7 will be shut and the big piston together with heavy load will be locked on the position of moving up. If the check valve 5 is opened, the oil on the bottom of big piston will flow back into the reservoir, which makes the big piston move down to its original position under deadweight.

From this operating principle of a hydraulic jack, it has been stated that the processing of draw-

第 1 章 液压与气压传动概论

1.1 液压与气压传动的研究内容

液压与气压传动是研究以有压流体（压力油或压缩空气）为能源介质，来实现各种机械传动和自动控制的学科。液压传动与气压传动实现传动和控制的方法基本相同，它们都是利用各种控制元件组成能够实现特定功能的基本回路，再由若干基本回路有机组合成能完成一定控制功能的传动系统，从而进行能量的传递、转换与控制。因此，要研究液压与气压传动及其控制技术，首先要了解传动介质的基本物理性能及其静力学、动力学特性；还要了解组成系统的各类液压与气动元件的结构、工作原理、工作性能以及由这些元件所组成的各种控制回路的性能和特点，并在此基础上进行液压与气压传动控制系统的设计。

1.2 液压传动的工作原理

物理学中有一条著名的定律，称为帕斯卡定律。这个定律指出："在密闭容器内，施加于静止液体上的压力将以等值同时传到液体中各点"。这一基本原理，是由法国物理学家伯雷斯·帕斯卡在 1620 年发现的，它奠定了液压传动科学的理论基础。

液压传动的工作原理以图 1-1 所示液压千斤顶为例来说明。

当向上抬起杠杆 2 时，小液压缸 1 的小活塞向上运动，下腔容积增大形成局部真空，单向阀 7 关闭，油箱 6 中的油液在大气压作用下经吸油管顶开单向阀 8 进入小液压缸下腔。当向下压杠杆时，小液压缸下腔容积减小，油液受挤压，压力升高，关闭单向阀 8，顶开单向阀 7，油液经排油管进入大液压缸 4 的下腔，推动大活塞上移顶起重物 3。如此不断上下扳动杠杆，则不断有油液进入大液压缸下腔，使重物逐渐举升。如杠杆停止动作，大液压缸下腔油液压力将使单向阀 7 关闭，大活塞连同重物一起被自锁不动，停止在举升位置。如打开截止阀 5，大液压缸下腔通油箱，大活塞在自重作用下向下移，恢复到原始位置。

从液压千斤顶的工作原理中可以看出：小液压缸 1 与单向阀 7、8 一起完成吸油与排油，将杠杆的机械能转换为油液的压力能输出，称为（手动）液压泵。大液压缸 4 将油液的压力能转换为机械能输出，抬起重物，称为（举升）液压缸。在这里大、小液压缸组成了最简单的液压传动系统，实现了力和运动的传递。

实现力和运动的传递，其工作介质——液压油应具备两个重要的性质：其一，几乎是不可压缩的；其二，具有力的放大作用。

1. 力的传递

如图 1-1 所示，设大液压缸活塞面积为 A_2，作用在其活塞上的负载力为 F_2，则该力在液压缸中所产生的液体压力为

ing oil into and out from small actuator 1 is performed by the small actuator 1, non-return valves 7 and 8, which is transforming mechanical energy to hydraulic pressure energy, and this setup is called hand-operated hydraulic pump. The big piston and big actuator 4 called lifting hydraulic actuator is transforming the hydraulic pressure energy to mechanical energy and moving up the heavy load. Both small and big hydraulic actuators constitute the simplest hydraulic transmission system, which accomplishs the two transmissions: power transmission and motion transfer.

To accomplish the power transmission and motion transfer, there are two properties of oil that constitute the basic elements of hydraulics. The first property is that oil is nearly incompressible. The second oil may be used to multiply forces.

1. Power transmission

Let's note the piston area of actuator 4 as A_2, the total thrust on it as F_2, then the hydraulic pressure in the actuator p_2

$$p_2 = \frac{F_2}{A_2} \tag{1-1}$$

According to Pascal's law, here that is $p_1 = p_2 = p$, the oil displacement pressure is called the hydraulic system pressure.

It is important here to overcome the heavy load for moving the actuator, and taking the thrust on the small piston of pump as F_1

$$F_1 = pA_1 = p_1 A_1 = p_2 A_1 = \frac{A_1}{A_2} F_2 \tag{1-2}$$

Where A_1 is the piston area of pump (small actuator).

The heavier the load is, the higher the hydraulic pressure is needed under a given A_1 and A_2, which illuminates the close relationship between the hydraulic system and outside load. Thus the moving piston can do work on an outside load.

We can get the first important law here that the working pressures in the hydraulic system depend on the outside load.

2. Motion transfer

The volume of displacement is equal to the volume drawn into it when the compressibility, leaking and deform action are neglected . Let S_1, S_2 note respectively piston moving positions in the pump and actuator, and then

$$S_1 A_1 = S_2 A_2 \tag{1-3}$$

Dividing by the moving time t on the two sides in the above formula, we have

$$q_1 = v_1 A_1 = v_2 A_2 = q_2 \tag{1-4}$$

Where v_1 and v_2 are noted respectively the average speed of hydraulic pump and hydraulic actuator; q_1 is the oil outflow rate from the pump and q_2 is the oil inflow rate of the actuator.

It is shown that the motion by the confined volume equals the change in hydraulic system, i. e. , moving speed of piston is adjusted by changing the flow rate into actuator. And we can get the second important law here that the motion speed of piston in actuator depends on the flow-in rate, and is independent of the outside load.

The discussion above states that the liquid pressure is related to the outside load, and motion

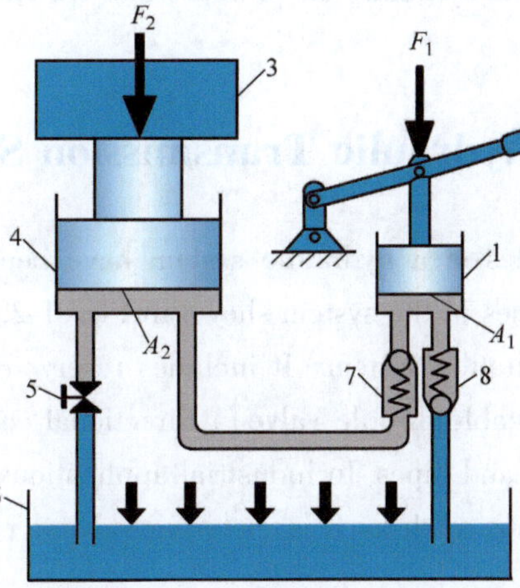

Fig. 1-1　Operating principle of hydraulic jack（液压千斤顶工作原理图）

1—Small actuator（小液压缸）　2—Lever（杠杆）　3—Heavy load（重物）

4—Big actuator（大液压缸）　5—Check valve（截止阀）　6—Reservoir（油箱）

7，8—Non-return valve（单向阀）

$$p_2 = \frac{F_2}{A_2} \tag{1-1}$$

根据帕斯卡原理，液压泵的排油压力 p_1 应等于液压缸中的液体压力，即 $p_1 = p_2 = p$，液压泵的排油压力又称为系统压力。

为了克服负载力使液压缸活塞运动，作用在液压泵活塞上的作用力 F_1 应为

$$F_1 = pA_1 = p_1A_1 = p_2A_1 = \frac{A_1}{A_2}F_2 \tag{1-2}$$

式中，A_1 为小液压缸活塞面积。

在 A_1、A_2 一定时，负载力 F_2 越大，系统中的压力 p 也就越高，所需的作用力 F_1 也就越大，即系统压力与外负载密切相关。

由此得出液压传动工作原理的第一个重要特征：液压传动中工作压力取决于外负载。

2. 运动的传递

如果不考虑液体的可压缩性、漏损和缸体、管路的变形，液压泵排出的液体体积必然等于进入液压缸的液体体积。设液压泵活塞位移为 S_1，液压缸活塞位移为 S_2，则有

$$S_1A_1 = S_2A_2 \tag{1-3}$$

上式两边同除以运动时间 t 得

$$q_1 = v_1A_1 = v_2A_2 = q_2 \tag{1-4}$$

式中，v_1、v_2 分别为液压泵活塞和液压缸活塞的平均运动速度；q_1、q_2 分别为液压泵输出的平均流量和输入液压缸的平均流量。

由上述可见，液压传动是靠密闭工作容积变化相等的原则实现运动（速度和位移）传递的。调节进入液压缸的流量即可调节活塞的运动速度，由此得出液压传动工作原理的第二个重要特征：活塞的运动速度只取决于输入流量的大小，而与外负载无关。

从上面的讨论还可以看出，与外负载力相对应的流体参数是流体压力，与运动速度相对应的流体参数是流体流量。因此，压力和流量是液压传动中两个最基本的参数。

speed to the liquid flow rate; thus hydraulic oil pressure and the flow rate are the two main parameters in the hydraulic systems.

1.3 Composition of Hydraulic Transmission System

Now that the basic principles of a hydraulic system have been described, let us look at the function each component performes in the system shown in Fig. 1-2. It reviews the operation of such a system, which is a simple hydraulic system. It includes reservoir 8, filter 7, hydraulic pump 1, flow rate control device 2 (adjustable throttle valve), directional control valve 3, hydraulic cylinder 4, working load 5, relief valve 6 and pipes. In industrial applications there are many components that may be added to the basic system, such as flow metering valves, check valves, and heat exchangers.

First, the electric motor is started to drive hydraulic pump 1. Once hydraulic pump 1 is in operation, it brings into play three direct actions on the fluid:

1) A partial vacuum is created in hydraulic pump 1 inlet pipe.

2) Atmospheric pressure acting on the surface of the fluid in the reservoir forces the fluid into the inlet pipe to fill the partial vacuum.

3) The mechanical action of the pump forces the fluid from the inlet to the outlet of the pump and through the hydraulic system.

When hydraulic pump 1 is operating driven by a motor, the hydraulic oil flows from reservoir 8 pass filter 7 into the inlet of pump. Fluid pressure is affected by the design of the closed-center directional control valve. For the system to perform work, the directional control valve 3 must be connected to a cylinder or motor. Using a directional control valve, there will usually be high pressure on the inlet side of the control valve and no pressure value of any significance on the outlet side, which is connected to the cylinder. When the control valve is shifted to either extreme position (left position), the oil flows via flow rate control device 2 (adjustable throttle valve), directional control valve 3 into the left chamber of the hydraulic cylinder 4 to move the piston in the cylinder to the right-direction; and the oil in the right chamber of the hydraulic cylinder 4 will flow through the directional control valve 3, and back to the reservoir usually only meeting atmospheric pressure. When the directional control valve 3 is shifted to the reverse position, the system pressure will drop to whatever pressure is necessary to reverse movement of the cylinder piston assembly. The oil in the un-pressurized chamber of the cylinder will return to the reservoir through the directional control valve 3. This flow rate control valve 2 can be a variable restrictor or orifice, reducing the orifice size will lessen the volume of oil permitted to flow to the actuator. This will reduce actuator speed. Thus flow rate control valve 2 is used to regulate the speed of an actuator.

Pressure will rise and pump flow will again be diverted to the tank through the relief valve 6 at its setting. Working pressure in the hydraulic system depends on the outside load. The maximum load that can be placed on a system is determined by the relief valve 6 setting. The setting values would be the sum of maximum load and total pressure loss in the system. When the cylinder reaches the end of travel, the flow of fluid is restricted and pressure builds up to the relief setting. It cannot

1.3　液压传动系统的组成

以图 1-2 所示的典型液压系统为例，说明其组成和各种元件在系统中的作用。该系统由油箱 8、过滤器 7、液压泵 1、流量控制阀 2、换向阀 3、液压缸 4、工作台 5、溢流阀 6 和油管组成。

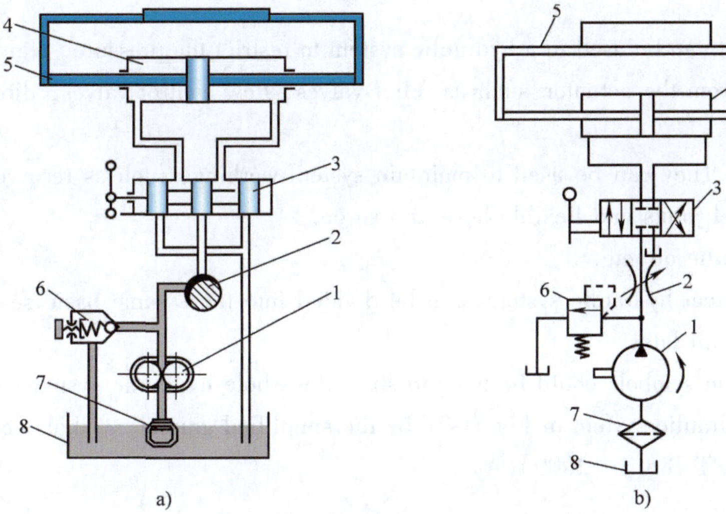

Fig. 1-2　Principle of a typical hydraulic system（典型液压系统原理图）

a）Partial configuration（半结构图示法）　b）Simplified graphics symbols（图形符号表示法）

1—Hydraulic pump（液压泵）　2—Flow rate control valve（流量控制阀）

3—Directional control valve（换向阀）　4—Hydraulic cylinder（液压缸）　5—Working load（工作台）

6—Relief valve（溢流阀）　7—Filter（过滤器）　8—Reservoir（油箱）

电动机带动液压泵 1 旋转，液压泵 1 开始工作，将对液压油带来三方面的作用：

1）在液压泵 1 的进口产生局部真空。

2）作用在液压油箱中液压油表面的大气压驱使液压油进入液压泵的进口以填充局部真空空间。

3）液压泵的机械作用使液压油从液压泵的进口流入而从液压泵的出口流出，此时获得一定压力的液压油流进液压系统。

当电动机带动液压泵 1 旋转，从油箱 8 经过滤器 7 吸油，换向阀 3 处于图示位置时，液压泵的全部压力油经溢流阀 6 流回油箱，压力油进不到液压缸 4，工作台 5 停止不动。当换向阀 3 的阀芯处于左位时，压力油经流量控制阀 2、换向阀 3 进入液压缸 4 的左腔，推动活塞向右运动。液压缸右腔的油液经换向阀 3 流回油箱。改变换向阀 3 阀芯的工作位置，使之处于右端位置时，液压缸活塞反向运动。

调节流量控制阀 2 的开口大小就可以改变进入液压缸的流量，从而控制液压缸活塞的运动速度。液压泵排出的多余油液经溢流阀 6 流回油箱。液压缸的工作压力取决于负载。液压泵的最大工作压力由溢流阀 6 调定，其调定值应为液压缸的最大工作压力及系统中油液流经

go any higher because the fluid can escape over the relief valve when the relief valve setting is reached. Thus the relief valve protects the system from excessive pressure.

From an operational standpoint, any hydraulic system can be divided into five logical segments:

(1) Power input segment (energy portion) It usually consists of pumps. In effect, the pump converts a mechanical force to hydraulic power and supply hydraulic oil to the system.

(2) The power output segment (actuator portion) It usually consists of rotary and oscillatory motors or linear cylinders, while it converts the hydraulic power back to mechanical force to do work.

(3) Control elements It can be used in a hydraulic system to restrict the pressure, regulate the volume of oil directed to or from the actuator, such as relief valves, flow control valves, directional control valves and so on.

(4) Assistant elements They can be used to maintain system working, such as reservoir, filter, tubing connectors, swivel joints and flexible hose and so on.

(5) Working oil hydraulic oil, etc.

Because almost all practical hydraulic systems can be divided into these same basic segments, they will be considered in detail later.

Usually simplified graphic symbols could be used to show the whole hydraulic system. Fig. 1-2b shows the principle of hydraulic system in Fig. 1-2a by the simplified graphic symbols according to the Chinese standard (GB/T 786. 1—2009).

1.4 Features of Hydraulic and Pneumatic Transmission

1. The advantages of hydraulic transmission

1) A hydraulic system can produce higher power than electrical equipment under the same volume. The hydraulic equipment system has smaller volume, light, high power consistency and compact configuration at a given power. The volume and weight of a hydraulic motor are about 12% of an electric motor.

2) Hydraulic equipment has a good working stability. It is because of light, less inertia, quick response; the hydraulic equipment can realize celerity start-up, brake and frequent change in motion direction.

3) The hydraulic transmission can reach a wide range of speed regulation (with the range of 2000:1), and the speed can also be regulated during the work processing.

4) The hydraulic transmission can easily realize automation and the pressure, flow rate and the flow direction can be regulated and controlled. If we combine it with electric, electron or pneumatic control systems, a more complex transmission system with remote control can be realized.

5) The hydraulic system can protect from over-load easily, which cannot be done by electricity or machine equipment.

6) Because of standardization, series, and all-purpose application; the hydraulic system is easier in design, fabrication and application.

7) The hydraulic system is easier than machine equipment in doing line motion.

阀和管道的压力损失的总和。因此，系统的工作压力不会超过溢流阀的调定值，溢流阀对系统还起着过载保护的作用。

从上面的例子可以看出，液压传动系统主要由以下五个部分组成：

（1）功率输入装置（能源装置） 把机械能转换成流体压力能的装置，如各类液压泵，为系统提供压力油。

（2）功率输出装置（执行元件） 把流体的压力能转换成机械能输出的装置，如作直线运动的液压缸和作回转运动的液压马达。

（3）控制元件 对系统中流体压力、流量和流动方向进行控制或调节的装置，如溢流阀、流量控制阀与换向阀等。

（4）辅助元件 保证系统正常工作所需的上述三种以外的装置。如油箱、过滤器、管件等。

（5）工作介质 液压油等。

图 1-2a 所示的液压系统图是一种半结构式工作原理图，它的直观性强，容易理解，但绘图费事。为了简化液压系统的表示方法，世界各国通常采用图形符号来绘制系统原理图。常用液压与气动元件图形符号（摘自 GB/T 786.1—2009）见附录 B。图 1-2b 就是按此图形符号标准绘制的图 1-2a 所示液压系统的原理图。

1.4 液压与气压传动的特点

1. 液压传动的优点

1）在同等体积下，液压装置比电气装置产生更高的动力。在同等功率下，液压装置体积小，重量轻，功率密度大，结构紧凑。液压马达的体积和重量只有同等功率电动机的 12% 左右。

2）液压装置工作比较平稳。由于重量轻、惯性小、反应快，液压装置易于实现快速起动、制动和频繁的换向。

3）液压装置能在大范围内实现无级调速（调速比可达 2000:1），它还可以在运行过程中进行调速。

4）液压传动易于自动化，它对液体压力、流量或流动方向易于进行调节或控制。当将液压控制和电气控制、电子控制或气动控制结合起来使用时，整个传动装置能实现复杂的顺序动作，也能方便地实现远程控制。

5）液压装置易于实现过载保护，这是电气传动装置和机械传动装置无法办到的。

6）由于液压元件已实现了标准化、系列化和通用化，液压系统的设计、制造和使用都比较方便。

7）用液压传动实现直线运动远比用机械传动简单。

2. 液压传动的缺点

1）由于流体流动的阻力损失和泄漏是不可避免的，所以液压传动在工作过程中常有较多的能量损失。

2）工作性能易受温度变化的影响，因此不宜在很高或很低的温度条件下工作。

3）为了减小泄漏，液压元件的制造精度要求较高，因而价格较贵。

4）出现故障时不易找出原因。

2. The shortages of hydraulic transmission

1) Leak. Oil Leaks are inevitable because of the loss in fluid flow resistance. So more energy loss exists in a hydraulic transmission.

2) Working temperature. The working temperature has strong effect on the working property of a hydraulic system because of the viscosity-temperature character of hydraulic oil. It is suitable for working in a proper temperature.

3) Cost. The cost is high because of the needs in high precision fabricate for hydraulic elements.

4) It is difficult to find the reasons of fault.

3. The advantages of pneumatic transmission

1) The air can be obtained and expelled from the atmosphere. It cannot bring pollution to the environment.

2) It is of low viscosity and lower pressure loss in pipes. The pressure air is convenient for convergence supply and remote transportation.

3) It is of low working pressure (usually 0.3-0.8MPa). A lower material and fabricate precision is required for the pneumatic transmission elements.

4) The pneumatic transmission has a simple servicing. The air pipe is not easy to be jammed.

5) Safety. The pneumatic system can protect from over-load easily.

4. The shortcomes of pneumatic transmission

1) It is because of air compressibility. The working stabilities for pneumatic transmission system are poorer than those of hydraulic transmission system.

2) Because of lower working pressure and small sizes in configuration, the push force of pneumatic transmission is usually much lower.

3) Lower transmission efficiency.

To sum up, the strong-points of hydraulic and pneumatic transmission have taken the main advantages, and the shortages have been overcome and improved by technical renovation.

1.5 The Development History and Application of Hydraulic and Pneumatic Transmission

The fundamental law underlying the whole science of hydraulics was discovered by Blaise Pascal, a French physicist, in the seventeenth century. But it was not until the end of the 18 century that man found ways to make the snugly fitting parts required in hydraulic systems and other modern equipment. Since then progress has been rapid.

Hydraulic transmission has been experiencing the process as below.

The 17th and 18th centuries were a productive period in the development of hydraulic theory. Torricelli studied fluid motion in the early 17th century. Late in that century, Sir Isaac Newton conducted studies on viscosity and the resistance of submerged bodies in a moving fluid. The key achievements of the period occurred in the middle of the 18th century when Daniel Benoulli developed the theory of transmission of energy in fluid streams and Blaise Pascal, at about the same time, established the principle of hydrostatic pressure transmission.

3. 气压传动的优点

1）空气可以从大气中取得，同时，用过的空气可直接排放到大气中去，处理方便，万一空气管路有泄漏，除引起部分功率损失外，不致产生不利于工作的严重影响，也不会污染环境。

2）空气的粘度很小，在管道中的压力损失较小，因此压缩空气便于集中供应和远距离输送。

3）因压缩空气的工作压力较低（一般为 0.3～0.8MPa），因此，对气动元件的材料和制造精度要求较低。

4）气动系统维护简单，管道不易堵塞。

5）使用安全，并且便于实现过载保护。

4. 气压传动的缺点

1）由于空气具有可压缩的特性，因此运动速度的平稳性不如液压传动。

2）因为工作压力较低和结构尺寸不宜过大，因而气压传动装置的总推力一般不可能很大。

3）传动效率较低。

总的说来，液压与气压传动的优点是主要的，而它们的缺点通过技术进步和多年的不懈努力，已得到克服或得到了很大的改善。

1.5 液压与气压传动的发展及应用概况

虽然在 17 世纪中叶法国物理学家伯雷斯·帕斯卡提出了静压传递原理，但是在 18 世纪末才开始找到应用在液压系统和其他现代装备中合适的元件。从那以后，液压技术得到迅速的发展。

17、18 世纪是液压基础理论建立的最兴旺的时期。其中，在 17 世纪初期，意大利数学和物理学家托里切利研究流体运动原理；17 世纪后期，艾萨克牛顿研究物体在流动的液体中的粘性和阻力问题；18 世纪中叶是最关键的时期，主要的成就有丹莱尔·苈某里发展了流体能量传递原理，同时伯雷斯·帕斯卡建立并提出了静压传递原理。从此静压传递原理（即帕斯卡原理）为流体传动（液压、气压传动）奠定了理论基础。

静压传递原理在 18 世纪后期得到广泛应用。18 世纪末，英国制造了世界上第一台水压机。

在上述理论基础上，纳维推导流体运动方程，到了 19 世纪初期斯托克斯也独立发现相同的方程并进一步发展了纳维的流体运动方程。

近代液压、气压传动是由 19 世纪崛起并蓬勃发展的石油工业推动起来的，最早实践成功的液压传动装置是舰艇上的炮塔转位器，其后才在机床上应用。第一次世界大战期间引入基于液压原理的新武器。在 20 世纪 30 年代初期和后期在大型自动化工业中引入液压制动。1940 年开始使用拖拉机以增强农机设备的机动性和效率。第二次世界大战期间，由于军事工业和装备迫切需要反应迅速、动作准确、输出功率大的液压传动及控制装置，促使液压技术迅速发展；战后，液压技术很快转入民用工业。在机床、工程机械、冶金机械、塑料机械、农林机械、汽车、船舶等行业得到了广泛的应用和大幅度的发展。近几年液压传动应用到大型挖掘和建筑施工的设备中，所涉及的总的动力常比最大型的航空系统所需的动力

This principle was first used in the latter part of the 18th century. The first hydraulic pressure machine was manufactured by England in the late 18th century. The fundamentals of fluid theory were established by the above work and refinements were added by Navier who derived the mathematics of motion in liquids including equations for fluid flow with friction. This was early in the 19th century. It was followed by the work of Stokes, who independently discovered the same equations and further extended the work of Navier.

Recently hydraulic and pneumatic pressure transmission technology has been developed with a large scale petrolic industry in the 19th century, and the barbette displace was the first one successful using hydraulic equipment, and then hydraulic machine tool. In World War Ⅰ many new machines based on the principles of hydraulics had been used. The great automotive industry introduced hydraulic brakes in the early thirties and hydraulic transmissions in the late thirties. The tractor industry began using hydraulics in 1940 to increase the flexibility and utility of farm equipment. In World War Ⅱ because of the demand transmission and control equipments in fast reaction, precision action and high output powers boosted development in hydraulic technology. After the War, the hydraulic development turned into civil industry, such as machine tool, engineering, metallurgy, plastic machine,farm machine, vehicle and watercraft. In more recent years, the role of leadership in hydraulic power application has been taken over largely by some of the larger earthmoving and construction equipment manufacturers. The total power involved is often greater than that required in even the largest aircraft systems.

With the development of higher automation of hydraulic machines and increasing use of hydraulic and pneumatic elements, the scaled elements and integrated hydraulic system with miniaturization is inevitable. Especially in recent years hydraulic and pneumatic transmission is combined closely with the sensor and micro-electronics technology. It has been emerging amounts of new valves such as hydraulic-electricity proportional valves, digital valves, hydraulic and electro-hydraulic servo cylinders and the integrative elements, which will lead the hydraulic and pneumatic technology to the development of higher pressure, higher speed, larger power, lower energy wastage and noise, longevity and high integration. Computer aided design (CAD) and test (CAT) and practical control technology used in hydraulic and pneumatic system will be the trend. Nowadays the application of hydraulic transmission system has become one of the important indications of industry level for a country. In developed countries, 95% of engineering machine, 90% of numerical control center and more than 95% of automation assembly lines use the hydraulic transmission systems.

The examples of application hydraulic and pneumatic are shown in Tab. 1-1.

还高。

随着液压机械自动化程度的不断提高，液压、气动元件应用数量急剧增加，元件小型化、系统集成化是必然的发展趋势。特别是近十年来，液压和气动技术与传感技术、微电子技术密切结合，出现了许多诸如电液比例控制阀、数字阀、电液伺服液压缸等机（液）电一体化元器件，使液压技术在高压、高速、大功率、节能高效、低噪声、使用寿命长、高度集成化等方面取得了重大进展。无疑，液压元件和液压系统的计算机辅助设计（CAD）、计算机辅助试验（CAT）和计算机实时控制也是当前液压和气动技术的发展方向。现今采用液压传动的程度已成为衡量一个国家工业水平的重要标志之一。如发达国家生产的95%的工程机械、90%的数控加工中心、95%以上的自动生产线都采用了液压传动。

液压与气压传动系统在工业中的应用见表 1-1。

Tab. 1-1　Examples of application of hydraulic and pneumatic transmission technology in industries
（液压与气压传动在各类机械行业中的应用举例）

Fields（行业名称）	Examples（应用举例）
Engineering machine（工程机械）	Grab，loading machine，bulldozer，shovel machine，etc（挖掘机、装载机、推土机、铲运机等）
Mine machine（矿山机械）	Charge，digger，elevator，hydraulic support，etc（凿岩机、开掘机、提升机、液压支架等）
Architecture machine（建筑机械）	Pile driver，jack，flat machine，etc（打桩机、液压千斤顶、平地机等）
Metallurgy machine（冶金机械）	Rolling mill，press machine，etc（轧钢机、压力机等）
Fabrication（机械制造）	Tool machine，numeric-control machining center，automatic assembly line，air-spanner，press machine，punch，model-forge machine，air hammer，sand-model press machine，feed machine，die-casting machine，tension machine，etc（机床、数控加工中心、自动线、气动扳手、压力机、模锻机、空气锤、砂型压实机、加料机、压铸机等）
Light industries（轻工机械）	Packer，injection-plastic machine，rubber-sulfuration machine，food packager，vacuum-plating machine，loom，printing and dyeing machine，etc（打包机、注射机、橡胶硫化机、食品包装机、真空镀膜机、织布机、抛砂机、印染机等）
Automobile industry（汽车工业）	High altitude operating car，lift，redirector，etc（高空作业车、汽车式起重机、转向器等）
Water project（水利工程）	Dam，strobe，ship machine，ship-rudder，etc（水坝、船闸水闸启闭机、船用机械船舵液压操纵等）
Farming industry（农林机械）	Fertilizer packager，combine harvester，tractor，farming suspension system，etc（化肥包装机、联合收割机、拖拉机、农机悬挂系统等）

Chapter 2 Fundamental Hydraulic Fluid Mechanics

2.1 Performances of the Hydraulic Oil

2.1.1 Main performances

1. Density (kg/m^3)

The ratio of the mass of a given amount of a substance to the volume that this amount occupies. If this volume is denoted as V, the mass as m and the density ρ

$$\rho = m/V \tag{2-1}$$

The density is one of the main parameters of fluid. It changes with the temperature and the pressure but is neglectable in this context. Usually density for hydraulic oil is $900kg/m^3$.

2. Compressibility

It refers to deformation that results from shear stresses. All fluids compress as the pressure increases, resulting in an increase in density. A common way to describe the compressibility of a fluid is by the following definition of the bulk modulus of elasticity K

$$K = -\frac{\Delta p}{\Delta V}V \tag{2-2}$$

The bulk modulus, also called the coefficient of compressibility, is defined as the ratio of the change in pressure (Δp) to relative change in volume ($\Delta V/V$) while the temperature remains constant.

3. Viscosity

Viscosity can be thought of as the internal stickiness of a fluid. Under the action of an external force, the cohesion between fluid molecules would stop the fluid flow, which is called friction force. This is referred to as the viscosity of fluid. Needless to say, the viscosity is an extremely important fluid property in our study of fluid flows and is a condition of hydraulic oil.

The sketch of viscosity is illustrated by Fig. 2-1. The experiments have proved that friction force between the two fluid molecules can be described as

$$F_f = \mu A \frac{du}{dy}$$

Where μ is viscosity coefficient, also kinematic viscosity.

If the τ is described as the friction force per unit area A. Then

$$\tau = \frac{F_f}{A} = \mu \frac{du}{dy} \tag{2-3}$$

第 2 章　液压流体力学基础

2.1 液压油的性质

2.1.1 主要性质

1. 密度（kg/m^3）

单位体积液体的质量称为该液体的密度。

体积为 V、质量为 m 的液体的密度 ρ 为

$$\rho = m/V \tag{2-1}$$

密度是液体的一个重要物理参数，随着温度或压力的变化，其密度也会发生变化，但变化量一般很小，可以忽略不计。一般液压油的密度为 $900kg/m^3$。

2. 可压缩性

液体受压力作用后其体积缩小的性质称为液体的可压缩性。通常用体积弹性模量 K 来描述液体的可压缩性，有

$$K = -\frac{\Delta p}{\Delta V}V \tag{2-2}$$

简而言之，可压缩性定义为压力的变化量 Δp 与体积相对变化量 $\Delta V/V$ 的比值，显然，体积弹性模量与压力单位相同。它表示产生单位体积相对变化所需要的压力增量。液压油的 K 值很大，为 $(1.2 \sim 2) \times 10^3 MPa$，当压力变化不大时，油液的可压缩性可忽略不计。但液压油混入空气时，其可压缩性便显著增加，进而影响液压系统的工作性能，故应尽力防止空气混入油液中。

3. 粘性

液体在外力作用下流动时，分子间的内聚力要阻碍分子相对运动而产生一种内摩擦力，这一特性称为液体的粘性。液体只有在流动时才会呈现出粘性，静止的液体是不会呈现粘性的。粘性是液体的重要物理特性，也是选择液压用油的依据。

粘性使流体内部各处的速度不相等，以图 2-1 为例，若两平行平板间充满液体，当上平板以速度 u_0 相对于静止的下平板向右移动时，由于受液体粘性的作用，使紧贴于下平板的液体层速度为零，紧贴于上平板的液体层速度为 u_0，而中间各层液体的速度从上到下近似呈线性递减的规律分布。

实验结果表明，液体流动时相邻液层间的内摩擦力 F_f 与液层接触面积 A、液层间的速度梯度 du/dy 成正比，即

$$F_f = \mu A \frac{du}{dy}$$

It is the Newtoniau's law of inner friction. It means that the viscosity of fluid is the friction on per unit of area fluid under per unit of velocity gradient.

(1) Dynamic viscosity or absolute viscosity (Pa · s) The dynamic viscosity μ can be gotten from the formula (2-3), i. e.

$$\mu = \frac{\tau}{\frac{du}{dy}}$$

It means that if $\frac{du}{dy} = 1$, then $\mu = \tau$, i. e., the inner friction force τ on per unit area of fluid between contiguity surfaces, is called dynamic viscosity or also absolute viscosity.

(2) Kinematic viscosity (m^2/s) The ratio of absolute viscosity to the oil mass density is called the kinematic viscosity v in analysis and calculation of hydraulic system, i. e.

$$v = \frac{\mu}{\rho} \tag{2-4}$$

(3) Relative viscosity Viscosity tested by special viscometer under a given condition is also called conditional viscosity. The unit of relative viscosity is different by the testing conditions, such as °E used China, SSU used in America, R used in England and so on.

(4) Temperature-viscosity The relative viscosity °E is tested by the °E viscometer. 200 cm^3 of tested fluid (such as hydraulic oil) is put into a container of the viscometer which has a ϕ2. 8mm small hole at the end of it (as shown in Fig. 2-2b), and then the time (t_1) of 200 mL tested fluid deadweight flowing through the equal small hole under a set of temperature will be recorded; meanwhile, the time (t_2) of 200cm^3 of steam water under a temperature of 20℃ deadweight flowing through the equal small hole will be also recorded, as shown in Fig. 2-2a. The ratio of t_1 to t_2 is called °E viscosity value of this tested fluid. Usually the set temperatures are 20℃, 50℃ and 100℃. So there are the °E_{20}, °E_{50} and °E_{100}.

Take the note °E_t to describe the viscosity

$$°E_t = t_1/t_2 \tag{2-5}$$

The conversion formula between the °E_t and kinematic viscosity (m^2/s) is

$$v = (7.31\,°E_t - \frac{6.31}{°E_t}) \times 10^{-6} \tag{2-6}$$

The viscosity is strongly dependent on temperature of liquids in which cohesive forces play a dominant role. Note that the viscosity of liquid decreases with increased temperature. The change of oil temperature will affect the performance and leakage of the oil, thus the viscosity-temperature changes. For the viscosity less than 15 °E and the temperature 30-150℃, the viscosity-temperature formula is described as follows

$$v_t = v_{50}(\frac{50}{t})^n \tag{2-7}$$

Where v_t ($10^{-6}m^2/s$) is kinematic viscosity at a set of temperature; v_{50} is the kinematic viscosity when the temperature is 50℃; and n is index as shown in Tab. 2-1.

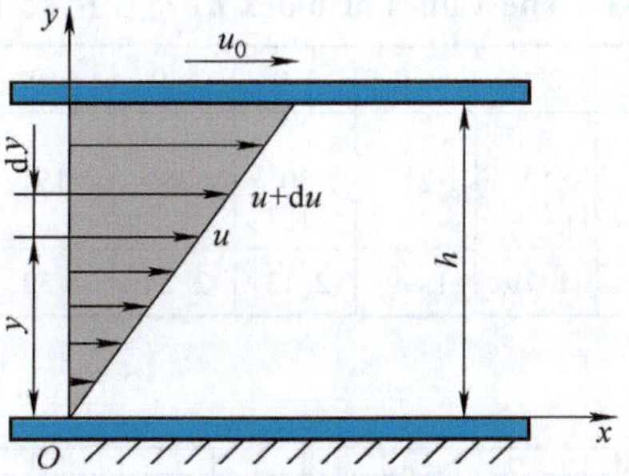

Fig. 2-1　Sketch of viscosity（液体粘性示意图）

式中，μ 为比例系数，称为粘度系数或动力粘度。

若以 τ 表示液层间在单位面积上的内摩擦力，则

$$\tau = \frac{F_f}{A} = \mu \frac{du}{dy} \tag{2-3}$$

这就是牛顿液体的内摩擦定律。

由此可知，液体的粘度是指在单位速度梯度下流动时单位面积上产生的内摩擦力。粘度是衡量粘性的指标。常用的粘度有三种，即动力粘度、运动粘度和相对粘度。

（1）动力粘度 μ　由式（2-3）得出

$$\mu = \frac{\tau}{\frac{du}{dy}}$$

动力粘度 μ 的物理意义是：当速度梯度 $du/dy = 1$ 时，则 $\mu = \tau$，即单位速度梯度下接触液层间单位面积上的内摩擦力，称为动力粘度，又称绝对粘度，其单位是 Pa · s。

（2）运动粘度 υ　在液压系统的实际计算和分析中常使用绝对粘度 μ 与该液压油密度 ρ 的比值来表示，该比值称为运动粘度 υ，即

$$\upsilon = \frac{\mu}{\rho} \tag{2-4}$$

其单位是 m^2/s。

（3）相对粘度　相对粘度又称条件粘度。它是采用特定的粘度计在规定的条件下测出来的液体粘度。根据测量条件的不同，各国采用的相对粘度的单位也不同。例如：我国采用恩氏粘度（$°E$），美国用国际赛氏粘度（SSU），英国采用雷氏粘度（R）等。

恩氏粘度由恩氏粘度计测定，即将 200mL 的被测液体装入底部有 ϕ2. 8mm 小孔的恩氏粘度计的容器中（图 2-2b），在某一特定温度 t 时，测定液体在自重作用下流过小孔所需的时间 t_1，和同体积的蒸馏水在 20℃时流过同一小孔（图 2-2a）所需的时间 t_2 之比值，便是该液体在 t 时的恩氏粘度。一般以 20℃、50℃、100℃作为测定恩氏粘度的标准温度，由此而得来的恩氏粘度分别用 $°E_{20}$、$°E_{50}$ 和 $°E_{100}$ 表示。

恩氏粘度用符号 $°E_t$ 表示

$$°E_t = t_1/t_2 \tag{2-5}$$

恩氏粘度和运动粘度的换算关系式为

$$\upsilon = (7.31\ °E_t - \frac{6.31}{°E_t}) \times 10^{-6} \tag{2-6}$$

Tab. 2-1 **The values of index n(特征指数 n 的数值)**

$°E_{50}$	1.2	1.5	1.8	2.0	3.0	4.0	5.0	6.0	7.0	8.0	9.0	10.0	15.0
$v_{50}/(10^{-6}m^2\cdot s^{-1})$	2.5	6.5	9.5	12	21	30	38	45	52	60	68	76	113
n	1.39	1.59	1.72	1.79	1.99	2.13	2.24	2.32	2.42	2.49	2.52	2.56	2.75

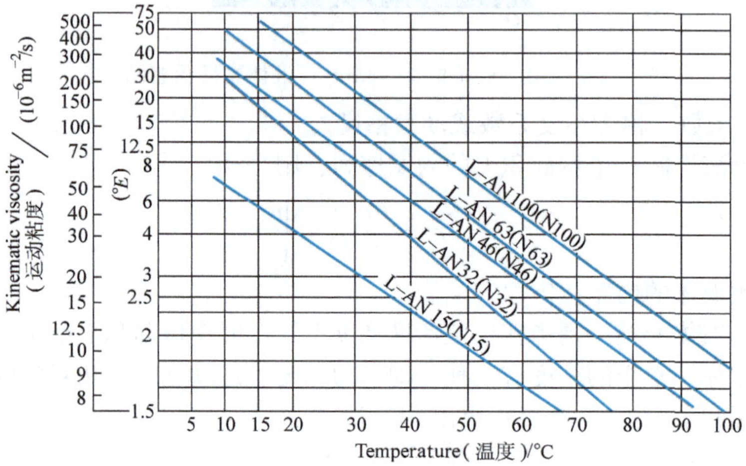

Fig. 2-3 The viscosity-temperature of homemade oils
(几种国产油液粘温图)

(5) Viscosity-pressure The distance between molecules will be reduced and cohesion (viscosity) will be increased with pressurization. Under lower pressure (<5MPa), the viscosity of hydraulic oil is not strongly affected by tiny changes in pressure. However, pressure up to 10MPa will increase fluid viscosity sufficiently to require consideration in engineering calculations. The viscosity can be computed by

$$v_p = v_0 e^{bp} \approx v_0(1 + 0.03p) \tag{2-8}$$

Where v_p is kinematic viscosity at pressure p($10^{-6}m^2/s$); v_0 is kinematic viscosity at an atmospheric pressure($10^{-6}m^2/s$); b is viscosity-pressure coefficient, usually $b = 0.002$-0.003 for common hydraulic oil and p is pressure of oil (MPa).

(6) Other performances Besides above, there are also other performances for hydraulic oil, which cover physical and chemical, such as anti-inflammability, anti-oxygenation, anti-concreting, anti-foam and anti-corrosion, etc. The performances are different for the different products of oils. In this case you can refer to the manual of oil products.

2.1.2 The requests and choice of hydraulic oil

In the hydraulic transmission system, the oil plays two roles of transmission energy and lubrication on the surfaces of working interaction. So it will affect the hydraulic system, such as the reliability, stability, and efficiency. The requests for the hydraulic fluids are: have an appropriate viscosity, good in property of favorable viscosity-temperature, a good lubricity, chemical and environ-

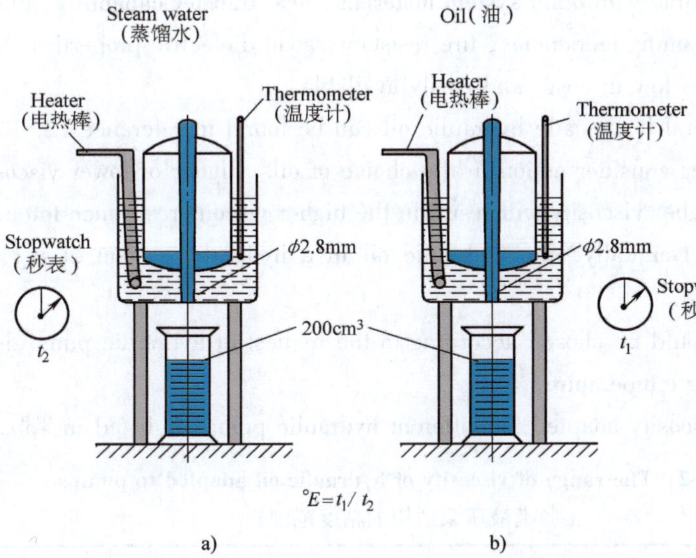

Fig. 2-2 Principle of °E viscometer (恩氏粘度计)

其单位是 m^2/s。

(4) 温度对粘度的关系 温度对油液粘度的影响较大。温度增高，油液粘度会显著降低。油液粘度的变化直接影响液压系统的性能和泄漏量，因此希望粘度随温度的变化越小越好。常用粘温特性描述这种关系，粘温特性好，表示粘度随温度升高而下降的量相对少一些。

对于粘度不超过 15 $^\circ E_t$ 的液压油，当温度在 30～150℃范围内时，可用下述近似公式计算温度为 t 时的运动粘度

$$\upsilon_t = \upsilon_{50}\left(\frac{50}{t}\right)^n \tag{2-7}$$

式中，υ_t 为温度 t 时油液的运动粘度（$10^{-6}m^2/s$）；υ_{50} 为温度 50℃时油液的运动粘度（$10^{-6}m^2/s$）；n 为与油液粘度有关的特性指数，见表 2-1。

(5) 压力对粘度的影响 油液随压力增大其分子间距缩小，内聚力增大，粘度也随之增大。在压力不高（<5MPa）且变化不大时，其影响可忽略不计。但压力较高或变化较大（≥10MPa）时，则需要考虑压力对粘度的影响，它们之间存在如下关系

$$\upsilon_p = \upsilon_0 e^{bp} \approx \upsilon_0(1 + 0.03p) \tag{2-8}$$

式中，υ_p 为压力为 p 时的运动粘度（$10^{-6}m^2/s$）；υ_0 为压力为一个大气压时的运动粘度（$10^{-6}m^2/s$）；b 为粘度压力系数，对一般液压油 b＝0.002～0.003；p 为油液压力（MPa）。

(6) 其他特性 液压油还有其他一些物理化学性质，如抗燃性、抗氧化性、抗凝性、抗泡沫性、抗乳化性、缓蚀性、润滑性、导热性、稳定性以及相容性（主要指对密封材料、软管等不侵蚀、不溶胀的性质）等，这些性质对液压系统的工作性能有重要影响。对于不同品种的液压油，这些性质的指标是不同的，具体应用时可查油类产品手册。

2.1.2 对液压油的要求和选用

液压系统中的工作油液具有双重作用，一是作为传递能量的介质，二是作为润滑剂润滑运动零件的工作表面，因此油液的性能会直接影响液压传动的性能，如工作的可靠性、工况

mental stabilities, compatible with other system materials, heat transfer capability, high bulk modulus, low volatility, low foaming tendencies, fire resistant, good dielectric properties, nontoxic or allergenic, non-malodorous, low in cost, and easily available.

The qualities of several homemade hydraulic oil can be found in reference 12.

The viscosity is a key consider action in the choice of oil. Higher or lower viscosity will affect the hydraulic system. Higher viscosity will result in the higher viscous resistance force, and at lower viscosity larger leak oil. Generally, the hydraulic oil in a hydraulic system at $v_{40} = (10\text{-}60) \times 10^{-6}\mathrm{m^2/s}$ is recommended.

The hydraulic oil should be chosen according to the request of hydraulic pump due to the high load, speed and operating temperature.

The hydraulic oil viscosity adapted for different hydraulic pumps is listed in Tab. 2-2.

Tab. 2-2 The range of viscosity of hydraulic oil adapted to pumps
(各类液压泵适用的粘度范围)

Types(液压泵类型)		Viscosities(运动粘度) $/10^{-6}\mathrm{m^2 \cdot s^{-1}}$		Types(液压泵类型)	Viscosities(运动粘度) $/10^{-6}\mathrm{m^2 \cdot s^{-1}}$	
		5-40℃①	40-80℃①		5-40℃①	40-80℃①
Vane Pumps (叶片泵)	$P < 7\mathrm{MPa}$	30-50	40-75	Gear pumps(齿轮泵)	30-70	95-165
	$P \geq 7\mathrm{MPa}$	50-70	50-90	Radial piston pumps(径向柱塞泵)	30-50	65-240
Screw pumps(螺杆泵)		30-50	40-80	Axial piston pumps(轴向柱塞泵)	30-70	70-150

① 5-40℃,40-80℃ are described as the temperatures of hydraulic system.
5～40℃,40～80℃是指液压系统温度。

2.2 Hydrostatics

Hydrostatics studies the laws when fluid is at rest.

2.2.1 Characteristics of hydrostatics

1. Hydrostatics

Pressure is the amount of force on per unit of area. Static pressure: the action force in normal on per unit of area. It is intituled pressure in physics and action force in engineering usually.

2. Characteristics

1) In any homogeneous fluid system at rest, the pressure increases with the depth of the fluid.

2) Pressure at any point in a homogeneous fluid system at rest acts perpendicularly on the surfaces in contact with the fluid. For curved surfaces, the pressure at any point acts in the direction normal to the surface at that point.

2.2.2 The basic formula of hydrostatics

1. The basic formula of bydrostatics

The acting pressures on the fluid at rest. In a container they include the liquid weight, and the

的稳定性、系统的效率及零件的使用寿命等。在选用液压油时应满足下列几项要求：粘度合适，粘温特性好，润滑性良好，化学稳定性和环境稳定性好，与系统元件的材料的兼容性好，具有良好的热交换性，高的体积弹性模量；低挥发性，低的泡沫性和高的抗燃性，具有良好的绝缘性能，无毒，无过敏性，低成本，方便使用。

几种常用的国产液压油的主要质量指标可查阅参考文献 12。

选择液压用油首先要考虑的是粘度问题。在一定条件下，选用的油液粘度太高或太低，都会影响系统的正常工作。粘度高的油液流动时产生的阻力较大，克服阻力所消耗的功率较大，而此功率损耗又将转换成热量使油温上升。粘度太低，会使泄漏量加大，导致系统的容积效率下降。一般液压系统的油液粘度 υ_{40} 在（10～60）$\times 10^{-6}\mathrm{m^2/s}$ 之间，更高粘度的油液应用较少。

在液压系统中，液压泵因承受压力大、转速高、温升大、润滑要求严格，所以常根据液压泵的要求选择液压油的粘度。

各类液压泵适用的粘度范围见表 2-2。

2.2　液体静力学

液体静力学是研究液体处于静止状态下的力学规律以及这些规律的应用。

2.2.1　静压力及其特性

1. 液体的静压力

静止液体在单位面积上所受的法向力称为静压力。液体静压力在物理学上称为压强，在工程实际应用中习惯称为压力。

2. 液体静压力的特性

1）液体静压力垂直于其承压面，其方向和该面的内法线方向一致。

2）静止液体内任一点所受到的静压力在各个方向上都相等。

2.2.2　静压力基本方程式

1. 静压力基本方程式

在重力作用下的静止液体所受的力，除了液体重力，还有液面上作用的外加压力，其受力情况如图 2-4a 所示。

如果忽略沿 x 方向的力，对于计算离液面深度为 h 的某一点压力，可取出一个底面包含该点的垂直小液柱为研究体，如图 2-4b 所示。设小液柱底面积为 ΔA，高为 h，其体积为 ΔAh，则液柱的重力为 $\rho gh\Delta A$，并作用于液柱的重心上。由于液柱处于平衡状态，所以液柱所受各力存在如下关系

$$p\Delta A = p_0\Delta A + \rho gh\Delta A \tag{2-9}$$

等式两边同除以 ΔA，则得

$$p = p_0 + \rho gh \tag{2-10}$$

式（2-10）即为静压力基本方程式。由式（2-10）可知，重力作用下的静止液体，其压力分布有如下特征：

1）静止液体内任一点处的压力由两部分组成：一部分是液面上的压力 p_0，另一部分是该

force on the fluid surface, shown in Fig. 2-4a.

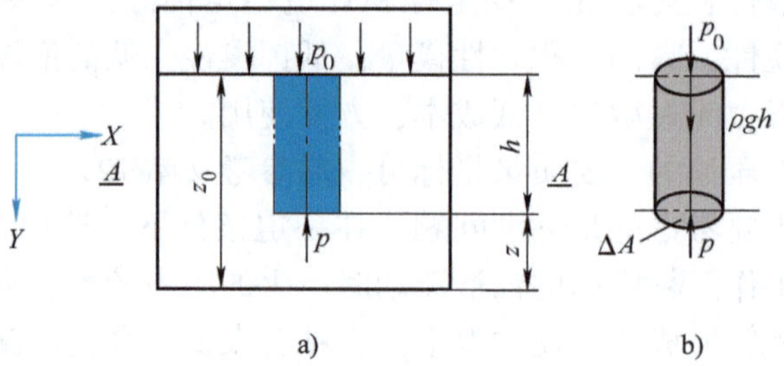

Fig. 2-4 The distribution of forces in a container with rest fluid
(静止液体内压力分布规律)

If the pressure in an x axial direction is ignored, then to compute the pressure at a depth h point, we can consider a control fluid volume in a direction perpendicular to line A-A, shown in 2-4a and 2-4b. If the end area of the volume as ΔA, depth h, then the volume is $\Delta A \cdot h$, so weight is $\Delta A \cdot \rho gh$. The total balance force formula is

$$p\Delta A = p_0\Delta A + \rho gh\Delta A \tag{2-9}$$

Formula (2-9) is divided by ΔA, then

$$p = p_0 + \rho gh \tag{2-10}$$

The formula (2-10) is the basic equation for hydrostatic. It states the distribution status of hydrostatics as follows:

1) The pressure p on a rest fluid contained involves two parts: one is the acting pressure on the fluid surface p_0; the other is the weight of the volume itself; if the p_0 = atmospheric pressure p_a, then

$$p = p_a + \rho gh \tag{2-11}$$

2) The pressure p is increased with the depth h.

3) Isotonic pressure surface, that is, the pressures are all equal at the surface composed by all points at given depth h, such as at the line of A-A.

4) Conservation of energy

$$\frac{p_0}{\rho g} + z_0 = \frac{p}{\rho g} + z = \text{constant} \tag{2-12}$$

Where, $p_0/(\rho g)$ is pressure energy at per unit mass fluid.

Formula (2-12) states that the total energy is conservation at any given point.

2. The definition and units of pressure

(1) Absolute pressure It is based on the absolute vacuum, such as formula (2-11)。

(2) Relative gauge pressure Based on the atmosphere, such as p-p_a = ρgh. The pressures measured by a pressure gauge are all relative pressure due to the objects all equilibriums loaded in nature.

(3) Vacuum (negative pressure) When the pressure is lower than atmosphere.

The relationship of three pressures is shown in Fig. 2-5.

The units of pressure and relations between different pressures can be described:

点以上液体自重所形成的压力。当液面上只受大气压 p_a 作用时，液体内任一点处的压力为

$$p = p_a + \rho g h \tag{2-11}$$

2）静止液体内的压力随液体深度呈直线规律递增。

3）离液面深度相同处各点的压力均相等，而压力相等的所有点组成的面称为等压面。

4）能量守恒，即

$$\frac{p_0}{\rho g} + z_0 = \frac{p}{\rho g} + z = \text{常量} \tag{2-12}$$

式中，$p_0/(\rho g)$ 为静止液体中单位质量液体的压力能；z_0，z 为单位质量液体的势能。

式（2-12）的物理意义为静止液体中任一质点的总能量保持不变，即能量守恒。

2. 压力的表示方法及单位

（1）绝对压力　当压力以式（2-11）表示时，称为绝对压力，其值是以绝对真空为基准进行度量的。

（2）相对压力　实际中超过大气压力的压力 $p - p_a = \rho g h$ 称为相对压力或表压力，其值是以大气压为基准进行度量的。因大气中的物体受大气压的作用是自相平衡的，所以用压力表测得的压力值是相对压力。

（3）真空度　当绝对压力低于大气压时，绝对压力不足于大气压力的那部分压力值，称为真空度。此时相对压力为负值。

绝对压力、相对压力和真空度的关系如图 2-5 所示。

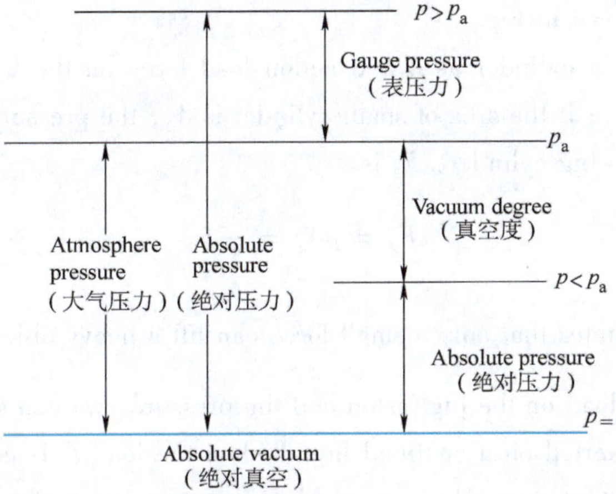

Fig. 2-5　Absolute, relative and vacuum pressure（绝对压力、相对压力和真空度）

压力的单位以及各种表示法之间的换算关系如下：

1Pa（帕）$=1\ \mathrm{N/m^2}$，1bar（巴）$=1\times10^5\mathrm{Pa}=1\times10^5\mathrm{N/m^2}$，1at（工程大气压）$=1\mathrm{kgf/cm^2}=9.8\times10^4\mathrm{N/m^2}$，$1\mathrm{mH_2O}$（米水柱）$=9.8\times10^3\mathrm{N/m^2}$，1mmHg（毫米汞柱）$=1.33\times10^2\mathrm{N/m^2}$。

例 2-1　如图 2-6 所示，容器内充满油液。已知：油的密度 $\rho=900\mathrm{kg/m^3}$，活塞上的作用力 $F=1000\mathrm{N}$，活塞面积 $A=1\times10^{-3}\mathrm{m^2}$，忽略活塞的质量。计算活塞下方深度为 $h=0.5\mathrm{m}$ 处的静压力。

解： 根据公式，在 h 深处 p 点的压力为

$$p = p_0 + \rho g h$$

1Pa = 1 N/m^2; 1bar = 1×10^5Pa = $1 \times 10^5 N/m^2$; 1at = $1kgf/cm^2$ = $9.8 \times 10^4 N/m^2$; $1mH_2O$ = $9.8 \times 10^3 N/m^2$; 1mmHg = $1.33 \times 10^2 N/m^2$.

Example 2-1 The oil is full in a container. For a given condition, the density of oil $\rho = 900kg/m^3$; the action force on this piston surface $F = 1000N$; the area of piston $A = 1 \times 10^{-3} m^2$, if the mass of piston is neglected, try to calculate the static pressure p at $h = 0.5m$, as shown in Fig. 2-6.

Solution: According to formula (2-10), the pressure at the depth h can be calculated, $p = p_0 + \rho gh$, and here $p_0 = \frac{F}{A} = \frac{1000}{1 \times 10^{-3}}Pa = 10^6 Pa$, and then

$$p = p_0 + \rho gh = (10^6 + 900 \times 9.8 \times 0.5) Pa$$
$$= 1.0044 \times 10^6 \ N/m^2 \approx 10 \times 10^5 \ Pa$$

From this example, we can see that the weight of fluid oil (ρgh) is very small, so pressure is equal nearly at all points in the container, i. e. , $p \approx p_0 = 10^6 Pa$. This is an important result used in the later chapters.

2.2.3 The principle of pascal

Pressure exerted on a confined liquid is transmitted undiminished in all directions and acts with equal force on all equal areas. The principle of Pascal is also called as the static transmission principle. Its application is shown in Fig. 2-7.

Set the area of the big cylinder as A_1 and action load force on the big piston as F_1, then, the pressure on it is $p = F_1/A_1$; if the area of small cylinder is A_2, the pressure on the small piston is also p, in order to hold the big cylinder, F_2 is

$$F_2 = pA_2 = \frac{A_2}{A_1}F_1 \tag{2-13}$$

The above formula states that only a small force can lift a heavy object due to $\frac{A_2}{A_1} < 1$. From the relationship between the load on the big piston and the pressure, we can see that pressure cannot be created in the pressure exerted on a confined liquid when the load F_1 is equal to zero no matter how high F_2 is. It is an important principle in the hydraulic transmission that the pressure exerted on a confined liquid depends on the external load.

2.2.4 Effect of fluid pressure on curved surfaces

Effect of fluid pressure on a container surface has two statuses:

1) When the wall is plane, the total of effect force on it is equal to the pressure of fluid multiply the effect areas and the direction of effect point is perpendicular to the wall surface.

2) When the wall is a curved surface, the effect force on a point, such as x direction, the useful pressure is exerted on the projection of the curved surfaces A_x in the plane of action. That is $F_x = pA_x$. Let's consider an example of cylinder.

Example 2-2 Fig. 2-8 shows a cylindrical member of inside radius r and length l. Calculation:

其中

$$p_0 = \frac{F}{A} = \frac{1000}{1 \times 10^{-3}}\text{Pa} = 10^6\text{Pa}$$

则深度为 h 处的液体压力为

$$p = p_0 + \rho g h = (10^6 + 900 \times 9.8 \times 0.5)\text{Pa}$$

$$= 1.0044 \times 10^6\text{N/m}^2 \approx 10 \times 10^5\text{Pa}$$

由本例可见，相比液体所受的外界压力，液体自重所产生的那部分静压力 ρgh 很小，计算中可以忽略不计，因而认为整个静止液体内部的压力是近似相等的。在以后的有关章节中分析计算压力时，都采用这一结论。

2.2.3 帕斯卡原理

在密闭容器内，施加于静止液体的压力可以等值地传递到液体中各点，这就是帕斯卡原理，也称为静压传递原理。帕斯卡原理应用实例如图 2-7 所示。图中大缸活塞面积为 A_1，作用在活塞上的负载为 F_1，液体所形成的压力 $p = F_1/A_1$。由帕斯卡原理知：小活塞处的压力也为 p，若小活塞面积为 A_2，则为防止大活塞下降，在小活塞上应施加的力为

$$F_2 = pA_2 = \frac{A_2}{A_1}F_1 \tag{2-13}$$

由上式可知：由于 $A_2/A_1 < 1$，所以用一个很小的推力 F_2，就可以推动一个比较大的负载 F_1。从负载与压力的关系还可以发现，当大活塞上的负载 $F_1 = 0$ 时，不考虑活塞自重和其他阻力，则不论怎样推动小液压缸的活塞，也不能在液体中形成压力，这就是液压传动中一个很重要的概念——液体内的压力是由外负载决定的。

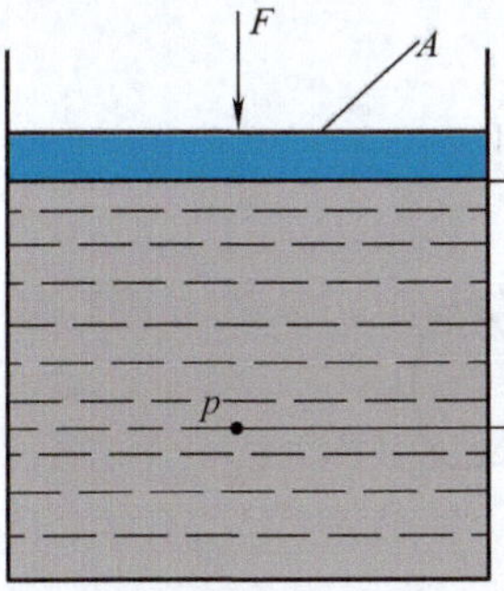

Fig. 2-6 Calculation of fluid static pressure
（流体静压力计算）

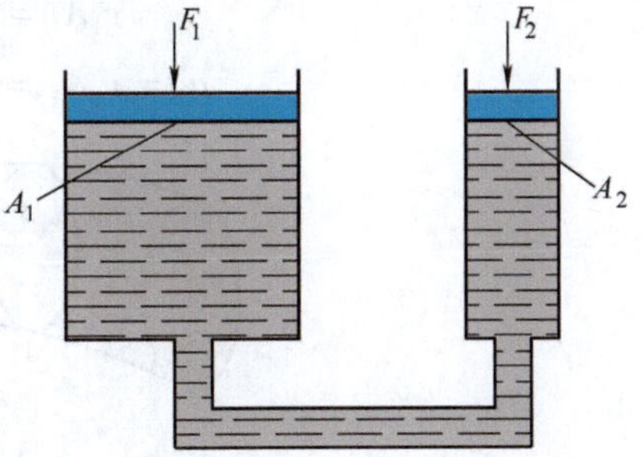

Fig. 2-7 Example of application of Pascal principle
（帕斯卡原理应用实例）

2.2.4 静压力对固体壁面的作用力

液体和固体壁面接触时，固体壁面将受到液体静压力的作用。

1）当固体壁面为一平面时，液体压力在该平面上的总作用力 F 等于液体压力 p 与该平面面积 A 的乘积，其作用方向与该平面垂直。

2）当固体壁面为一曲面时，液体压力在该曲面 x 方向上的总作用力 F_x 等于液体压力 p 与曲面在该方向投影面积 A_x 的乘积，即 $F_x = pA_x$，此式适用于任何曲面。下面以液压缸缸筒的受力情况为例加以证明。

The effect force F_x on the right segment of the cylinder at x direction.

Solution: Consider a tiny control surface $dA = l ds = lr d\theta$, and then the component of forces in the x direction as $dF_x = dF\cos\theta = p dA \cos\theta = plr\cos\theta d\theta$. Integral to give the effect force F_x on the right segment of the cylinder at x direction as follows, where the $A_x = 2rl$, is the projection of the curved surfaces in the x direction of a plane of action.

$$F_x = \int_{-\frac{\pi}{2}}^{\frac{\pi}{2}} dF_x = \int_{-\frac{\pi}{2}}^{\frac{\pi}{2}} plr\cos\theta d\theta = 2plr = pA_x$$

2.3 Hydrodynamics

Hydrodynamics in this section studies the laws of velocity and pressure for a flowing fluid. The equations of continuity, Bernoulli and momentum are basic motion equations that describe the dynamics laws in flowing fluid. A series of relationships among velocity, flow and pressure can be developed from the basic equations of motion for a falling body, which are the theoretical foundation in hydraulic transmission.

2.3.1 Equation of continuity—conservation of mass

Fig. 2-9 shows a tube with different across-section areas; the flow is stable, and the symbols A_1, v_1, ρ_1, A_2, v_2, ρ_2 describe the across-section area, velocity and density at 1 and 2 respectively. The mass is equal on two across-section area according to the conservation of mass, i. e.

$$\rho_1 v_1 A_1 = \rho_2 v_2 A_2 \tag{2-14}$$

For incompressible flow, $\rho_1 = \rho_2$, i. e.

$$v_1 A_1 = v_2 A_2 \tag{2-15}$$

or

$$q = v_1 A_1 = v_2 A_2 = vA = \text{constant} \tag{2-16}$$

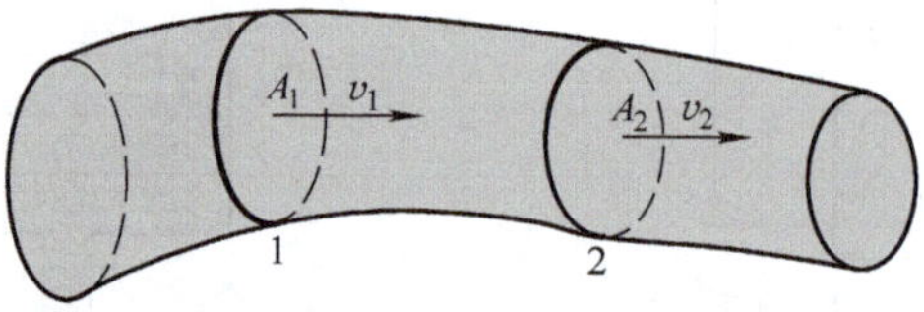

Fig. 2-9 Sketch of conservation mass

(液流连续性方程推导简图)

Formula (2-16) is the equation of flow continuity. It states that the flow rate is invariable when incompressible fluid flow flows through any across-section area. So the fluid velocity changes inversely to the flow across-section area.

2.3.2 Bernoulli equation—conservation of energy

1. Ideal equation of Bernoulli

The assumptions: inviscid flow (no shear stresses), steady flow, constant density and without energy loss.

The total energy on any across-section area is equal according to the equation of Bernoulli—

例 2-2 图 2-8 所示液压缸缸筒半径为 r，长度为 l。求液压油液对缸筒右半壁内表面在 x 方向上的作用力 F_x。

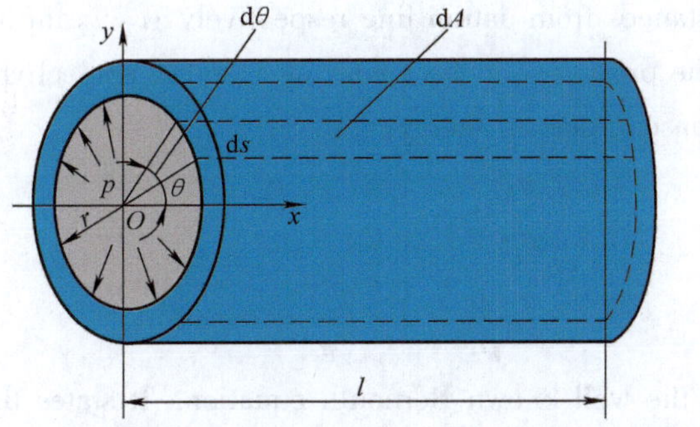

Fig. 2-8 Effect force on the inner surface of the cylinder
(压力油作用在缸筒内壁上的力)

解： 在右半壁面上取一微小面积 $dA = lds = lrd\theta$，则压力油作用在 dA 上的力 $dF = pdA$ 在 x 方向的分力 $dF_x = dF\cos\theta = pdA\cos\theta = plr\cos\theta d\theta$。

对上式积分，得右半壁面在 x 方向的作用力

$$F_x = \int_{-\frac{\pi}{2}}^{\frac{\pi}{2}} dF_x = \int_{-\frac{\pi}{2}}^{\frac{\pi}{2}} plr\cos\theta d\theta = 2plr = pA_x$$

式中，A_x 为缸筒右半壁面在 x 方向的投影面积，$A_x = 2rl$。

2.3 液体动力学

液体动力学是研究液体流动时流速和压力变化规律的。流动液体的连续性方程、伯努利方程、动量方程是描述流动液体力学规律的三个基本方程式。这些内容构成了液压传动技术的理论基础。

2.3.1 流量连续性方程——质量守恒定律

图 2-9 所示为一不等截面管，液体在管内作恒定流动。任取 1、2 两个通流截面，设其面积分别为 A_1 和 A_2，两个截面中液体的平均流速和密度分别为 v_1、ρ_1 和 v_2、ρ_2，根据质量守恒定律，在单位时间内流过两个截面的液体质量相等，即

$$\rho_1 v_1 A_1 = \rho_2 v_2 A_2 \tag{2-14}$$

不考虑液体的压缩性，有 $\rho_1 = \rho_2$，则得

$$v_1 A_1 = v_2 A_2 \tag{2-15}$$

或写为

$$q = vA = \text{常量} \tag{2-16}$$

式 (2-16) 是液流的流量连续性方程，它说明恒定流动中流过各截面的不可压缩流体的流量是不变的。因而流速和通流截面的面积成反比。

2.3.2 伯努利方程——能量守恒定律

1. 理想液体的伯努利方程

假设理想液体因无粘性，又不可压缩，因此在管内作稳定流动时没有能量损失。根据能

Conservation of energy.

Fig. 2-10 is the sketch of Bernoulli equation. Consider any two across-section areas A_1, A_2, z_1, and z_2 describe their distances from datum line respectively; v_1, v_2 the velocities at the center A_1 and A_2 respectively; p_1, p_2 the pressures at the center of A_1 and A_2 respectively.

According to the conservation of energy,

$$\frac{p_1}{\rho g} + z_1 + \frac{v_1^2}{2g} = \frac{p_2}{\rho g} + z_2 + \frac{v_2^2}{2g}$$

or

$$\frac{p}{\rho g} + z + \frac{v^2}{2g} = \text{constant} \tag{2-17}$$

Formula (2-17) is the well-known Bernoulli equation. It states that ideal fluid flow in a tube includes three forms of energies: pressure energy, potential energy, and kinetic energy. These three energies can be transferred between each other, but the total energy is always invariable, which is called energy conservation.

2. Real equation of Bernoulli

In many hydraulic systems practical fluid flow through in a tube, pump and several elements, the energies can be lost due to the viscous and friction resistance, so if total loss of h_w occurs in the two across-section areas A_1 and A_2. In addition, the real velocity is a nonuniform distribution in flow field and we can set a kinetic correction factor α to offset this. The coefficient α is defined by

$$\alpha = \frac{\frac{1}{2}\int_A u^2 \rho u \mathrm{d}A}{\frac{1}{2}\rho A v v^2} = \frac{\int_A u^3 \mathrm{d}A}{v^3 A} \tag{2-18}$$

It is the ratio of real kinetic energy to average kinetic energy. Here $\alpha = 1.1$ when it is turbulent flow, and $\alpha = 2$ when laminar flow, but in practice usually set the $\alpha = 1$. Thus, after introducing the energy loss and kinetic correction factor α, the following modifying equation of formula (2-17) is regarded as the real or practical equation of Bernoulli,

$$z_1 + \frac{p_1}{\rho g} + \frac{\alpha_1 v_1^2}{2g} = z_2 + \frac{p_2}{\rho g} + \frac{\alpha_2 v_2^2}{2g} + h_w \tag{2-19}$$

Note that when formula (2-19) is used in engineering calculation:

1) The choice of across-section area 1 and 2 should be selected along the streamline direction of fluid flow which has a steady flow status.

2) The position z and pressure p should be two values at the same point, usually one can choose the center point of this across-section area.

3. Application example of the equation of Bernoulli

Example 2-3 Venturi meter

The Venturi meter shown reduces the pipe diameter from 0.1m to *a* minimum of 0.05m as shown in Fig. 2-11. Calculate the flow rate and the mass flux assuming ideal conditions.

Solution: The control volume is (double dot dash line shown in Fig. 2-11) selected as shown so that at the entrance and exit correspond to the sections where the pressure information of the manometer can be applied. The manometer's reading is interpreted as follows

量守恒定律，同一管道每一截面的总能量都是相等的。

在图 2-10 中任取两个截面 A_1 和 A_2，它们距基准水平面的距离分别为 z_1 和 z_2，断面平均流速分别为 v_1 和 v_2，压力分别为 p_1 和 p_2。根据能量守恒定律有

$$\frac{p_1}{\rho g}+z_1+\frac{v_1^2}{2g}=\frac{p_2}{\rho g}+z_2+\frac{v_2^2}{2g}$$

或

$$\frac{p}{\rho g}+z+\frac{v^2}{2g}=\text{常量} \tag{2-17}$$

以上两式即为理想液体的伯努利方程，其物理意义为：在管内作稳定流动的理想流体具有压力能、势能和动能三种形式的能量，在任一截面上这三种能量可以互相转换，但其总和不变，即能量守恒。

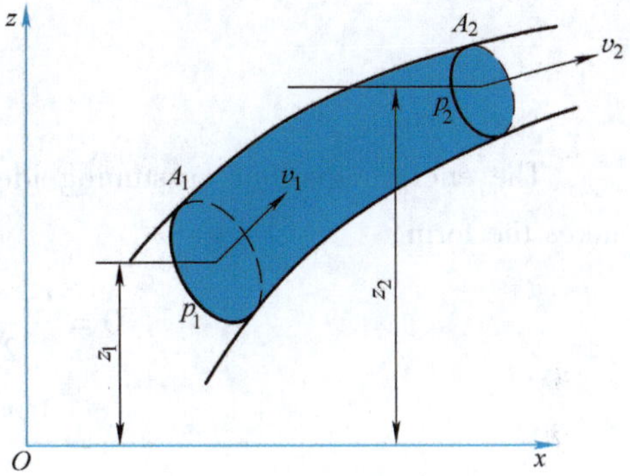

Fig. 2-10 Sketch of Bernoulli equation

（伯努利方程推导简图）

2. 实际液体伯努利方程

实际液体在管道内流动时，由于液体存在粘性，会产生内摩擦力，消耗能量；由于管道形状和尺寸的变化，液流会产生扰动，消耗能量。因此，实际液体流动时存在能量损失。设单位质量液体在两截面之间流动的能量损失为 h_w。

另外，因实际流速 u 在管道通流截面上的分布不是均匀的，为方便计算，一般用平均流速替代实际流速计算动能。显然，这将产生计算误差。为修正这一误差，便引进了动能修正系数 α，它等于单位时间内某截面处的实际动能与按平均流速计算的动能之比，其表达式为

$$\alpha=\frac{\frac{1}{2}\int_A u^2\rho u\mathrm{d}A}{\frac{1}{2}\rho A v v^2}=\frac{\int_A u^3\mathrm{d}A}{v^3 A} \tag{2-18}$$

动能修正系数 α 在湍流时取 1.1，在层流时取 2。实际计算时常取 $\alpha=1$。

在引进了能量损失 h_w 和动能修正系数 α 后，实际液体的伯努利方程表示为

$$z_1+\frac{p_1}{\rho g}+\frac{\alpha_1 v_1^2}{2g}=z_2+\frac{p_2}{\rho g}+\frac{\alpha_2 v_2^2}{2g}+h_w \tag{2-19}$$

在利用上式进行计算时必须注意的是：

1）截面 1、2 应顺流向选取，且选在流动平稳的通流截面上。

2）z 和 p 应为通流截面的同一点上的两个参数，为方便起见，一般将这两个参数定在通流截面的轴心处。

3. 伯努利方程应用举例

例 2-3 文丘利流量计

图 2-11 所示为一文丘利流量计。其中管子直径从 0.1m 减小到 0.05m。计算在理想状态下的体积流量和质量流量。

解： 根据题意取控制体（如图 2-11 虚线框所示）为研究对象，并且在控制体的进、出两个截面上含有压力数据信息。

$$p_a = p_b$$

$$p_1 + \rho g(z + 1.2) = p_2 + \rho g z + \rho_{Hg} g \times 1.2 = p_2 + \rho g z + 13.6 \rho g \times 1.2$$

Here z is the distance from the pipe centerline to the top of the mercury column. The manometer then gives

$$\frac{p_1 - p_2}{\rho g} = 15.12\text{m}$$

Continuity allows us to relate v_2 to v_1 by

$$v_1 A_1 = v_2 A_2$$

$$v_2 = \frac{A_1}{A_2} v_1 = 4v_1$$

The energy equation, assuming ideal conditions (no losses and uniform flow) with $h_w = 0$, takes the form

$$0 = \frac{v_2^2 - v_1^2}{2g} + \frac{p_2 - p_1}{\rho g} + (z_2 - z_1)$$

$$= \frac{16v_1^2 - v_1^2}{2g} - 15.12$$

So $$v_1 = 4.45\text{m/s}$$

The flow rate is

$$q_V = A_1 v_1 = \pi \times 0.05^2 \times 4.5 = 0.035\text{m}^3/\text{s}$$

The mass flux is

$$q_m = \rho q_V = 1000 \times 0.035 = 35\text{kg/s}$$

Example 2-4 Try to analyse the condition of a pump drawing into oil from a reservoir by the equation of Bernoulli. The setup is shown in Fig. 2-12. Set the absolute pressure on the drawing input pump orifice (2-2 across-section area) as p_2, the pressure at 1-1 across-section area, i. e., the hydraulic oil surface, as p_1. Here $p_1 = p_a$, p_a is atmosphere pressure; and the distance from pump orifice to hydraulic oil surface of the reservoir is h.

Solution: Take 1-1 across-section area as the datum plane, and consider the kinetic correction factor $\alpha_1 = \alpha_2 = 1$, and then the equation of Bernoulli for 1-1 and 2-2 across-section areas can be described as follows

$$\frac{p_1}{\rho g} + \frac{\alpha_1 v_1^2}{2g} = \frac{p_2}{\rho g} + \frac{\alpha_2 v_2^2}{2g} + h + h_w$$

Where $p_1 = p_a$, $v_1 = 0$ is the velocity at 1-1; v_2 is the velocity at 2-2; h_w is the energy loss of whole pipeline, so the above formula can be simplified as

$$\frac{p_a}{\rho g} = \frac{p_2}{\rho g} + h + \frac{v_2^2}{2g} + h_w$$

The pressure (vacuum or negative pressure) of pump orifice is

$$p_a - p_2 = \rho g h + \frac{1}{2}\rho v_2^2 + \rho g h_w = \rho g h + \frac{1}{2}\rho v_2^2 + \Delta p$$

From this we can see that pressure of input pump orifice involves three parts: ①pressure needed

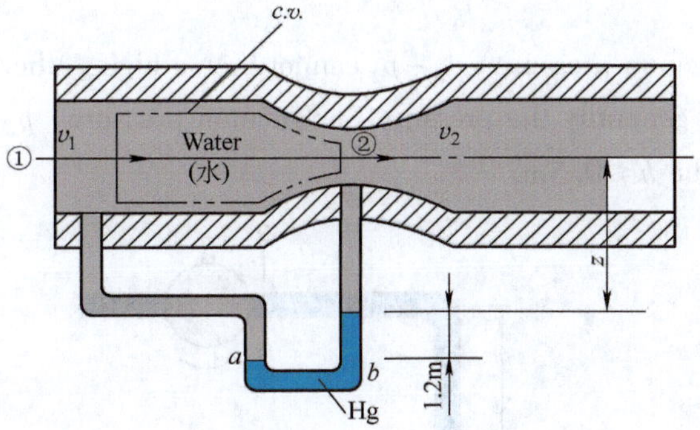

Fig. 2-11 Venturi meter（文丘利流量计）

那么，此压力计的读数为

$$p_a = p_b$$

$$p_1 + \rho g(z + 1.2) = p_2 + \rho gz + \rho_{Hg} g \times 1.2 = p_2 + \rho gz + 13.6\rho g \times 1.2$$

这里 z 表示水银柱的顶部到管子中心线的距离，则

$$\frac{p_1 - p_2}{\rho g} = 15.12\text{m}$$

根据质量连续性有

$$v_1 A_1 = v_2 A_2$$

$$v_2 = \frac{A_1}{A_2} v_1 = 4v_1$$

假设在理想条件下，能量损失 $h_w = 0$，即

$$0 = \frac{v_2^2 - v_1^2}{2g} + \frac{p_2 - p_1}{\rho g} + (z_2 - z_1) = \frac{16v_1^2 - v_1^2}{2g} - 15.12$$

求得

$$v_1 = 4.45\text{m/s}$$

因此体积流量为

$$q_V = A_1 v_1 = \pi \times 0.05^2 \times 4.5\text{m}^3/\text{s} = 0.0350\text{m}^3/\text{s}$$

质量流量为

$$q_m = \rho q_V = 1000 \times 0.035\text{kg/s} = 35.0\text{kg/s}$$

例 2-4 应用伯努利方程分析液压泵正常吸油的条件。如图 2-12 所示液压泵从油箱吸油，设液压泵吸油口处（2-2 截面）的绝对压力为 p_2，油箱液面（1-1 截面）压力 p_1 为大气压 p_a，泵吸油口至油箱液面高度为 h。

解： 取油箱液面（1-1 截面）为基准面，对 1-1，2-2 两截面列伯努利方程（动能修正系数取 $\alpha_1 = \alpha_2 = 1$）有

$$\frac{p_1}{\rho g} + \frac{\alpha_1 v_1^2}{2g} = \frac{p_2}{\rho g} + \frac{\alpha_2 v_2^2}{2g} + h + h_w$$

式中，p_1 等于大气压；v_1 为油箱液面流速，可视为零；v_2 为吸油管速；h_w 为吸油管路的能

that results in velocity v_2;②pressure needed that lifts oil surface to a given height h;③the pressure loss in the pipeline.

The pressure (or vacuum pressure) $p_a - p_2$ cannot be too high; otherwise it will produce cavitation and make noise. So generally the pressure (or vacuum pressure) $p_a - p_2$ is limited to less than 0.3×10^5(Pa) and set the $h \leqslant 0.5$m.

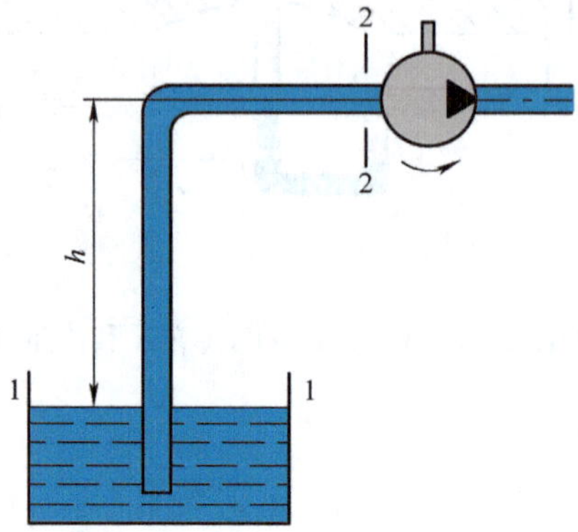

Fig. 2-12 Setup of hydraulic pump(液压泵从油箱吸油示意图)

2.3.3 Equation of momentum—conservation of momentum

In any system, of which Fig. 2-13 and Fig. 2-14 are two examples, the rate of change of momentum in the system equals the net applied external force

$$F = m\frac{dv}{dt} \tag{2-20}$$

The equation looks the same as the relationship

$$F = ma \tag{2-21}$$

Known as Newtonian's second law to all engineering students. Assume a frictionless, incompressible liquid in a cylindrical passage as shown in Fig. 2-14. The flow rate is being accelerated uniformly with time:

$$q_2 = q_1 + \int_0^t \frac{dq}{dt} \tag{2-22}$$

The force balance is, from formula(2-20)

$$p_1A - p_2A = \frac{d}{dt}(\rho ALv) \tag{2-23}$$

Because $q = Av$, so

$$p_1 - p_2 = \frac{\rho L}{A}\frac{dq}{dt} \tag{2-24}$$

Formula (2-24) says that a fluid in a passage cannot be accelerated (undergo a change in momentum) without having a pressure differential across the section of the fluid. This is important later for studying the stability of certain forms of valves.

A similar reasoning applies to the flow of fluid through curved passages as shown in Fig. 2-15. Because velocity is a vector quantity it contains elements of both speed and direction. Therefore a

量损失。

代入已知条件，上式可简化为

$$\frac{p_a}{\rho g}=\frac{p_2}{\rho g}+h+\frac{v_2^2}{2g}+h_w$$

即液压泵吸油口的真空度为

$$p_a-p_2=\rho gh+\frac{1}{2}\rho v_2^2+\rho gh_w=\rho gh+\frac{1}{2}\rho v_2^2+\Delta p$$

由此可知，液压泵吸油口的真空度由三部分组成：①产生一定流速 v_2 所需的压力；②把油液提升到高度 h 所需的压力；③吸油管的压力损失。

为保证液压泵正常工作，液压泵吸油口的真空度不能太大。若真空度太大，在绝对压力 p_2 低于油液的空气分离压 p_g 时，溶于油液中的空气会分离析出形成气泡，产生气穴现象，出现振动和噪声。为此，必须限制液压泵吸油口的真空度小于 0.3×10^5Pa，具体措施除增大吸油管直径、缩短吸油管长度、减少局部阻力以降低 $\frac{1}{2}\rho v_2^2$ 和 Δp 两项外，一般对液压泵的吸油高度 h 进行限制，通常取 $h\leqslant0.5$m。若将液压泵安装在油箱液面以下，则 h 为负值，对降低液压泵吸油口的真空度更为有利。

2.3.3 动量方程——动量守恒

在图2-13、图2-14或图2-15所示的任意一个系统中，作用在物体上全部外力的矢量和应等于物体在力作用方向上的动量变化率，即

$$F=m\frac{\mathrm{d}v}{\mathrm{d}t} \tag{2-20}$$

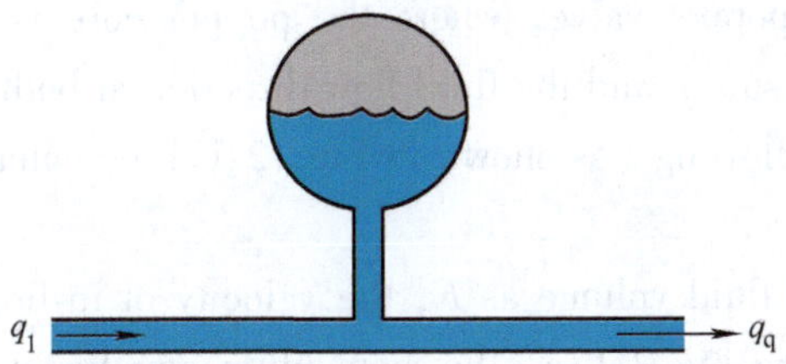

Fig. 2-13 Sketch of oil flow through a pipeline with a pressure vessel
（带有压力容器的管道流动示意图）

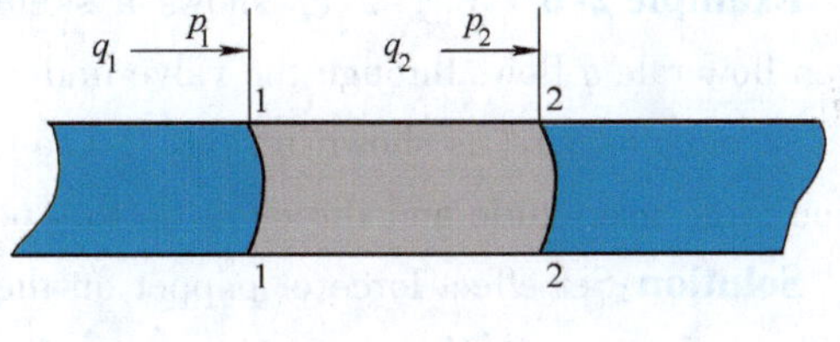

Fig. 2-14 Sketch of oil flow through a pipeline
（液压油在一个管道中流动的示意图）

上式与牛顿第二定律相似，即

$$F=ma \tag{2-21}$$

假设无摩擦和不可压缩的液体在图2-14所示的流道中流动时，取1-1和2-2截面所包括的区域为控制体，则在2-2截面上的流量等于1-1截面的流量 q_1 加上随时间均匀地变化（增加）的增量，即

$$q_2=q_1+\int_0^t\frac{\mathrm{d}q}{\mathrm{d}t} \tag{2-22}$$

由式（2-20）可以得出力的平衡公式

change in the direction of flow, for instance, hydraulic oil may flow through a bent tubing, will cause a change in momentum as defined in formula (2-20). The forces can be resolved into a component F_x which is axial to the inlet direction and a component F_y which is normal to the inlet direction.

$$F_x = \rho qv(1 - \cos\beta) \text{ and } F_y = \rho qv\sin\beta \qquad (2\text{-}25)$$

For a 90° bend $F_x = F_y$.

Example 2-5 Fig. 2-16 shows a sketch of a spool valve. When oil fluid flows through the valve, try to calculate: the axial effect force of oil fluid on the spool surface.

Solution: The control volume is selected as shown such that the entrance and exit correspond to the sections as arrows shown in Fig. 2-16. If it is steady flow and then the effect force of spool on the control volume (fluid) can be calculated by momentum equation

$$F = \rho q(v_2\cos\theta_2 - v_1\cos\theta_1)$$

Because $\theta_2 = 90°$, thus $F = -\rho qv_1\cos\theta_1$. The effect direction is on the left; but the reaction force of the fluid on the spool is $F' = -F = \rho qv_1\cos\theta_1$ and the direction is right. Therefore the action force of fluid on the spool makes the spool to tend to be closed.

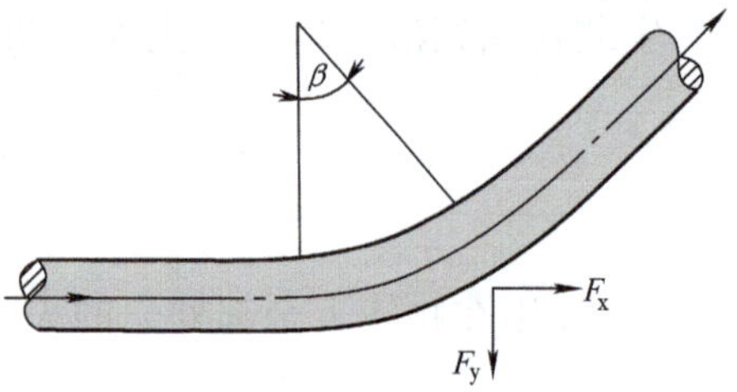

Fig. 2-15 Sketch of oil through curved passages
(液压油流过弯曲管道的示意图)

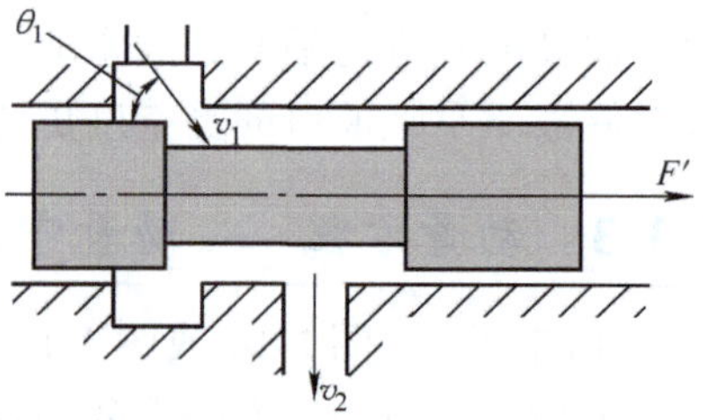

Fig. 2-16 Hydraulic dynamic on the spool valve
(滑阀上的液动力)

Example 2-6 Fig. 2-17 shows a sketch of a poppet valve, where the poppet core is 2α. When flow rate q flows through the valve under the pressure p and the fluid flow direction at both statuses of out-flowing (as shown in Fig. 2-17a) and in-flowing (as shown in Fig. 2-17b), calculate action force magnitude and direction on this poppet core.

Solution: Set effect force of poppet on the control fluid volume as F, the velocity of in-flowing v_1 and out-flowing v_2. However, for status of out-flowing Fig. 2-17a, the momentum equation is established along the streamline direction

$$p\frac{\pi}{4}d^2 - F = \rho q(v_2\cos\theta_2 - v_1\cos\theta_1)$$

Because of $v_1 \ll v_2$, neglect v_1, i. e., $v_1 = 0$, if $\theta_2 = \alpha, \theta_1 = 0$, the above formula becomes

$$F = p\frac{\pi}{4}d^2 - \rho qv_2\cos\alpha$$

Similarly, we can get the result for status Fig. 2-17b as follows

$$p\frac{\pi}{4}(D^2 - d_2{}^2) - p\frac{\pi}{4}(D^2 - d^2) - F = \rho q(v_2\cos\theta_2 - v_1\cos\theta_1)$$

Here $\theta_1 = 90°, \theta_2 = \alpha$ and then

$$F = \frac{\pi}{4}p(d^2 - d_2{}^2) - \rho qv_2\cos\alpha$$

$$p_1 A - p_2 A = \frac{\mathrm{d}}{\mathrm{d}t}(\rho A l v) \tag{2-23}$$

由于流量 $q = Av$，因此

$$p_1 - p_2 = \frac{\rho l}{A}\frac{\mathrm{d}q}{\mathrm{d}t} \tag{2-24}$$

式（2-24）表明沿流体流动的截面，在没有压差的情况下，流道中的液体将不会产生动量的变化。这点对研究某一种特定形式的阀的稳定性问题是一个重要的理论基础。

类似地，可推导流体流过弯曲管道（图 2-15）的情况。由于速度是具有大小和方向的矢量，因此流动中的液体在流动方向发生改变时，如液压油在一个弯曲的流道中流动时，动量将发生改变，其改变量可以采用式（2-20）计算。而作用力可以分成进口沿轴线方向分量 F_x 和沿进口的法向分量 F_y，即

$$F_x = \rho q v(1 - \cos\beta) \text{ 和 } F_y = \rho q v \sin\beta \tag{2-25}$$

对于 90°弯曲的弯管，$F_x = F_y$。

例 2-5　当液流通过图 2-16 所示的滑阀时，求：液流对阀芯的轴向作用力。

解： 取阀进、出口之间的液体为控制体积。设液流作恒定流动，则在此控制体积内液体上的力按照动量守恒原理应为

$$F = \rho q(v_2\cos\theta_2 - v_1\cos\theta_1)$$

式中，θ_1、θ_2 分别为液流流经滑阀时进、出口流束与滑阀轴线之间的夹角，称之为液流速度方向角。

因 $\theta_2 = 90°$，故得

$$F = -\rho q v_1 \cos\theta_1\text{，方向向左}$$

而液体对阀芯的轴向作用力为

$$F' = -F = \rho q v_1 \cos\theta_1\text{，方向向右}$$

即这时液流有一个力图使阀口关闭的液动力。

例 2-6　图 2-17 所示锥阀的锥角为 2α。液体在压力 p 的作用下以流量 q 流经锥阀，当液流方向是外流式（图 2-17a）和内流式（图 2-17b）时，求：作用在阀芯上的液动力的大小和方向。

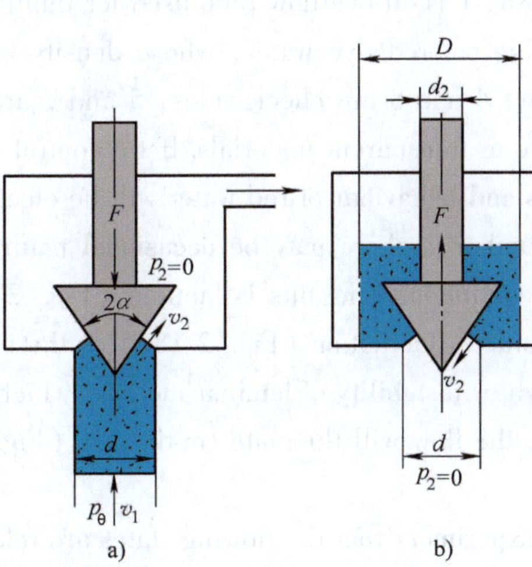

Fig. 2-17　Hydraulic dynamic on the poppet valve（锥阀上的液动力）

For the two cases above the fluid action pressures on the poppet are all equal to F. The action directions are shown in Fig. 2-17a and Fig. 2-17b respectively. And both calculations mentioned above show that the fluid dynamic $\rho q v_2 \cos\alpha$ acted on the poppet is all negative value and action directions are the same as those of F shown in Fig. 2-17a and Fig. 2-17b respectively. So for the Fig. 2-17a the fluid dynamic pressure makes the poppet orifices tend to be closed, and for the Fig. 2-17b tend to be opened.

From these two examples, we can see that action force direction of fluid dynamic pressure on poppet or spool should be considered according to the detail status and we could not consider all tend spool orifice to be closed in any conditions.

2.4 Characteristics of Fluid Flow in Pipeline

When a continuity viscous fluid flows through variable section such as an abruptly bend pipe or orifice, impact between fluid and wall will occur. In this case, fluid will lose parts of energy in order to overcome resistance. This can be presented by introducing the pressure loss h_w and kinetic correction factor α, i. e. , in the above mentioned real fluid Bernoulli's equation

$$z_1 + \frac{p_1}{\rho g} + \frac{\alpha_1 v_1^2}{2g} = z_2 + \frac{p_2}{\rho g} + \frac{\alpha_2 v_2^2}{2g} + h_w$$

The term of h_w includes two parts: ①pressure losses along parallel pipes; ②minor (or local) losses. The minor losses occur in some local places, such as valves entrance or exit orifice, sudden expansion, bends, elbows, or gradual expansions or contractions pipe.

The pressure loss $h_w$and kinetic correction factor α are all related with the states of fluid flow.

2.4.1 States of fluid flow and Reynolds number

Reynolds, a British physical scientist, had done lots of observation to discover that there are three main states of flow, i. e. , laminar, transition and turbulent in a pipe. The Reynolds test is shown in Fig. 2-18. In Fig. 2-18a, 1 is an overflow pipe used for limiting water level; 2 is a supply pipe;3 is a small reservoir flowing colored dye water, whose density is the same as water; 6 is a large pipe reservoir flowing water; 4 and 8 are check vales, 5 and 7 are all pipes flowing red water from 3. The whole setup is made of transparent materials. If we control the velocity by adjusting the check valve 8, the flowing states and behaviour of red water will be clearly seen in big pipe 7. Under a lower velocity (or Reynolds number), there may be occasional natural disturbances which damp out quickly and then become a continuous line, this is laminar (Fig. 2-18b). If transition occurs, there will be sharp bursts of turbulent fluctuation (Fig. 2-18c) as the increasing velocity (or Reynolds number) causes a breakdown or instability of laminar motion, which is called transition. At sufficiently large velocity (or Re), the flow will fluctuate continually (Fig. 2-18d) and that is termed fully turbulent.

It has also been proved by experiment that the flowing states are related with average velocity v, pipe diameter d and kinematic viscosity ν flowing in a pipe. Reynolds number, a dimensionless number, consists of these three parameters, i. e.

解： 设阀芯对控制体的作用力为 F，流入速度为 v_1，出流速度为 v_2。对于图 2-17a 所示的情况，控制体取在阀口下方（图中阴影部分），沿液流方向列出动量方程

$$p\frac{\pi}{4}d^2 - F = \rho q(v_2\cos\theta_2 - v_1\cos\theta_1)$$

因为 $v_1 \ll v_2$，可忽略 v_1，$\theta_2=\alpha$，$\theta_1=0°$，则代入整理后得

$$F = p\frac{\pi}{4}d^2 - \rho q v_2\cos\alpha$$

对于图 2-17b 所示的情况，将控制体取在上方。同理，列出动量方程为

$$p\frac{\pi}{4}(D^2 - d_2{}^2) - p\frac{\pi}{4}(D^2 - d^2) - F = \rho q(v_2\cos\theta_2 - v_1\cos\theta_1)$$

因为，$\theta_1=90°$，$\theta_2=\alpha$ 则

$$F = \frac{\pi}{4}p(d^2 - d_2{}^2) - \rho q v_2\cos\alpha$$

在上述两种情况下，液流对锥阀作用力的大小都等于 F，而作用方向各自与图示方向相反。由上述两个 F 的计算式可以看出，其中作用在锥阀上的液动力项 $\rho q v_2\cos\alpha$ 均为负值，亦即此力的作用方向应与图示方向一致。因此，在图2-17a所示情况下，液动力力图使锥阀关闭；可是在图 2-17b 所示情况下，却欲使之打开。所以不能笼统地认为，阀上稳态液动力的作用方向是固定不变的，必须对具体情况作具体分析。

2.4 管道中液流的特性

由于流动的液体具有粘性，以及液体流动时突然转弯和通过阀口会产生相互撞击和出现漩涡等，液体流动时必然会产生阻力。为了克服阻力，液体流动时需要损耗一部分能量。这种能量损失可以用 h_w 和动能修正系数 α 来描述，即用上节所叙述的实际流体的伯努利方程表示

$$z_1 + \frac{p_1}{\rho g} + \frac{\alpha_1 v_1^2}{2g} = z_2 + \frac{p_2}{\rho g} + \frac{\alpha_2 v_2^2}{2g} + h_w$$

其中 h_w 由两部分组成：①沿程压力损失；②局部压力损失。局部压力损失发生在液压流过的局部位置，如液体流过进、出阀口，突然的扩大管，弯管或渐扩或渐缩的管道等。

h_w 和动能修正系数 α 均与液体在管道中的流动状态有关。

2.4.1 流体的流态与雷诺数

液体在管道中流动时存在两种流动状态：层流和湍流。两种流动状态可通过雷诺实验进行观察，如图 2-18 所示。

实验装置如图 2-18a 所示，溢流管 1 用来保持液面高度不变，2 为供水管，在容器 3 和 6 中分别装满了密度与水相同的红色液体和水，4 和 8 是阀门，5 和 7 分别为小管和大管。整套试验装置由透明材料制作而成。大管 7 中流体流动速度可通过调节阀门 8 的开度得到控制，从而可明显地从大管 7 中红色液体流动状态判别为：当管中流速小到一定值时，红色液体在大管 7 中呈一条明显的直线，将小管 5 的出口上下移动，则红色直线也上下移动，层次分明不相混杂（图 2-18b），液体的这种流动状态称为层流。当流速逐渐增大到某一值时，可以看到红线上下波动呈波纹状（图 2-18c）。表明层流状态被破坏，液流开始出现紊乱。

$$Re = \frac{vd}{\upsilon} \tag{2-26}$$

The Reynolds number was observed to be a ratio of the inertial force to the viscous force. Hence, when this ratio becomes large, it is expected that inertial forces may dominate the viscous forces. Re_{cr} is a critical value between laminar and turbulence usually determined by experimental data. The familiar critical Reynolds numbers based on different pipe materials are shown in Tab. 2-3.

Tab. 2-3 Familiar critical Reynolds numbers based on different pipe materials (常见液流管道的临界雷诺数)

Pipes(管道)	Re_{cr}	Pipes(管道)	Re_{cr}
Smooth metal pipe(光滑金属圆管)	2320	Smooth pipe with eccentric annulari gap (光滑的偏心环状缝隙)	1000
Hosepipe(橡胶软管)	1600 – 2000	Column valve orifice(圆柱形滑阀阀口)	260
Smooth pipe with concentric annular gap (光滑的同心环状缝隙)	1100	Poppet valve orifice(锥阀阀口)	20 – 100

For flow in noncircular ducts, the Reynolds number is calculated by

$$Re = \frac{4vR}{\upsilon} \tag{2-27}$$

Here R is hydraulic radius, defined by

$$R = \frac{A}{x} = \frac{cross\text{-}sectional\ area}{wetted\ perimeter} \tag{2-28}$$

It equals to the ratio of cross-sectional area to wetted perimeter.

2.4.2 Losses along circle parallel pipe

The losses due to viscosity in equal diameter pipe are referred to as losses in parallel pipe, which will change with the different flowing states.

1. Losses in parallel pipe at laminar flow

(1) Velocity profile in a laminar pipe flow This volume can be considered a section as shown in Fig. 2-19. If we set radius as r, length as l, action pressure on two surfaces as p_1 and p_2, friction action on the inside surface of pipe as F_f, then a force balance in the x-direction yields,

$$(p_1 - p_2)\pi r^2 = F_f = -2\pi r l \mu \frac{du}{dr} \tag{2-29}$$

The velocity gradient is negative because of velocity decreasing as the radii increasing. If we set $\Delta p = p_1 - p_2$, and $du = -\frac{\Delta p}{2\mu l} r dr$, integrate it and under the boundary of $u = 0$ at $r = R$, we obtain

$$u = \frac{\Delta p}{4\mu l}(R^2 - r^2) \tag{2-30}$$

It says that velocity profile in a laminar pipe flow along radii direction is a parabola profile and the maximum velocity is at the axis center $r = 0$, $u_{max} = \frac{\Delta p}{4\mu l}R^2$.

若大管 7 中液体流速继续增大，红线消失（图 2-18d)，表明液流完全紊乱，这时流动状态称为湍流。

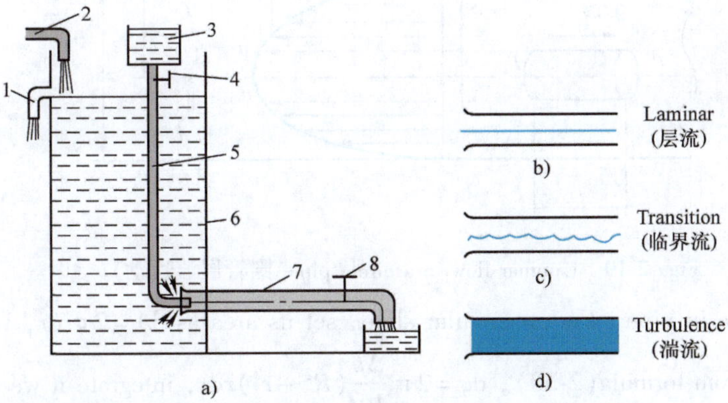

Fig. 2-18　Setup of Reynolds test（雷诺试验装置）

1—Overflow pipe（溢流管）　2—Supply pipe（供水管）　3，6—Reservoir（容器）
4，8—Check vale（阀门）　5—Small pipe（小管）　7—Big pipe（大管）

实验结果还证明，液体在圆管中的流动状态不仅与管内的平均流速 v 有关，还和管道内径 d、液体的运动粘度 υ 有关。而决定流动状态的，是这三个参数所组成的一个称为雷诺数 Re 的无量纲数，即

$$Re = \frac{vd}{\upsilon} \tag{2-26}$$

式（2-26）中的雷诺数 Re 的物理意义为：惯性力与粘性力之比。如果这个比值大表明惯性力占优势。记 Re_{cr} 为判别液流由层流转变为湍流的临界雷诺数，常见液流管道的临界雷诺数由实验求得，见表 2-3。

对于非圆截面的管道来说，Re 可由下式计算

$$Re = \frac{4vR}{\upsilon} \tag{2-27}$$

式中，R 为通流截面的水力半径，它等于液流的有效面积 A 和它的湿周（有效截面的周界长度）x 之比，即

$$R = \frac{A}{x} \tag{2-28}$$

水力半径大，意味着液流和管壁接触少，阻力小，通流能力大，即使通流截面积小时也不易堵塞。

2.4.2　沿程压力损失

液体在等直径管中流动时因粘性摩擦而产生的损失，称为沿程压力损失。液体的沿程压力损失也因流体流动状态的不同而有所区别。

1. 层流时的沿程压力损失

（1）通流截面上的流速分布规律　如图 2-19 所示，液体在等直径水平圆管中作层流运动。在液流中取一段与管轴相重合的微小圆柱体作为研究对象，设其半径为 r，长度为 l，作用在两端面的压力分别为 p_1 和 p_2，作用在侧面的内摩擦力为 F_f。液流在作匀速运动时受

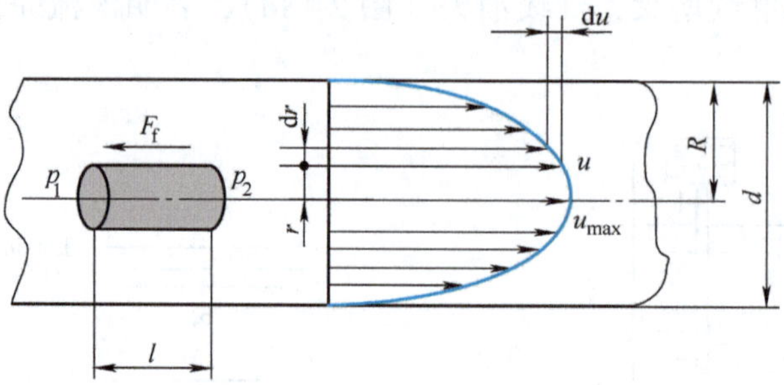

Fig. 2-19 Laminar flow in a circle pipe(圆管层流运动)

(2) The flow rate in pipe For an annular slice, set its area as $dA=2\pi r dr$, volume flow rate $dq=udA=2\pi r dru$, from formula(2-30), $dq=2\pi\frac{\Delta p}{4\mu l}(R^2-r^2)rdr$, integrate it we obtain

$$q=\int_0^R 2\pi\frac{\Delta p}{4\mu l}(R^2-r^2)rdr=\frac{\pi R^4}{8\mu l}\Delta p=\frac{\pi d^4}{128\mu l}\Delta p \tag{2-31}$$

(3) Average velocity in pipe According to the definition of average velocity, we obtain

$$v=\frac{q}{A}=\frac{1}{\pi R^2}\frac{\pi R^4}{8\mu l}\Delta p=\frac{\Delta p}{8\mu l}R^2=\frac{d^2}{32\mu l}\Delta p \tag{2-32}$$

Formula(2-32) says that the average velocity is 1/2 of the maximum velocity.

(4) Losses along circle parallel pipe From formula (2-32), the loss is

$$\Delta p_\lambda=\Delta p=\frac{32\mu l v}{d^2} \tag{2-33}$$

The formula presents that the loss along circle parallel pipe is proportional to length l, velocity v, viscosity μ, and inversely proportional to the diameter squared. The formula (2-33) can be also written as

$$\Delta p_\lambda=\frac{64}{\frac{vd}{\upsilon}}\frac{l}{d}\frac{\rho v^2}{2}=\frac{64}{Re}\frac{l}{d}\frac{\rho v^2}{2}=\lambda\frac{l}{d}\frac{\rho v^2}{2} \tag{2-34}$$

Where λ is the resistance coefficient along a circle pipe, In theory, $\lambda=\frac{64}{Re}$, but in a practical case, $\lambda=\frac{75}{Re}$ for a metal pipe, $\lambda=\frac{80}{Re}$ for a hosepipe because influence of temperature needs to be considered.

2. Losses in parallel pipe at turbulence flow

When turbulence flow has happened, particles in fluid have behaved ruleless motion and changeable speed and direction with time. Until now, the law can be only probed by experiment.

The experiment has shown that resistance coefficient λ is related with Reynolds number Re and relative roughness in pipe wall Δ/d, i. e., $\lambda=f(Re,\Delta/d)$, here Δ is absolute roughness in pipe wall. However, water-power slippery pipe happened when Reynolds number were more than 2320 ($Re>2320$) due to less affection of roughness despite turbulence. The resistance coefficient λ can be calculated by experimental formula as follows for water-power slippery pipe

力平衡，故有

$$(p_1 - p_2)\pi r^2 = F_f = -2\pi r l \mu \frac{du}{dr} \tag{2-29}$$

因流速 u 随 r 的增大而减小，故 du/dr 为负值，所以加一负号。令 $\Delta p = p_1 - p_2$，$du = -\frac{\Delta p}{2\mu l} r dr$，对此式进行积分，并应用边界条件，当 $r = R$ 时，$u = 0$，得

$$u = \frac{\Delta p}{4\mu l}(R^2 - r^2) \tag{2-30}$$

可见管内液体质点的流速在半径方向上按抛物线规律分布。最小流速发生在管壁上，即 $r = R$ 处，$u_{min} = 0$；最大流速发生在轴线上，即 $r = 0$ 处，$u_{max} = \frac{\Delta p}{4\mu l} R^2$。

（2）通过管道的流量　对于微小环形通流截面面积 $dA = 2\pi r dr$，所通过的流量 $dq = u dA = 2\pi r dr u$，所以，$dq = 2\pi \frac{\Delta p}{4\mu l}(R^2 - r^2) r dr$，于是积分得

$$q = \int_0^R 2\pi \frac{\Delta p}{4\mu l}(R^2 - r^2) r dr = \frac{\pi R^4}{8\mu l}\Delta p = \frac{\pi d^4}{128\mu l}\Delta p \tag{2-31}$$

（3）管道内的平均流速　根据平均流速的定义，可得

$$v = \frac{q}{A} = \frac{1}{\pi R^2}\frac{\pi R^4}{8\mu l}\Delta p = \frac{R^2}{8\mu l}\Delta p = \frac{d^2}{32\mu l}\Delta p \tag{2-32}$$

将式（2-32）与 $u_{max} = \frac{R^2}{4\mu l}\Delta p$ 比较可知，平均流速 v 为最大流速 u_{max} 的一半。

（4）沿程压力损失　从式（2-32）中求出的 Δp 表达式即为沿程压力损失

$$\Delta p_\lambda = \Delta p = \frac{32\mu l v}{d^2} \tag{2-33}$$

由上式可知，液流在直管中作层流流动时，其沿程压力损失与管长、流速、粘度成正比，而与管径的平方成反比。适当变换式（2-33）可写成如下形式

$$\Delta p_\lambda = \frac{64}{\frac{vd}{\nu}}\frac{l}{d}\frac{\rho v^2}{2} = \frac{64}{Re}\frac{l}{d}\frac{\rho v^2}{2} = \lambda \frac{l}{d}\frac{\rho v^2}{2} \tag{2-34}$$

式中，λ 为沿程阻力系数，理论值 $\lambda = 64/Re$，考虑实际流动中的油温变化不匀等问题，因而在实际计算时，对金属管取 $\lambda = 75/Re$，橡胶软管 $\lambda = 80/Re$。

由于在液压传动中，液体自重和位置变化对压力的影响很小可以忽略，所以在水平管的条件下推导的式（2-34）同样适用于非水平管。

2. 湍流时的沿程压力损失

湍流时，液体质点作无规则的相互混杂运动，其运动速度的大小和方向都随时间而变，是一种很复杂的流动，目前主要还是靠实验来探索其某些规律。实验证明，湍流沿程阻力系数 λ 除与雷诺数有关外，还与管壁的表面粗糙度值有关，即 $\lambda = f(Re, \Delta/d)$，这里 Δ 为管壁的绝对表面粗糙度值，Δ/d 为相对表面粗糙度值。对光滑圆管，当 $Re > 2320$ 以后，尽管属于湍流，但由于油液的粘度使贴近管壁处还会有一层薄的流层，它能够减弱管壁表面粗糙度值的影响，这种情况称为水力光滑管。此时可以认为 λ 只与 Re 有关，而与管壁表面粗糙

$$\lambda = 0.3164Re^{-0.25} \qquad (2320 < Re < 10^5)$$

or $$\lambda = 0.032 + 0.212Re^{-0.237} \qquad (10^5 < Re < 3\times10^6) \tag{2-35}$$

The effect of roughness in pipe wall on loss along pipe has become a main factor when Reynolds number is increasing gradually, which is called completely roughness pipe, and λ can be calculated by experimental formula as follows

$$\lambda = (2gld/2\Delta + 1.74)^{-2} \qquad (Re > 900d/\Delta) \tag{2-36}$$

Here Δ is related with material of pipe, such as steel tube 0.04mm, copper pipe 0.0015-0.01mm, aluminum 0.0015-0.06 mm and hosepipe 0.03mm. On the other hand, the velocity is well distributed at turbulence flow with the maximum velocity $u_{max} = (1\text{-}1.3)v$. In this case, losses in parallel pipe can be calculated by formula (2-34).

2.4.3 Minor losses in pipe system

In a pipe system, fluid flows through pipe entrance or exit; sudden variety across-sections, such as expansion or contraction; bends, elbows, tees, and other fittings; valves, open or partially closed; gradual expansions or contractions, turbulence flow will occur to give resistance and energy loss. Energy losses are called minor losses.

Usually the minor losses Δp_ξ can be calculated by

$$\Delta p_\xi = \xi\frac{\rho v^2}{2} \tag{2-37}$$

Where ξ is the minor resistance coefficient (can be found from the manual); ρ is density (kg/m^3), and v is average velocity (m/s).

In a hydraulic system where oil flows through different valves, it is difficult for us to calculate the minor losses by formula (2-37) because of the different types of spool. Thus in this case, one should obtain the values of pressure losses Δp_s under a given rating flow rate q_s from a hydraulic handbook, and then calculate the flow rate except the rating rate by pressure loss formula

$$\Delta p_\xi = \Delta p_s\left(\frac{q}{q_s}\right)^2 \tag{2-38}$$

Where q is the real flow rate that oil fluid flows through the valve.

The total energy losses in a whole hydraulic system can be summed after calculating out several section's losses by

$$\sum \Delta p = \sum \Delta p_\lambda + \sum \Delta p_\xi = \sum \lambda\frac{l}{d}\frac{\rho v^2}{2} + \sum \xi\frac{\rho v^2}{2} \tag{2-39}$$

2.5 Flow Rate and Pressure Features of Orifice

This section presents the flow rate formulas for flow through a orifice and clearance, which are the main theoretical foundation used in studying and designing of the hydraulic system leaking.

2.5.1 Thin wall orifice

Thin wall orifice defined as the radio of flow length l to diameter of orifice d is less than 0.5 as

度值无关，λ 值可用下列经验公式计算

$$\lambda = 0.3164Re^{-0.25} \quad (2320 < Re < 10^5)$$

或

$$\lambda = 0.032 + 0.212Re^{-0.237} \quad (10^5 < Re < 3\times10^6) \tag{2-35}$$

当 Re 继续增大时，管壁表面粗糙度值对沿程压力损失的影响便转化为主要因素，而 Re 则不起作用了，这种情况称完全粗糙管。这时沿程阻力系数 λ 可按下式计算

$$\lambda = (2gld/2\Delta + 1.74)^{-2} \quad (Re > 900d/\Delta) \tag{2-36}$$

管壁表面粗糙度值 Δ 和管道材料有关，对钢管取 0.04mm，铜管取 0.0015 ~ 0.01mm，铝管取 0.0015 ~ 0.06mm，橡胶软管取 0.03mm。另外湍流中的流速分布是比较均匀的，其最大流速 $u_{max} = (1 \sim 1.3)\ v$。

这样，湍流时沿程压力损失的计算，就可采用层流时的计算公式，即式（2-34）。

2.4.3 局部压力损失

局部压力损失产生的原因：液体流经管道的进、出口，突然变化的截面，弯头、接头、阀口，全开或部分开的阀口，渐扩或渐缩管道等处时，液体流速的大小和方向将急剧发生变化，因而会产生漩涡，并发生强烈的湍动现象，于是产生流动阻力，由此造成的压力损失称为局部压力损失。

局部压力损失 Δp_ξ 的计算一般按下式进行

$$\Delta p_\xi = \xi \frac{\rho v^2}{2} \tag{2-37}$$

式中，ξ 为局部阻力系数（具体数值可查阅有关手册）；ρ 为液体密度（kg/m^3）；v 为液体的平均流速（m/s）。

因阀芯结构较复杂，故按式（2-37）计算液体流过各种液压阀的局部压力损失较困难，这时可在产品目录中查出阀在额定流量 q_s 下的压力损失 Δp_s。当流经阀的实际流量不等于额定流量时，通过该阀的压力损失 Δp_ξ 可用下式计算

$$\Delta p_\xi = \Delta p_s \left(\frac{q}{q_s}\right)^2 \tag{2-38}$$

式中，q 为通过阀的实际流量。

在求出液压系统中各段管路的沿程压力损失和各局部压力损失后，整个液压系统的总压力损失应为所有沿程压力损失和所有局部压力损失之和，即

$$\sum \Delta p = \sum \Delta p_\lambda + \sum \Delta p_\xi = \sum \lambda \frac{l}{d}\frac{\rho v^2}{2} + \sum \xi \frac{\rho v^2}{2} \tag{2-39}$$

式（2-39）适用于两相邻局部障碍之间距离大于管道内径 10 ~ 20 倍的场合，否则计算出来的压力损失值比实际数值小。这是因为如果局部障碍距离太小，通过第一个局部障碍后的流体尚未稳定就进入第二个局部障碍，这时的液流湍动更强烈，阻力系数要高于正常值的 2 ~ 3 倍。

2.5　孔口及缝隙的流量压力特性

本节主要介绍液流经过小孔及缝隙的流量公式。在研究节流调速及分析计算液压元件的泄漏时它们是重要的理论基础。

shown in Fig. 2-20, usually the orifice is sharp edged.

When hydraulic oil flows out from a small orifice to a large chamber, the flowing fluid form a contraction across-section C-C and then expansion will occur because of inertia, in which part or all of the velocity energy is converted into pressure energy. If the ratio of diameter D before this orifice to the diameter d after it, i. e. , $D/d \geqslant 7$, it is called complete contraction; while $D/d < 7$, it is called incomplete contraction. For the orifice before and after section 1-1 and 2-2, if the kinetic correction factor $\alpha = 1$, then Bernoulli equation is

$$\frac{p_1}{\rho g} + \frac{v_1^2}{2g} = \frac{p_2}{\rho g} + \frac{v_2^2}{2g} + \sum h_\xi \tag{2-40}$$

Where $\sum h_\xi$ is the minor losses caused by orifice and include two parts: fluid flow through suddenly contraction $\sum h_{\xi 1}$ and suddenly expansion sections $\sum h_{\xi 2}$. Here $h_{\xi 1} = \xi v_c{}^2/(2g)$ can be looked up from hydraulic manual, and $h_{\xi 2} = (1 - A_c/A_2) v_c^2/(2g)$, since it is considered the $A_c \ll A_2$, thus $\sum h_\xi = h_{\xi 1} + h_{\xi 2} = (\xi + 1) v_c^2/(2g)$, because of $A_1 = A_2, v_1 = v_2$, so

$$v_c = \frac{1}{\sqrt{\xi + 1}} \sqrt{\frac{2}{\rho}(p_1 - p_2)} = C_v \sqrt{\frac{2\Delta p}{\rho}} \tag{2-41}$$

Where $C_v = \dfrac{1}{\sqrt{\xi + 1}}$ is the speed coefficient, which describes the effects of minor losses.

Further, the fluid flow rate that flows through this orifice is

$$q = A_c v_c = C_c A_0 v_c = C_c C_v A_0 \sqrt{\frac{2\Delta p}{\rho}} = C_d A_0 \sqrt{\frac{2\Delta p}{\rho}} \tag{2-42}$$

Where A_0 is the across-section area of this orifice; the section contraction coefficient $C_c = A_c/A_0$; C_d is flow rate coefficient, $C_d = C_v C_c$; C_d is determined usually by experiment.

In the case of complete contraction, i. e. , $Re \leqslant 10^5$, C_d can be calculated by

$$C_d = 0.964 Re^{-0.05} \tag{2-43}$$

In the case of $Re > 10^5$, C_d is considered a constant of $C_d = 0.6$-0.61; further in the case of incomplete contraction, C_d can be selected by Tab. 2-4, and C_d will be increased to 0. 7 to 0. 8 because of the main pipe wall effect on the fluid.

Tab. 2-4 Flow rate coefficients in incomplete contraction

(不完全收缩时流量系数 C_d 的值)

$\frac{A_0}{A}$	0. 1	0. 2	0. 3	0. 4	0. 5	0. 6	0. 7
C_d	0. 602	0. 615	0. 634	0. 661	0. 696	0. 742	0. 804

This is the reason of low resistance losses when fluid flows along the length of the pipe in thin orifice. It has less sensitivity to temperature, and thin orifice is thus usually used to throttle adjustor.

Poppet and spool valve orifices are similar to the thin orifice, so both are all used to the hydraulic component orifices. Although obeys the formula (2-42), C_d and A_0 should be changed for the dif-

2.5.1　薄壁小孔

当小孔的通流长度 l 与孔径 d 之比 $l/d \leqslant 0.5$ 时，称为薄壁小孔，如图 2-20 所示。一般薄壁小孔的孔口边缘都做成刃口形式。

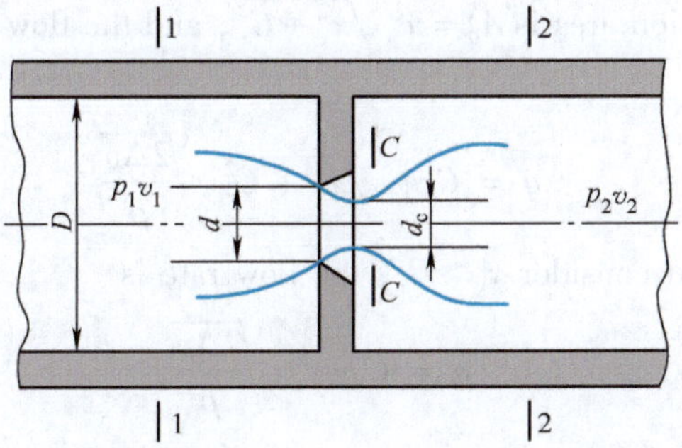

Fig. 2-20　Fluid flow through orifice（通过薄壁小孔的液流）

当液流经过管道由小孔流出时，由于液体的惯性作用，使通过小孔后的液流形成一个收缩截面 C-C，然后再扩散，这一收缩和扩散过程产生很大的能量损失。当孔前通道直径与小孔直径之比 $D/d \geqslant 7$ 时，液流的收缩作用不受孔前通道内壁的影响，这时的收缩称为完全收缩；当 $D/d < 7$ 时，孔前通道对液流进入小孔起导向作用，这时的收缩称为不完全收缩。

现对孔前、孔后通道截面 1-1 和 2-2 列伯努利方程，并设动能修正系数 $\alpha = 1$，则有

$$\frac{p_1}{\rho g} + \frac{v_1^2}{2g} = \frac{p_2}{\rho g} + \frac{v_2^2}{2g} + \sum h_\xi \tag{2-40}$$

式中，$\sum h_\xi$ 为液流流经小孔的局部能量损失，它包括两部分：液流经截面突然缩小时的 $h_{\xi1}$ 和突然扩大时的 $h_{\xi2}$。$h_{\xi1} = \xi \dfrac{v_c{}^2}{2g}$，$h_{\xi2} = \dfrac{\left(1 - \dfrac{A_c}{A_2}\right) v_c^2}{2g}$。因为 $A_c \ll A_2$，所以 $\sum h_\xi = h_{\xi1} + h_{\xi2} = (\xi + 1)\dfrac{v_c^2}{2g}$。又因为 $A_1 = A_2$ 时，$v_1 = v_2$，将这些关系代入伯努利方程，得出

$$v_c = \frac{1}{\sqrt{\xi + 1}} \sqrt{\frac{2}{\rho}(p_1 - p_2)} = C_v \sqrt{\frac{2\Delta p}{\rho}} \tag{2-41}$$

式中，C_v 为速度系数，$C_v = \dfrac{1}{\sqrt{\xi + 1}}$，反映了局部阻力对速度的影响。

所以经过薄壁小孔的流量为

$$q = A_c v_c = C_c A_0 v_c = C_c C_v A_0 \sqrt{\frac{2\Delta p}{\rho}} = C_d A_0 \sqrt{\frac{2\Delta p}{\rho}} \tag{2-42}$$

式中，A_0 为小孔截面积；C_c 为截面收缩系数，$C_c = \dfrac{A_c}{A_0}$；C_d 为流量系数，$C_d = C_v C_c$。

流量系数 C_d 的大小一般由实验确定，在液流完全收缩的情况下，$Re \leqslant 10^5$ 时，C_d 可由下式计算

$$C_d = 0.964 Re^{-0.05} \tag{2-43}$$

ferent orifices.

Take Fig. 2-21 as an example. It is a cylinder spool valve orifice; A is a valve seat, B is a spool core, and the diameter of the spool core is d, the radial clearance is C_r; when spool core moves left a distance of x_v relative to valve seat, the effective wide is $\sqrt{x_v^2 + C_r^2}$, circumferential length is $w = \pi d$, and the across-section area is $A_0 = w\sqrt{x_v^2 + C_r^2}$, and the flow rate that flows through the orifice is calculated below

$$q = C_d w \sqrt{x_v^2 + C_r^2}\sqrt{\frac{2\Delta p}{\rho}} \tag{2-44}$$

Here if neglect C_r, and consider $x_v \gg C_r$, the flow rate is

$$q = C_d w x_v \sqrt{\frac{2\Delta p}{\rho}} \tag{2-45}$$

The flow rate coefficient in the above two equations can be obtained by Fig. 2-22, in which the Reynolds number can be calculated as follows

$$Re = \frac{4vR}{\upsilon} = \frac{4v}{\upsilon}\sqrt{x_v^2 + C_r^2} \tag{2-46}$$

When $Re > 10^3$, C_d is usually a constant, $C_d = 0.67\text{-}0.74$.

For a hydraulic valve whatever flowing in or out, θ is the angle between streamline and spool line and is called speed direction angle, it is usually $\theta = 69°$.

The poppet valve orifice is shown in Fig. 2-23, in which A is the valve seat, and B is the poppet core. When poppet moves up a distance of x_v relative to the valve seat, the average diameter of $d_m = (d_1 + d_2)/2$, relatively, the across-section area at the average diameter is $A_0 = \pi d_m x_v \sin\alpha$. If we set the pressure different between the inlet and outlet of this valve orifice as Δp, then the flow rate is

$$q = C_d A_0 \sqrt{\frac{2\Delta p}{\rho}} = C_d \pi d_m x_v \sin\alpha \sqrt{\frac{2\Delta p}{\rho}} \tag{2-47}$$

Here the flow rate coefficient C_d can be obtained by Fig. 2-24. We can see from Fig. 2-24 that C_d changes between 0.77 and 0.82.

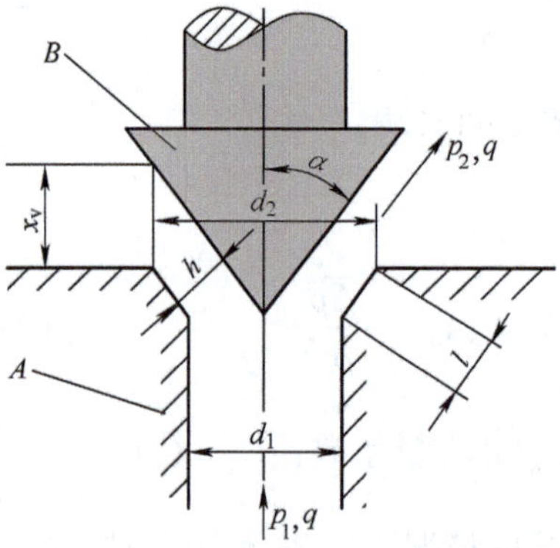

Fig. 2-23 Orifice shape of poppet valve
(锥阀阀口)

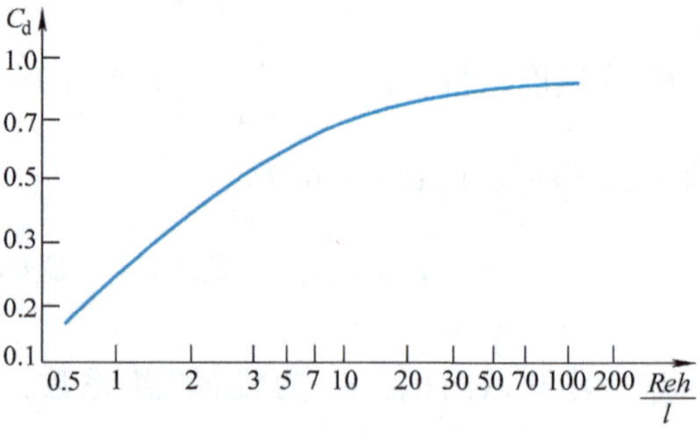

Fig. 2-24 Flow coefficient C_d of poppet valve orifice
(锥阀阀口的流量系数 C_d)

当 $Re>10^5$ 时，C_d 可以认为是不变的常数，计算时按 $C_d=0.60\sim0.61$ 选取。液流不完全收缩时，C_d 可按表 2-4 来选择。这时由于管壁对液流进入小孔起导向作用，C_d 可增至 0.7～0.8。

薄壁小孔因其沿程阻力损失非常小，通过小孔的流量对油温的变化不敏感，因此薄壁小孔多被用作调节流量的节流器使用。

锥阀阀口和滑阀阀口因为比较接近于薄壁小孔，所以常用做液压阀的可调节孔口。它们的流量计算满足式（2-42），但流量系数 C_d 和孔口的截面积 A_0 随着孔口的不同应作具体计算。

如图 2-21 所示的是圆柱滑阀阀口，图中 A 为阀座，B 为阀芯，设阀芯直径为 d，阀芯与阀座间半径间隙为 C_r，当阀芯相对于阀座向左移动一个距离 x_v 时，阀口的有效宽度为 $\sqrt{x_v^2+C_r^2}$，令 w 为阀口的周向长度，$w=\pi d$，则阀口的通流截面面积 $A_0=w\sqrt{x_v^2+C_r^2}$，由式（2-42）可求出通过阀口的流量为

$$q = C_d w \sqrt{x_v^2 + C_r^2}\sqrt{\frac{2\Delta p}{\rho}} \tag{2-44}$$

当 $x_v \gg C_r$ 时，略去 C_r 不计，就有

$$q = C_d w x_v \sqrt{\frac{2\Delta p}{\rho}} \tag{2-45}$$

以上两式的流量系数 C_d 由图 2-22 查出。图中的雷诺数按下式计算

$$Re = \frac{4vR}{\upsilon} = \frac{4v}{\upsilon}\sqrt{x_v^2 + C_r^2} \tag{2-46}$$

图 2-22 中虚线 1 表示 $x_v=C_r$ 时的理论曲线，虚线 2 表示 $x_v \gg C_r$ 时的理论曲线，实线则表示实验测定的结果。当 $Re>10^3$ 时，C_d 一般为常数，其值在 0.67～0.74 之间。阀口棱边圆滑或有很小的倒角时 C_d 比锐边时大，一般在 0.8～0.9 之间。

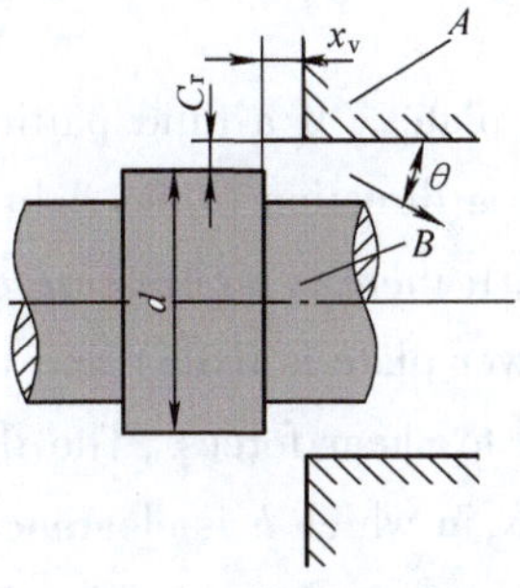

Fig. 2-21　Sketch of cylinder spool orifice
（圆柱滑阀阀口示意图）

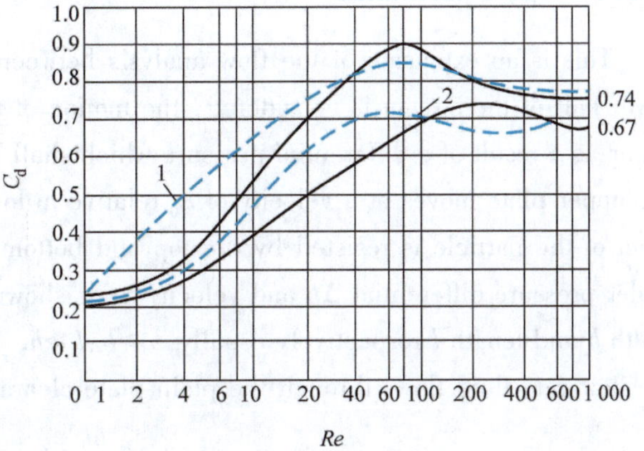

Fig. 2-22　Flow coefficient on the orifice of spool valve
（滑阀阀口的流量系数）

液流流经阀口时，不论是流入还是流出，其流束与滑阀轴线间总保持着一个角度 θ，称为速度方向角，一般 $\theta=69°$。

图 2-23 所示为锥阀阀口，图中 A 为阀座，B 为阀芯，当锥阀阀芯向上移动 x_v 距离时，阀座平均直径 $d_m=(d_1+d_2)/2$ 处的通流截面积 $A_0=\pi d_m x_v \sin\alpha$，这时如果锥阀前后的压差

2.5.2 Stubby orifice and slot orifice

The stubby orifice is defined as $0.5 < l/d \leqslant 4$; and $l/d > 4$ as slot orifice.

The flow rate equation for the stubby orifice is the same as formula (2-42), but the flow rate coefficient C_d must be obtained from the curve in Fig. 2-25. It has shown in Fig. 2-25 that the Reynolds number Re reaches high values; and the C_d is almost a constant. The Stubby orifice is usually used as a fixed throttler because it is easy to be fabricated.

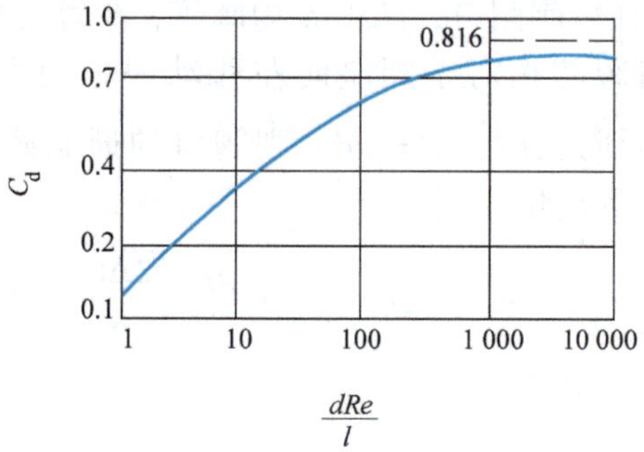

Fig. 2-25 Flow rate coefficients in stubby orifice
(短孔的流量系数)

The flow rate equation for slot orifice obeys the formula (2-31), i. e., $q = \frac{\pi d^4}{128\mu l}\Delta p$, here the inference of viscosity leads to laminar flow state. From formula (2-31), flow rate is proportional to pressure different Δp and inversely proportional to viscosity μ, thus it is dependent on viscosity.

2.5.3 Plate clearance

This is an extension of the flow analysis between two parallel plates. If a fluid particle in the lower half of the stream is considered, the motion of the particle in a direction to the right can only occur as a result of a differential pressure which shall be termed Δp. If there is no pressure differential, upper plate moves at a velocity of u_0 relative to lower plate (lower plate is fixed) and the movement of the particle is resisted by the top and bottom surfaces due to shear forces. The fluid flows under pressure differential Δp and velocity u_0 as shown in Fig. 2-26, in which h is clearance height, width b and length l respectively usually, $b \gg h$, $l \gg h$. The control volume is shown in Fig. 2-26, and the flow rate fluid flows through the plain plate clearance is

$$q = \frac{bh^3}{12\mu l}\Delta p + \frac{u_0}{2}bh \tag{2-48}$$

The formula (2-48) has two statuses:

1) Fluid flow at pressure differential ($\Delta p \neq 0$), and $u_0 = 0$, the formula (2-48) becomes as

$$q = \frac{bh^3}{12\mu l}\Delta p \tag{2-49}$$

2) Fluid flow by viscosity shear, i. e., $u_0 \neq 0$, and $\Delta p = 0$, the formula (2-48) changes to

为 Δp，则通过的流量为

$$q = C_d A_0 \sqrt{\frac{2\Delta p}{\rho}} = C_d \pi d_m x_v \sin\alpha \sqrt{\frac{2\Delta p}{\rho}} \tag{2-47}$$

流量系数 C_d 可在图 2-24 中查出。由图 2-24 可知，当雷诺数较大时，C_d 变化很小，其值在 0.77 ~0.82 之间。

2.5.2　短孔和细长孔

当长径比为 $0.5 < l/d \leqslant 4$ 时，称为短孔；当 $l/d > 4$ 时，称为细长孔。短孔的流量表达式同式（2-42），但流量系数 C_d 应按图 2-25 中的曲线来查。由图 2-25 可知，雷诺数较大时，C_d 基本稳定在 0.8 左右。由于短孔加工比薄壁孔容易得多，因此短管常用作固定节流器。

流经细长孔的液流，由于粘性的影响，流动状态一般为层流，所以细长孔的流量可用液流流经圆管的流量公式，即式（2-31）计算。

从式（2-31）可看出，液流经过细长孔的流量和孔前后压差 Δp 成正比，而和液体粘度 μ 成反比，因此流量受液体温度影响较大，这和薄壁小孔是不同的。

2.5.3　平板缝隙

研究两平行平板缝隙间流体流动。如果液体受到压差 Δp 的作用，液体会流动。如果没有压差 Δp 的作用，而两平行平板之间有相对运动，即一平板固定，另一平板以速度 u_0 运动时，由于液体存在粘性，液体也会被带着移动，在上、下平板邻近的液体受到剪切作用而引起流动。液体通过平行平板缝隙时的最一般的流动情况，是既受压差 Δp 的作用，又受平行平板相对运动的作用，其计算图如图 2-26 所示。

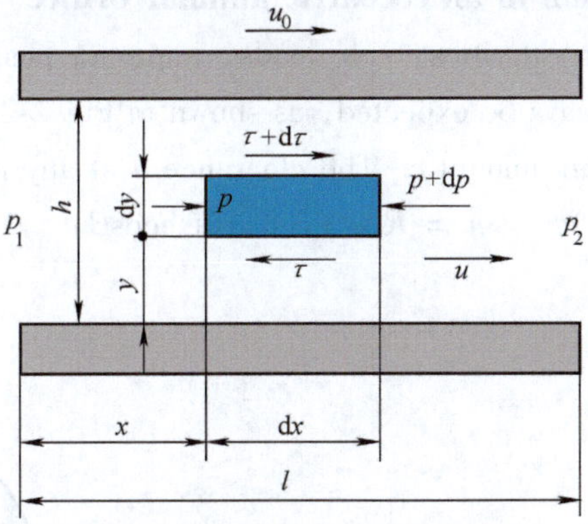

Fig. 2-26　Flow in parallel plain clearance
（平行平板缝隙间的液流）

图中 h 为缝隙高度，b 和 l 分别为缝隙宽度和长度，一般 $b \gg h$，$l \gg h$。在液流中取一个微元体 dx 和 dy（宽度方向取单位长），其左、右两端面所受的压力分别为 p 和 $p + dp$，上、下两面所受的切应力分别为 $\tau + d\tau$ 和 τ，由此得通过平行平板缝隙的流量为

$$q = \frac{bh^3}{12\mu l}\Delta p + \frac{u_0}{2}bh \tag{2-48}$$

式（2-48）是所有缝隙流的通用公式。下面分别有两种特殊情况：

$$q = \frac{u_0}{2}bh \tag{2-50}$$

From formula (2-48) and formula (2-49), the flow increases as the cube of the clearance under pressure differential. This means that the inner clearance is playing an important role for leakage through such a flow path in hydraulic system elements.

2.5.4 Cylinder annular clearance

The clearance surface occurs for some relative motion elements in hydraulic system such as between piston and piston orifice, cylinder spool valve and seat. There are concentric annular orifice, eccentric annular orifice and conical annular orifice.

1. The flow rate equation in a concentric annular orifice

Fig. 2-27 shows a sketch of concentric clearance flow, we set the diameter of the cylinder as d, clearance h, circumferential length (clearance length) l. Let's consider annular clearance expanded along the length direction is the same as a plain plate clearance, so substituting $b = \pi d$ into formula (2-48), and the flow rate equation in a concentric annular orifice can be obtained

$$q = \frac{\pi d h^3}{12\mu l}\Delta p \pm \frac{\pi d h u_0}{2} \tag{2-51}$$

If the motion direction of cylinder is the same as the direction of pressure differential, the symbol in formula (2-51) chooses "+", otherwise "−". Under the condition of $u_0 = 0$, the flow rate is

$$q = \frac{\pi d h^3}{12\mu l}\Delta p \tag{2-52}$$

2. The flow rate equation in an eccentric annular orifice

Because of machining irregularities, side loads, improper positioning and other effects, some degree of eccentricity must always be expected, as shown in Fig. 2-28. A piston of radius r is eccentric in a bore of radius R by an amount e. The clearance h at any point is

$$h = R - (r\cos\alpha + e\cos\beta) \tag{2-53}$$

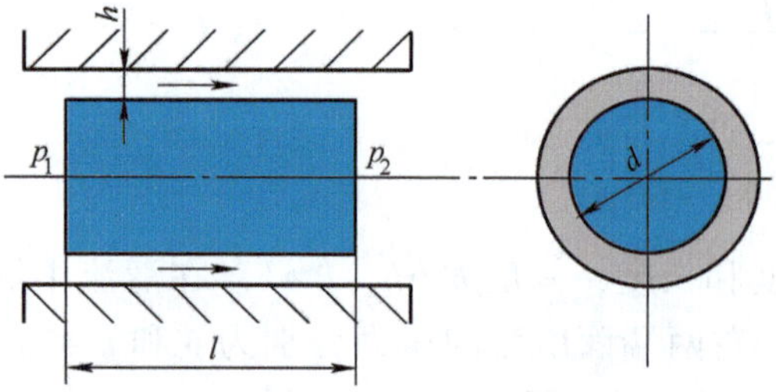

Fig. 2-27 Sketch of concentric clearance flow
(同心环缝隙流动)

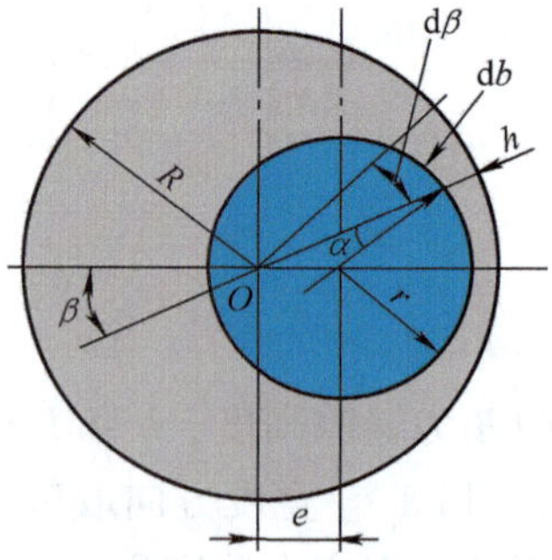

Fig. 2-28 Eccentric annular orifice
(偏心环形缝隙)

1）当平行平板间没有相对运动，即 $u_0=0$ 时，通过的液流纯由压差引起，称为压差流动，其流量为

$$q=\frac{bh^3}{12\mu l}\Delta p \tag{2-49}$$

2）当平行平板两端不存在压差时，通过的液流纯由平板运动引起，称为剪切流动，其流量值为

$$q=\frac{u_0}{2}bh \tag{2-50}$$

从式（2-48）、式（2-49）可看出，在压差作用下，流过固定平行平板缝隙的流量与缝隙的三次方成正比，这说明液压元件内缝隙的大小对其泄漏量的影响是很大的。

2.5.4　环形缝隙

在液压元件中，某些相对运动零件，如柱塞与柱塞孔、圆柱滑阀阀芯与阀体孔之间的间隙为圆柱环形间隙。根据二者是否同心又分为同心圆柱环形间隙和偏心环形间隙。如二者中任一零件具有锥度，则形成圆锥环形间隙。

1. 流经同心圆柱环形缝隙的流量

图 2-27 所示为同心环形缝隙的流动。设圆柱体直径为 d，缝隙值为 h，缝隙长度为 l。如果将环形缝隙沿圆周方向展开，就相当于一个平行平板缝隙。因此只要把 $b=\pi d$ 代入式（2-48），就可得同心环形缝隙的流量公式，即

$$q=\frac{\pi dh^3}{12\mu l}\Delta p\pm\frac{\pi dhu_0}{2} \tag{2-51}$$

当圆柱体移动方向和压差方向相同时取正号，方向相反时取负号。若无相对运动，即 $u_0=0$，则同心环形缝隙流量公式为

$$q=\frac{\pi dh^3}{12\mu l}\Delta p \tag{2-52}$$

2. 流经偏心圆柱环形缝隙的流量

由于加工和安装误差，出现偏心，如图 2-28 所示。一个柱塞半径 r 与柱塞孔半径 R 的安装位置产生偏心量 e，在任意角度 β 处的缝隙量为 h，根据图 2-28 可计算为

$$h=R-(r\cos\alpha+e\cos\beta) \tag{2-53}$$

对于很小的间隙量，由于 α 值很小并且 $\cos\alpha\approx1$，则式（2-53）可以写成

$$\begin{aligned}h&=R-(r+e\cos\beta)=R-r-e\cos\beta\\&=h_0-e\cos\beta=h_0(1-\varepsilon\cos\beta)\end{aligned} \tag{2-54}$$

式中，h_0 为内、外圆同心时半径方向的缝隙，即 $h_0=R-r$；ε 为相对偏心率，$\varepsilon=e/h_0$。

因缝隙很小，$R\approx r=d/2$，可把微小圆弧 $\mathrm{d}b$ 所对应的环形缝隙间的流动近似地看成是平行平板缝隙的流动。将 $\mathrm{d}b=r\mathrm{d}\beta$ 代入式（2-48），则流量增量为

$$\mathrm{d}q=ch^3r\mathrm{d}\beta+\frac{u_0h}{2}r\mathrm{d}\beta \tag{2-55}$$

其中，$c=\dfrac{\Delta p}{12\mu l}$。

将式（2-54）代入式（2-55）

For very small clearances, α is very small and $\cos\alpha \approx 1$. The equation can then be written as

$$h = R - (r + e\cos\beta) = R - r - e\cos\beta$$
$$= h_0 - e\cos\beta = h_0(1 - \varepsilon\cos\beta) \tag{2-54}$$

Where h_0 is an eccentric clearance when it is in concentric condition; ε is called the relative eccentricity $\varepsilon = e/h_0$.

Because of small clearance, $R \approx r = d/2$, so the flow situation of circle clearance flow relative to circle arc db can be considered as plates clearance flow. Substitute db = rdβ into formula (2-48), the incremental flow is

$$dq = ch^3 r d\beta + \frac{u_0 h}{2} r d\beta \tag{2-55}$$

Here $c = \dfrac{\Delta p}{12\mu l}$.

Substitute formula (2-54) into formula (2-55)

$$dq = ch_0^3(1 - \varepsilon\cos\beta)^3 r d\beta + \frac{u_0 h_0}{2}(1 - \cos\beta) r d\beta \tag{2-56}$$

$$q = \int_0^{2\pi} ch_0^3(1 - \varepsilon\cos\beta)^3 r d\beta + \frac{u_0}{2}\int_0^{2\pi} h_0(1 - \varepsilon\cos\beta) r d\beta \tag{2-57}$$

Integrating

$$q = \pi d h_0^3 C\left(1 + \frac{3\varepsilon^2}{2}\right) \pm \frac{\pi d h_0 u_0}{2}$$

or

$$q = \frac{\pi d h_0^3}{12\mu l}\Delta p\left(1 + \frac{3\varepsilon^2}{2}\right) \pm \frac{\pi d h_0 u_0}{2} \tag{2-58}$$

This analysis states that when the eccentricity is maximum, $e = h_0$ and $\varepsilon = 1$. The flow is then 2.5 times greater than it would be indicated by the use of formula (2-51). In practical terms, formula (2-58) states that the use of needle valves or other devices dependent upon annular orifices as precise metering orifices is dangerous unless extreme care to assume concentricity is taken.

3. The flow rate flow and pressure through a conical annular clearance

Because of machining irregularities, such as piston or bore, valve core or seat core, some degree of conic must always be expected, as shown in Fig. 2-29. A piston of diameter d_0 is the degree of conic in a bore of diameter d by an amount h_1 and h_2. If $h_1 < h_2$ it is called inverse degree of conic as shown in Fig. 2-29a; otherwise sequence degree of conic as shown in Fig. 2-29b. The degree of conic existence in valve core will not only affect the flow rate but also the pressure distribution in a clearance fluid.

The half-angle of the cone is set as θ, the valve core moves right at a speed of u_0. Take a control volume as shown in Fig. 2-29a, here h_1, p_1 and h_2, p_2 are denoted as the clearance height and pressure between input and output respectively, and h and p as clearance height h and pressure at left distance x respectively, meanwhile clearance height is considered as nearly constant at a tiny volume wide dx; thus for the status of Fig. 2-29a, substituting $-\Delta p/l = dp/dx$ into formula (2-

$$dq = ch_0^3(1-\varepsilon\cos\beta)^3 r d\beta + \frac{u_0 h_0}{2}(1-\cos\beta) r d\beta \tag{2-56}$$

则

$$q = \int_0^{2\pi} ch_0^3(1-\varepsilon\cos\beta)^3 r d\beta + \frac{u_0}{2}\int_0^{2\pi} h_0(1-\varepsilon\cos\beta) r d\beta \tag{2-57}$$

对式（2-57）积分，可得流量公式

$$q = \pi d h_0^3 C\left(1+\frac{3\varepsilon^2}{2}\right) \pm \frac{\pi d h_0 u_0}{2}$$

或

$$q = \frac{\pi d h_0^3}{12\mu l}\left(1+\frac{3\varepsilon^2}{2}\right)\Delta p \pm \frac{\pi d h_0 u_0}{2} \tag{2-58}$$

正负号意义同前。

由式（2-58）可以看出，当偏心距 $e = h_0$ 时，即 $\varepsilon = 1$ 时（最大偏心状态），其通过的流量是同心环形缝隙流量的 2.5 倍。因此在液压元件中，有配合的零件应尽量使其同心，以减小缝隙泄漏量。

3. 流经圆锥环形间隙的流量及压力分布

对于阀芯与阀体孔同心的情况。当柱塞或柱塞孔、阀芯或阀体孔因加工误差带有一定锥度时，两相对运动零件之间的间隙为圆锥环形间隙，其间隙大小沿轴线方向变化，将出现图 2-29 所示的倒锥和顺锥两种情况。

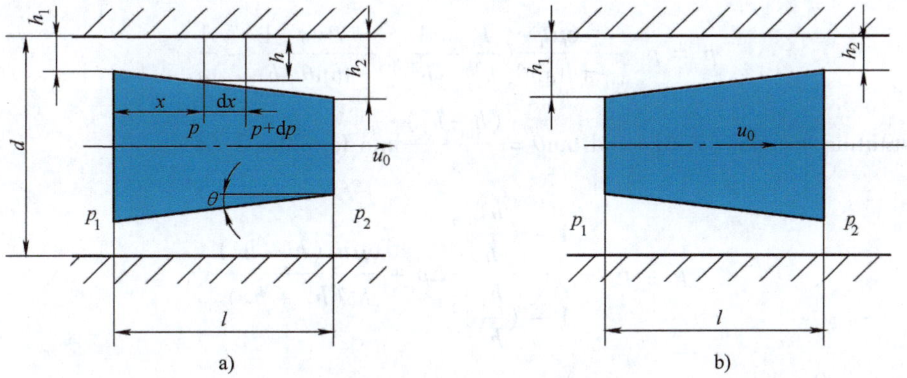

Fig. 2-29　Fluid flow through a conical annular clearance（环形圆锥缝隙的液流）

a）inverse cone（倒锥）　b）Sequence cone（顺锥）

倒锥：图 2-29a 所示的阀芯大端为高压，液流由大端流向小端；顺锥：图 2-29b 所示的阀芯小端为高压，液流由小端流向大端。

阀芯存在锥度不仅影响流经间隙的流量，而且影响缝隙中的压力分布。

下面分别对上述两种情况进行分析。设圆锥半角为 θ，阀芯以速度 u_0 向右移动，进、出口处的缝隙和压力分别为 h_1、p_1 和 h_2、p_2，并设距左端面 x 距离处的缝隙为 h，压力为 p，则液体在微小单元 dx 处流动，由于 dx 值很小可认为 dx 段内缝隙宽度不变。

对于图 2-29a 所示的流动情况，由于 $-\Delta p/l = dp/dx$，将其代入同心环形缝隙流量式（2-51），得

$$q = -\frac{\pi d h^3}{12\mu}\frac{dp}{dx} + \frac{\pi d u_0 h}{2} \tag{2-59}$$

51), we obtain

$$q = -\frac{\pi d h^3}{12\mu}\frac{dp}{dx} + \frac{\pi d u_0 h}{2} \tag{2-59}$$

Because $h = h_1 + x\tan\theta$, substituting $dx = dh/\tan\theta$ into formula (2-59) and then,

$$dp = -\frac{12\mu q}{\pi d\tan\theta}\frac{dh}{h^3} + \frac{6\mu u_0}{\tan\theta}\frac{dh}{h^2} \tag{2-60}$$

Integrating and substituting $\tan\theta = (h_2 - h_1)/l$ into formula (2-59)

$$\Delta p = p_1 - p_2 = \frac{6\mu l}{\pi d}\frac{(h_1 + h_2)}{(h_1 h_2)^2}q - \frac{6\mu l}{h_1 h_2}u_0 \tag{2-61}$$

We obtain the flow rate as

$$q = \frac{\pi d}{6\mu l}\frac{(h_1 h_2)^2}{(h_1 + h_2)}\Delta p + \frac{\pi d h_1 h_2}{(h_1 + h_2)}u_0 \tag{2-62}$$

When $u_0 = 0$, flow rate is

$$q = \frac{\pi d}{6\mu l}\frac{(h_1 h_2)^2}{(h_1 + h_2)}\Delta p \tag{2-63}$$

Integrating formula (2-61) and substituting the boundary condition at $h = h_1$, $p = p_1$, we obtain the pressure distribution in this clearance flowing

$$p = p_1 - \frac{6\mu q}{\pi d\tan\theta}\left(\frac{1}{h_1^2} - \frac{1}{h^2}\right) + \frac{6\mu u_0}{\tan\theta}\left(\frac{1}{h_1} - \frac{1}{h}\right) \tag{2-64}$$

Substituting formula (2-62) and $\tan\theta = \frac{(h - h_1)}{x}$ into formula (2-64),

$$p = p_1 - \frac{1 - \left(\frac{h_1}{h}\right)^2}{1 - \left(\frac{h_1}{h_2}\right)^2}\Delta p + \frac{6\mu u_0 (h - h_2)}{h^2 (h_1 + h_2)}x \tag{2-65}$$

When $u_0 = 0$, we have

$$p = p_1 - \frac{1 - \left(\frac{h_1}{h}\right)^2}{1 - \left(\frac{h_1}{h_2}\right)^2}\Delta p \tag{2-66}$$

The formula (2-62) and formula (2-65) give expression to flow rate and pressure distribution for the status of inverse degree of conic.

For the status of Fig. 2-29b, the sequence degree of conic the flow rate formula is the same as the formula (2-62), the status of inverse degree of conic, but the pressure distribution when $u_0 = 0$ is

$$p = p_1 - \frac{\left(\frac{h_1}{h}\right)^2 - 1}{\left(\frac{h_1}{h_2}\right)^2 - 1}\Delta p \tag{2-67}$$

由于 $h = h_1 + x\tan\theta$，$dx = dh/\tan\theta$，代入式（2-59）并整理后得

$$dp = -\frac{12\mu q}{\pi d\tan\theta}\frac{dh}{h^3} + \frac{6\mu u_0}{\tan\theta}\frac{dh}{h^2} \tag{2-60}$$

对上式进行积分，并将 $\tan\theta = (h_2 - h_1)/l$ 代入得

$$\Delta p = p_1 - p_2 = \frac{6\mu l}{\pi d}\frac{(h_1 + h_2)}{(h_1 h_2)^2}q - \frac{6\mu l}{h_1 h_2}u_0 \tag{2-61}$$

将上式移项可求出环形圆锥缝隙的流量公式

$$q = \frac{\pi d}{6\mu l}\frac{(h_1 h_2)^2}{(h_1 + h_2)}\Delta p + \frac{\pi d h_1 h_2}{(h_1 + h_2)}u_0 \tag{2-62}$$

当阀芯没有运动，即 $u_0 = 0$ 时，流量公式为

$$q = \frac{\pi d}{6\mu l}\frac{(h_1 h_2)^2}{(h_1 + h_2)}\Delta p \tag{2-63}$$

环形圆锥缝隙中压力的分布可通过对式（2-60）积分，并将边界条件 $h = h_1$，$p = p_1$ 代入得

$$p = p_1 - \frac{6\mu q}{\pi d\tan\theta}\left(\frac{1}{h_1^2} - \frac{1}{h^2}\right) + \frac{6\mu u_0}{\tan\theta}\left(\frac{1}{h_1} - \frac{1}{h}\right) \tag{2-64}$$

将式（2-62）代入式（2-64），并将 $\tan\theta = (h - h_1)/x$ 代入得

$$p = p_1 - \frac{1 - \left(\frac{h_1}{h}\right)^2}{1 - \left(\frac{h_1}{h_2}\right)^2}\Delta p + \frac{6\mu u_0 (h - h_2)}{h^2 (h_1 + h_2)}x \tag{2-65}$$

当在压差流的情况下（$u_0 = 0$ 时），则有

$$p = p_1 - \frac{1 - \left(\frac{h_1}{h}\right)^2}{1 - \left(\frac{h_1}{h_2}\right)^2}\Delta p \tag{2-66}$$

式（2-62）和式（2-65）分别为倒锥安装时流经圆锥环形间隙的流量及压力分布公式。

对于图 2-29b 所示的顺锥情况，其流量计算公式和倒锥安装时流量计算公式相同，但其压力分布（在 $u_0 = 0$ 时），则为

$$p = p_1 - \frac{\left(\frac{h_1}{h}\right)^2 - 1}{\left(\frac{h_1}{h_2}\right)^2 - 1}\Delta p \tag{2-67}$$

或

$$p = p_1 - \frac{1 - \left(\frac{h_1}{h}\right)^2}{1 - \left(\frac{h_1}{h_2}\right)^2}\Delta p \tag{2-68}$$

4. 液压卡紧现象及卡紧力

如果阀芯在阀套内的安装出现偏心距 e（图 2-30），则阀芯与阀套之间的间隙 h 将分别

or
$$p = p_1 - \frac{1-(\frac{h_1}{h})^2}{1-(\frac{h_1}{h_2})^2}\Delta p \tag{2-68}$$

4. Hydraulic lock and force

If there is an eccentricity "e" between spool core and seat due to setting, as shown in Fig. 2-30, and the gap "h" between spool core and seat (shown in Fig. 2-30c) should be changed by two directions of circumference β and x axes respectively. As we know, from the formula (2-54), $h = h_1 - e\cos\beta$ at $x = 0$ and $h = h_2 - e\cos\beta$ at $x = l$. However, the value of h at any point is

$$\begin{cases} h = h_1 - e\cos\beta + x\tan\theta (\text{for inverse order}) \\ h = h_1 - e\cos\beta - x\tan\theta (\text{for in order}) \end{cases} \tag{2-69}$$

Let us consider two special positions, i. e. , $\beta = 0$ and $\beta = \pi$ (Fig. 2-30 c), it is clear that the value of h at $\beta = 0$ is more than that at $\beta = \pi$. So some results can be educed from the formula (2-66) and formula (2-68):

1) For the case of eccentric with inverse order conical annular clearance (Fig. 2-30a): since $h_1 < h < h_2$ and also $h_1/h < 1$; from the formula (2-66), the pressure at point of $\beta = 0$ is less than that at $\beta = \pi$, i. e. ,$p|_{\beta=0} < p|_{\beta=\pi}$ (Fig. 2-30c), thus the side force (radial direction) F will be created and it will enlarge the eccentric value e (Fig. 2-30a) to make the spool core locked on the wall of seat. From here, the conclusion can be obtained that the hydraulic lock will be produced when the fluid flows through an eccentric gap with inverse order conical annular clearance.

2) For the case of eccentric with sequence order conical annular clearance (Fig. 2-30b): since $h_2 < h < h_1$ and also $h_1/h > 1$; from the formula (2-68), the pressure at point of $\beta = 0$ is more than that at $\beta = \pi$, i. e. ,$p|_{\beta=0} > p|_{\beta=\pi}$ (Fig. 2-30c), thus the side force (radial direction) F will be created and it will reduce the eccentric value e (Fig. 2-30b) to make the spool core centered in the seat. From here, the conclusion can be obtained that the eccentric will be eliminated of oneself when the fluid flows through an eccentric gap with sequence order conical annular clearance.

In fact, the hydraulic lock is objective and what we can do is to try to eliminate it. There are several methods to reduce the hydraulic lock and one of them is to balance notches on the spool core. The balance notches are used to make the all equal hydraulic pressure in the circumference direction of the spool core. The depth and width of it are usually 0. 3-1. 0mm. It has been proved experimentally that the hydraulic lock force will be reduced consumedly to 5% of original state when it has 3 balance notches. And the hydraulic lock force (radial direction force) can be reduced to 2. 7% of original state and almost in a concentric circle when it has 7 balance notches.

随圆周方向β和轴向长度方向x而异。从式（2-54）可知，在 $x=0$ 处，$h=h_1-e\cos\beta$；在$x=l$处，$h=h_2-e\cos\beta$；而对于任意一点处的 h 值为

$$\begin{cases} h = h_1 - e\cos\beta + x\tan\theta \text{（对于倒锥情况）} \\ h = h_1 - e\cos\beta - x\tan\theta \text{（对于顺锥情况）} \end{cases} \tag{2-69}$$

下面取两个特殊位置，即$\beta=0$ 和$\beta=\pi$（图 2-30c）处来分析，显然在$\beta=0$ 处的 h 值小于$\beta=\pi$ 处的 h 值。同时，从式（2-66）和式（2-68）可以看出：

1）对于带有偏心的环形圆锥缝隙的液流（倒锥情况，如图 2-30a 所示），由于 $h_1<h<$

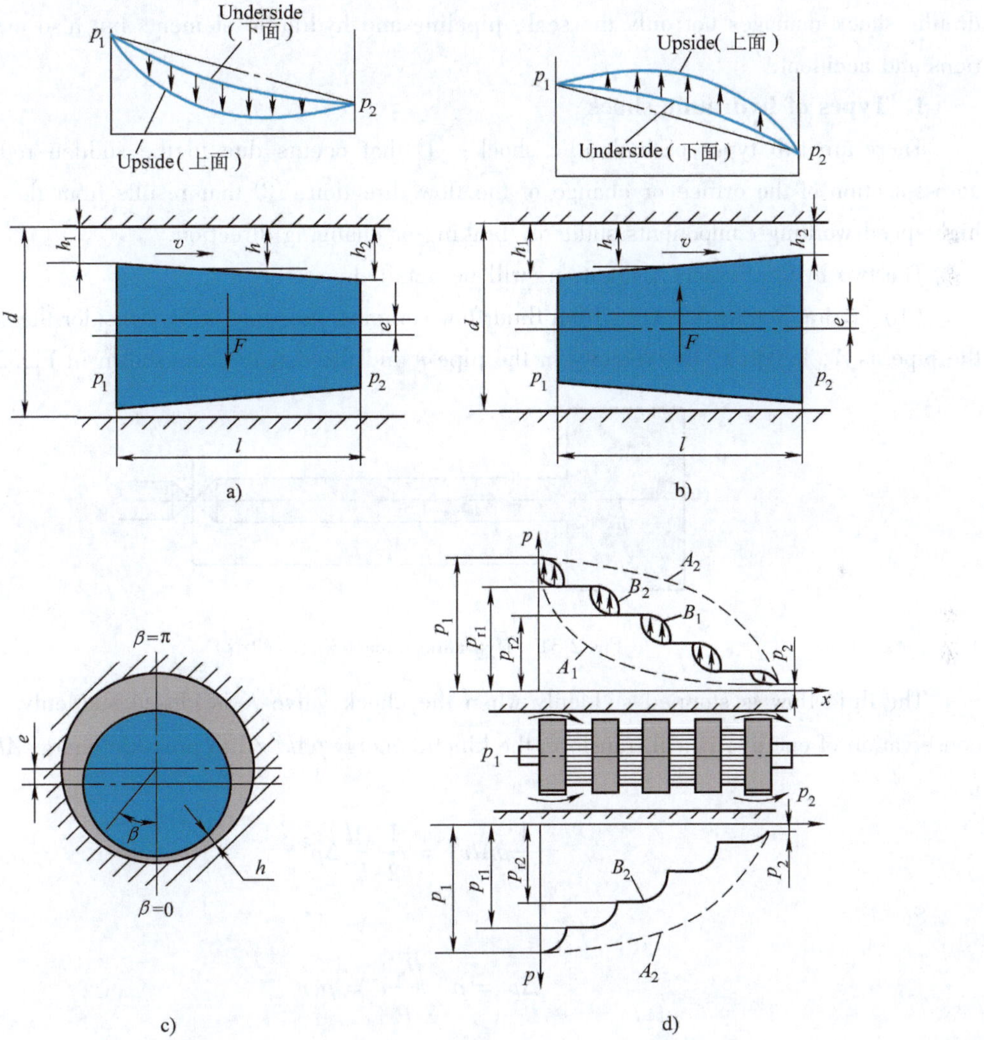

Fig. 2-30　Fluid flow through a conical annular clearance with eccentric
（带有偏心的环形圆锥缝隙的液流）
a）Eccentric with inverse order conical annular（倒锥）
b）Eccentric with in order conical annular（顺锥）
c）Section figure at any point（任意一点处的截面图形）
d）Spool core notched balance pressure（圆柱体上开均压槽）

2.6 Hydraulic Shock and Cavitation

In hydraulic systems, hydraulic shock and cavitation may occur. Thus it is important to understand the reasons in order to control by some measures.

2.6.1 Hydraulic shock

In hydraulic pipeline systems, pressure of hydraulic oil can suddenly become higher than several times those of normal working conditions, which is called pressure peak or hydraulic shock. hydraulic shock damages not only the seal, pipeline and hydraulic elements but also makes error actions and accident.

1. Types of hydraulic shock

There are two types of hydraulic shock: ① that occurs due to the sudden reduction of the across-section of the orifice or change of the flow direction; ② that results from the inertia of the high speed working components suddenly braking or changing direction.

The two hydraulic shock situations will be detailed.

(1) Hydraulic shock results from fluid flow stopping suddenly We consider the section area of the pipe as A, length l, the velocity in the pipe v and the density ρ as shown in Fig. 2-31.

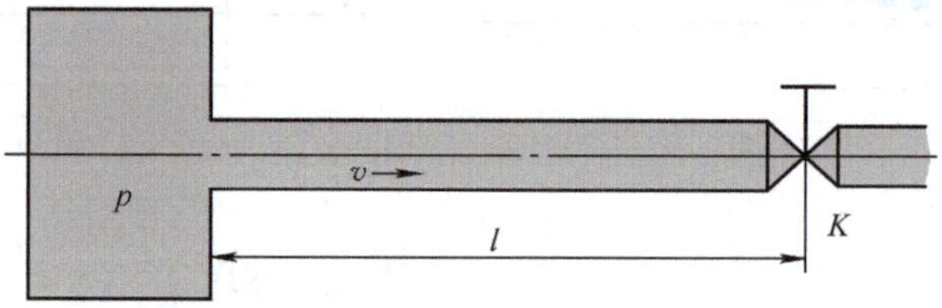

Fig. 2-31 Hydraulic impact (液压冲击)

The fluid flow is stopped suddenly when the check valve K is closed suddenly. According to conservation of energy, it will transform the kinetic energy $\rho Alv^2/2$ to pressure energy $Al\Delta p^2/(2K')$, i. e.

$$\frac{1}{2}\rho Alv^2 = \frac{1}{2}\frac{Al}{K'}\Delta p^2$$

So

$$\Delta p = \rho\sqrt{\frac{K'}{\rho}}v = \rho cv \tag{2-70}$$

Where Δp (Pa) is the value of pressure increased when the hydraulic shock occurs; c (m/s) is the spread speed of shock wave, $c = \sqrt{K'/\rho}$.

For a given hydraulic oil and pipe material, from the formula (2-70) we know the values of ρ and c are two fixed values. So enlarging the section area of pipe is the only way to reduce the value of Δp due to lower velocity value v. If the velocity v is limited within 4.5m/s, the value of Δp is less than 5.0MPa.

h_2，$h_1/h<1$，因此从式（2-66）看，在$\beta=0$处的压力$p|_{\beta=0}$小于$\beta=\pi$处的压力$p|_{\beta=\pi}$（图 2-30c），这样就产生了的侧压力（径向力）F使偏心距e增大（图 2-30a），结果将使阀芯压向阀套的壁面。由此可见，圆柱滑阀阀芯或阀套有锥度且间隙向流动方向增大（倒锥）时，将产生液压卡紧现象。

2）对于带有偏心的环形圆锥缝隙的液流（顺锥情况，如图 2-30b 所示），由于$h_2<h<h_1$，$h_1/h>1$，因此从式（2-68）看，$\beta=0$处的压力$p|_{\beta=0}$大于$\beta=\pi$处的压力$p|_{\beta=\pi}$（图 2-30c），这样就产生了的侧压力（径向力）F使偏心距e减小直到趋于同心（图 2-30b）。由此可见，圆柱滑阀阀芯或阀套有锥度且间隙向流动方向减小（顺锥）时，不产生液压卡紧现象。

实际上液压卡紧力是客观存在的，我们只能尽量采取措施减小它。已有几种办法可以减小液压卡紧力，其中简单且行之有效的方法是在圆柱体上开均压槽（图 2-30d），使槽内液体压力在圆周方向处处相等。均压槽的深度和宽度一般为 0.3～1.0mm。实验证明，当开三条均压槽时，液压侧向力（径向力）可减少到原来的 5% 左右，当开七条时，可减少到 2.7% 左右，从而大大地减小液压卡紧力。

2.6 液压冲击和气穴现象

在液压传动中，液压冲击和气穴现象都会给液压系统的正常工作带来不利影响，因此需要了解这些现象产生的原因，并采取相应的措施以减小其危害。

2.6.1 液压冲击

在液压系统中，因某些原因液体压力在一瞬间会突然升高，产生很高的压力峰值，这种现象称为液压冲击。液压冲击的压力峰值往往比正常工作压力高好几倍，瞬间压力冲击不仅引起振动和噪声，而且会损坏密封装置、管道和液压元件，还会使某些液压元件（如压力继电器、顺序阀等）产生误动作，造成设备事故。

1. 液压冲击的类型

液压系统中的液压冲击按其产生的原因不同可分为：①因液流通道迅速关闭或液流迅速换向使液流速度的大小或方向发生突然变化时，液流的惯性导致的液压冲击；②高速运动的部件突然制动或换向时，因运动部件的惯性引起的液压冲击。

下面分别根据这两种情况进行分析。

（1）液体突然停止运动时产生的液压冲击　图 2-31 中，设管道的截面积为A，长度为l，管道中液流的流速为v，密度为ρ。当管道的末端突然关闭时，液体立即停止运动。根据能量守恒定律，液体的动能$\rho Alv^2/2$转化为液体的压力能$Al\Delta p^2/(2K')$，即

$$\frac{1}{2}\rho Alv^2 = \frac{1}{2}\frac{Al}{K'}\Delta p^2$$

所以

$$\Delta p = \rho\sqrt{\frac{K'}{\rho}}v = \rho cv \tag{2-70}$$

式中，Δp为液压冲击时压力的升高值（Pa）；K'为液体的等效体积模量（Pa）；c为冲击波在管中的传播速度（m/s），$c=\sqrt{K'/\rho}$。

Spread speed of shock wave in pipe can be calculated by

$$c = \sqrt{\frac{K'}{\rho}} = \frac{\sqrt{K/\rho}}{\sqrt{1 + \frac{d\,K}{\delta\,E}}} \tag{2-71}$$

Where K is the bulk modulus of elasticity; d the inner diameter of pipe; δ the wall thickness of pipe and E the modulus of elasticity(Pa). Usually the value of spread speed of shock wave c is between 890 and 1270 m/s.

Formula (2-70) is only used to the closed pipe, i. e. , the time t of the check valve closed is less than the time t_c (time of critical close), i. e.

$$t < t_c (t_c = 2l/c) \tag{2-72}$$

Formula (2-72) is intituled the whole hydraulic shock situation. The pressure peak value at non-whole hydraulic shock is less than the whole hydraulic shock and it can be calculated by

$$\Delta p = \rho c v \frac{t_c}{t} \tag{2-73}$$

If the check valve is partly closed, the speed of fluid drops from v to v', i. e. ,the speed changed valve $\Delta v = v - v'$, we replace v with Δv in formula (2-70) and formula (2-73) and the pressure valve Δp increased in this case can be calculated. The maximum pressure p_{max} can be obtained by

$$p_{max} = p + \Delta p \tag{2-74}$$

Where p is an operation pressure.

(2) Hydraulic shock results from the motion parts braked If a motion part, total mass $\sum M$, the time in reducing speed is Δt with speed reducing Δv during the braking, then the system shock pressure Δp can be calculated approximatively according to the conservation of momentum

since $$\Delta p A \Delta t = \sum M \Delta v$$

So $$\Delta p = \frac{\sum M \Delta v}{A \Delta t}$$

Where A is effective working area(m^2).

Example 2-7 If the inner diameter of pipe is $d = 200$mm, the wall thickness $\delta = 10$mm, the initialization speed $v = 2$m/s, pressure $p = 2.0$MPa, the bulk modulus of elasticity of fluid $K = 2.0 \times 10^3$ MPa, the modulus of elasticity of pipe material $E = 2.0 \times 10^5$ MPa. Try to calculate the maximum pressure drop Δp when the check valve is closed suddenly.

Solution: Firstly calculate the spread speed of shock wave c. Set the density of fluid as $\rho = 900$ (kg/m^3), and from the formula (2-71)

$$c = \sqrt{\frac{K'}{\rho}} = \frac{\sqrt{K/\rho}}{\sqrt{1 + \frac{d\,K}{\delta\,E}}} = \frac{\sqrt{2 \times 10^9 / 900}}{\sqrt{1 + \frac{200 \times 2 \times 10^9}{10 \times 2 \times 10^{11}}}} \text{m/s} = 1360.8\text{m/s}$$

So $$\Delta p = \rho c v = 900 \times 1360.8 \times 2\text{N/m}^2 = 24.5 \times 10^5\text{Pa}$$

2. Measures to reduce the hydraulic shock

The measures are:

由式（2-70）可知，对于一定的某种油液和管道材质来说，ρ 和 c 均为定值，因此降低 Δp 的有效办法是通过加大管道的通流截面积以降低 v 值。一般若将 v 限制在 4.5m/s 以内，Δp 一般不会超过 5.0MPa。

液压冲击波在管中的传播速度 c 可按下式计算

$$c = \sqrt{\frac{K'}{\rho}} = \frac{\sqrt{K/\rho}}{\sqrt{1 + \frac{d}{\delta}\frac{K}{E}}} \tag{2-71}$$

式中，K 为液压油的体积模量（Pa）；d 为管道内径（m）；δ 为管道壁厚（m）；E 为管道材料的弹性模量（Pa）。

冲击波在管道中液压油内的传播速度 c 一般为 890 ~ 1270 m/s。

式（2-70）仅适用于管道瞬间关死的情况，亦即阀门的关闭时间 t 小于压力波来回依次所需的时间 t_c（临界关闭时间）的情况，即

$$t < t_c \ (t_c = 2l/c) \tag{2-72}$$

凡满足式（2-72）的称为完全冲击，否则便是非完全冲击。非完全冲击时引起的压力峰值比完全冲击时的低，按下式计算

$$\Delta p = \rho c v \frac{t_c}{t} \tag{2-73}$$

如果阀门部分关闭，使液流流速从 v 降到 v'，即冲击前后的稳态流速变化值为 $\Delta v = v - v'$，这种情况下只要在式（2-70）和式（2-73）中以 Δv 代替 v，便可求得相应条件下的压力升高值 Δp。知道了 Δp，便可求得出现冲击后管道中的最大压力

$$p_{max} = p + \Delta p \tag{2-74}$$

式中，p 为正常工作压力（N/m^2）。

（2）运动部件制动时产生的液压冲击　设总质量为 $\sum M$ 的运动部件在制动时的减速时间为 Δt，速度的减小值为 Δv，则根据动量定理可近似地求得系统中冲击压力 Δp，即

$$\Delta p A \Delta t = \sum M \Delta v$$

所以

$$\Delta p = \frac{\sum M \Delta v}{A \Delta t}$$

式中，A 为液压缸的有效工作面积（m^2）。

例 2-7　已知某管道的内径 $d = 200$mm，壁厚 $\delta = 10$mm，液体在管中初始速度 $v = 2$m/s，压力 $p = 2.0$MPa，液体的体积模量 $K = 2.0 \times 10^3$MPa，管道材料的弹性模量 $E = 2.0 \times 10^5$MPa。当阀门突然关闭时，试求最大压力升高值 Δp。

解　先计算冲击波传播速度 c。设液体的密度 $\rho = 900$（kg/m^3），由式（2-71）可得

$$c = \sqrt{\frac{K'}{\rho}} = \frac{\sqrt{K/\rho}}{\sqrt{1 + \frac{d}{\delta}\frac{K}{E}}} = \frac{\sqrt{2 \times 10^9 / 900}}{\sqrt{1 + \frac{200 \times 2 \times 10^9}{10 \times 2 \times 10^{11}}}} \text{m/s} = 1360.8 \text{m/s}$$

所以

$$\Delta p = \rho c v = 900 \times 1360.8 \times 2 \text{N/m}^2 = 24.5 \times 10^5 \text{Pa}$$

2. 减小液压冲击的措施

分析前面各式中 Δp 的影响因素，可以归纳出减小液压冲击的主要措施有：

1) Prolong the time closing check and motion components. Permit also the two chambers of hydraulic actuator going each other at neutral position of the direction valve in order to reduce the hydraulic shock.

2) Valve orifice of motion working-piece is designed accurately making speed uniformity change.

3) Expand properly the diameter of pipe to make the flow speed less than or equal to the value commended.

4) Try to shorten the length of pipe to decrease the transmission time. Try to change the complete shock to incomplete shock.

5) Utilize rubber tube or accumulator at point of hydraulic shock.

2.6.2 Cavitation

1. The principle and harm of cavitation

If the pressure at a point is reduced far enough, hydraulic oil will vaporize and vapour cavities will be formed in the liquid, which will make air separated from oil and result in large numbers of air bubble. This phenomenon is called cavitation. The pressure at which vapourisation commences is called the vapour pressure of the liquid. This vapour pressure is quite dependent on the temperature of the oil.

The phenomenon of cavitation happens easily in suction of pump and valve orifice. The suction of pump will produce lower enough pressure due to small diameter and high resistance force; or oil could not be filled completely into pump due to excessive high rotational speed. On the other hand, if the valve orifice designed is so small in diameter that result in very high flow speed, which will result in lower pressure at the point in terms of the energy conservation. All mentioned above will result in the phenomenon of cavitation.

It is well known that the air bubble resulting from the phenomenon of cavitation will be ruptured quickly at high oil pressure and cohere again into fluid accompanying the vacuum forming, high pressure oil entrance, and it will result in high temperature and shock. The bubble having oxygen will cause acidification and cauterization. The phenomenon of cauterization resulting from cavitation is called air-corrosion which will damage working-pieces of hydraulic machinery and also shorten the machine's life.

2. Measures to decrease the cavitation

The measures are:

1) Reducing the pressure drop, usually the special pressure is controlled $p_1/p_2<3.5$.

2) When design it is better for trying to avoid narrow, elbow or sudden direction changed.

3) Try to reduce the suction height of pump. The resistance of suction oil for pump should be less, such as using bigger inner diameter, less elbow or sudden direction changed.

4) Good seal parts to avoid air entrance.

5) Resist corrosive material used try to keep the element rigidity and increase exterior roughness for withstanding corrosion.

1）延长阀门关闭和运动部件制动换向的时间，可采用换向时间可调的换向阀；还可以从阀结构设计上做到当换向阀移到中位时，使液压缸两腔瞬时互通，以减小液压冲击。

2）正确设计阀口，使运动部件制动时速度变化比较均匀。

3）适当增大管径，使液流流速小于或等于推荐值，以限制或降低流速且可以减小压力冲击波传播速度。

4）尽量缩短管道长度，可以减小压力波的传播时间，使完全冲击改变为不完全冲击。

5）采用橡胶软管或在容易发生冲击的地方设置限压阀或蓄能器。

2.6.2 气穴现象

1. 气穴现象的机理及危害

在液流中，如果某点处压力低于当时温度下空气分离压时，溶于油中的空气就会分离出来形成气泡；当继续降低到饱和蒸汽压时，油液将汽化产生大量气泡，这种现象称为气穴现象。汽化压力称为蒸汽压，它受油温的影响。

最容易产生气穴处是泵的吸油腔和阀口。泵的吸油腔，常因吸油面过低、吸油管径过小、吸油管路的其他阻力过大，以致吸油腔压力过低或泵的转速过高，吸油腔未能完全充满油液。如果阀口通流截面很小，流速升得很高，根据能量守恒，速度的升高也将使该处的压力很低，这些原因都将导致气穴的产生。

当气穴产生的气泡进入高压区时，在周围高压油的冲击下将迅速破裂，并又凝结成液体而使体积减小，形成局部真空，周围高压油液瞬间补进，使局部产生高温和压力冲击，以及液体游离出来的空气含有氧气，具有较强的酸化作用，造成金属表面剥落、腐蚀。这种因气穴造成的对金属表面的侵蚀作用称为气蚀，它将缩短零件的使用寿命。

2. 减少气穴现象的措施

为减少气穴现象和减小气蚀的危害，一般采取如下一些措施：

1）减小阀孔或其他元件通道前后的压差，一般使压力比 $p_1/p_2<3.5$。

2）在设计液压元件和管路时，尽量避免设计成狭窄油道或急转弯油道，以防止产生低压区。

3）尽量降低液压泵的吸油高度。泵的吸油腔阻尼应尽量小，如采用内径较大的吸油管、容量较大的过滤器，少用弯头。

4）各元件的连接处要密封可靠，以防止空气进入。

5）对容易产生气蚀的地方，其零件如泵的配油盘等应采用抗腐蚀能力强的金属材料，并增加零件的机械强度，减小表面粗糙度值，以增强其抗腐蚀能力。

Chapter 3 Hydraulic Pumps

3.1 Introduction

Hydraulic pumps convert mechanical energy into hydraulic energy. They supply fluid to the components in the hydraulic system.

3.1.1 Operating principle

The operating principle of every pump in a hydraulic system depends upon the alternative change of their working volumes. The components of a single piston pump are shown in Fig. 3-1. The pump consists of two check valves 1 and 2, a spring 3, a cylinder 4, a piston 5 and an eccentric wheel 6. An oil tight chamber is formed between the piston and the cylinder bore. There are two different pumping actions: oil suction and delivery.

1) Suction oil. As the eccentric wheel driven by a startup motor rotates in a clockwise direction, the piston moves downwards under spring pressure. Then the oil tight chamber increases and a vacuum is formed, which causes liquid to be forced through check valve 1 by atmospheric pressure on the surface of the oil in the reservoir (check valve 2 is shut off). This pumping action ends when the geometry center of the eccentric wheel reaches the lowest point O_1'.

2) Delivery oil. The eccentric wheel continues to rotate after the suction action, the piston moves upwards together with the eccentric wheel's moving up. Accordingly, oil is forced out through check valve 2 under pressure as a result of the oil tight chamber's decreasing. The delivery action will not end until the geometry center of the eccentric wheel reaches the highest point O_1''. Note that check valve 1 is shut off during this pumping action.

In this way, the piston moves up and down repeatedly with the eccentric wheel's continuous rotation, oil is thus forced during the first half cycle and discharged out during the second half cycle.

Note the piston diameter as d, the offset of the eccentric wheel as e, the longest stroke the piston can reach as s and the discharged oil volume V, then

$$s = 2e$$

$$V = (\pi d^2/4)s = (\pi d^2/2)e$$

Where V is the discharged oil volume per revolution for single piston pumps; V is also called the delivery, which only depends upon the geometry dimensions d and e.

As described above, the basic conditions that constitute a hydraulic pump can be concluded as follows:

1) Oil tight chamber. The size of this oil tight working chamber will change periodically. It will connect with the inlet port when the size increases and with the outlet port when decreases.

第3章 液 压 泵

3.1 概述

液压泵是一种能量转换装置，它将机械能转换为流体压力能，是液压传动系统中的动力元件，为系统提供压力油。

3.1.1 工作原理

液压系统中的各种液压泵，其工作原理都是依靠密闭容积的变化来进行工作的。图3-1所示为容积式单柱塞泵工作原理。其由单向阀1和2、弹簧3、缸体4、柱塞5和偏心轮6等组成，柱塞与缸体孔之间形成密闭容积空间。当原动机带动偏心轮顺时针方向旋转时，柱塞在弹簧力的作用下向下运动，柱塞与缸体孔组成的密闭容积逐渐增大，形成局部真空，油箱中的油液在大气压的作用下经单向阀1进入其内（此时单向阀2关闭），这一过程称为吸油，在偏心轮的几何中心转到最下点 O_1' 时终止。吸油过程终了，偏心轮继续旋转，柱塞随偏心轮向上运动，柱塞与缸体孔组成的密闭容积逐渐减小，油液受挤压经单向阀2排出（此时单向阀1关闭），这一过程称为排油，到偏心轮的几何中心转到最上点 O_1'' 时终止。偏心轮连续旋转，柱塞上、下往复运动，泵半个周期吸油，半个周期压油。

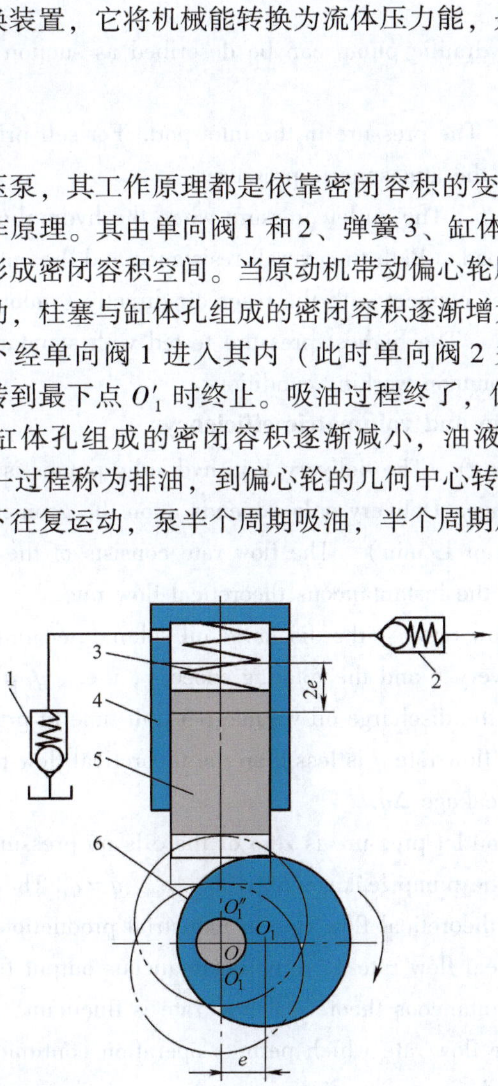

Fig. 3-1 Operating principle displacement pumps with single pistion（容积式单柱塞泵的工作原理图）

1，2—Check valve（单向阀） 3—Spring（弹簧） 4—Cylinder（缸体）

5—Piston（柱塞） 6—Eccentric wheel（偏心轮）

2) Flow-deploying. The corresponding mechanism is used to ensure that the inlet port is separate from the outlet port. There are valve type orifice distributing, port plate and axial type distributing, etc. For the pump shown in Fig. 3-1, flow-deploying action is accomplished by check valves 1 and 2, which is also called valve type orifice distributing.

3) The absolute oil pressure in the reservoir should be equal or more than the atmospheric pressure, which is absolutely necessary external condition for a capacity pump. Thus the reservoir should be connected to atmospheric pressure or charged pressure air.

3.1.2 Main performance parameters in hydraulic pumps

1. Pressure

The pressures in a hydraulic pump can be described as suction pressure, working pressure p and rated pressure p_s.

(1) Suction pressure The pressure in the inlet port. For self-priming suction pumps, the suction pressure is lower than the atmospheric pressure.

(2) Working pressure p The outlet pressure when the hydraulic pump is in operation, which depends upon the overall loads. Both the overall resistance and the output pressure will reduce to zero if the discharge pipelines connect with the reservoir directly (unloaded operation).

(3) Rated pressure p_s The highest pressure tested with standard method when the pump rotates continuously under common working conditions.

2. Delivery, flow rate and volumetric efficiency

(1) Delivery V (cm^3/r) The delivery V in hydraulic pumps is the discharged oil volume per revolution without oil leakage. Delivery only depends upon the geometry dimension of the pump.

(2) Flow rate (m^3/s or L/min) The flow rate consists of the average theoretical flow rate, the practical flow rate and the instantaneous theoretical flow rate.

Average theoretical flow rate q_t: the discharge oil volume per unit time in theory, which is proportional to the pump delivery V and the rotating speed n, i. e., $q_t = nV$.

Practical flow rate q: the discharge oil volume per unit time in practice. If the outlet pressure is beyond zero, the practical flow rate q is less than the theoretical flow rate q_t ($q = q_t - \Delta q$) due to the existence of the flow rate leakage Δq.

Note here: When the outlet pressure is zero or there is no pressure difference between the inlet and the outlet pressures, the pump leakage $\Delta q = 0$, i. e., $q = q_t$. The flow rate under this condition is considered equal to the theoretical flow rate in industrial production.

Instantaneous theoretical flow rate q_{sh}: any instantaneous output flow rate in theory of hydraulic pump, generally, the instantaneous theoretical flow rate is fluctuant, i. e., $q_{sh} \neq q_t$.

Rated flow rate q_s: the flow rate which permits operation continuously under the rated pressure and the rated rotating speed.

3. Power and efficiency

(1) Input power P_i The mechanical power of the driving shaft in hydraulic pumps. Note the input torque as T, the angular speed as ω, then $P_i = T\omega$.

(2) Output power P_o The output hydraulic power of the hydraulic pump, i. e., the product of

设柱塞直径为 d，偏心轮偏心距为 e，则柱塞向上最大行程 $s=2e$，排出的油液体积 $V=(\pi d^2/4)s=(\pi d^2/2)e$。对单柱塞泵 V 即为泵每一转所排出的油液体积，称为泵的排量，排量只与几何尺寸（d 和 e）有关。

根据以上分析，构成液压泵的基本条件是：

1）有一个（或若干个）密封的容积，且密封容积有周期的变化，密封容积由小变大时吸油，由大变小时压油。

2）具有相应的配流机构，利用该配流机构将吸油口和压油口隔开。配流机构的形式有阀式配流、盘式配流和轴式配流等。图 3-1 所示的泵是通过单向阀 1 和 2 实现这一要求的，因此称之为阀式配流机构。

3）油箱内体积的绝对压力必须等于或大于大气压力。这是容积式液压泵吸入油液的外部条件。因此，为保证泵正常吸油，油箱必须与大气相通，或采用密封的充压油箱。

3.1.2 液压泵的主要性能参数

1. 压力

（1）吸入压力　泵进口处的压力，自吸泵的吸入压力低于大气压力。

（2）工作压力　液压泵工作时的出口压力，其大小取决于负载总和。若排油管路直接接回油箱，则总阻力为零，泵排出压力为零，泵的这一工况称为卸荷（或卸载）。

（3）额定压力　是指在正常工作条件下，按试验标准连续运转的最高压力。

2. 排量、流量和容积效率

（1）排量 V　是指在没有泄漏情况下，液压泵每转一转理论上应排出的油液体积，排量的大小仅与泵的几何尺寸有关。

（2）流量　液压泵的流量又分为平均理论流量、实际流量、瞬时理论流量。

1）平均理论流量 q_t。液压泵在单位时间内理论上排出的油液体积，它正比于泵的排量 V 和转速 n，即 $q_t=nV$。常用的单位为 m^3/s 或 L/min。

2）实际流量 q。液压泵在单位时间内实际排出的油液体积。当泵的出口压力不等于零时，因存在泄漏流量 Δq，因此实际流量 q 小于理论流量 q_t，即 $q=q_t-\Delta q$。

在此，需要指出，当泵的出口压力等于零或进、出口压力差等于零时，泵的泄漏量 $\Delta q=0$，即 $q=q_t$。工业生产中将此时的流量等同于理论流量。

3）瞬时理论流量 q_{sh}。液压泵任一瞬时理论输出的流量。一般液压泵的瞬时理论流量是波动的，即 $q_{sh}\neq q_t$。

4）额定流量 q_s。液压泵在额定压力、额定转速下允许连续运行的流量。

3. 功率和效率

（1）输入功率 P_i　驱动液压泵轴的机械功率为泵的输入功率，若输入转矩为 T、角速度为 ω，则 $P_i=T\omega$。

（2）输出功率 P_o　液压泵输出的液压功率，即平均实际流量 q 和工作压力 p 的乘积，$P_o=pq$。

（3）容积效率 η_V　液压泵的实际流量 q 与理论流量 q_t 的比值称为液压泵的容积效率，可表示为 $\eta_V=q/q_t=(q_t-\Delta q)/q_t$。

容积损失产生的原因：①高压腔的泄漏；②吸油阻力大，粘度大，泵转速太高导致吸油时油液不能全部充满工作腔。

the average flow rate q and the working pressure p, $P_o = pq$.

(3) Volumetric efficiency η_V The ratio of the practical flow rate q to the theoretical flow rate q_t, it can be expressed as: $\eta_V = q/q_t = (q_t - \Delta q)/q_t$.

The reasons for the loss of the volume efficiency: ①the leakage in the high-pressure chamber; ②the large resistance in the inlet port, high viscidity and high rotating speed can prevent the working chamber from full of oil.

(4) Mechanical efficiency η_m η_m is reflected by the loss of the hydraulic torque. The reasons for this are: ①the torque loss due to mechanical friction between the components with relative motion; ②the friction loss due to viscosity.

(5) Overall efficiency η and mechanical efficiency η_m The overall efficiency is the ratio of the output power P_o to the input power P_i, i. e.

$$\eta = \frac{P_o}{P_i} = \frac{pq}{T\omega} = \eta_V \eta_m$$

Where η_m is the mechanical efficiency of the pump.

A hydraulic pump with good performance should possess high overall efficiency, not just high volumetric efficiency.

4. Rotating speed

(1) Rated rotating speed n_s the highest rotating speed for normal continuous and long-time operation under rated pressure.

(2) Highest rotating speed n_{max} the highest rotating speed for short-time operation under rated pressure, which is higher than n_s.

(3) Lowest speed n_{min} the lowest speed of the pump under normal operation.

(4) Range of rotating speed the speed range between n_{min} and n_{max}.

3.1.3 Performance curves

The curves are acquired by tests under given medium, rotating speed and oil temperature, which reflect the relationships between the pump working pressure and its volumetric efficiency η_V (or the practical flow rate q), overall efficiency η, input power P_i.

Fig. 3-2 shows a performance curve of a hydraulic pump, in which the abscissa represents working pressure p of the hydraulic pump and the ordinate is the volumetric efficiency η_V (or the practical flow rate q), overall efficiency η and input power P_i.

The volumetric efficiency η_V (or the practical flow rate q) increases inversely with the working pressure p and $\eta_V = 100\%$, $q = q_t$ (the theoretical flow rate) when $p = 0$; while the overall efficiency η increases as p and reaches the highest value when p closes to the rated pressure of the hydraulic pump.

For some hydraulic pumps with certain rotating speed ranges or variable displacement capacity, the format of the general characteristic curves shown in Fig. 3-3 is commonly used to reflect all their performances. The abscissa represents the working pressure p of the pump and the ordinate represents the pump's flow rate q, rotating speed n or delivery V. Fig. 3-3 also shows the equal efficiency curve, and the equal power curve.

（4）机械效率 η_m　表现在液压转矩上的损失。原因是：①液压泵相对运动件之间的机械摩擦而引起的摩擦转矩损失；②粘性引起的摩擦损失。

（5）总效率 η　液压泵的输出功率 P_o 与输入功率 P_i 之比为总效率，即

$$\eta = \frac{P_o}{P_i} = \frac{pq}{T\omega} = \eta_V \eta_m$$

式中，η_m 为液压泵的机械效率。

一台性能良好的液压泵应要求其总效率最高，而不仅仅是容积效率最高。

4. 转速

（1）额定转速 n_s　在额定压力下，能连续长时间正常运转的最高转速，称为液压泵的额定转速。

（2）最高转速 n_{max}　在额定压力下，超过额定转速允许短时间运行的最高转速。

（3）最低转速 n_{min}　正常运转所允许的液压泵的最低转速。

（4）转速范围　最低转速与最高转速之间的转速为液压泵工作的转速范围。

3.1.3 特性曲线

液压泵的性能曲线是在特定的介质、转速和温度下通过试验得出的。它表示液压泵的工作压力与容积效率 η_V（或实际流量 q）、总效率 η 和输入功率 P_i 的关系。

图 3-2 所示为液压泵的性能曲线。曲线的横坐标为液压泵的工作压力 p，纵坐标为液压泵的容积效率 η_V（或实际流量 q），总效率 η 和输入功率 P_i。

由图示性能曲线可看出：液压泵的容积效率 η_V（或实际流量 q）随泵的工作压力升高而降低，压力为零时容积效率 $\eta_V = 100\%$，实际流量等于理论流量。液压泵的总效率 η 随泵的工作压力升高而升高，接近液压泵的额定压力时总效率 η 最高。

对某些工作转速在一定范围的液压泵或排量可变的液压泵，为了揭示液压泵整个工作范围的全性能特性，一般用图 3-3 所示的通用特性曲线表示。曲线的横坐标为泵的工作压力 p，纵坐标为泵的流量 q、转速 n 或排量 V，图中绘制有泵的等效率曲线与等功率曲线。

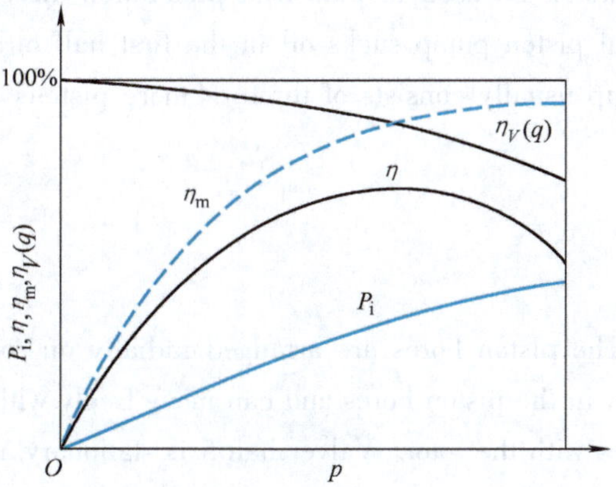

Fig. 3-2　Performance curves of hydraulic pumps
（液压泵的性能曲线）

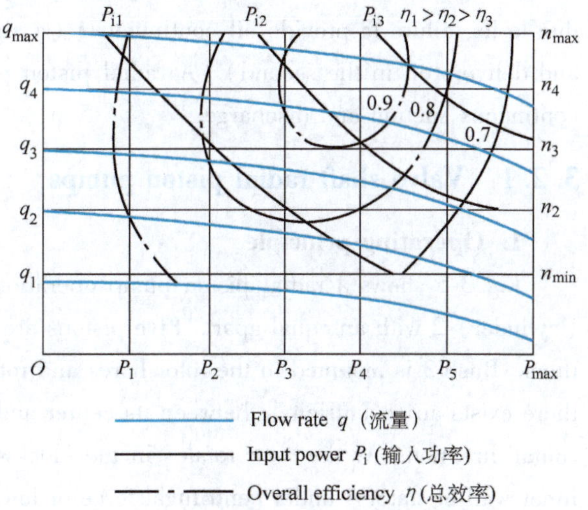

Fig. 3-3　General performance curves of hydraulic pumps
（液压泵的通用性能曲线）

3.1.4 Classification

Hydraulic pumps could be classified by the forms of main moving components and their moving manners: gear, vane, piston and screw pumps.

1) Gear pumps could be further classified in two broad categories, external or internal mesh.

2) Vane pumps could be further divided into double-acting vane pumps, single-acting vane pumps and cam-rotor vane pumps.

3) Piston pumps consist of radial piston pumps and axial piston pumps.

4) Screw pumps include single-screw pumps, two-screw pumps and three-screw pumps.

Hydraulic pumps may be further classified according to the types of delivery: fixed delivery and variable delivery. Variable delivery can be accomplished in single-acting vane pumps, radial or axial piston pumps.

3.1.5 Graphic symbols

The graphic symbols of hydraulic pumps are shown in Fig. 3-4.

3.2 Piston Pumps

The oil suction and delivery actions of a piston pump depend upon the size change of the chamber produced by the reciprocating movement of the pistons within their cylinder bores. For piston pumps, the circular pistons and their corresponding cylinder bores can get high precision match-up and ensure good performance of seal. High volumetric and overall efficiencies can be acquired even in operation under high pressure. The piston pumps can be divided into radial and axial types according to their pistons' arrangements and their different moving directions with respect to the transmission shafts. The signal piston pump shown in Fig. 3-1 is a radial type. The piston is located along the radial direction. However, signal piston pump cannot be used in industrial production directly due to its failure to provide oil continuously (a signal piston pump sucks oil in the first half circle and delivers oil in the second). A radial piston pump usually consists of three or more pistons for continuous suction and discharge.

3.2.1 Valve shaft radial piston pumps

1. Operating principle

Fig. 3-5 shows a radial piston pump operation. The piston bores are arranged radially on rotor (cylinder) 2 with an equal apart. Five pistons are set in the piston bores and can move freely within them. Bush 3 is mounted in the rotor bores and rotates with the rotor. Valve shaft 5 is stationary and there exists an eccentricity *e* between its center and that of the stator 4. The stator moves in the horizontal direction. As the rotor rotates in the clockwise direction, the pistons are pressed against the inner wall of stator 4 under centrifugal force or low pressure oil and then forced outward in the upper semicircle, then a partial vacuum is formed, drawing the oil in the reservior into chamber *b* through orifice *a* of the valve shaft; while the pistons in the next semicircle are pushed inward by the inner

3.1.4 分类

液压泵按主要运动构件的形状和运动方式分为齿轮泵、叶片泵、柱塞泵和螺杆泵：

1）齿轮泵分为外啮合齿轮泵和内啮合齿轮泵。

2）叶片泵分为双作用叶片泵、单作用叶片泵和凸轮转子叶片泵。

3）柱塞泵分为径向柱塞泵和轴向柱塞泵。

4）螺杆泵分为单螺杆泵、双螺杆泵和三螺杆泵。

液压泵按排量能否改变分为定量泵和变量泵，其中变量泵可以是单作用叶片泵、径向柱塞泵和轴向柱塞泵。

3.1.5 图形符号

液压泵的图形符号如图 3-4 所示。

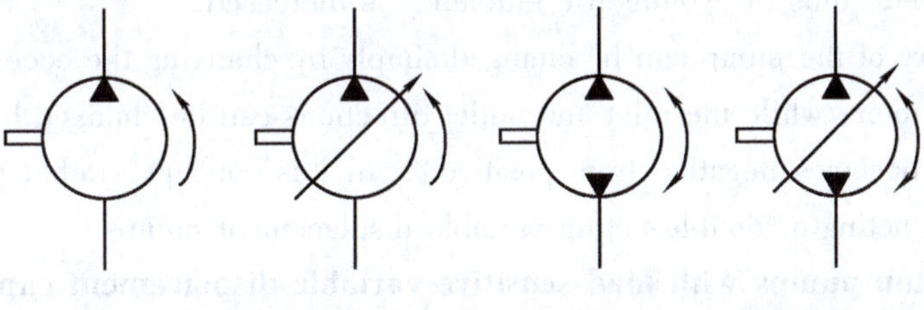

Fig. 3-4 Graphic symbols of hydraulic pumps（液压泵的图形符号）

a）Check fixed delivery hydraulic pump（单向定量液压泵）

b）Check variable delivery hydraulic pump（单向变量液压泵）

c）Reversible fixed delivery pump（双向定量液压泵）

d）Reversible variable delivery pump（双向变量液压泵）

3.2 柱塞泵

柱塞泵是依靠柱塞在缸体孔内作往复运动时产生的容积变化进行吸油和压油的。由于柱塞和缸体内孔都是圆柱表面，容易得到高精度的配合，密封性能好，在高压下工作仍能保持较高的容积效率和总效率。根据柱塞的布置和运动方向与传动主轴相对位置的不同，柱塞液压泵可分为径向柱塞泵和轴向柱塞泵两类。图 3-1 所示的单柱塞泵其柱塞沿径向放置被称为径向柱塞泵，并且单个柱塞因其半个周期吸油、半个周期排油，供油不连续而不能直接用于工业生产。为使柱塞泵能够连续地吸油和压油，柱塞数必须大于 3。

3.2.1 配流轴式径向柱塞泵

1. 工作原理

图 3-5 所示为配流轴式径向柱塞泵。在转子 2 上径向均匀排列着柱塞孔，孔中装有柱塞 1，柱塞可在柱塞孔中自由滑动。衬套 3 固定在转子孔内并随转子一起旋转。配流轴 5 固定不动，其中心与定子中心有偏心量 e，移动定子可改变偏心量 e。当转子顺时针方向转动时，柱塞在离心力或在低压油的作用下压紧在定子 4 的内壁上，当柱塞转到上半周时柱塞向外伸出，径向孔内的密闭工作容积不断增大，产生局部真空，油箱中的油液经配油轴上的 a 孔进入 b 腔；当柱塞转到下半周时，定子表面将柱塞往里推，密闭工作容积不断减小，将 c 腔的

wall of the stator, which reduces the working volume of the oil tight chamber and forces a quantity of liquid in chamber *c* out of the cylinder through orifice *d* on the valve shaft. In each radial bore, the pistons suck and deliver oil once per revolution of the rotor.

The pump's delivery is expressed by

$$V = \frac{\pi d^2}{2} ez \tag{3-1}$$

Where d is the piston diameter; e is the eccentricity between the stator and cylinder case (rotor); z is the number of the pistons.

2. Structure features

1) The inlet and outlet ports of the valve shaft are separated by a center land and balancing grooves are set in their opposite directions. The hydraulic radial forces acting on the valve shaft are balanced out by the balancing grooves. Not only the wear of its sliding surface but also leakage via clearance are reduced, thus the volumetric efficiency is increased.

2) The delivery of the pump can be changed simply by changing the eccentricity e of the stator with respect to the rotor; while the inlet and outlet directions can be changed by changing the direction of e (i. e., e becomes negative from positive). In this concept, radial piston pumps can be changed into signal-acting or double-acting variable displacement pumps.

3. Radial piston pumps with load-sensitive variable displacement capacity

This type of pump gets its name from the fact that the output pressure of the pump depends upon the load. The components of this pump are identified in Fig. 3-6. The output oil with a pressure of p_1 (working pressure) enters the actuator through control valve V_2, whose output pressure p_2 depends upon the load on the actuator. Two paths of oil (with the pressures of p_1 and p_2) connect with the two ends of three-ported valve V_1 respectively. When valve V_1 is balanced out, the pressure difference of valve V_2 is

$$p_1 - p_2 = F_t/A$$

Where A is the effective area of the valve core cover for valve V_1 and F_t is the spring force in the right end of its valve core.

If F_t is considered constant, then $p_1 - p_2$ is also constant (0.2-0.3MPa). i. e., a certain flow is put out for a certain size of orifice in valve V_2. There is an eccentricity of the stator with respect to the rotor. The moving direction and displacement of the stator depend upon the pressure difference of the left and right variable displacement pistons. To meet the requirements of flow rate and pressure change in actuators, the output flow rate and pressure of the pump must be adjusted. This can be accomplished by the stator's displacement, which will change the eccentricity e of the stator.

For this type of pumps, if the output flow rate q is constant, the pressure difference of valve V_2 ($\Delta p = p_1 - p_2$) will increase by reducing the orifice of valve V_2. Then the valve core of three-ported valve V_1 moves in the right due to uneven forces, opening valve orifices a and c. Immediately this happens the pressurized oil in the left variable piston cylinder opens to the reservoir and the pressure p_3 falls down. The stator moves in the left due to uneven force, reducing the eccentricity e, and then the output flow rate q and the pressure difference of valve V_2 will be reduced. The three-ported valve

油从配流轴上的d孔向外压出。转子每转一周，柱塞在每个径向孔内吸、压油各一次。泵的排量为

$$V = \frac{\pi d^2}{2} ez \tag{3-1}$$

式中，d 为柱塞直径；e 为定子与缸体（转子）之间的偏心距；z 为柱塞数。

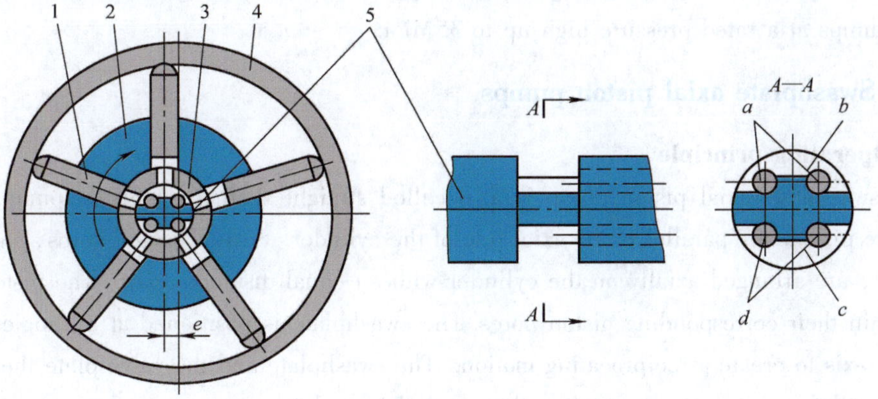

Fig. 3-5 Valve shaft radial piston pump（配流轴式径向柱塞泵）

1—Piston（柱塞） 2—Rotor（转子） 3—Bush（衬套）

4—Stator（定子） 5—Valve shaft（配流轴）

2. 结构特点

1）配流轴上的吸、压油窗口由中间隔墙分开，同时对应的方向开有平衡油槽，使作用在配流轴上的液压径向力实现了平衡，既减小了滑动表面的磨损，又减小了间隙泄漏，提高了容积效率。

2）改变定子与转子的偏心量 e，可改变泵的排量；改变偏心量 e 的方向（即使偏心量 e 由正值转为负值）时，泵的吸、压油方向改变。因此径向柱塞泵可做成单向或双向变量泵。

3. 负载敏感变量径向柱塞泵

负载敏感变量泵因其输出压力取决于负载而得名，其原理图如图3-6所示。液压泵的出口压力油 p_1（工作压力）经控制元件 V_2 后进入执行元件，V_2 的出口压力 p_2 由执行元件的负载决定，因压力油 p_1 和 p_2 被分别引到三通阀 V_1 的阀芯两端，当 V_1 的阀芯受力平衡时，V_2 前后压差为

$$p_1 - p_2 = F_t / A$$

式中，A 为 V_1 阀芯端面的有效作用面积；F_t 为阀芯右端弹簧力。

若视 F_t 不变，则 $p_1 - p_2$ 为定值（0.2～0.3MPa），即对应于 V_2 一定的开口

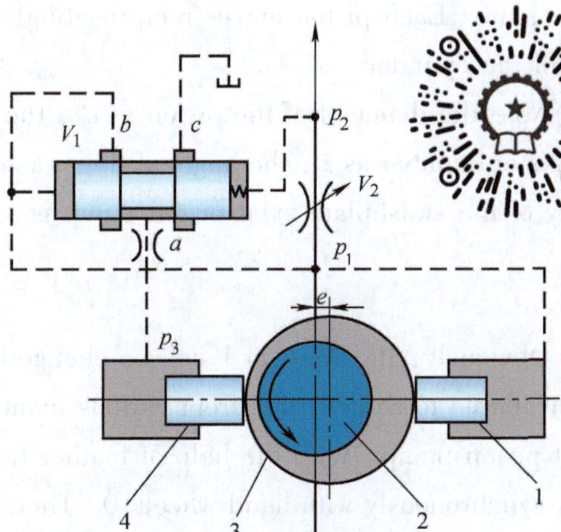

Fig. 3-6 Operating principle of load-sensitive variable discharges of radial piston pump

（负载敏感变量径向柱塞泵原理图）

1—Right variable piston（右变量活塞） 2—Rotor（转子）

3—Stator（定子） 4—Left variable piston（左变量活塞）

V_1 will again balance out when the pressure difference returns to its original value. At this moment, the valve core of valve V_2 returns to its home position and valve orifices a and c are cut off again. No further oil flows into the left variable piston cylinder and the stator stays at a new position. A certain amount of flow rate on request is put out with respect to the orifice of valve V_2. Similarly, when the orifice of valve V_2 increases, the eccentricity and the output flow rate will be increased.

Recent improvements in the design of radial piston pumps have resulted in the widespread use of these pumps at a rated pressure high up to 35MPa.

3.2.2 Swashplate axial piston pumps

1. Operating principle

The swashplate axial piston pump is also called straight shaft axial piston pump. The center lines of the pistons are parallel to the axial line of the cylinder. Several piston bores, each fitted with one piston, are arranged axially on the cylinder with an equal distance apart. The pistons can move freely within their corresponding piston bores. The swashplate is positioned at an angle β relative to the piston axis to create a reciprocating motion. The swashplate and the valve plate themselves keep stationary and the pistons are pressed on the swashplate under low pressure oil or spring force. There are two waist shaped ports on the valve plate that are separated from each other by the transition region. The width of this transition is equal to or slight longer than the waist shaped ports at the bottom of the cylinder to prevent the connection of these two waist shaped ports (the inlet and outlet ports).

As the cylinder rotates with the transmission shaft in a direction shown in the Fig. 3-7, the pistons in the upper semicircle are forced outward gradually under low pressure oil. The oil tight working volume in the cylinder bores is on the increase and a vacuum is formed, drawing the oil into port a of the valve plate. While the pistons in the next semicircle are pushed inward by the swashplate gradually, reducing the oil tight working volume and forcing the oil out through port b of the same valve plate. Each piston moves reciprocatingly once with one suction and delivery action per revolution of the cylinder.

Note the diameter of the piston as d, the diameter of the cylinder bore distribution circle as D, the piston number as z, the angle of the swashplate with respect to the piston axis as β, then the delivery of the swashplate axial piston pump is

$$V = \frac{\pi d^2}{4} Dz\tan\beta \tag{3-2}$$

Obviously, the delivery V can be changed by varying the angle of the swashplate. Several means are available to achieve this from various manufacturers. Fig. 3-7 shows a manual variable-displacement piston pump. With the help of feather key, the piston 7 moves up and down when screw 9 rotates synchronously with hand wheel 10. Then the swashplate swings about its rotation center through pin 6, thus changing the swashplate angle with respect to the piston axis. As the largest swashplate angle (shown in Fig. 3-7), the pistons have the longest stroke within the cylinder barrel, i. e., $\beta = \beta_{max}$, $s = s_{max}$.

Note the vertical distance between the pin (force applying point) and the swashplate gyration centerline as L, then

面积，泵输出一定的流量，定子具有一定的偏心，且定子的移动量大小及方向取决于左变量活塞缸与右变量活塞缸的压差。定子的移动会改变定子的偏心量，从而改变泵输出流量及压力，以适应执行元件的流量及压力变化。

调节控制元件 V_2，如减小其开口面积，则在泵输出流量 q 不变时，V_2 前后压差 $\Delta p = p_1 - p_2$ 将增大，三通滑阀 V_1 的阀芯受力平衡被破坏，阀芯右移，开启阀口 a 和 c，左变量活塞缸的压力油与油箱连通，压力 p_3 下降，定子受力平衡被破坏，定子向左移动，偏心量 e 减小，泵输出的流量 q 减小，控制元件 V_2 前后压差减小，当压差恢复到原来值时，三通滑阀 V_1 阀芯受力重新平衡，阀芯回到中位，阀口 a 和 c 被切断，左变量活塞缸封闭，定子稳定在新的位置，泵输出与控制元件 V_2 开口面积相适应的流量，以满足执行元件的流量需求。若增大控制元件 V_2 的开口面积，类似上面的分析，定子偏心量将增大，泵输出的流量增加。

由于结构上的一些改进，图 3-5 所示径向柱塞泵的额定压力可达 35MPa，加之变量方式灵活，且可以实现双向变量，因此应用日益广泛。

3.2.2 斜盘式轴向柱塞泵

1. 工作原理

斜盘式轴向柱塞泵又称直轴式轴向柱塞泵，该液压泵的柱塞中心线平行于缸体的轴线。如图 3-7 所示，缸体上均匀分布着几个轴向排列的柱塞孔，柱塞可在孔内沿轴向滑动，斜盘的中心线与缸体中心线斜交成一个 β 角，以产生往复运动。斜盘和配油盘固定不动。柱塞可在低压油或弹簧作用下压紧在斜盘上。在配油盘上有两个腰形窗口，它们之间由过渡区隔

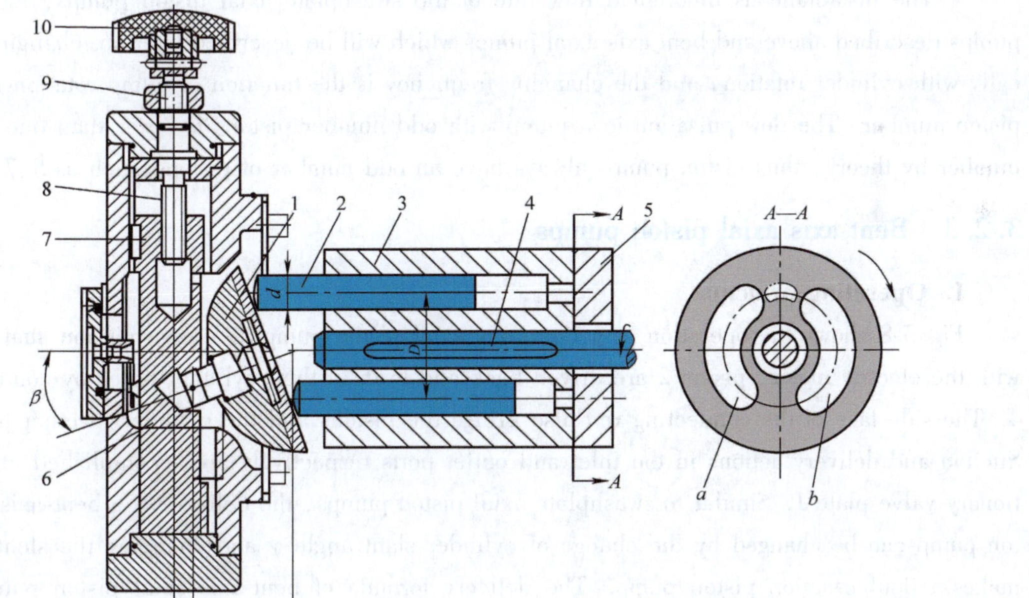

Fig. 3-7 Swashplate axial piston pump（斜盘式轴向柱塞泵）

1—Swashplate（斜盘） 2—Piston（柱塞） 3—Cylinder（缸体）
4—Transmission shaft（传动轴） 5—Valve plate（配流盘） 6—Pin（轴销）
7—Adjusting piston（变量柱塞） 8—Bolt（螺钉） 9—Screw（丝杠） 10—Hand wheel（手轮）
a—Inlet port（吸油窗口） b—Outlet port（压油窗口）

$$\tan\beta_{max} = s_{max}/L$$

Because the displacement of the pin equals that of the adjusting piston, then $\tan\beta = s/L$, substituting it into formula(3-2), we have

$$V = \frac{\pi d^2}{4} Dz \frac{s}{L} \tag{3-3}$$

From formula (3-3), we can see that the pump delivery is proportional to the piston's displacement. To limit the hydraulic side forces on the pistons, the swashplate largest angle β_{max} should be less than 18°.

2. Structure features

1) For piston pumps, the machining accuracy between the pistons and the cylinder bores is easy to acquire; The piston pumps can reach high volumetric efficiency and high rated pressure (high up to 32MPa).

2) To avoid pressure pulsation at the bottom of pistons (tight volumes) where oil pressures are changed from suction to delivery, damping grooves (or holes) are set in the front of the suction and delivery ports of valve plate, or the valve plate is placed with an angle along the cylinder rotating direction.

3) Leakage past the clearance among the pistons, piston shoes and cylinder block face are carried through the pump body and a case drain connection to the reservoir, which carries away the heat generated and prevents any heat buildup and overheating within the pump.

4) The instantaneous theoretical flow rate of the swashplate axial piston pumps, radial piston pumps described above and bent axis axial pumps which will be described later are changing periodically with cylinder rotation, and the changing frequency is the function of pump rotation speed and piston number. The flow pulsation in a pump with odd number pistons is lower than one with even number by theory, thus piston pumps always have an odd number of pistons, such as 5, 7 or 9.

3.2.3 Bent-axis axial piston pumps

1. Operating principle

Fig. 3-8 shows the operation of a bent-axis axial piston pump. As transmission shaft 5 rotates with the electric motor, piston 2 are driven back and forth in their cylinder bores by connecting rod 4. The side face of the connecting rod also brings the pistons and the cylinder together to revolve. Suction and delivery actions in the inlet and outlet ports respectively are accomplished through stationary valve plate 1. Similar to swashplate axial piston pumps, the delivery of a bent-axis axial piston pump can be changed by the change of cylinder slant angle γ and changing the slent direction makes a double-action piston pump. The delivery formula of bent-axis axial piston pumps is the same as that of swashplate axial piston pumps.

2. Constant power variable displacement bent-axis axial piston pump

Here is a bent-axis axial piston pump with the function of constant power variable displacement. Fig. 3-9a shows its working principle.

(1) Variable displacement process (composed by components 1-4) the upper chamber of adjusting piston 13 connects with the pump's outlet port. The oil enters into the oil chamber of control pis-

开，不能连通。过渡区宽度等于或稍大于缸体底部窗口宽度，以防止吸油区和压油区连通。

当传动轴以图示方向带动缸体转动时，位于左半圆的柱塞在低压油的作用下逐渐向外伸出，使缸体孔内密闭工作腔容积不断增大，产生局部真空，将油液从配油盘配油窗口 a 吸入；位于右半圆的柱塞被斜盘推着逐渐向里缩入，使密闭工作腔容积不断减小，将油液经配油盘配油窗口 b 压出。缸体旋转一周，每个柱塞往复运动一次，完成一次吸油和压油动作。若柱塞直径为 d，缸体柱塞孔分布圆直径为 D，柱塞数为 z，斜盘倾角为 β，则斜盘式轴向柱塞泵的排量

$$V = \frac{\pi d^2}{4} Dz\tan\beta \tag{3-2}$$

显然，改变斜盘的倾角 β 可以改变泵的排量。斜盘式轴向柱塞泵的变量方式可以有多种，图 3-7 所示为手动变量泵。当旋转手轮 10 带动丝杠 9 旋转时，因导向平键的作用，变量柱塞 7 将上下移动并通过轴销 6 使斜盘绕其回转中心摆动，改变倾角大小。图示位置斜盘倾角 $\beta = \beta_{\max}$，轴销距水平轴线的位移 $s = s_{\max}$。若轴销距斜盘回转中心的力臂为 L，则可得 $\tan\beta_{\max} = s_{\max}/L$，又由于轴销随同变量活塞一起移动，因此轴销的位移即变量活塞的位移 s，于是有 $\tan\beta = s/L$，代入公式（3-2），则有

$$V = \frac{\pi d^2}{4} Dz \frac{s}{L} \tag{3-3}$$

泵的排量与变量活塞的位移成正比。为使柱塞所受的液压侧向力不致过大，斜盘的最大倾角 $\beta_{\max}$ 一般小于 18°。

2. 结构特点

1）在构成吸压油腔密闭容积的三对运动摩擦副中，柱塞与缸体柱塞孔之间的圆柱环形间隙加工精度易于保证；缸体与配流盘、滑履与斜盘之间的平面缝隙采用静压平衡，间隙磨损后可以补偿，因此轴向柱塞泵的容积效率较高，额定压力可达 32MPa。

2）为防止柱塞底部的密闭容积在吸、压油腔转换时因压力突变而引起压力冲击，一般在配流盘吸、压油窗口的前端开设减振槽（孔），或将配流盘顺缸体旋转方向偏转一定角度放置。

3）泵内压油腔的高压油经三对运动副的间隙泄漏到缸体与泵体之间的空间后，再经泵体上方的泄漏油口直接引回油箱。这不仅可保证泵体内的油液为零压，而且可随时将热油带走，保证泵体内的油液不致过热。

4）斜盘式轴向柱塞泵以及前面介绍过的径向柱塞泵和后面将介绍的斜轴式轴向柱塞泵的瞬时理论流量随缸体的转动而周期性变化，其变化频率与泵的转速和柱塞数有关。由理论推导柱塞数为奇数时的脉动小于偶数，因此柱塞泵的柱塞取为奇数，一般为 5、7 或 9。

3.2.3 斜轴式轴向柱塞泵

1. 工作原理

图 3-8 所示为斜轴式轴向柱塞泵，当传动轴 5 随电动机一起转动时，连杆 4 推动柱塞 2 在缸体 3 中作往复运动，同时连杆的侧面带动柱塞连同缸体一起旋转。通过固定不动的配流盘 1 的吸油窗口、压油窗口进行吸油、压油。与斜盘式轴向柱塞泵类似，可通过改变缸体的倾斜角度 γ 来改变泵的排量；通过改变缸体的倾斜方向来构成双向变量轴向柱塞泵。斜轴式轴向柱塞泵排量公式与斜盘式完全相同。

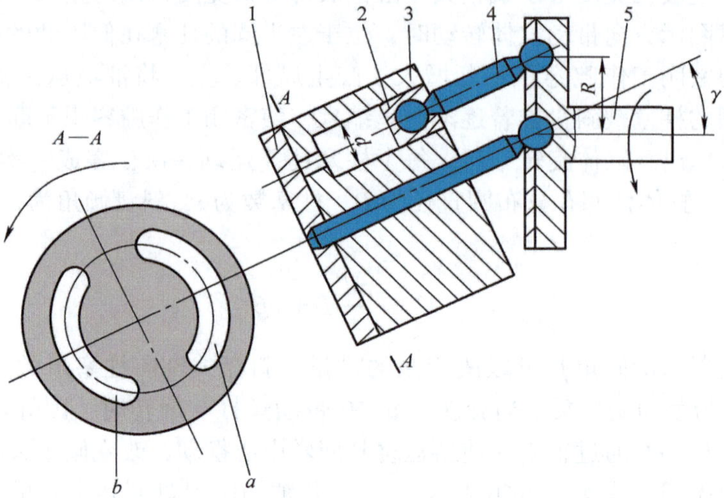

Fig. 3-8 Bent-axis axial piston pump (斜轴式轴向柱塞泵)
1—Valve plate(配流盘) 2—Piston(柱塞) 3—Cylinder(缸体)
4—Connecting rod(连杆) 5—Transmission shaft(传动轴)
a—Inlet port(吸油窗口) *b*—Outlet port(压油窗口)

ton 7 through stationary damping orifice 6. Variable springs 9 and 10 form a dual-spring with the distance s_p between the installation height of inner spring 10 and the spring plate. Spring 12 is located at the down end of servo-valve 11. When the hydraulic pressure acting on the control piston 7 exceeds the resultant set force of springs 9 and 12, control piston 7 will push servo-valve 11 core to move downwards. Then oil ports *a* and *b* connect with each other and the pressure oil enters into the down chamber of the adjusting piston 13. The variable piston will move upwards because the area of the down chamber is larger than that of the upper chamber, causing valve plate 4 and cylinder block 1 to swing about the point *O* through pin 5, which in turn reduces the swing angle γ of the cylinder.

(2) Variable displacement with servo constant power process (composed by components 5-13)

With the reduction of swing angle γ, variable displacement piston 13 moves up and provides a feedback for compression spring 9 through pin 5. Then control piston 7 and servo-valve core 11 move up to their home position, closing the oil ports *a* and *b*. The hydraulic pressure acting on the control piston and the spring force are balanced out. The pump puts out a certain amount of flow because the adjusting piston and the cylinder are kept at a certain position. When the stroke of the adjusting piston reaches s_p, inner spring 10 works, i. e., the hydraulic pressure acting on the control piston and the resultant force of the springs 9, 10 and 12 are balanced out.

(3) Characteristic curves The constant power variable characteristic curve is illustrated in Fig. 3-9b. The slope of line 1 is regulated by the stiffness of spring 9, while the slope of line 2 is determined by the resultant stiffness of springs 9 and 10. The precompress value of spring 12 is used to control the horizontal position of curves *BCD*. As can be seen from this curve, the hydraulic pressure acting on the control piston becomes larger with the increase of the pump's output pressure, while the servo variable displacement process makes its output flow increase inversely with the output

2. 恒功率变量轴向柱塞泵

下面介绍一种斜轴式轴向柱塞泵，它具有恒功率变量功能。图 3-9a 所示为其工作原理图。其工作原理如下：

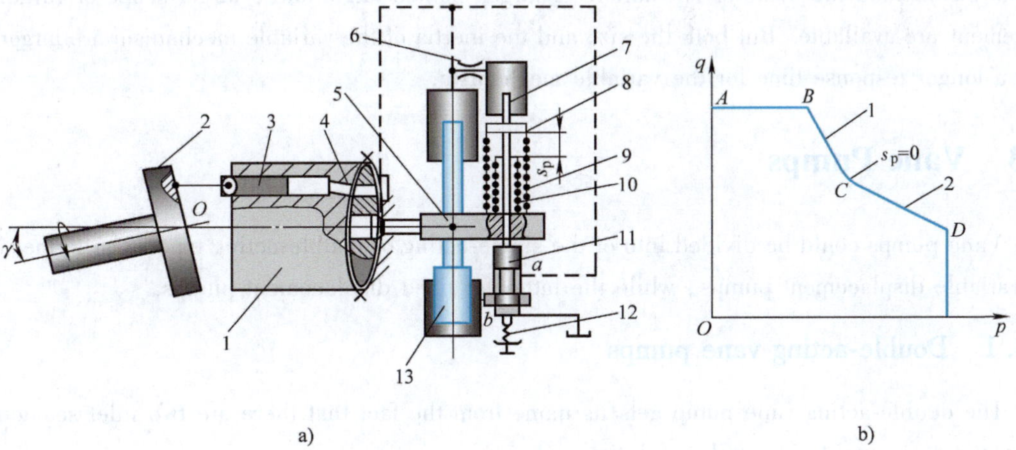

Fig. 3-9 Constant power variable displacement bent-axis axial piston pump

（恒功率变量斜轴式轴向柱塞泵）

a) Operating principle（工作原理） b) Performance curve（特性曲线）

1—Cylinder（缸体） 2—Swashplate（斜盘） 3—Piston（柱塞） 4—Valve plate（配流盘）

5—Pin（拨销） 6—Damping orifice（阻尼孔） 7—Control piston（控制柱塞） 8—Spring plate（弹簧座）

9，10，12—Spring（弹簧） 11—Servo valve（伺服阀） 13—adjusting piston（变量柱塞）

（1）变量过程（由零件 1 ~ 4 组成） 变量柱塞 13 的上腔油室常通泵的压油腔，同时经固定阻尼孔 6 进入控制柱塞 7 的油腔。变量弹簧 9 和 10 为双弹簧，其中内弹簧 10 的安装高度与弹簧座之间相距 s_p，弹簧 12 位于伺服阀 11 的下端。当作用在控制柱塞 7 的液压力大于弹簧 9 和弹簧 12 的预压缩力之和时，控制柱塞 7 推动伺服阀阀芯 11 向下移动，接通油口 a 与 b，压力油进入变量柱塞 13 下腔。因变量柱塞 13 下腔作用面积大于上腔作用面积，导致变量柱塞 13 向上运动，通过拨销 5 带动配流盘 4 和缸体 1 一起绕 O 点摆动，减小缸体摆角 γ。

（2）伺服恒功率变量过程（由零件 5 ~ 13 组成） 在上述缸体摆角 γ 减小的同时，由于变量柱塞 13 向上运动通过拨销 5 反馈压缩弹簧 9 并使控制柱塞 7 和伺服阀阀芯 11 上移复位，关闭油口 a 和 b。此时，作用在控制活塞上的液压力与弹簧力平衡，变量柱塞稳定在一定位置，缸体具有一定的摆角，泵输出一定的流量。当变量柱塞上移的行程等于 s_p 时，内弹簧 10 参与工作，即作用在控制柱塞上的液压力与弹簧 9、10、12 的合力相平衡，变量柱塞上移行程等于 s_p，为变量特性曲线上的拐点。

（3）特性曲线 恒功率变量特性曲线如图 3-9b 所示，图中直线 1 的斜率由外弹簧 9 的刚度决定，直线 2 的斜率由内、外弹簧 9 与 10 的合成刚度决定，弹簧 12 的预压缩量则用来使曲线 BCD 在水平方向平移。由曲线可以看到，随着泵的出口压力升高，控制活塞上所受液压力将进一步增大，由伺服变量过程使泵输出的流量随压力增大而减小。因为泵的出口压力 p 与输出流量 q 的乘积近似为常数，因此称这种变量方式为恒功率变量泵。

与斜盘式泵相比较，斜轴式泵由于柱塞和缸体所受的径向作用力较小，允许的倾角较

pressure. The product of output pressure p and flow rate q is almost constant, pump with such way of delivery is also named constant power variable displacement pump.

Compared with the swashplate piston pump, in a bent-axis axial piston pump, the radial force on the pistons and the cylinder is smaller. A larger slanted angle and a larger scope of variable displacement are available. But both the size and the inertia of the variable mechanism are larger, causing a longer response time for the variable mechanism.

3.3 Vane Pumps

Vane pumps could be divided into of the single-acting or double-acting type. The former is used for variable displacement pumps, while the latter for fixed displacement pumps.

3.3.1 Double-acting vane pumps

The double-acting vane pump gets its name from the fact that there are two inlet segments and two outlet segments during each revolution.

1. Operating principle

The principal parts of the pump include a transmission shaft 9, a rotor 13, a stator 5, left and right valve plates 2 and 6, front and rear pump housings 3 and 7, etc. (Fig. 3-10). The stator is anchored securely in the housing of the pump, which does not move. The rotor is slotted and is driven by a shaft. Each slot of the rotor serves the purpose of holding a flat, rectangular vane. The vanes are free to move radially in the slot. As the rotor turns, centrifugal force ejects the vanes outward to contact and follow the inner wall of the stator. By this action, the vanes divide the enclosed area by the rotor, the inner ring of the stator, the left and right valve plates into four chambers (Fig. 3-11a). The

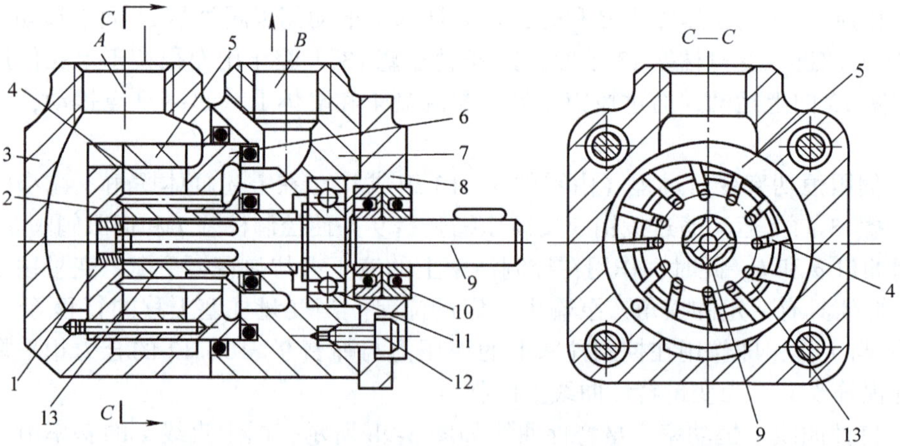

Fig. 3-10 Double-acting vane pump(双作用叶片泵结构)

1,11—Bearing(轴承) 2,6—Left and right valve plate(左、右配流盘)

3,7—Front and rear pump housing(壳体) 4—Vane(叶片) 5—Stator(定子)

8—End cover(端盖) 9—Transmission shaft (传动轴) 10—Dustproof ring(防尘圈)

12—Screw(螺钉) 13—Rotor(转子)

大，所以变量范围较大。由于靠摆动缸体来改变流量，故其体积和变量机构的惯量较大，变量机构动作的响应速度较低。

3.3 叶片泵

叶片泵分为单作用叶片泵和双作用叶片泵两种，前者用作变量泵，后者为定量泵。

3.3.1 双作用叶片泵

双作用叶片泵因转子旋转一周，叶片在转子叶片槽内滑动两次，完成两次吸油和两次压油而得名。

1. 工作原理

图 3-10 所示为双作用叶片泵的结构图，主要零件包括传动轴 9、转子 13、定子 5、左配流盘 2、右配流盘 6、叶片 4 和前壳体 3、后壳体 7 等。转子上开有叶片槽且由轴驱动，叶片可在叶片槽内径向自由滑动。当传动轴带动转子旋转时，位于转子叶片槽内的叶片在离心力的作用下向外甩出，紧贴定子内表面随转子旋转。由定子的内环、转子的外圆和左、右配流盘组成的密闭容积（图 3-11a）被叶片分割为四部分。因为存在半径差，因此随着转子顺时针方向旋转（图 3-11a），由叶片 1 和 3、叶片 5 和 7 所分割的两部分密闭容积逐渐减小；由叶片 7 和 1、叶片 3 和 5 所分割的两部分密闭容积逐渐增大。容积减小时受挤压的油液经配流盘上的压油窗口 9 和 11 排出，容积增大时形成局部真空，油箱的油液在大气压作用下经配流盘的吸油窗口 10 和 12 吸油，传动轴每转一转排出的油液体积即双作用叶片泵的排量

$$V = 2\pi B(R^2 - r^2) - \frac{2zBS(R-r)}{\cos\theta} \tag{3-4}$$

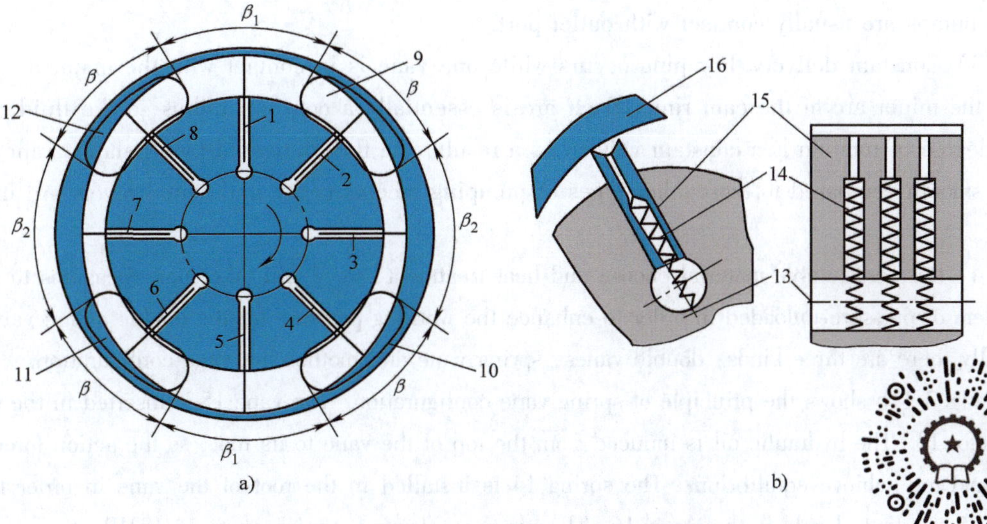

Fig. 3-11 Operating principle（泵工作原理图）

a）Operating principle of double-acting vane pump（双作用叶片泵的工作原理图）

b）Operating principle of spring vane configuration（弹簧叶片式泵工作原理图）

1，2，3，4，5，6，7，8，15—Vane（叶片） 9，11—Outlet port（压油窗口）

10，12—Inlet port（吸油窗口） 13—Rotor（转子） 14—Spring（弹簧） 16—Stator（定子）

inner ring of the stator is oval and the radii are different for each arc. If the rotor turns in a clockwise direction, the two oil tight chambers enclosed by vanes 1 and 3, 5 and 7 decrease in volume; while that enclosed by vanes 7 and 1, 3 and 5 are on the increase. As each chamber is decreased in volume, the oil is pressurized and ejected from the pump through outlet port 9 and 11 on the valve plate. And as each chamber is increased in volume, the oil is drawn into the pump under atmospheric pressure through inlet ports 10 and 12 on the same valve plate.

The delivery V for double-acting vane pump is defined as the output oil volume per revolution of the shaft, which is expressed by the formula

$$V = 2\pi B(R^2 - r^2) - \frac{2zBS(R-r)}{\cos\theta} \tag{3-4}$$

Where R and r are the radii of the major arc and the minor arc of the inner ring of the stator respectively; B is the rotor width; S is the vane thickness; z is the number of vanes; θ is the slanted angle of the vane slot.

The second part in the right side of formula (3-4) is the effect of the vane thickness on delivery, which will decrease the theoretical delivery of pump.

2. Structure features

1) Simplicity. Because the two inlet segments and the two outlet segments are designed symmetrically on the valve plate, the pressure forces acting on the rotor and the stator are equal and opposite, completely canceling each other out. As a result, a small radial force acts on the bearings. Consequently, the life of this type of pump is exceptionally good.

2) Intra vane feature. Many of the balanced vane pumps also have an intra vane feature which helps to keep the vane in continuous contact with the inner ring during the rises and falls while also minimizing vane-tip contact forces. To reach this, the vane-tips of the vane grooves in double-acting vane pumps are usually connect with outlet port.

3) Constant delivery. Pumping occurs while one vane is in contact with the major arc and one with the minor arc of the cam ring. Each arc is essentially a constant radius so the fluid delivered per degree of rotation is a constant value. As a result, the flow delivered by a balance vane pump is very smooth compared to most other types of pumping mechanisms, and sound levels are inherently low.

4) Besides suitable material choose and heat treatment, we would take some measures to solve the problem of pressure-unloaded in order to enhance the working pressure for the double action vane pump. Usually there are three kinds: double vanes, spring vane and mother-son vane configuration.

Fig. 3-11b shows the principle of spring vane configuration. The vane 15 is inserted in the groove of the rotor 13. The hydraulic oil is induced from the top of the vane to its root, so the action forces on the root and top achieve equilibrium. The spring 14 is installed in the root of the vane in order to ensure vane 15 touched closely to the stator 16. The pressure of pump can reach to 14-18MPa.

3.3.2 Single-acting vane pumps

The single-acting vane pump gets its name from the fact that there are only one inlet segment and a separate outlet segment during each revolution.

式中，R、r分别为定子圆弧段的大、小半径；B为转子的宽度；S为叶片的厚度；z为叶片数；θ为叶片槽相对于径向的倾斜角。

式（3-4）右边第二项为叶片厚度对排量的影响，它使泵的理论排量减小。

2. 结构特点

1）配流盘的两个吸油窗口和两个压油窗口对称布置，因此作用在转子和定子上的液压径向力平衡，轴承承受的径向力小，寿命长。

2）为保证叶片在转子叶片槽内自由滑动并始终紧贴定子内环，双作用叶片泵一般采用叶片槽根部全部通压油腔的办法。

3）由于双作用叶片泵的吸、压油经常是在定子的大半径圆弧和小半径圆弧段进行的，每个圆弧段半径不变，因此转子每转一周的排量是个定值。与其他泵相比，双作用叶片泵运行更平稳，产生的噪声较小。

4）为了提高双作用叶片泵的工作压力，除了对有关零件选用合适的材料和进行热处理外，结构上采取必要的措施解决叶片卸荷问题，使叶片压向定子的作用力减小。通常有双叶片式，弹簧叶片式和母子叶片式等结构。

图3-11b所示为弹簧叶片式结构的工作原理图。转子13的叶片槽内装有叶片15。叶片根部的油从叶片的顶部引入，叶片顶部和根部所受的作用力基本平衡。为了保证叶片15和定子16的表面能紧密接触，在叶片根部装有弹簧14。定子表面所受的力仅为弹簧力，叶片泵的压力可达14~18MPa。

3.3.2 单作用叶片泵

单作用叶片泵的转子每转一周，吸、压油各一次，故称为单作用。

1. 工作原理

单作用叶片泵的工作原理如图3-12所示。与双作用叶片泵相同，单作用叶片泵也是由转子2、定子1、叶片3和配油盘（图中未画出）等零件组成的。与双作用叶片泵明显不同之处是，定子的内表面是圆形的，转子与定子之间有一偏心量e，配油盘只开一个吸油窗口和一个压油窗口。当转子转动时，由于离心力作用，叶片顶部始终压在定子内圆表面上。这样，两相邻叶片间就形成了密闭容腔。显然，当转子按图示方向旋转时，图中下半部A的

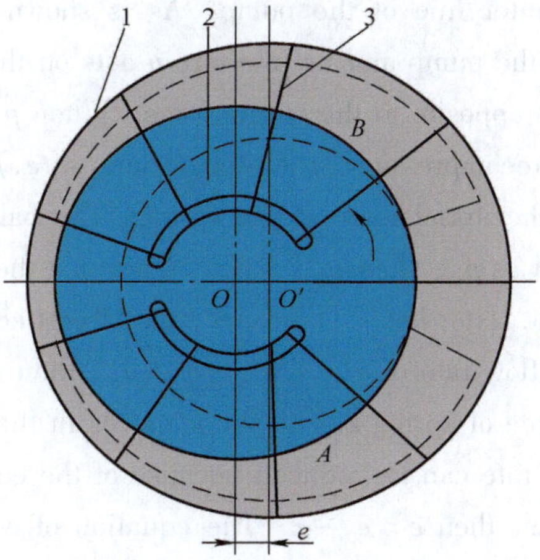

Fig. 3-12 Operating principle of single-acting vane pump（单作用叶片泵工作原理图）

1—Stator（定子） 2—Rotor（转子） 3—Vane（叶片） A—Inlet port（吸油窗口） B—Outlet port（压油窗口）

1. Operating principle

A typical construction of single-acting vane pump is shown in Fig. 3-12. Similar to double-acting vane pumps, the principal components are: a rotor 2, a stator 1, vane 3 and a valve plate (ignored here), etc. Unlike double-acting vane pumps, the inner ring of the stator is circle. There is an eccentricity e in the rotor with respect to the stator and there is only one inlet port and a separate outlet port on the valve plate. When the rotor rotates, the vane-tips are kept pressed on the inner ring of the stator under centrifugal force, forming an oil tight chamber between two adjacent vanes. Obviously, when the rotor rotates in the direction shown in the figure, the down chamber serves as an inlet port A and the up chamber as an outlet port B. Their volumes change for oil suction and delivery. The oil tight regions are shown in this figure; each region accomplishes suction and delivery once per revolution of the rotor. single-acting vane pumps have another name of non-unload type vane pumps. The delivery V for single-acting vane pump is expressed by

$$V = 4BzRe\sin\frac{\pi}{z} \tag{3-5}$$

Where B is the axial width of the rotor (or the width of the vanes); z is the number of the vanes; R is the radius of the inner ring of the stator; e is the eccentricity of the stator with respect to the rotor.

As described above, the inlet and outlet ports can be changed by changing the direction of the stator eccentricity with respect to the rotor. The delivery of the pump can be changed simply by changing its amplitude. If $e = 0$, the oil tight chamber won't change in volume and the pump fails to work. Similarly, the vane grooves are usually set backwards with a slant angle of 20°-30° (unlike the double-acting type) to ensure the smooth movement of the vanes and reduce their wear damage.

2. Pressure-limiting variable vane pumps

A typical construction of pressure-limiting variable vane pumps is shown in Fig. 3-13. The rotor is stationary, while the stator can move in the horizontal direction. Fig. 3-14 shows the stator is pushed to the left by pressure control spring 2, producing an eccentricity e_0 of its center O_2 with respect to the rotor's center O_1. This eccentricity can be adjusted by eccentricity adjusting screw 6. The largest flow rate q_{max} is certain after the adjusting screw 6 is adjusted. The inlet and outlet ports are symmetrical about the center line of the pump. As is shown in Fig. 3-14a, output oil flows through the inner pipeline of the pump and its pressure p acts on the small piston area A. The force $F = pA$ acting on the piston is opposite to the spring force. When $pA = Kx_0$ (K is the stiffness of the spring, and x_0 is the spring precompression at an eccentricity of e_0), the hydraulic pressure on the pistons is balanced out with the initial force of the spring. The balancing pressure p is the highest pressure for this pump, noted as p_B, then $p_BA = Kx_0$. When p (the pressure of this system) $< p_B$, then $pA < Kx_0$, at this state the stator keeps stationary, the largest eccentricity e_0 is constant, and the pump puts out the maximum flow rate of q_{max}. When $p > p_B$, then $pA > Kx_0$, indicating that the oil pressure is larger than the force of spring 2. The stator moves in the right, and the spring is further compressed. The output flow rate can be reduced because of the eccentricity e. Note the additional compression of the spring as x, then $e = e_0 - x$. The equation of equilibrium for the stator is $pA = K(x_0 + x)$. By substituting the formula $p_BA = Kx_0$ into the above equation, then $e = e_0 - A(p - p_B)/K$ (when $p > p_B$), which indicates the relationship between the eccentricity e and the working pressure p

容腔是吸油腔，上半部B的容腔是压油腔，它们容积的变化分别对应着吸油和压油过程。封油区如图中部所示。由于在转子每转一周的过程中，每个密封容腔完成吸油、压油各一次。单作用式叶片泵的转子受不平衡液压力的作用，故又被称为非卸荷式叶片泵，其排量

$$V = 4BzRe\sin\frac{\pi}{z} \tag{3-5}$$

式中，B 为转子的轴向宽度（叶片宽度）；z 为叶片数；R 为定子内圆半径；e 为定子与转子之间的偏心距。

上述工作原理表明，改变定子与转子偏心距的方向也就改变了泵的吸、压油口，即原来的吸油口变成压油口，原来的压油口变成吸油口；改变上述偏心距的大小意味着改变了泵的排量。当偏心量为零时，密封容腔不会有容积变化，因此也就不具备液压泵的工作条件了。

2. 限压式变量叶片泵的变量原理

图3-13所示为YBX型外反馈限压式变量叶片泵的结构图，图3-14a所示为其原理图。转子的中心 O_1 固定，定子可左右移动，在限压弹簧2的作用下，定子被推向左端，使定子中心 O_2 与转子中心 O_1 有一初始偏心距 e_0，此偏心距大小可由偏心调节螺钉6调节，它决定了螺钉本次调节后泵的最大流量 q_{max}。如图3-14a所示，泵工作时，出口压力 p 经泵内通道作用在小柱塞面积 A 上，因此柱塞上的作用力 F（$F=pA$）与弹簧作用力方向相反。当 $pA=Kx_0$（K 为弹簧刚度，x_0 是偏心量为 e_0 时弹簧的预压缩量）时，柱塞所受的液压力与弹簧初始力平衡，此时的压力 p 称为该泵的限定压力，用 p_B 表示，则 $p_BA=Kx_0$；当系统压力 $p<p_B$ 时，则 $pA<Kx_0$，这表明定子不动，最大偏心距 e_0 保持不变，泵也保持最大流量 q_{max}。当系统压力 $p>p_B$ 时，则 $pA>Kx_0$，这表明压力油的作用力大于弹簧2的作用力，使定子向右移动，弹簧被压缩，偏心距 e 减小，泵的流量也随之减少。当偏心距变化时弹簧增加的压缩量为 x，则偏心距 $e=e_0-x$，此时定子受力平衡方程为 $pA=K(x_0+x)$，将 $p_BA=Kx_0$ 代入得：$e=e_0-A(p-p_B)/K$（当 $p>p_B$ 时），此式表明当液压系统压力 p 超过泵的限定压力 p_B 时偏

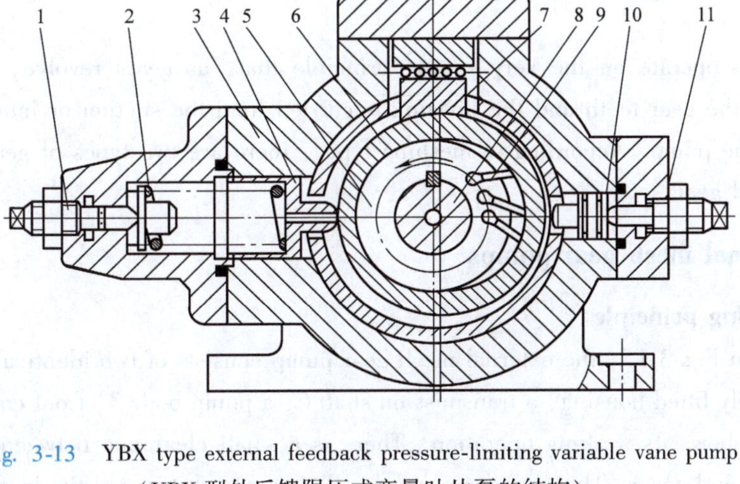

Fig. 3-13 YBX type external feedback pressure-limiting variable vane pump
（YBX型外反馈限压式变量叶片泵的结构）

1—Preload adjusting screw（预紧力调节螺钉） 2—Pressure control spring（限压弹簧） 3—Pump body（泵体）
4—Spring plate（弹簧座） 5—Rotor（转子） 6—Stator（定子） 7—Sliding block（滑块）
8—Transmission shaft（传动轴） 9—Vanes（叶片） 10—Feedback piston（反馈柱塞）
11—Adjusting screw for maximum flow rate（最大流量调节螺钉）

of the pump when the hydraulic system pressure p surpasses the highest pressure p_B. Obviously, the higher the working pressure p, the smaller the eccentricity e and the smaller the output flow rate of the pump.

The pressure-flow characteristics curve is shown in Fig. 3-14b. Point B is an inflexion point, $p_B = Kx_0/A$; and the pressure reaches the highest at point C, $p_C = K(x_0 + e_{max})/A$. At AB section, the hydraulic pressure on the control piston is lower than the set spring force, the eccentricity of the stator $e = e_{max}$ and the delivery of the pump reaches the maximum(AB section is slightly tilted downwards due to the practical flow reduction with the increasing pressure and the pump's leakage). Following point B, the output flow of the pump reduces with the increase of the output pressure, as can be seen from BC section. The slope of BC section depends upon the stiffness of spring. The output flow reduces to zero at point C. Adjusting the pressure set screw can change the precompress value x_0 of the spring, as well as the pressure at point B. Curve BC moves in the horizontal direction, while adjusting the maximum flow screw in the right of the stator will change the highest eccentricity e_{max} and the highest flow rate of the pump. Curve AB moves in the longitudinal direction. The flow rate is equal to zero when the output pressure increases to p_C at point C and won't increase any more, so the highest pressure of the pump is p_C.

This type of variable displacement pump is controlled by the output pressurized oil which acts on the pistons, so it is also called external feedback pressure-limiting variable displacement vane pump.

As described above, for pressure-limiting or constant power variable pumps and load-sensitive variable pump systems, the delivery (flow rate) is controlled automatically based on the feedback of the system pressure, so they are also called pressure-compensated variable pumps.

3.4 Gear Pumps

Gear pumps operate on the very simple principle that, as gears revolve, oil trapped in the spaces between the gear teeth and the housing is carried from the suction or intake side to the discharge side of the pump. According to meshing types, there are two types of gear pumps: external gear and internal gear.

3.4.1 External mesh gear pumps

1. Operating principle

As shown in Fig. 3-15, the external mesh gear pump consists of two identical gears 4 and 8 enclosed in a closely fitted housing, a transmission shaft 6, a pump body 3, front cover 5 and rear cover 1. Fig. 3-16 shows its working operation. There is a small clearance between the gear end face and the housing end cover. The clearance between the gear teeth top and the housing surface is also quite small and the gear housing can be separated as left and right oil tight chambers. As the gears rotate, a vacuum is formed as the teeth un-meshed, which causes liquid to be forced in through the inlet port. Fluid is displaced as the gear teeth meshed at the outlet side and is forced out of the pump into the hydraulic system. In this way, as the gears rotate continuously, the pump carries flu-

心距e和泵的工作压力p之间的关系，即工作压力p越高，偏心距e越小，泵的流量也就越小。

在图3-14b所示的特性曲线中，B点为拐点，对应的压力$p_B=Kx_0/A$；C点为极限压力，$p_C=K(x_0+e_{max})/A$。在AB段，作用在控制活塞上的液压力小于弹簧的预压缩力，定子偏心距$e=e_{max}$，泵输出最大流量。因为随着压力增高，泵的泄漏量增加，泵的实际输出流量减小，因此线段AB略为向下倾斜。拐点B之后，泵输出流量随出口压力的升高而自动减小，如曲线BC所示，曲线BC的斜率与弹簧的刚度有关。到C点，泵的输出流量为零。

调节图3-14中的压力调节螺钉可以改变弹簧的预压缩量x_0，即改变特性曲线中拐点B的压力p_B的大小，曲线BC沿水平方向平移。调节定子左边的最大流量调节螺钉，可以改变定子的最大偏心距e_{max}，即改变泵的最大流量，曲线AB上下移动。由于泵的出口压力升至C点的压力p_C时，泵的流量等于零，压力不会再增加，因此泵的最高压力为p_C。

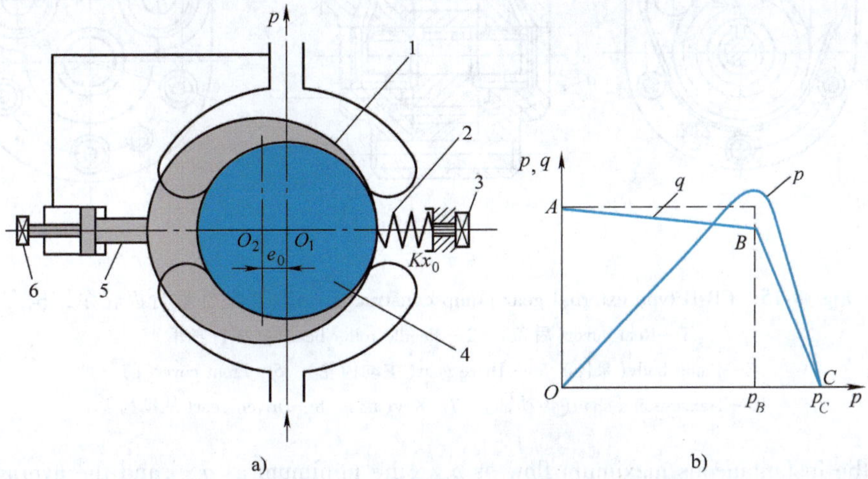

Fig. 3-14 Pressure-limiting variable vane pump（限压式变量叶片泵）

a）Operation（原理图） b）Pressure-flow characteristics curve（特性曲线）

1—Stator（定子） 2—Pressure control spring（限压弹簧） 3—Preload adjusting screw（预紧力调节螺钉）
4—Rotor（转子） 5—Piston（柱塞） 6—Eccentricity adjusting screw（偏心调节螺钉）

这种变量泵是由出油口引出的压力油作用在柱塞上来控制变量的，故称为外反馈限压式变量叶片泵。

综上所述，限压式变量泵以及负载敏感变量泵、恒功率变量泵都是通过系统压力（压差）的反馈作用来自动调节泵的排量（流量）的，因此又总称为压力补偿变量泵。

3.4 齿轮泵

齿轮泵是利用齿轮啮合原理工作的，根据啮合形式不同分为外啮合齿轮泵和内啮合齿轮泵两种。

3.4.1 外啮合齿轮泵

1. 工作原理

外啮合齿轮泵（图3-15）由几何参数完全相同的主动齿轮4和从动齿轮8、传动轴6、

ids through the housing and forces them out by a meshing action, this process repeats itself each revolution of the pump. There is no unique flow-deploying mechanism for the inlet and outlet ports separated from each other by the meshing gear teeth and the pump housing.

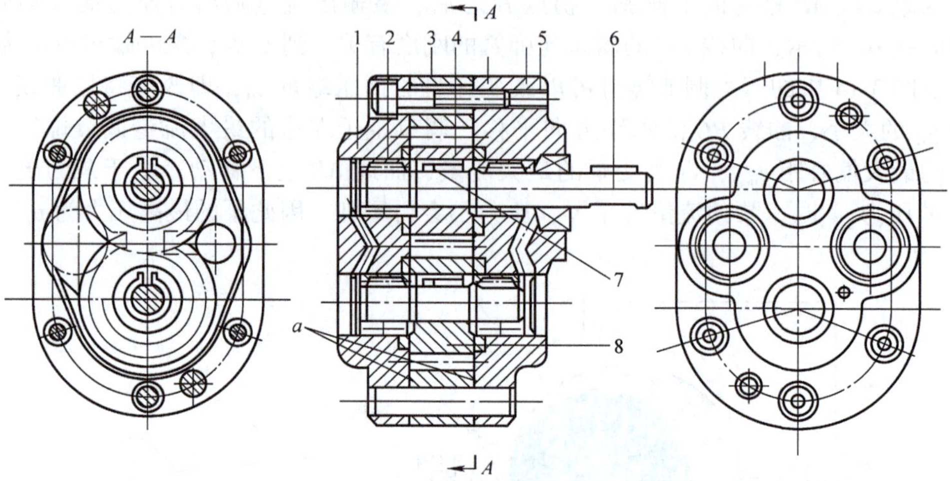

Fig. 3-15 CB-B type external gear pump construction(CB-B 型外啮合齿轮泵结构图)

1—Rear cover(后盖) 2—Needle roller bearing(滚针轴承)

3—Pump body(泵体) 4—Drive gear(主动齿轮) 5—Front cover(前盖)

6—Transmission shaft(传动轴) 7—Key(键) 8—Driven gear(从动齿轮)

Note the instantaneous maximum flow as q_{max}, the minimum as q_{min}, and the average as q_p, then the instantaneous flow fluctuating coefficient δ_Q is expressed by

$$\delta_Q = \frac{q_{max} - q_{min}}{q_p} \tag{3-6}$$

δ_Q increases inversely with the number of gears.

The delivery can be obtained from the formula that describes the area of tooth valley for a gear: $A = \pi m^2$, i. e.

$$V = 2\pi z m^2 B \tag{3-7}$$

Where z and m represent the teeth number and the modulus of the gear respectively, B is the width of the gear teeth.

From formula (3-7), we can see that the delivery V is proportional to the gear teeth number (z) and the square of its modulus (m^2). So, when the diameter of gear pitch circle $D_j = mz$ is given, the delivery of the pump can be increased by increasing the modulus m or decreasing the gear teeth number z. It is for this reason that the number of gear teeth is usually fewer, and the gear should be modified to avoid root undercut due to a small number of gear teeth.

2. Structure features

(1) Noise reduction The flow pulsation is a source of noise. If considering the design (such as setting two sets of staggered gears on the same shaft) to avoid pulsing flow due to the trapping of oil between the meshing gear teeth, very favorable low noise can be achieved.

泵体3、前盖5及后盖1等主要零件组成。图3-16所示为外啮合齿轮泵的工作原理图。由于齿轮端面与壳体端盖之间的缝隙很小，齿轮齿顶与壳体内表面的间隙也很小，因此可以看成将齿轮泵壳体内分隔成左、右两个密闭容腔。当齿轮按图示方向旋转时，右侧的齿轮逐渐脱离啮合，密闭工作容积逐渐增大，形成局部真空，油箱中的油液在大气压力的作用下，经吸油管进入吸油腔，将齿谷空间充满，并随着齿轮的转动，把油液带到左侧压油腔。左侧压油腔轮齿逐渐进入啮合，密封工作容积逐渐减小，把油液挤压出去，输送到压力管路中。齿轮不断地旋转，泵的吸、压油口便不断地吸油和压油。在齿轮泵的啮合过程中，啮合点沿啮合线把吸油区和压油区自然分隔开来，因此没有单独的配油机构。

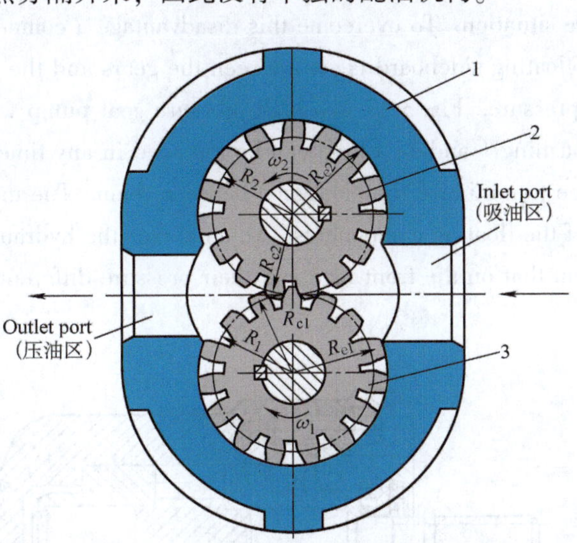

Fig. 3-16 External gear pump operation（外啮合齿轮泵工作原理图）

1—Pump body（泵体） 2—Drive gear（主动齿轮） 3—Driven gear（从动齿轮）

若瞬时最大流量为 q_{max}，最小流量为 q_{min}，平均流量为 q_p，则 δ_Q 表示泵的瞬时理论流量脉动系数，有

$$\delta_Q = \frac{q_{max} - q_{min}}{q_p} \tag{3-6}$$

δ_Q 值随齿数增多而减小。

齿轮泵的排量可根据轮齿齿谷的面积 $A = \pi m^2$ 得到，即

$$V = 2\pi z m^2 B \tag{3-7}$$

式中，z 为齿数；m 为齿轮模数；B 为齿宽。

由式（3-7）可以看到，齿轮泵的排量 V 与齿轮模数 m 的平方成正比，与齿数 z 的一次方成正比。因此，在齿轮节圆直径 $D_j = mz$ 一定时，增大模数 m，减少齿数 z 可以增大泵的排量，因为这一原因，齿轮泵的齿数一般较少，为避免因齿数少而产生根切，需对齿轮进行修正。

2. 结构特点

（1）降低齿轮泵的噪声　齿轮泵产生噪声的一个主要根源来自流量脉动，为减小齿轮泵的瞬时理论流量脉动，可同轴安装两套齿轮，每套齿轮之间错开半个齿距，两套齿轮之间用一平板相互隔开，组成共同吸油和压油的两个分离的齿轮泵。由于两个泵的脉动错开了半

(2) Leakage Here the leakage is the inner leakage of a pump, i. e., a part of oil in the inlet port is displaced to the outlet port with the gears' rotation. Obviously, the leakage will reduce the volumetric efficiency. The main leakage of an external pump is the gear end cover leakage, which takes up about 70%-75% of the total leakage. To reduce the leakage of the end cover is the main path to increase the volumetric efficiency of gear pumps.

(3) Clearance As it is shown in Fig. 3-15, the surface that is clearance formed by the front and rear covers and the gears can not be too small due to the technology limitation of machining and assembly. Instead, the clearance will be larger as the gears wear out, thus pumps of this type only find use in low pressure situation. To overcome this disadvantage a compensatory component such as a floating bushing or a floating sideboard is set between the gears and the front and rear covers in the gear pumps with high pressure. Fig. 3-17 is a high pressure gear pump with floating bushing. These components, floating bushing 1 and 2, which can be replaced in any time after wearing out, cooperate with the gear surface to minimize the clearance between them. The method aims to lead pressurized oil into the back of the floating components, which makes the hydraulic pressure acting force on the back face larger than that on the front face and their pressure differential is undertaken by an extremely thin oil film.

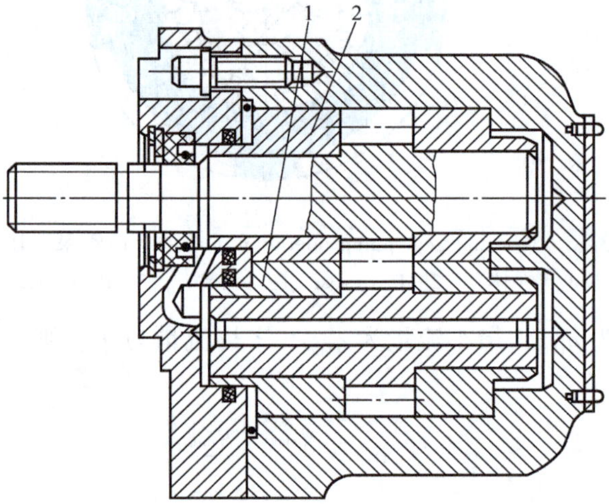

Fig. 3-17 The high pressure gear pump with floating bushing(采用浮动轴套的中高压齿轮泵)
1,2—Floating bushing(浮动式轴套)

(4) Unbalanced hydraulic radial force The resultant hydraulic pressure on the gears and the shaft is unbalanced due to the pressure difference between the inlet and outlet ports, see Fig. 3-18a. The unbalanced hydraulic radial force can be expressed by

$$F = K\Delta pBD_e \tag{3-8}$$

Where K is the coefficient: for the drive shaft, $K = 0.75$, for the driven shaft, $K = 0.85$; Δp is the pressure difference between the inlet and outlet ports; B is the gear width; D_e is the diameter of the tip circle.

As described above, when the size of the pump is given, the higher the oil pressure, the larger the unbalanced radial force. The hydraulic radial force acting on the gear shaft not only affects the

个周期，各自的脉动量相互抑制，因此，总的脉动量大大减小。

（2）泄漏 这里所说的泄漏是指液压泵的内部泄漏，即一部分压力油从压油腔流回吸油腔，没有输送到系统中去。显然，泄漏降低了液压泵的容积效率。外啮合齿轮泵的泄漏主要是齿轮端面泄漏，这部分泄漏量约占总泄漏量的70%～75%。减小端面泄漏是提高齿轮泵容积效率的主要途径。

（3）间隙 图3-15所示的齿轮泵，一方面由前、后盖与齿轮端面形成的端面间隙因加工工艺和装配工艺的限制，间隙值不可能很小；另一方面磨损后间隙会越来越大，因此只适用于低压场合。为使齿轮泵能在高压下工作，并且具有较高的容积效率，需要从结构上采取措施对端面间隙进行自动补偿。通常采用的端面间隙自动补偿装置有浮动轴套式和弹性侧板式两种，其原理都是引入压力油使轴套或侧板紧贴齿轮端面，压力越高，贴得越紧，因而可自动补偿端面损失和减小间隙。图3-17所示为采用浮动轴套的中高压齿轮泵。浮动式轴套1和2是浮动安装的，轴套左侧的空腔均与泵的压油腔相通。当泵工作时，浮动式轴套1和2受左侧油压作用向右移动，将齿轮两侧面压紧，从而自动补偿了端面的间隙。

（4）液压径向不平衡力 在齿轮泵中，由于在压油腔和吸油腔之间存在着压差，液体压力的合力作用在齿轮和轴上，是一种径向不平衡力，如图3-18a所示。径向不平衡力的大小为

$$F = K\Delta pBD_e \tag{3-8}$$

式中，K为系数，对于主动齿轮，$K=0.75$，从动齿轮，$K=0.85$；Δp为泵进、出口压差；D_e为齿顶圆直径。

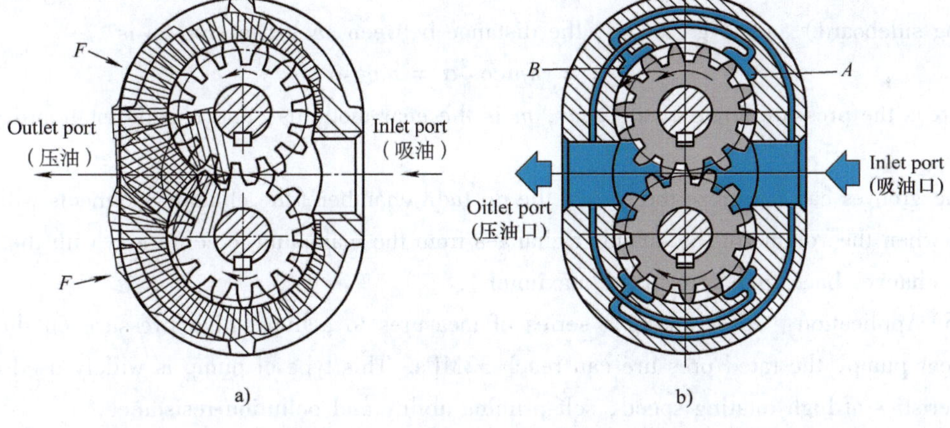

Fig. 3-18 Unbalanced force（不平衡力）

a) Unbalanced hydraulic radial force（液压径向不平衡力） b) Balancing grooves（平衡槽）

由此可见，当泵的尺寸确定以后，油液压力越高，径向不平衡力就越大。其结果是加速轴承的磨损，增大内部泄漏，甚至造成齿顶与壳体内表面的摩擦。减小径向不平衡力的方法有：

1）开压力平衡槽（图3-18b）。在盖板上开设平衡槽A、B，使它们分别与低、高压腔相通，产生一个与吸油腔和压油腔对应的液压径向力，起平衡作用。但这种方法会增加内泄漏，较少使用。

life of the bearings, but also distorts the gear shaft, resulting in the gear tip's scraping against the inner contour of the pump body.

Measures to balance the hydraulic radial force:

1) We can set balancing grooves A and B on the cover board (Fig. 3-18b), making them connect with the inlet and outlet ports respectively. Another hydraulic radial force will be produced in the opposite direction to balance the previous radial force. Note that this measure will increase inner leakage-level, so it finds few applications.

2) Another method is to reduce the area of the outlet ports and thus the high pressure oil action area on the gear.

(5) Phenomenon of surrounded oil and unloading measures To achieve a steady gear transmission, the gear overlap coefficient ε of the pump should be greater than 1 (usually $\varepsilon = 1.05\text{-}1.10$), i. e. , the next pair of gear teeth begin to mesh before the previous pair unmesh. An oil tight chamber is formed between the two pairs of teeth when they mesh at the same time, see Fig. 3-19. As the gears rotate, the size of the chamber constantly changes from zero to the maximum and then back to zero during one complete revolution. As the size begins to increase, a vacuum is formed without oil compensation, which will cause cavitation and noise. As the gears continue to rotate, the size of the chamber decreases and the oil is forced out through the gap, which not only increases pressure and makes the gear bearing undertake periodic pressure pulsation, but also increases the temperature of the oil. The action described above is called phenomenon of surrounded oil, which will largely reduce the pump's life and should be diminished from the pump system. To achieve this, most gear pumps are designed with relief grooves on the front and rear cover boards or floating bearing sets (floating sideboard), see Fig. 3-19d, the distance between each two grooves is

$$a = \pi m \cos^2\alpha = t_0 \cos\alpha \tag{3-9}$$

Where α is the pressure angle of the gear; m is the gear modulus; and t_0 is the standard gear base pitch.

The grooves can minimize the size of the oil tight chamber (the chamber connects with the outlet port when the volume of the chamber changes from the maximum to zero, and with the inlet port when it changes back from zero to the maximum).

(6) Application After taking a series of measures to achieve high pressure on the external mesh gear pump, the rated pressure can reach 32MPa. This type of pump is widely used due to its characteristics of high rotating speed, self-priming ability and pollution-resistance.

3.4.2 Internal mesh gear pumps

Fig. 3-20 shows the operation of a internal mesh gear pump. A pinion is driven by a shaft, and the crescent-shaped block acts as a division between suction and discharge. As the driving pinion meshes with and drives the internal gear, the size of the top oil tight chamber increases as the tops of teeth unmesh and begin to draw oil. The size of the bottom oil tight chamber decreases as the bottoms of teeth begin to mesh and then oil is forced out.

The greatest advantages of internal gear pumps are: no surrounded oil, low sound level and lower flow pulsation compared with external gear pumps. The rated pressure can reach 30MPa with

2）缩小压油腔。通过减小高压油在齿轮上的作用面来减小径向不平衡力。

（5）困油现象与卸荷措施　为了使齿轮平稳地啮合运转，根据齿轮啮合原理，齿轮的重叠系数应大于1，即存在两对轮齿同时进入啮合的情况。因此，就有一部分油液困在两对轮齿所形成的封闭容腔之内，这个封闭容腔先随齿轮转动逐渐减小，以后又逐渐增大。减小时会使被困油液受挤压而产生高压，并从缝隙中挤出，导致油液发热，同时也使轴承受到不平衡负载的作用；封闭容腔的增大会造成局部真空，使溶于油液中的气体分离出来，产生气穴，这就是齿轮泵的困油现象。其封闭容积的变化如图3-19所示。困油现象使齿轮泵产生强烈的噪声和气蚀，影响、缩短其工作的平稳性和寿命。消除困油的方法，通常是在两端盖板上开卸荷槽，见图3-19d中的虚线。当封闭容积减小时，通过右边的卸荷槽与压油腔相通，而封闭容积增大时，通过左边的卸荷槽与吸油腔相通，两卸荷槽的间距必须确保在任何时候都不使吸、排油相通。两卸荷槽之间的距离为

$$a = \pi m \cos^2\alpha = t_0 \cos\alpha \qquad (3\text{-}9)$$

式中，α 为齿轮压力角；m 为齿轮模数；t_0 为标准齿轮的基节。

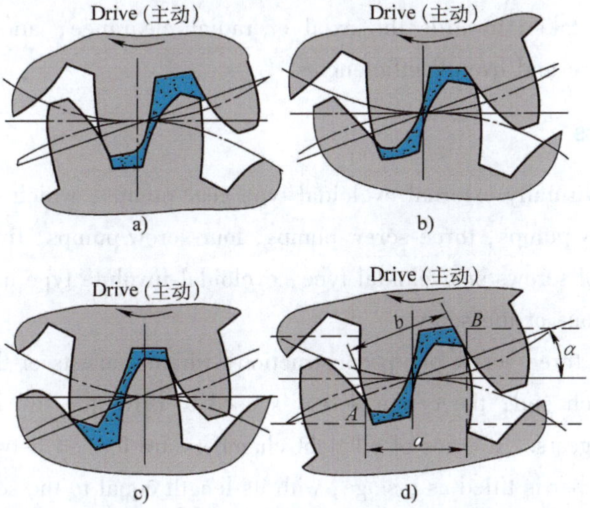

Fig. 3-19　Phenomenon of surrounded oil and its eliminating measure
（齿轮泵的困油现象及消除措施）

在开设卸荷槽后，可使闭死容积限制为最小，即容积由大变小时与压油腔相通，容积由小变大时与吸油腔相通。

（6）应用　外啮合齿轮泵在采取了一系列的高压化措施后，额定压力可达32MPa。由于它具有转速高、自吸能力好、抗污染能力强等一系列优点，因此得到了广泛的应用。

3.4.2　内啮合齿轮泵

图3-20所示为内啮合齿轮泵的工作原理，一对相互啮合的小齿轮和内齿轮与侧板所围成的密闭容积被轮齿啮合线和月牙板分隔成两部分。当传动轴带动小齿轮按图示方向旋转时，内齿轮同向旋转，图中上半部轮齿脱开啮合，所在的密闭容积增大，为吸油腔；下半部轮齿进入啮合，所在的密闭容积减小，为压油腔。

内啮合齿轮泵的最大优点是：无困油现象，流量脉动较外啮合齿轮泵小，噪声低。采用

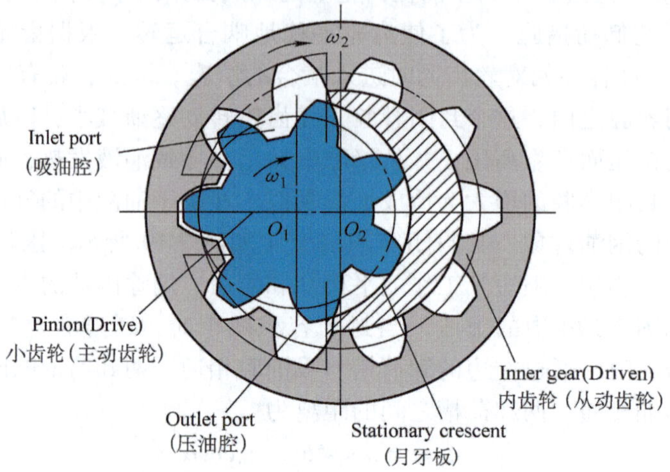

Fig. 3-20 Operating principle of inner gear pump(内啮合齿轮泵工作原理)

compensating measures taken towards the axial or radial clearance, and thus the pump can be achieve higher volumetric and overall efficiencies.

3. 4. 3 Screw pumps

Screw pumps are virtually external cycloidal type gear pumps, which can be divided as single-screw pumps, two-screw pumps, three-screw pumps, four-screw pumps, five-screw pumps, etc. according to the number of screws or cycloidal type, cycloidal-involute type and circular type based on the different cross sections of the screw.

Fig. 3-21 shows a three-screw pump construction, which consists of three parallel screws, oppositely threaded on each end; the center screw is used to drive the two outer idler rotors directly without external timing gears. A series of oil tight chambers are formed between the three screws and the housing. Each chamber is titled as a stage, with its length equal to the screw pitch. As the driving shaft rotates in a clockwise direction, the size of the oil tight chamber in the left of the screws is formed gradually and then increases to serve as inlet port, while the chamber in the right of the screws diminishes gradually and then becomes outlet port. There is at least an intact oil tight chamber between the inlet and outlet ports. The more stages the screws have, the higher the rated pressure of the pump.

The greatest advantages of the screw pumps are that they possess low sound level, even output flow, especially fit for precise mechanics which require low pressure pulsation and steady flow. In addition, screw pumps have a good characteristic of self-priming which permits high rotating speed and large flow rate. Consequently, screw pumps find many applications in large-size hydraulic system as auxiliary pumps. They are also widely used to transmit liquid with high grade of viscosity such as crude oil.

Because of complex machining techniques, and strict requirement of machining precision, special machining equipment is needed to produce inner gear pumps and screw pumps. Therefore there is a limitation to their applications.

轴向和径向间隙补偿措施后，泵的额定压力可达30MPa，容积效率和总效率均较高。

3.4.3 螺杆泵

螺杆泵实质上是一种外啮合摆线齿轮泵，按其螺杆根数有单螺杆泵、双螺杆泵、三螺杆泵、四螺杆泵和五螺杆泵等；按螺杆的横截面分为摆线齿形、摆线—渐开线齿形和圆形齿形三种不同形式的螺杆泵。

图3-21所示为一种三螺杆泵的结构图，在壳体2内放置有三根平行且互为啮合的双头螺杆，中间为主动螺杆（凸螺杆），两侧为从动螺杆（凹螺杆）。三根螺杆的外圆与壳体之间形成多个密闭容积，每个密闭容积为一级，其长度约等于螺杆的螺距。当传动轴（与凸螺杆为一整体）如图示方向旋转时，左端螺杆密封空间逐渐形成，容积增大为吸油腔；右端螺杆密封空间逐渐消失，容积减小为压油腔。在吸油腔与压油腔之间至少有一个完整的密闭工作腔，螺杆的级数越多，泵的额定压力越高（每一级工作压差为2～2.5MPa）。

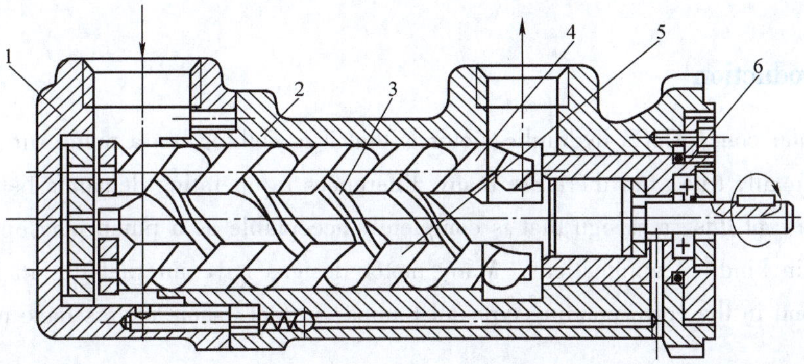

Fig. 3-21 Screw pump（螺杆泵）

1—Rear cover（后盖） 2—Housing（壳体） 3—Drive screw（protruding）主动螺杆（凸螺杆）

4，5—Driven screw（sunken）从动螺杆（凹螺杆） 6—Front cover（前盖）

螺杆泵与其他容积式液压泵相比，具有结构紧凑、体积小、重量轻、自吸能力强、运转平稳、流量无脉动、噪声小、对油液污染不敏感及工作寿命长等优点。目前常用在精密机床上和用来输送粘度大或含有颗粒物质的液体。螺杆泵的缺点是其加工工艺复杂，加工精度要求高，所以应用受到一定限制。

Chapter 4 Hydraulic Actuators

The function of hydraulic actuators is to translate hydraulic energy of fluid into energy of the machine. They will drive the mechanism motion path like a straight, swing or rotation. The straight or swing is referred to as hydraulic actuator or swing motion actuator, and the rotational motion is called hydraulic motor. The output parameters are force and speed or torque and rotating speed.

4.1 Hydraulic Motors

4.1.1 Introduction

The popular concept of a hydraulic motor is that it is nothing but a pump run backwards. This may be superficially true, but there are many differences in operating demands between pumps and motors. Because of this, a design that is completely acceptable as a pump may operate poorly as a motor in certain kinds of applications. Many motor designs have internal design features differing from those found in the corresponding types of pumps; in fact, some motors have no pump counterpart.

1. Characteristic parameters

(1) Power and overall efficiency The input power P_{Mi} of motor is

$$P_{Mi} = \Delta p q_M \tag{4-1}$$

The output power P_{Mo} of the motor is

$$P_{Mo} = T_{Mo} \cdot 2\pi n \tag{4-2}$$

The overall efficiency η_M of the motor is the ratio of P_{Mo} to P_{Mi}, i. e.

$$\eta_M = \frac{P_{Mo}}{P_{Mi}} = \eta_{Mm}\eta_{MV} \tag{4-3}$$

Where Δp is the pressure differential between the inlet and outlet ports of the motor; q_M is the flow rate put in the motor inlet port, $q_M = Vn$; V is the delivery of the motor; n is the rotating speed of the motor; T_{Mo} is actual output torque; ω is the angular speed, $\omega = 2\pi n$; η_{Mm} and η_{MV} are mechanical and volumetric efficiencies respectively.

(2) Torque and mechanical efficiency

1) The theoretic torque. The hydraulic power of pump will be equal to that of the motor without energy loss, then we can get the torque of the motor from formula (4-1) and formula (4-2) directly

$$\Delta p q_M = T_{Mt} \cdot 2\pi n$$

第 4 章　液压执行元件

液压执行元件是将流体的压力能转换为机械能的元件，它驱动机构作直线往复运动、摆动或旋转运动。液压执行元件分为两类：液压马达和液压缸。作直线往复运动的称为液压缸，作摆动的称为摆动液压马达，作旋转运动的称为液压马达。其输出为力与速度或转矩与转速。

4.1　液压马达

4.1.1　概述

从工作原理上讲，液压传动中的泵和马达都是靠工作腔密闭容器的容积变化而工作的，所以说泵可以作马达用，反之也一样，即泵与马达有可逆性。实际上由于二者工作状况不一样，为了更好发挥各自工作性能，在结构上存在某些差别，使之不能通用。

1. 特性参数

（1）功率与总效率　马达输入功率 P_{Mi} 为

$$P_{\mathrm{Mi}} = \Delta p q_{\mathrm{M}} \tag{4-1}$$

马达输出功率 P_{Mo} 为

$$P_{\mathrm{Mo}} = T_{\mathrm{Mo}} \cdot 2\pi n \tag{4-2}$$

马达的总效率 η_{M} 等于马达的输出功率 P_{Mo} 与输入功率 P_{Mi} 之比，即

$$\eta_{\mathrm{M}} = \frac{P_{\mathrm{Mo}}}{P_{\mathrm{Mi}}} = \eta_{\mathrm{Mm}}\eta_{\mathrm{M}V} \tag{4-3}$$

式中，Δp 为液压马达进出、口之间的压差；q_{M} 为输入液压马达的实际流量，$q_{\mathrm{M}} = Vn$，V 为液压马达的排量，n 为液压马达转速；T_{Mo} 为液压马达输出的实际转矩；ω 为角速度，$\omega = 2\pi n$；η_{Mm} 为马达的机械效率；$\eta_{\mathrm{M}V}$ 为马达的容积效率。

（2）马达的转矩和机械效率

1）理论转矩。如果不计损失，液压泵输出的液压功率应当全部转化为液压马达输出的机械功率，即可以从式（4-1）和式（4-2）得出液压马达的理论转矩 T_{Mt}

$$\Delta p q_{\mathrm{M}} = T_{\mathrm{Mt}} \cdot 2\pi n$$

$$T_{\mathrm{Mt}} = \frac{\Delta p q_{\mathrm{M}}}{2\pi n} \tag{4-4}$$

2）起动转矩。设 T_{Ms} 为马达从静止状态到开始起动时的起动输出转矩，ΔT_{Ms} 为开始起动时的静摩擦转矩损失。通常起动性能指标采用起动机械效率 η_{Ms} 表示，其表达式为

$$\eta_{\mathrm{Ms}} = \frac{T_{\mathrm{Ms}}}{T_{\mathrm{Mt}}} = \frac{T_{\mathrm{Mt}} - \Delta T_{\mathrm{Ms}}}{T_{\mathrm{Mt}}} \tag{4-5}$$

$$T_{\mathrm{Mt}} = \frac{\Delta p q_{\mathrm{M}}}{2\pi n} \tag{4-4}$$

2) Startup torque. If we define T_{Ms} as the startup torque, i. e., the actual motor output torque from begining to start up, and ΔT_{Ms} as the torque loss at the static state. The startup performance should be described by startup mechanical efficiency, denoted as η_{Ms}

$$\eta_{\mathrm{Ms}} = \frac{T_{\mathrm{Ms}}}{T_{\mathrm{Mt}}} = \frac{T_{\mathrm{Mt}} - \Delta T_{\mathrm{Ms}}}{T_{\mathrm{Mt}}} \tag{4-5}$$

3) Mechanical efficiency. If we define the T_{Mo} as the actual output torque after startup and ΔT_{Mo} as the torque loss at the dynamic state. However, the practical output torque T_{Mo} is also less than the theoretical torque T_{Mt} due to the energy loss produced by the relative movement of each components and the relative movement between the fluid and the components, thus

$$T_{\mathrm{Mo}} = T_{\mathrm{Mt}} - \Delta T_{\mathrm{Mo}} \tag{4-6}$$

The mechanical efficiency of the motor, noted as η_{Mm}

$$\eta_{\mathrm{Mm}} = \frac{T_{\mathrm{Mo}}}{T_{\mathrm{Mt}}} = \frac{T_{\mathrm{Mt}} - \Delta T_{\mathrm{Mo}}}{T_{\mathrm{Mt}}} \tag{4-7}$$

Here $\eta_{\mathrm{Ms}} < \eta_{\mathrm{Mm}}$ comes from formulae (4-5) and (4-7) because the static friction coefficient is larger than that of dynamic state and also $\Delta T_{\mathrm{Ms}} > \Delta T_{\mathrm{Mo}}$.

(3) Flow rate and volumetric efficiency The practical flow rate q_{M} is the flow rate that flows into the motor inlet port under a certain pressure. Note here that a leakage Δq exists due to the clearance in motor. So, to reach the required rotating speed, the practical q_{M} should be

$$q_{\mathrm{M}} = q_{\mathrm{Mt}} + \Delta q \tag{4-8}$$

Where q_{Mt} is the theoretical flow rate without leakage in the motors.

The volumetric efficiency η_{MV} is the ratio of theoretical flow rate q_{Mt} to the practical flow rate q_{M}

$$\eta_{MV} = \frac{q_{\mathrm{Mt}}}{q_{\mathrm{M}}} = \frac{q_{\mathrm{M}} - \Delta q}{q_{\mathrm{M}}} = 1 - \frac{\Delta q}{q_{\mathrm{M}}} \tag{4-9}$$

(4) Delivery and rotating speed The motor delivery is defined as the needed oil volume per revolution of the motor output shaft without leakage when the volumetric efficiency equals 1. There are two types of motors, namely, fixed delivery when the motor delivery V is constant and variable delivery when the motor delivery V is variable. The motor rotating speed n is

$$n = \frac{q_{\mathrm{Mt}}}{V} = \frac{q_{\mathrm{M}}\eta_{MV}}{V} \tag{4-10}$$

(5) Working and rated pressures The practical pressure of input oil is called working pressure in motors, whose value depends on the load on motors. The differential between the input pressure and the output pressure is called motor pressure differential. While the rated pressure is the highest pressure that ensures the motor's continuous and normal motion according to test standard.

2. Classification

Motors could be classified according to their working performances: High-speed hydraulic motors whose rated rotating speed is beyond 500r/min; Low-speed hydraulic motors whose rated rotating speed is below 500r/min.

High-speed hydraulic motors may be further divided into gear motors, vane motors, axial piston

3）机械效率。设 T_{Mo} 为起动后马达实际输出转矩，ΔT_{Mo} 为马达中各种零件间相对运动和流体与零件相对运动而产生的能量损失，则有

$$T_{Mo} = T_{Mt} - \Delta T_{Mo} \tag{4-6}$$

马达的机械效率 η_{Mm} 为

$$\eta_{Mm} = \frac{T_{Mo}}{T_{Mt}} = \frac{T_{Mt} - \Delta T_{Mo}}{T_{Mt}} \tag{4-7}$$

由于开始起动时的静摩擦因数大于起动后的动摩擦因数，使得起动时能量损失 ΔT_{Ms} 大于起动后的能量损失 ΔT_{Mo}。因此从式（4-5）和式（4-7）可以看出起动机械效率 η_{Ms} 小于机械效率 η_{Mm}。

（3）流量与容积效率　液压油以一定的压力流入液压马达入口的流量称为马达的实际流量，用 q_M 表示。由于马达存在间隙，会产生泄漏 Δq，为达到要求转速，则输入马达的实际流量 q_M 必须为

$$q_M = q_{Mt} + \Delta q \tag{4-8}$$

式中，q_{Mt} 为马达理论流量。

马达的理论流量 q_{Mt} 与实际流量 q_M 之比为马达的容积效率 η_{MV}，即

$$\eta_{MV} = \frac{q_{Mt}}{q_M} = \frac{q_M - \Delta q}{q_M} = 1 - \frac{\Delta q}{q_M} \tag{4-9}$$

（4）马达的排量与转速　马达的排量 V 是在容积效率等于 1 时，即没有泄漏的情况下，使马达输出轴旋转一周所需要油液的体积。马达排量 V 不可变的称为定量马达，可变的称为变量马达。马达的转速 n 为

$$n = \frac{q_{Mt}}{V} = \frac{q_M \eta_{MV}}{V} \tag{4-10}$$

（5）工作压力与额定压力　马达输入油液的实际压力称为马达的工作压力，其大小取决于马达的负载。马达进口压力与出口压力的差值称为马达的压差。

按试验标准规定，能使马达连续正常运转的最高压力称为马达的额定压力。

2. 分类

按照工作特性马达可分为两大类：额定转速在 500r/min 以上的为高速液压马达，额定转速在 500r/min 以下的为低速液压马达。高速液压马达有齿轮马达、叶片马达、轴向柱塞马达、螺杆马达等。低速液压马达有单作用连杆型径向柱塞马达和多作用内曲线径向柱塞马达等。按其排量能否改变可分为定量马达和变量马达。

液压马达的图形符号如图 4-1 所示。

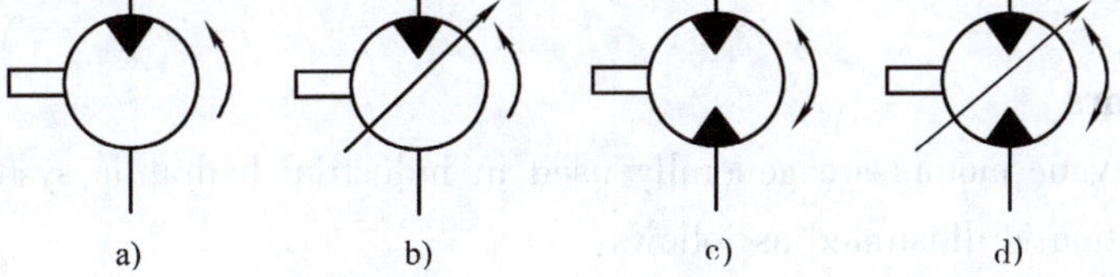

Fig. 4-1　Graphics symbols of motor（液压马达图形符号）

a）One-way fixed displacement motor（单向定量液压马达）　b）One-way variable displacement motor（单向变量液压马达）
c）Double-way fixed delivery motor（双向定量液压马达）　d）Double-way variable delivery motor（双向变量液压马达）

motors, screw motors, etc. Low-speed hydraulic motors include single-acting connecting rod type radial piston motors and incurve multiple-acting radial piston motors.

Motors can also be classified as fixed delivery and variable delivery according to their delivery types.

The graphic symbols of motor are shown in Fig. 4-1.

4.1.2 Operating principle of hydraulic motors

1. Axial piston motors

Fig. 4-2 shows the working principle of an axial piston motor. Valve plate 4 and swashplate 1 are stationary, while motor shaft 5 and cylinder 2 rotate together. When pressurized oil enters into the piston bores of cylinder 2 through valve plate 4, pistons 3 extend out under oil pressure against the swashplate 1, which then produces a normal reaction force F on pistons 3, this force can be further decomposed into axial branch force F_x and vertical branch force F_y. F_x is balanced out with the pressure acting on the pistons, while F_y develops a torque with respect to the cylinder center, which causes a counter clockwise rotation of motor shaft. Note that the instantaneous overall torque created by axial piston motor is fluctuant and motor shaft 5 will rotate clockwise if the pressurized oil is put from the other direction. The swashplate angle α or the delivery not only decides the mortor's torque, but also its rotating speed. The larger the angle α, the higher the torque and the lower the speed.

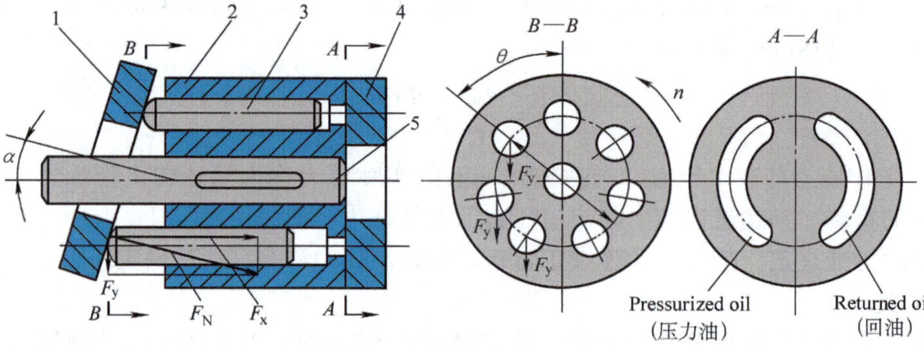

Fig. 4-2 Operating principles of axial piston motor(轴向柱塞马达的工作原理)
1—Swashplate(斜盘) 2—Cylinder(缸体) 3—Piston(柱塞)
4—Control lens(配油盘) 5—Motor's shaft(马达轴)

The delivery calculation for an axial piston motor is exactly the same as that for an axial piston pump.

2. Vane motors

Double-acting vane motors are generally used in industrial hydraulic systems, here a double-acting design operation is illustrated as follows:

Fig. 4-3 shows a vane motor operation. When high pressure oil is put into a working region enclosed by vanes 4 and 8 from the inlet port, pressurized oil acts on the two sides of vane 4 and 8 and develops no torque. Vanes 1,5,3 and 7 undertake high pressure of oil on one side and low pressure

4.1.2 液压马达的工作原理

1. 轴向柱塞马达

轴向柱塞马达的工作原理如图 4-2 所示，配油盘 4 和斜盘 1 固定不动，马达轴 5 与缸体 2 相连接并一起旋转。当压力油经配油盘 4 的窗口进入缸体 2 的柱塞孔时，柱塞 3 在压力油作用下外伸，紧贴斜盘 1，斜盘 1 对柱塞 3 产生一个法向反力 F，此力可分解为轴向分力 F_x 和垂直分力 F_y。F_x 与柱塞上的液压力相平衡，而 F_y 则使柱塞对缸体中心产生一个转矩，带动马达轴逆时针方向旋转。轴向柱塞马达产生的瞬时总转矩是脉动的。若改变马达压力油输入方向，则马达轴 5 按顺时针方向旋转。斜盘倾角 α 的改变，即排量的变化，不仅影响马达的转矩，而且影响它的转速。斜盘倾角越大，产生转矩越大，转速越低。

轴向柱塞马达的排量公式与轴向柱塞泵完全相同。

2. 叶片液压马达

常用叶片液压马达为双作用式，下面以双作用式来说明其工作原理。

叶片液压马达的工作原理如图 4-3 所示。当高压油 p 从进油口同时进入工作区段的叶片 4 和 8 之间的容积时，叶片 4 和 8 的两侧均受压力油 p 作用不产生转矩，而叶片 1 和 5、3 和 7 都有一侧受高压油 p 的作用，另一侧受低压油 p_t 的作用。由于叶片 1 和 5 伸出面积大于叶片 3 和 7 的伸出面积，所以产生使转子顺时针方向转动的转矩。由图可知，当改变进油方向时，即高压油 p 进入叶片 2 和 6 之间容积时，叶片带动转子逆时针方向转动。

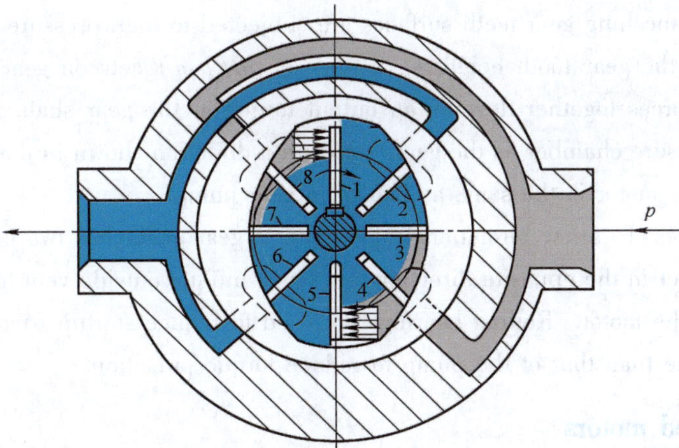

Fig. 4-3 Operating principle of balanced vane motor（叶片液压马达的工作原理）

1，2，3，4，5，6，7，8—Vanes（叶片）

叶片液压马达的排量公式与双作用叶片泵相同。

结构特点：为了适应马达正反转要求，叶片液压马达的叶片为径向放置，叶片倾角 $\theta=0$。为了使叶片底部始终通入高压油，在高、低油腔通入叶片底部的通路上装有梭阀。为了保证叶片液压马达在压力油通入后，高、低压腔不致串通并能正常起动，在叶片底部安装了预紧弹簧。叶片液压马达体积小，转动惯量小，反应灵敏，能适应较高频率的换向。但泄漏较大，低速时不够稳定。它适用于转矩小、转速高、机械性能要求不很严格的场合。

on the other, which develops a torque for clockwise rotation at the motor shaft due to the larger extended area of vanes 1 and 5 than that of vanes 3 and 7. Similarly, the rotor will turn with the vanes in the opposite direction simply by changing direction of the input flow, i. e., high pressure oil enters into the chamber enclosed by vanes 2 and 6.

The delivery calculation for a double-acting vane motor is exactly the same as that for a double-acting vane pump.

Structure features: The vanes in a vane motor are positioned radially with a zero vane angle ($\theta = 0$). A shuttle valve is set at the path between high/low pressure oil chambers and the vane bottom to ensure high pressure oil at the vane bottom. To ensure a normal startup, a spring with preload is mounted at the vane bottom to prevent the connection of high and low pressure chambers when high pressure oil enters into the vane motor. Vane type motors are small in size and thus the moment of inertia, making a short reaction time and applicable to frequent direction-changing occasions. However vane type motors find more applications in small torque, high speed, low mechanical performance request occasions due to its large leakage and inadequate stabilization at low speed.

3. Gear motors

Fig. 4-4 shows an external gear motor operation. C is the meshing point of gear Ⅰ and gear Ⅱ; h is the overall height of the gear. The distances between C and the tooth roots of gear Ⅰ and gear Ⅱ are a and b respectively, and the gear width is noted as B. When high pressure oil p enters into the high pressure chamber of the motor, all gear teeth there are subjected to high pressure oil, but only part of the two meshing gear teeth surfaces are subjected to high pressure oil. Because either a or b is smaller than the gear tooth height h, a force of $pB(h-a)$ acts on gear Ⅰ and $pB(h-b)$ on gear Ⅱ. This two forces together develop an output torque at the gear shaft. The oil is forced out through the low pressure chamber as the gear rotates in a direction shown in the Figure. The delivery calculation for a gear motor is the same as that for a gear pump.

Structure features: To allow birotational operation, a gear motor has two identical inlet and outlet ports which are set in the opposite direction and has a unique outside vent to guide the leakage in the bearings out of the motor. Rolling bearings are used to reduce startup friction torque. The teeth of the motor are more than that of the pump to reduce torque pulsation.

4.1.3 Low-speed motors

Low-speed motors are usually radial piston motors. They possess high pressure and large delivery to meet the needs of low speed rotation and large torque. Because of their large size and large moment of inertia, they are not suitable for the applications which require rapid response and frequent direction changing.

Low-speed motors may be further classified based on how they function: single-acting motors which provide a push in one direction; double-acting motors which provide push in both directions; multiple-acting motors which provide push in many directions.

1. Single-acting connecting rod type radial piston motors

Fig. 4-5 shows the operation of a signal-acting connecting rod type hydraulic motor. The motor is composed of a case, a connecting rod, a piston assembly, a crank, a valve shaft, etc. Five cylin-

3. 齿轮液压马达

外啮合齿轮液压马达的工作原理如图4-4所示，C为A、B两齿轮的啮合点，h为齿轮的全齿高。啮合点C到两齿轮Ⅰ、Ⅱ的齿根距离分别为a和b，齿宽为B。当高压油p进入马达的高压腔时，处于高压腔所有轮齿均受到压力油的作用，其中相互啮合的两个轮齿的齿面只有一部分齿面受高压油的作用。由于a和b均小于齿高h，所以在两个齿轮Ⅰ、Ⅱ上就产生大小为$pB(h\text{-}a)$和$pB(h\text{-}b)$的作用力。在这两个力的作用下，对齿轮产生输出转矩，随着齿轮按图示方向旋转，油液被带到低压腔排出。齿轮液压马达的排量公式同齿轮泵。

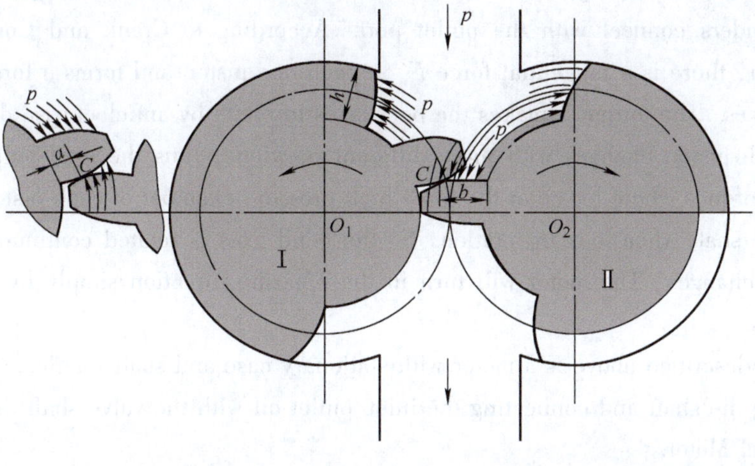

Fig. 4-4 Operating principle of external gear motor（外啮合齿轮液压马达的工作原理）

结构特点：适应正反转要求，进、出油口大小相等、具有对称性，有单独外泄油口将轴承部分的泄漏油引出壳体外；为了减小起动摩擦力矩，采用滚动轴承；为了减小转矩脉动，齿轮液压马达的齿数比泵要多。

4.1.3 低速液压马达

低速液压马达通常是径向柱塞式结构，为了获得低速和大转矩，采用高压和大排量，但它的体积和转动惯量很大，不能用于反应灵敏和频繁换向的场合。

低速液压马达按其每转作用次数，可分为单作用式和多作用式。若马达每旋转一周，柱塞作一次往复运动，则称为单作用式；若马达每旋转一周，柱塞作多次往复运动，则称为多作用式。

1. 单作用连杆式径向柱塞马达

图4-5所示为单作用连杆式液压马达的工作原理图，马达由壳体1、连杆3、活塞组件、曲轴4及配油轴6组成，壳体1内沿圆周呈放射状均匀布置了五个缸体，形成星形壳体；缸体内装有活塞2，活塞2与连杆3通过球铰连接，连杆大端做成鞍形圆柱瓦面紧贴在曲轴4的偏心圆上，其圆心为O_1，它与曲轴旋转中心O的偏心距$OO_1=e$，液压马达的配油轴6与曲轴4通过十字键连接在一起，随曲轴一起转动，马达的压力油经过配油轴通道，由配油轴分配到对应的活塞液压缸，在图中，液压缸的①、②、③腔通压力油，活塞受到压力油的作用；其余腔则与排油窗口接通；根据曲柄连杆机构运动原理，作用在柱塞上的切向分力F_t对曲轴旋转中心形成转矩T，使曲轴逆时针方向旋转。由于三个柱塞位置不同，所产生的转矩大小也不同，曲轴输出的总转矩等于高压腔相通的柱塞所产生的转矩之和。曲轴旋转时带

der blocks are arranged radially against the spherical surface of case 1 with an equal distance apart, forming a star-shaped cylinder block. The pistons 2 are set in the cylinder block and connect with connecting rods 3 through spherical hinges. The connecting rod larger heads with saddle-shape circular surface bear against the offset circular of crankshaft 4. The center of the crankshaft is O_1 and has an offset e with respect to the rotation center O of crankshaft, i. e. , $OO_1 = e$. Valve shaft 6 and crank 4 joint together through a cross key and rotates with the crankshaft. Pressurized oil is distributed to the corresponding piston cylinders by the valve shaft. As is shown in the Fig. 4-5, cylinders ①, ② and ③ connect with high pressure oil and the pistons are subjected to high pressure oil; while the other cylinders connect with the outlet port. According to Crank and Connecting Rod Mechanism operation, there is a tangential force F_t on each one piston and forms a torque T against the center of bend axes. The torque T makes the bend axes to rotate by anticlockwise direction. The torque acting on each piston changes with three different positions, thus the total output torque equals to the sum of torques where locate at the three high pressure chamber of each piston. The bend axes drives the valve shaft when it is in rotation. So the bend axes is rotated continuously with the flow-deploying state changes. The motor will turn in the opposite direction simply by changing the inlet and outlet ports.

What has been described above is a motor with stationary case and shaft rotation. The case will turn simply by fixing the shaft and connecting the inlet/outlet oil with the valve shaft directly, forming a so-called Wheel Motor.

The delivery for single acting radical-plunger type motors is expressed by

$$V = \frac{\pi d^2 ez}{2} \tag{4-11}$$

Where d is the diameter of the plunger; e is the eccentric distance of the crankshaft; z is the plunger number.

Structure features: They have the advantages of concise structure, working reliability, but are large in size and weight. They exist torque pulsation and poor stability at low speed. However, motors of this type have been improved recently by employing static-pressure supporting or static-pressure balance mechanism for the main friction pairs, allowing a lowest speed of 3r/min.

2. Incurve multiple-acting radial piston motors

As is shown in Fig. 4-6, the typical design consists of a stator 1, a rotor (cylinder) 2, pistons 3, valve shaft 6, etc. The inner surface of the stator 1 is composed by multiple identical curves with an equal distance apart. The number (six here) of the curve surfaces equals the action times of the motor. Two identical regions are divided at the concave bottom of each curve, one is the inlet region (working region), the other is the returned region. Several piston bores are arranged radially on rotor 2 with an equal distance apart (eight here). The piston heads contract with beam 4, which can move freely within the radial grooves of the cylinder block. Rolling wheels 5 on the journals at the two ends of the beam can roll around the inner surface of the stator. The inner bores of rotor 2 and the outer circular of valve shaft 6 match up to allow a sliding behavior of the shaft. An oil distribution port at each bottom of the piston bore is the same as that at valve shaft 6 in cylinder. The valve shaft is stationary and there are two oil distribution ports on it, each port having the same number as

动配油轴同步旋转，因此配流状态不断发生变化，从而保证曲轴连续旋转。如果进、排油口对换，液压马达也就反向旋转。

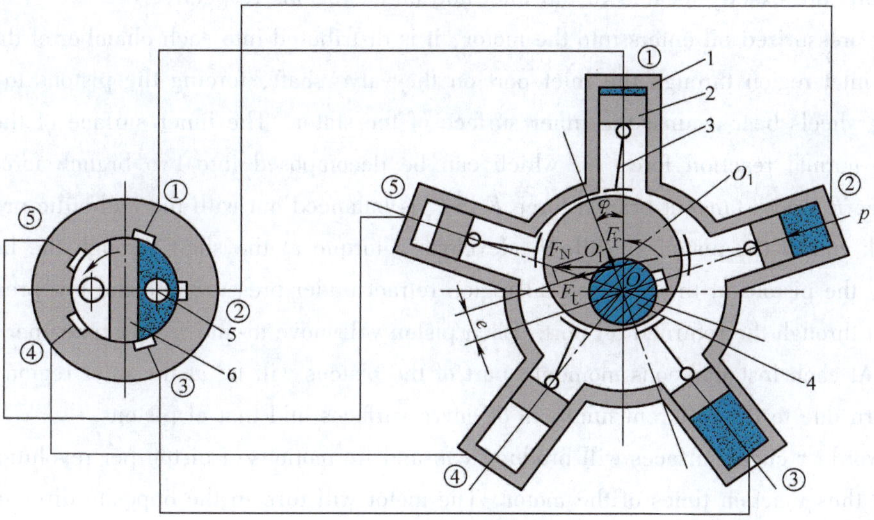

Fig. 4-5 Operating principle of crank and connecting rod type radical-plunger motor
（单作用连杆式径向液压马达的工作原理）
1—Case（壳体） 2—Piston（活塞） 3—Connecting rod（连杆）
4—Crank shaft（曲轴） 5—Inlet port（进油腔） 6—Valve shaft（配油轴）

以上讨论的是壳体固定、轴旋转的情况。如果将轴固定，进、排油直接通到配油轴中，就能达到外壳旋转的目的，构成了所谓的车轮马达。

单作用连杆式径向柱塞马达的排量 V 为

$$V = \frac{\pi d^2 ez}{2} \tag{4-11}$$

式中，d 为柱塞直径；e 为曲轴偏心距；z 为柱塞数。

结构特点：优点是结构简单、工作可靠。缺点是体积和重量较大、转矩脉动大、低速稳定性较差。近几年来因其主要摩擦副大多采用静压支承或静压平衡结构，其低速稳定性有很大的改善，最低转速可达 3r/min。

2. 多作用内曲线径向柱塞马达

多作用内曲线径向柱塞马达的典型结构如图 4-6 所示。它由定子 1、转子（缸体）2、柱塞 3 和配油轴 6 等组成。定子 1 的内表面由多段形状相同且均布的曲面组成，曲面的段数（本例为 6 段）就是马达的作用次数。每一曲面的凹部顶点处分为对称的两半，一半为进油区段（即工作区段），另一半为回油区段。转子 2 沿径向均布若干个柱塞孔（本例为 8 个）。柱塞头部与横梁 4 接触，横梁可在缸体的径向槽内滑动。安装在横梁两端轴颈上的滚轮 5 可沿定子内表面滚动。转子 2 的内孔和配油轴 6 外圆滑动配合，缸体中每个柱塞孔底部有一配油孔与配油轴 6 的配油窗口相同。配油轴固定不动，其圆周上有两组配油窗口，每组配油窗口的数目等于定子曲面的段数，一组配油窗口 A 与配油轴中心的进油孔相通，另一组配油窗口 B 与回油孔道相通，两组配油窗口的位置分别和定子内表面的进、回油区段位置严格对应。

that of stator curves. One group of oil distribution port A connects with the inlet port of the valve shaft center, while the other group B connects with the returned oil path. This two groups of oil distribution ports are exactly located in the inlet and outlet regions respectively.

When pressurized oil enters into the motor, it is distributed into each chamber of the piston bottom at the inlet region through the inlet port on the valve shaft, forcing the pistons to extend out. The rolling wheels bear against the inner surface of the stator. The inner surface of the stator then produces a normal reaction force F, which can be decomposed into two branch forces: a radial branch force F_r and a tangent branch force F_t. F_r is balanced out with the hydraulic pressure acting on the back end of the piston , while F_t develops a torque at the shaft through the beam. At the same time, the pistons at the returned oil region retract under pressure and the low pressure oil discharges out through the returned oil port. Each piston will move to-and-fro six times per revolution of the rotor. At each instantaneous moment, part of the pistons will be at the inlet region to make the cylinder turn due to the different numbers of curve surfaces and that of pistons.

In a word, x curve surfaces will produce x to-and-fro motions of piston per revolution of the cylinder, and thus x action times of the motor. The motor will turn in the opposite direction simply by changing the inlet and outlet ports of the motor. Most of the incurve motors are fixed delivery type motors.

The delivery of this type is expressed by

$$V = \frac{\pi d^2}{4} sxyz \tag{4-12}$$

Where d is the piston diameter; s is the piston stroke; x is the action times; y is the number of the piston rows; z is the number of pistons in each row.

Incurve multiple-acting radial piston motors have a low torque pulsation, balanced radial force, large startup torque and can move steadily at low speed. They find many applications in engineering, architecture, lifting and conveying machinery, coal mine, shipping, agriculture, etc.

4.2 Hydraulic Cylinders

4.2.1 Classification and speed-thrust characteristics

Cylinders may be further divided into piston type, plunger type and swing type based on their constructions or single-acting, double-acting according to their function. Some typical cylinders are described as follows:

1. Piston cylinders

There are double-rod type and signal-rod type piston cylinders in construction, stationary cylinder body type and stationary piston rod type in installation.

(1) Double-rod piston cylinders Fig. 4-7a shows a double-rod piston cylinder of stationary cylinder body type. It consists of two rods with the same diameter at each side of the piston. The inlet and outlet ports are located at the two ends of the cylinder respectively. The piston drives the workbench to move through the piston rod. The displacement range of the workbench equals treble effec-

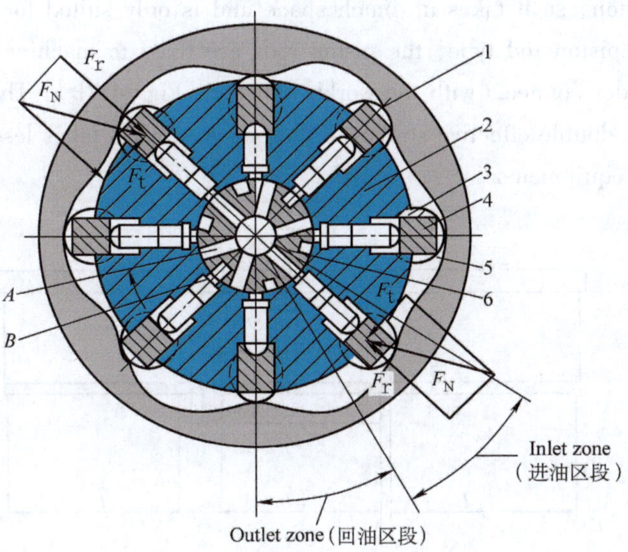

Fig. 4-6　Incurve multiple-acting radial piston motor
（多作用内曲线径向柱塞马达）
1—Stator（定子）　2—Rotor（转子）　3—Piston（柱塞）
4—Beam（横梁）　5—Rolling wheel（滚轮）　6—Valve shaft（配油轴）

当压力油输入马达后，通过配油轴上的进油窗口分配到处于进油区段的各柱塞底部油腔，液压油使柱塞顶出，滚轮顶紧在定子内表面上。定子表面给滚轮一个法向反力 F_N，这个法向反力 F_N 可分解为两个方向的分力，其中径向分力 F_r 与作用在柱塞后端的液压力相平衡，切向分力 F_t 通过横梁对转子产生转矩。同时，处于回油区段的柱塞受压缩进，低压油通过回油窗口排出。转子每转一周，每个柱塞往复移动六次。由于曲面数目和柱塞数不等，所以任一瞬时总有一部分柱塞处于进油区段，使缸体转动。

总之，若有 x 个导轨曲面，则缸体旋转一周，每个柱塞往复运动 x 次，马达作用次数就为 x。马达的进、回油口互换时，马达将反转。内曲线马达多为定量马达。

多作用内曲线径向柱塞马达的排量为

$$V = \frac{\pi d^2}{4} sxyz \tag{4-12}$$

式中，d 为柱塞直径；s 为柱塞行程；x 为作用次数；y 为柱塞排数；z 为每排柱塞数。

多作用内曲线径向柱塞马达具有转矩脉动小、径向力平衡、起动转矩大、能在低速下稳定运转等优点，故普遍应用于工程、建筑、起重运输、煤矿、船舶、农业等机械中。

4.2　液压缸

4.2.1　类型和速度与推力特性

液压缸的种类繁多，通常根据其结构特点分为活塞式、柱塞式和摆动式三大类；按其作用来分，有单作用式和双作用式。下面介绍几种常用的液压缸。

tive strokes of the piston, so it takes up much space and is only suited for small size equipments. While in a stationary piston rod type, the piston rods are fixed to machine tool with the help of a support and the cylinder connects with the workbench (see Fig. 4-7b). The displacement range of the workbench equals double effective strokes of the piston, thus it takes less space and is suited for large or medium size equipments.

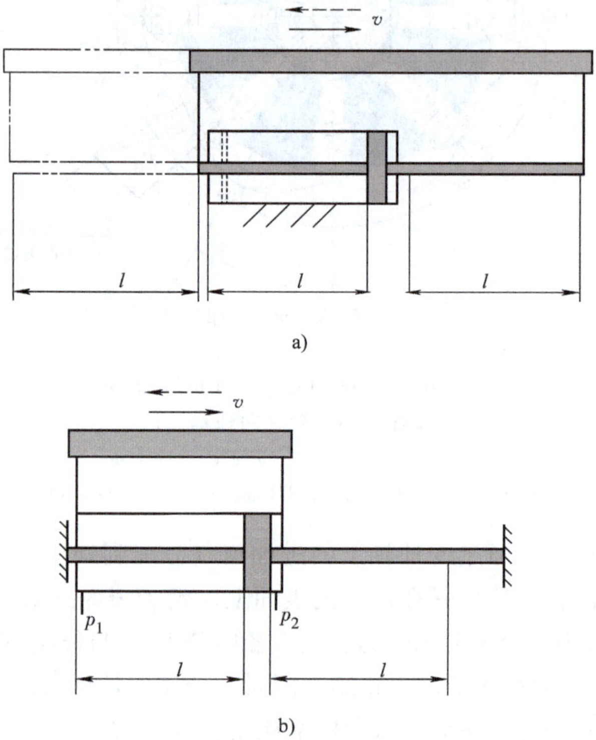

Fig. 4-7 Double-rod piston cylinder (双杆活塞缸)

a) Stationary cylinder body type(缸筒固定式) b) Stationary piston rod type(活塞固定式)

The two piston rods in the two sides of the double rod piston cylinder are equal in diameters. When input flow rates and pressures are constant, the output thrust F and speed v are equal in the two opposite directions. F and v can be calculated as follows

$$F = A(p_1 - p_2)\eta_m = \frac{\pi}{4}(D^2 - d^2)(p_1 - p_2)\eta_m \tag{4-13}$$

$$v = \frac{q\eta_V}{A} = \frac{4q\eta_V}{\pi(D^2 - d^2)} \tag{4-14}$$

Where A is the effective area of the cylinder; η_m and η_V are the mechanical and volumetric efficiencies of the cylinder respectively; D and d are the diameters of the piston and the piston rod respectively; q is the flow rate through the cylinder; p_1 and p_2 are the pressures of the inlet and outlet ports respectively.

(2) Single-rod piston cylinders As is shown in Fig. 4-8, the cylinder does not have a rod in the cap side of a single rod cylinder. Single-rod piston cylinders can also be divided into stationary cylinder body type and stationary piston rod type based on their installation. Unlike double-rod piston

1. 活塞式液压缸

活塞式液压缸可分为双杆式和单杆式两种结构形式，其安装形式有缸体固定式和活塞杆固定式。

（1）双杆活塞缸 图 4-7a 所示为缸筒固定式双杆活塞缸，其特点是活塞两端都有一根直径相等的活塞伸出。它的进、出油口位于缸筒两端。活塞通过活塞杆带动工作台移动，工作台移动范围等于活塞有效行程的三倍，占地面积大，因此仅适用于小型机床。图 4-7b 所示为活塞杆固定式。缸筒与工作台相连，活塞杆通过支架固定在机床上。此种安装形式，工作台的移动范围等于活塞有效行程 l 的 2 倍，因此占地面积小，常用于大中型设备中。

因双活塞杆液压缸的两端活塞杆直径相等，所以当输入流量和油液压力不变时，其往返运动速度和推力相等。则缸的推力 F 和运动速度 v 分别为

$$F = A(p_1 - p_2)\eta_m = \frac{\pi}{4}(D^2 - d^2)(p_1 - p_2)\eta_m \tag{4-13}$$

$$v = \frac{q\eta_V}{A} = \frac{4q\eta_V}{\pi(D^2 - d^2)} \tag{4-14}$$

式中，A 为液压缸的有效面积；η_m 为液压缸的机械效率；η_V 为液压缸的容积效率；D 为活塞直径；d 为活塞杆直径；q 为输入液压缸的流量；p_1 为进油腔压力；p_2 为回油腔压力。

（2）单杆活塞缸 单杆活塞缸如图 4-8 所示，活塞仅一端带活塞杆。单杆活塞缸也有缸筒固定和活塞杆固定的两种安装形式，两种安装方式的工作台移动范围均为活塞有效行程 l 的 2 倍。

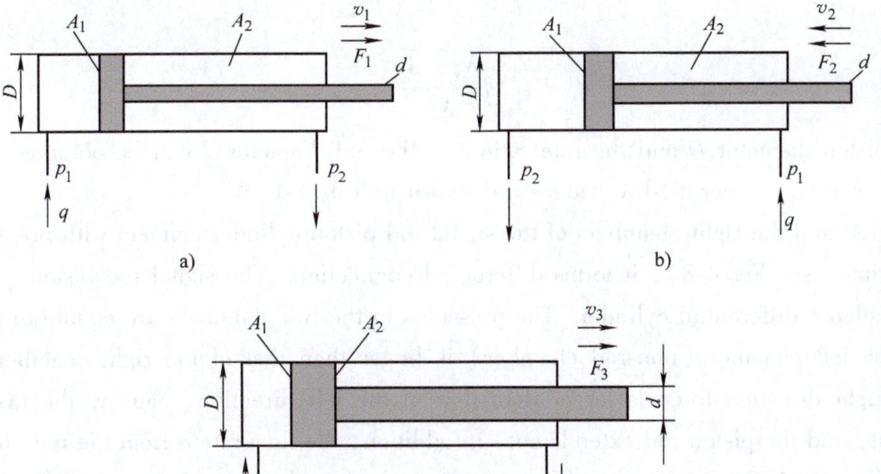

Fig. 4-8 Single-rod piston cylinder（单杆活塞缸）

a）Pressurized oil is put into the non-rod chamber（无杆腔进油）

b）Pressurized oil is put into the rod chamber（有杆腔进油） c）Differential connection（差动连接）

cylinders, the displacement range of the workbench is equal to double the piston effective stroke in both installation forms of this type.

The effective areas A_1 and A_2 in the left and right chambers are different. When the pressures of the inlet and outlet ports are p_1 and p_2 respectively, and the flow rates (q) in the two ports are equal, then the thrust and speed in the opposite directions are different.

When pressurized oil is put into the non-rod chamber (see Fig. 4-8a), the thrust F_1 applied on the piston and its speed v_1 are

$$F_1 = (A_1p_1 - A_2p_2)\eta_m = \frac{\pi}{4}[(p_1 - p_2)D^2 + p_2d^2]\eta_m \tag{4-15}$$

$$v_1 = \frac{q\eta_V}{A_1} = \frac{4q\eta_V}{\pi D^2} \tag{4-16}$$

When pressurized oil is put into the rod chamber (see Fig. 4-8b), the thrust F_2 applied on the piston and its speed v_2 are

$$F_2 = (p_1A_2 - p_2A_1)\eta_m = \frac{\pi}{4}[(p_1 - p_2)D^2 - p_1d^2]\eta_m \tag{4-17}$$

$$v_2 = \frac{q\eta_V}{A_2} = \frac{4q\eta_V}{\pi(D^2 - d^2)} \tag{4-18}$$

In engineering applications, the ratio of v_1 to v_2 is denoted as forth-back rate ratio λ_v, we obtain

$$\lambda_v = \frac{v_2}{v_1} = \frac{1}{1 - \left(\frac{d}{D}\right)^2}$$

$$d = D\sqrt{\frac{\lambda_v - 1}{\lambda_v}} \tag{4-19}$$

Given the piston diameter D and the rate ratio λ_v, the rod diameter d can be obtained. d increases with λ_v. A series of certified λ_v values are shown in Tab. 4-1.

If both the left and the right chambers of the signal rod piston cylinder connect with pressurized oil at the same time (see Fig. 4-8), it forms differential connection. The signal rod piston cylinder in this form is called a differential cylinder. The pressures in the two chambers are equal but the effective area of the left chamber (non-rod chamber) is larger than that of the right chamber (rod chamber). The right direction force is larger than that in the left direction, causing the piston to move in the right, and the piston rod extends out. In addition, the flow rate q from the rod chamber converges with the pump's flow rate q and then enters into the left chamber of the cylinder together. The piston moves faster by this action. This type of cylinder can only move in one direction (right in Fig. 4-8c). The thrust F_3 on the piston and its moving speed v_3 are

$$F_3 = p_1(A_1 - A_2)\eta_m = p_1\frac{\pi d^2}{4}\eta_m \tag{4-20}$$

$$v_3 = \frac{q\eta_V + q'}{A_1} = \frac{q\eta_V + \frac{\pi}{4}(D^2 - d^2)v_3}{\frac{\pi D^2}{4}} \tag{4-21}$$

单杆活塞缸因左、右两腔有效面积 A_1 和 A_2 不等，因此当进油腔和回油腔压力分别为 p_1 和 p_2，输入左、右两腔的流量均为 q 时，液压缸左、右两个方向的推力和速度不相同。

当无杆腔进油时（图4-8a），活塞的推力 F_1 和运动速度 v_1 分别为

$$F_1 = (A_1p_1 - A_2p_2)\eta_m = \frac{\pi}{4}[(p_1 - p_2)D^2 + p_2d^2]\eta_m \tag{4-15}$$

$$v_1 = \frac{q\eta_V}{A_1} = \frac{4q\eta_V}{\pi D^2} \tag{4-16}$$

当有杆腔进油时（图4-8b），活塞的推力 F_2 和活塞运动速度 v_2 分别为

$$F_2 = (p_1A_2 - p_2A_1)\eta_m = \frac{\pi}{4}[(p_1 - p_2)D^2 - p_1d^2]\eta_m \tag{4-17}$$

$$v_2 = \frac{q\eta_V}{A_2} = \frac{4q\eta_V}{\pi(D^2 - d^2)} \tag{4-18}$$

工程应用时将上列速度 v_1 和 v_2 的比值称为往返速比，以 λ_v 表示，于是得

$$\lambda_v = \frac{v_2}{v_1} = \frac{1}{1 - \left(\frac{d}{D}\right)^2}$$

$$d = D\sqrt{\frac{\lambda_v - 1}{\lambda_v}} \tag{4-19}$$

即已知活塞直径 D 和往复速比 λ_v，可求得活塞杆直径 d，而速比 λ_v 越大，活塞杆直径 d 越大。往复速比 λ_v 的推荐值见表4-1。

Tab. 4-1 λ_v recommendatory values（λ_v 推荐值）

Pressure（液压缸压力）/MPa	<10	12.5 – 20	>20
Speed ratio of reciprocating motion（往复速度比）λ_v	1.33	1.46 – 2	2

如果单杆活塞缸的左、右两腔同时通压力油（图4-8c），称为差动连接，差动连接的单杆活塞缸称为差动液压缸。差动液压缸虽然左、右两腔压力相等，但因为左腔（无杆腔）的有效面积大于右腔（有杆腔），使得活塞向右的作用力大于向左的作用力，因此活塞向右运动，活塞杆向外伸出；与此同时，液压缸有杆腔排出的流量 q' 与泵的流量 q 汇合进入液压缸的左腔，使活塞运动速度加快。对差动连接的液压缸，活塞只能一个方向运动（图4-8c所示为向右运动），作用在活塞上的推力 F_3 和活塞运动速度 v_3 分别为

$$F_3 = p_1(A_1 - A_2)\eta_m = p_1\frac{\pi d^2}{4}\eta_m \tag{4-20}$$

$$v_3 = \frac{q\eta_V + q'}{A_1} = \frac{q\eta_V + \frac{\pi}{4}(D^2 - d^2)v_3}{\frac{\pi D^2}{4}} \tag{4-21}$$

上式化简后得

$$v_3 = \frac{4q\eta_V}{\pi d^2} \tag{4-22}$$

The above two formulae can be simplified as

$$v_3 = \frac{4q\eta_V}{\pi d^2} \tag{4-22}$$

If $v_3 = v_2$, from formulae (4-18) and (4-22), we obtain: $D = \sqrt{2}d$.

2. Plunger cylinders

The piston cylinders discussed above are used widely. But the cylinder bores request high precise-machining and are hard to be machined at long stroke occasion, and their manufacturing cost will be increased. However, the plunger cylinder in Fig. 4-9 does not meet such troubles and is widely used in practical production, specifically, when double direction control is not needed.

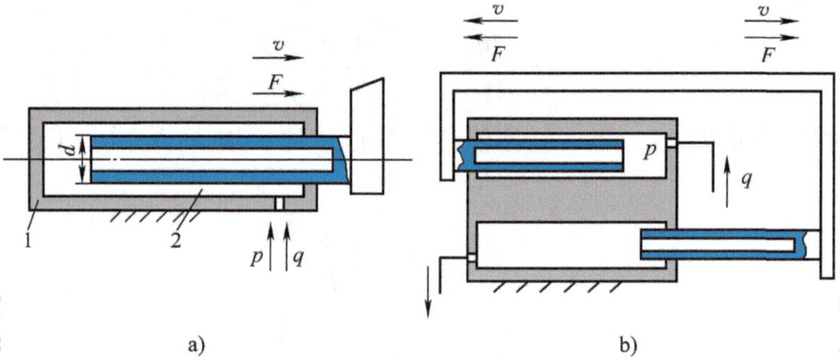

Fig. 4-9 Plunger cylinder(柱塞式液压缸)

1—Cylinder barrel(缸筒) 2—Piston(柱塞)

As is shown in Fig. 4-9a, the pistons do not contact the cylinder barrel, thus free from precise machining, possessing a good craft and low cost. Note that plunger type cylinder is a single-acting type and outside force such as weight or spring is needed to return the plunger to its home position. Two plunger cylinders are needed to accomplish bidirectional operation (Fig. 4-9b). The end cover of the plunger is subjected to oil pressure. The area of the end cover decides the output speed and thrust of the plunger cylinder. The plungers are usually quite thick and more heavier to guarantee enough thrust and stability. For this type of cylinder, uneven wear is easy to occur at horizontal installation, so plunger cylinders are usually suited for vertical installation, Hollow pistons are usually used to reduce the piston weight.

Note the plunger diameter as d, the input hydraulic oil flow rate as q, then the thrust F on the plunger and its speed v are

$$F = pA\eta_m = p\frac{\pi}{4}d^2\eta_m \tag{4-23}$$

$$v = \frac{q\eta_V}{A} = \frac{4q\eta_V}{\pi d^2} \tag{4-24}$$

3. Cylinders of other types

(1) Telescopic cylinders A considerable distance may be accomplished by using a telescoping cylinder (see Fig. 4-10, 4-11). The telescoping cylinder has multiple sleeves, one inside the other. It acts as though it were a one-piece cylinder until the outer section reaches its stop, then the next inner section extends to its stop, and so forth to the end of the stroke. Some telescoping cylinders are

如果要求差动液压缸活塞向右运动（差动连接）的速度与非差动连接时活塞向左运动的速度相等，即 $v_3=v_2$，则由式（4-18）、式（4-22）可知 $D=\sqrt{2}d$。

2. 柱塞式液压缸

前面所讨论的活塞式液压缸的应用非常广泛，但这种液压缸由于缸孔加工精度要求很高，当行程较长时，加工难度大，使得制造成本增加。在实际生产中，某些场合所用的液压缸并不要求双向控制，柱塞式液压缸正是满足了这种使用要求的一种价格低廉的液压缸。

如图 4-9a 所示，柱塞和缸筒内壁不接触，因此缸筒内孔不需精加工，工艺性好，成本低。柱塞式液压缸是单作用的，它的回程需要借助自重或弹簧等其他外力来完成，如果要获得双向运动，可将两柱塞式液压缸成对使用（图 4-9b）。柱塞式液压缸的柱塞端面是受压面，其面积大小决定了柱塞式液压缸的输出速度和推力，为保证柱塞式液压缸有足够的推力和稳定性，一般柱塞较粗，质量较大，水平安装时易产生单边磨损，故柱塞式液压缸适宜于垂直安装使用。为减轻柱塞的质量，常制成空心的。

当柱塞直径为 d，输入液压油流量为 q 时，柱塞上所产生的推力 F 和速度 v 分别为

$$F = pA\eta_{\mathrm{m}} = p\frac{\pi}{4}d^2\eta_{\mathrm{m}} \tag{4-23}$$

$$v = \frac{q\eta_V}{A} = \frac{4q\eta_V}{\pi d^2} \tag{4-24}$$

3. 其他形式液压缸

（1）伸缩式液压缸　伸缩式液压缸又称多套缸，它是由两级或多级活塞式液压缸套装而成的，前一级活塞缸的活塞是后一级活塞的缸筒。当通入压力油时，活塞有效面积最大的缸筒以最低油压力开始伸出，当行至终点时，活塞有效面积次之的缸筒在压力油的作用下开始伸出。各级伸出速度取决于外伸缸筒的有效面积，外伸缸筒有效面积越小，伸出速度越快。伸缩式液压缸可以获得很长的行程，缩回时轴向尺寸又很小。

图 4-10 和图 4-11 所示分别为双作用伸缩式液压缸和单作用伸缩式液压缸的工作原理图。它们的不同点主要是：单作用回程靠外力（例如重力），而双作用靠液压油作用回程。

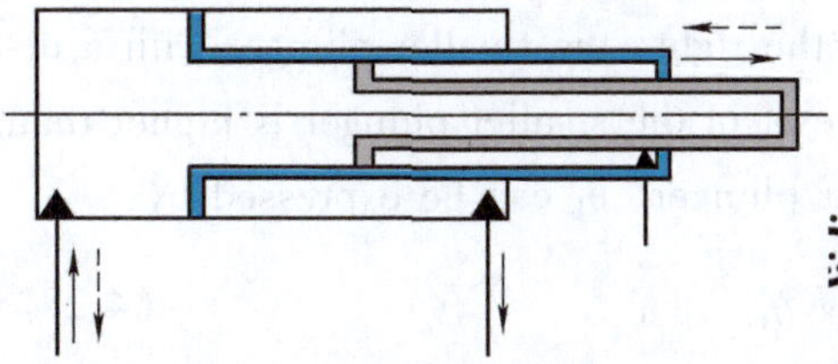

Fig. 4-10　Double-acting telescoping cylinder
（双作用伸缩液压缸）

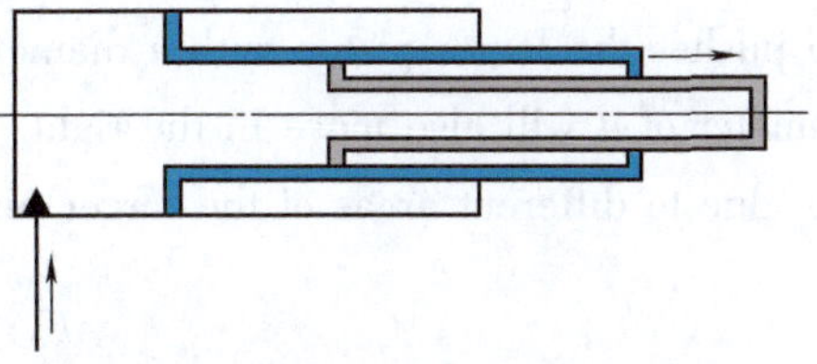

Fig. 4-11　Single-acting telescoping cylinder
（单作用伸缩液压缸）

伸缩式液压缸特别适用于工程机械及自动步进式输送装置。

（2）齿条活塞式液压缸　齿条活塞式液压缸由带有齿条杆的双作用活塞缸和齿轮齿条机构组成。如图 4-12 所示，压力油推动活塞往复运动，经齿条、齿轮机构变成齿轮轴往复旋动，从而带动工作部件作周期性的往复旋转运动，它多用于自动线、组合机床等转位或分度机构中。齿条活塞式液压缸传动轴输出转矩 T_{o} 及输出角速度 ω 分别为

$$T_{\mathrm{o}} = \Delta p\frac{\pi}{4}D^2\frac{D_{\mathrm{i}}}{2}\eta_{\mathrm{m}} = \Delta p\frac{\pi}{8}D^2D_{\mathrm{i}}\eta_{\mathrm{m}} \tag{4-25}$$

constructed so the inner sleeve moves first. The movement of the various sections depends upon the areas of pushing surface of each section. This type of cylinder gives long stroke with a short housing.

The cylinders shown in Fig. 4-10 and 4-11 are single-acting and double-acting telescoping cylinders respectively. Their principal difference lies in: single-acting type needs outside force (such as gravity) to return to its home position, while the double type accomplishes this by the hydraulic oil.

Telescoping cylinders are especially adaptable in engineering mechanics and automatic step feeding devices.

(2) Rack-piston cylinders As is shown in Fig. 4-12, a rack-piston cylinder consists of a double-acting piston cylinder with a rack shaft and a pinion and rack mechanism. With the help of the pinion and rack mechanism, the reciprocating motion of the piston translates into the reciprocating rotation of the gear shaft. The rack-piston cylinder finds many applications in changeable position or indexing mechanism such as automatic lines, combinational machine tools, etc. The output torque T_o and the angular speed ω of the rack-piston cylinder drive shaft are

$$T_o = \Delta p \frac{\pi}{4} D^2 \frac{D_i}{2} \eta_m = \Delta p \frac{\pi}{8} D^2 D_i \eta_m \tag{4-25}$$

$$\omega = \frac{q}{\frac{\pi}{4} D^2 \frac{D_i}{2}} \eta_V = \frac{8 q \eta_V}{\pi D^2 D_i} \tag{4-26}$$

Where Δp is the pressure differential of the left and right chambers in the cylinder; q is the flow rate put in the cylinder; D is the piston diameter; D_i is the diameter of the pinion pitch circle.

(3) Supercharge cylinders Supercharge cylinders are similar to the piston cylinders described above. But they do not convert the hydraulic energy into mechanical energy, instead, they only transfer the hydraulic energy to reach high pressure.

The cylinder shown in Fig. 4-13 is a complex supercharge cylinder composed by a piston cylinder and a plunger cylinder which are connected with each other. When the oil under low pressure p_a pushes the larger piston with a diameter of D to move in the right, the smaller plunger with a diameter of d will also move in the right. The output pressure p_b of the smaller plunger is higher than p_a due to different areas of the larger piston and the smaller plunger. p_b can be expressed by

$$p_b = p_a \left(\frac{D}{d} \right)^2 \eta_m = p_a K \eta_m \tag{4-27}$$

Where $K = D^2/d^2$ is called high pressure ratio, which represents the supercharge capacity of the supercharge cylinder.

It is not hard to see that supercharge is on the basis of reducing the effective flow $\left(q_b = \frac{q_a \eta_V}{K} \right)$. Supercharge cylinder as an intermediate in the oil, used on low pressure systems can meet the requirement of local high pressure.

4.2.2 Typical constructions and makeups

1. Typical constructions

The components of a single-rod cylinder are identified in Fig. 4-14. The principal parts of the

$$\omega = \frac{q}{\frac{\pi}{4}D^2 \frac{D_i}{2}}\eta_V = \frac{8q\eta_V}{\pi D^2 D_i} \tag{4-26}$$

式中，Δp 为液压缸左、右两腔压差；q 为进入液压缸的流量；D 为活塞的直径；D_i 为齿轮分度圆直径。

（3）增压器　增压器与前面介绍的活塞式液压缸相类似，但不是将液压能转换成机械能，而是液压能的传递，使之增压。

图 4-13 所示增压器为活塞缸与柱塞缸组成的复合缸。当低压油 p_a 推动直径为 D 的大活塞向右移动时，也推动与其连成一体的直径为 d 的小柱塞，由于大活塞与小柱塞面积不同，因此小柱塞缸输出的压力 p_b 要比 p_a 高，p_b 的大小可由下式求出

$$p_b = p_a\left(\frac{D}{d}\right)^2\eta_m = p_a K\eta_m \tag{4-27}$$

式中，$K = D^2/d^2$ 称为增压比，它表示增压缸的增压能力。

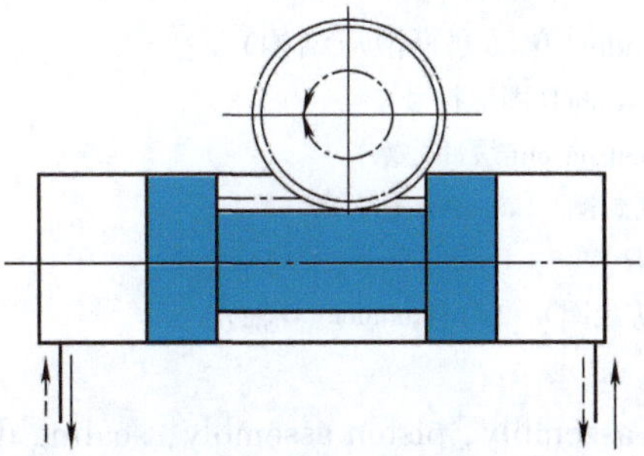

Fig. 4-12　Rack-piston cylinder（齿条活塞式液压缸）

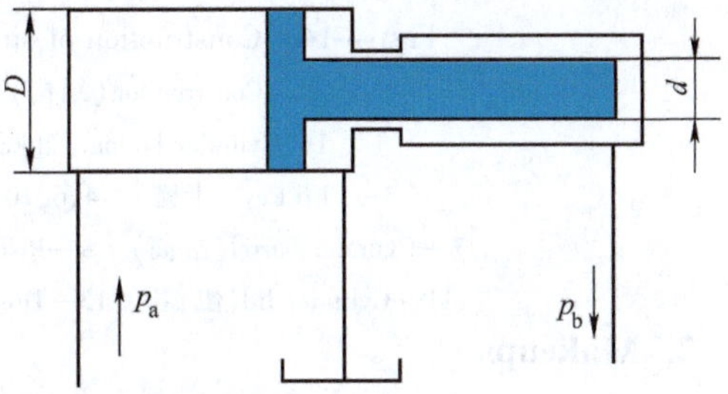

Fig. 4-13　Supercharge cylinder（增压器）

不难看出，增压能力是在降低有效流量的基础上得到的$\left(q_b = \frac{q_a\eta_V}{K}\right)$。增压器作为油路中的一个中间环节，用于使低压系统能满足局部高压油路要求。

4.2.2　典型结构和组成

1. 典型结构

图 4-14 所示为一种工程用的单活塞杆液压缸的结构图。它主要由缸底 1、缸筒 7、活塞 5、活塞杆 8、导向套 9 和缸盖 11 等主要零件组成。缸底与缸筒焊接成一体，缸盖与缸筒采用螺纹连接。为防止油液由高压腔向低压腔泄漏或向外泄漏，在活塞与活塞杆、活塞与缸筒、导向套与活塞杆之间均设有密封圈。为防止活塞快速退回到行程终端时撞击缸底，活塞杆后端设置了缓冲柱塞。为了防止脏物进入液压缸内部，在缸盖外侧还设有防尘圈。

2. 组成

液压缸按结构组成分为缸体组件、活塞组件、密封装置、缓冲装置和排气装置等。密封装置将在第 6 章单独叙述，下面介绍其他部分。

（1）缸体组件　缸体组件包括缸筒、缸盖和一些连接零件。缸筒可以用铸铁（低压时）和无缝钢管（高压时）制成。缸筒和缸盖的常见连接方式如图 4-15 所示。从加工的工艺性、外形尺寸和拆装是否方便不难看出各种连接方式的特点。图 4-15a 所示为法兰式连接，加工和拆装都很方便，只是外形尺寸大些。图 4-15b 所示为半环式连接，要求缸筒有足够的壁

cylinder are a cylinder bottom 1, a piston 5, a cylinder barrel 7, a piston rod 8, a guide sleeve 9, a cylinder lid 11, etc. The cylinder bottom and the cylinder barrel are welded together. Thread connection is employed for the cylinder lid and the cylinder barrel connection. Seals are set between the piston and the piston rod, the piston and the cylinder barrel, the guide sleeve and the piston rod to prevent oil leaking out or into the low pressure chamber from the high pressure chamber. A cushion piston is also arranged in the end of the piston rod to prevent the collision of the piston against the cylinder bottom when it retracts to the stroke end at a high speed. A dustproof ring is also mounted outside the cylinder lid to prevent dust from sneaking into the cylinder.

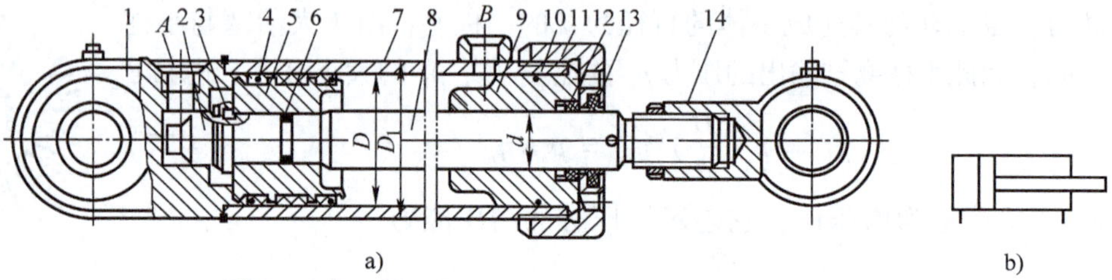

Fig. 4-14 Construction of single-rod cylinder(单活塞杆液压结构)

a) Construction(结构) b) Graphic symbol(图形符号)

1—Cylinder bottom(缸底) 2—Cushion piston(缓冲柱塞)

3—Clip key(卡键) 4,6,10,12—Seal(密封圈) 5—Piston(活塞)

7—Cylinder barrel(缸筒) 8—Piston rod(活塞杆) 9—Guide sleeve(导向套)

11—Cylinder lid(缸盖) 13—Dustproof ring(防尘圈) 14—Trunnion(耳轴)

2. Makeups

Cylinders can be further divided into cylinder block assembly, piston assembly, sealing device, cushion device and venting device.

(1) Cylinder block assembly The cylinder block assembly consists of a cylinder barrel, a cylinder lid, and other connecting components. The cylinder barrel can be made of cast iron (for low pressure occasions) or seamless steel tube (for high pressure occasions). The common connections of the cylinder barrel and cylinder lid are shown in Fig. 4-15. The feature of each type of connection is easy to see from their machining craft, size and removal/installation. Fig. 4-15a shows a flanged connection, which is easy to machine and remove or install, but with a relative large size. Fig. 4-15b shows a semi-ring connection, which requires that the cylinder barrel should be thick enough. Fig. 4-15c is threaded connection, whose size is small but inconvenient to remove or install, special tools are needed. Fig. 4-15d is a pull-rod type connection, which is easy to remove or install, but with a large size. Fig. 4-15e shows a welded connection, which is simple in construction, small in size, but may distort due to welding.

(2) Piston assembly Fig. 4-16 shows the pistons and rods connections. The most common connection types are thread connection and semi-ring connection. There also exist integral type, welding type, taper pin, etc.

Fig. 4-16a shows a threaded connection which is simple in construction and convenient to install or remove, but generally needs nut-loosenessproof mechanism. Fig. 4-16b shows a semi-ring connection, which has strong connection strength but complex construction and is inconvenient to remove or install. Semi-ring connection is usually used in high pressure or serious vibration occasions.

厚。图 4-15c 所示为螺纹式连接，外形尺寸小，但拆装不方便，要有专用工具。图 4-15d 所示为拉杆式连接，拆装容易，但外形尺寸大。图 4-15e 所示为焊接式连接，结构简单，尺寸小，但可能会因焊接有一些变形。

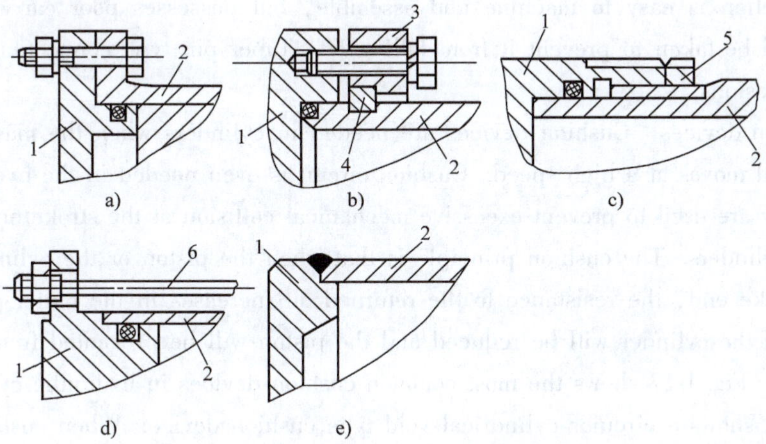

Fig. 4-15 Cylinder barrel and cylinder cover connections（缸筒与缸盖的连接方式）

a）Flange type（法兰式） b）Semi-ring type（半环式） c）Thread type（螺纹式）

d）Pull-rod type（拉杆式） e）Welding type（焊接式）

1—Cylinder cover（缸盖） 2—Cylinder barrel（缸筒） 3—Press plate（压板）

4—Semi-ring（半环） 5—Stopping nut（防松螺母） 6—Pull（拉杆）

（2）活塞组件 如图 4-16 所示，活塞与活塞杆的连接最常用的有螺纹式连接和半环式连接形式，除此之外还有整体式结构、焊接式结构、锥销式结构等。

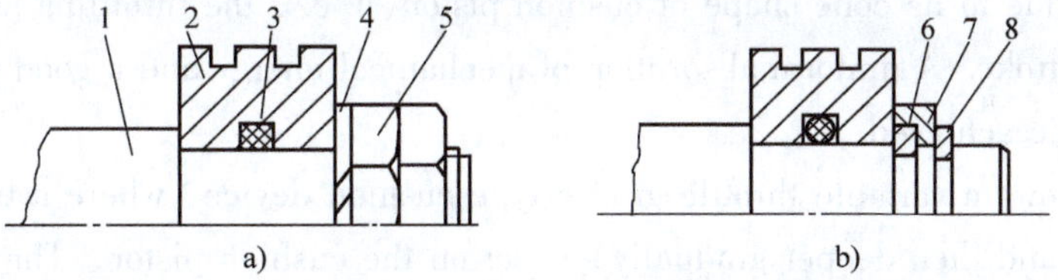

Fig. 4-16 Pistons and rods connections（活塞与活塞杆的连接）

a）Thread coupling（螺纹式连接） b）Semi-ring joint（半环式连接）

1—Piston rod（活塞杆） 2—Piston（活塞） 3—Seal（密封圈）

4—Against looseness gasket（防松垫圈） 5—Nut（螺母）

6—Clip key（半环） 7—Ring set（套环） 8—Spring collar（弹簧卡圈）

螺纹式连接如图 4-16a 所示，其结构简单，装拆方便，但一般需备螺母防松装置；半环式连接如图 4-16b 所示，其连接强度高，但结构复杂，装拆不方便，半环式连接多用于高压和振动较大的场合；整体式连接和焊接式连接结构简单，轴向尺寸紧凑，但损坏后需整体更换，对活塞与活塞杆比值较小、行程较短或尺寸不大的液压缸，其活塞与活塞杆可采用整体或焊接式连接；锥销式连接加工容易，装配简单，但承载能力小，且需要有必要的防脱落措施，在轻载情况下可采用锥销式连接。

（3）缓冲装置 当液压缸拖动负载的质量较大、速度较高时，一般应在液压缸中设缓冲装置，必要时还需在液压传动系统中设缓冲回路，以免在行程终端发生过大的机械碰撞，导致液压缸损坏。缓冲的原理是当活塞或缸筒接近行程终端时，在排油腔内增大回油阻力，

Integral connection and welded connection are simple in design, compact in axial size, but should be replaced totally once they suffer from damage. These two types of connections are usually used in the occasions with small ratio of piston to piston rod, short stroke, or small size of cylinders. Taper pin type connection is easy to machine and assemble, but possesses poor carrying capacity and measures should be taken to prevent it from falling off. Taper pin type connection can be used in light weight occasions.

(3) Cushion devices Cushion devices are needed in cylinders when the mass of the dragged load is large and moves at a high speed. Cushion circuit is even needed in the hydraulic systems if necessary. They are used to prevent excessive mechanical collision at the stroke end and damage to the hydraulic cylinders. The cushion principle is that when the piston or the cylinder barrel moves close to the stroke end, the resistance to the returned oil increases in the outlet port. By this the moving speed of the cylinder will be reduced and the piston will be prevented from barging against the cylinder lid. Fig. 4-17 shows the most common cushion devices in hydraulic cylinders.

Fig. 4-17a shows a circular cylindrical void type cushion device. When cushion piston enters into the inner bore of the cylinder lid, a cushion chamber is formed between the cylinder lid and the cushion piston. The enclosed oil can be discharged out only from the cylindrical void δ, which produces cushion pressure and reduces its speed. In this cushion device, the throttling area is constant, thus the cushion braking force is quite large at the first beginning of cushion process, but will reduce quickly. This type of cushion has the disadvantage of poor effect, but is still wildly used in a series of cylinders due to its simple design and low cost.

Fig. 4-17b shows a circular cone void type cushion device. The cushion void δ will change as the displacement due to its cone shape of cushion piston, i. e. , the throttling area reduces inversely with the cushion stroke. A uniform absorption of mechanical energy and a good performance of cushion behavior can be achieved.

Fig. 4-17c shows a variable throttle groove type cushion device, where a triangle throttle channel (first shallow and then deeper gradually) is set on the cushion piston. The throttle area decreases inversely with the cushion stroke, producing a smooth change of cushion pressure.

Fig. 4-17d shows an adjustable throttle orifice, in which the oil in the cushion chamber is discharged out from the small orifice. The cushion pressure can be changed simply by adjusting the area of the throttle orifice, which can meet the requirements of different load and speed occasions. In addition, when the piston moves in the opposite direction, the high pressure oil enters into the hydraulic cylinder through the non-return valve to prevent a slow or difficult startup due to inadequate thrust.

(4) Venting devices The hydraulic oil usually includes air or air will sneak into the hydraulic oil when the hydraulic cylinder stops, and the highest part of the cylinder will gather air. If the air in the oil is not eliminated in time, unsteady motion to cylinder, creep, vibration and even the hydraulic components erosion will occur due to oil oxidation. Venting devices are used to solve these problem, the common venting devices are shown in Fig. 4-18. The venting valve and the venting plug should be located at the highest point of the cylinder.

Note that not all cylinders need venting devices. Special venting devices are not necessary for those with low requirements. Just set the oil port at the highest point in the two ends of the cylinder

从而降低液压缸的运动速度，避免活塞与缸盖相撞。液压缸中常用的缓冲装置如图 4-17 所示。

图 4-17a 所示为圆柱形环隙式缓冲装置，当缓冲柱塞进入缸盖上的内孔时，缸盖和缓冲活塞间的封闭油液只能从环形间隙 δ 挤压出去，于是排油压力升高形成缓冲压力，使活塞的运动速度减慢从而实现减速缓冲。这种装置结构简单，制造成本低，但实现减速所需行程较长，适用于运动部件惯性大、运动速度不太高的场合。

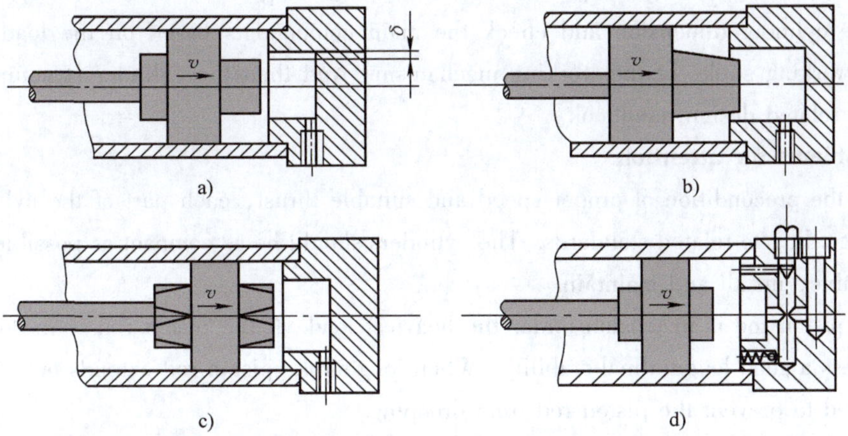

Fig. 4-17 Cylinder cushion device（液压缸的缓冲装置）

a）Circular cylindrical void（圆柱形环隙式） b）Circular cone void（圆锥形环隙式）

c）Variable throttle groove（可变节流槽式） d）Adjustable throttle orifice（可调节流孔式）

图 4-17b 所示为圆锥形环隙式缓冲装置，由于缓冲柱塞为圆锥形，所以缓冲环形间隙 δ 随位移量而改变，即节流面积随缓冲行程的增大而缩小，使机械能的吸收较均匀，其缓冲效果较好。

图 4-17c 所示为可变节流槽式缓冲装置，在缓冲柱塞上开有由浅入深的三角节流槽，节流面积随着缓冲行程的增大而逐渐减小，缓冲压力变化平缓。

图 4-17d 所示为可调节流孔式缓冲装置，它不但有凸台和凹腔结构，而且在缸盖中还装有针形节流阀和单向阀。当活塞上的凸台进入端盖内孔后，封闭在活塞与端盖间的油液只能从针形节流阀排出，调节节流孔的大小，可控制缓冲腔内缓冲压力的大小，以适应液压缸不同的负载和速度工况对缓冲的要求；同时当活塞反向运动时，高压油从单向阀进入液压缸内，活塞也不会因推力不足而产生起动缓慢或困难等现象。

（4）排气装置 由于液压油中混有空气或液压缸长时间停止使用时空气侵入，在液压缸的最高部位常会聚积空气，若不及时排除就会使缸的运动不平稳，引起爬行和振动，严重时会使油液氧化腐蚀液压元件。排气装置就是为解决此问题而设置的，常用的排气装置如图 4-18 所示。排气阀和排气塞都要安装在液压缸的最高部位，在液压缸排气时打开，让活塞全行程往复移动数次，排气完毕后关闭。

应当指出，并非所有的液压缸都应设置排气装置，对于要求不高的液压缸往往不设专门的排气装置，而是将通油口布置在缸筒两端的最高处，使缸中的空气随油液的流动而排走。对于速度稳定性要求较高以及较大型的液压缸，则必须设置。

barrel and the air in the cylinder will be vented out with the oil. As to the cylinders which require high stability of speed and those of relative large size, venting devices are needed.

4.2.3 Main dimension calculation

The cylinder structure dimension has a direct relationship with the working mechanism of the mainframe. After analyzing the working conditions of the mainframe hydraulic system, working out the load chart and calculating the pressure under each working condition (see Chapter 9), we can calculate the cylinder dimension and check the main components based on the load, the moving speed, the working stroke of the working mechanism. And then the cylinder structure can be designed (see related design handbook).

1. Problems for attention

1) On the precondition of proper speed and suitable thrust, each part of the cylinder must be designed according to related standards. The cylinders should be as compact as possible and convenient to machine, install and maintain.

2) The piston rod is in tension under the heaviest load. If the piston rod is in compression, it should possess a good longitudinal stability. When long stroke piston rod extends out, auxiliary support is needed to prevent the piston rod from drooping.

3) Only one end of the cylinder can be fixed for position for installation or fixing to ensure non-obstruction for cylinder's expanding with heat and contracting with cold.

4) Whether the cylinder system needs cushions, venting and dustproof devices or not must be determined based on the concrete working conditions.

2. Main dimension calculation

(1) Inner diameter D of the cylinder tube given the cylinder thrust F and the selected working pressure p or the moving speed and the input flow rate, the inner diameter D of the cylinder tube can be determined according to the related formulae of this chapter. D must be rounded to a nearest number according to standard GB/T 2348—1993.

(2) The diameter d of the piston rod d should meet the requirements of the cylinder speed or the speed ratio. It must be also rounded to a nearest number according to GB/T 2348—1993. Then the strength and stability of the structure can be checked.

(3) Cylinder tube length s s depends upon the longest working stroke, generally $s \leqslant 20D$.

(4) The shortest guide length H H is defined as the distance between the piston support surface center point and the center point of the guide sleeve sliding surface, see Fig. 4-19. If H is too short, poor stability will occur when the piston rod extends out to its full length. But the size of the cylinder will become larger if H is too long, so there should be a suitable length of H. Note the longest stroke of the cylinder as L, by experience, H can be estimated by

$$H \geqslant \frac{L}{20} + \frac{D}{2} \tag{4-28}$$

Generally the length of the guide sleeve sliding surface $A = (0.6\text{-}1.0)D$ when $D < 80\text{mm}$ and $A = (0.6\text{-}1.0)d$ when $D > 80\text{mm}$. The piston width $B = (0.6\text{-}1.0)D$. If the guiding length H is inadequate, an extra guiding spacing sleeve K can be used to add the length of H (see Fig. 4-19).

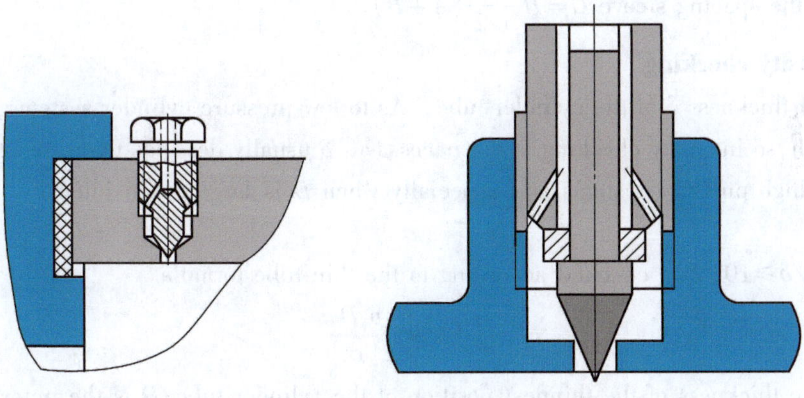

Fig. 4-18 Exhaust elements（排气装置）

a）Type of valve（阀式） b）Type of plug（塞式）

4.2.3 主要尺寸的确定

液压缸的结构尺寸与主机的工作机构有直接关系。在对主机液压系统进行工况分析、编制负载图、确定各工况工作压力之后（详见第 9 章），根据工作机构负载、运动速度、工作行程等确定液压缸的尺寸和结构，对主要零件进行验算，最后进行液压缸的结构设计，具体设计时还需参考有关设计手册。

1. 应注意的问题

1）在保证所获得速度和推力的前提下，应尽可能使液压缸各部分结构按有关标准来设计，尽量做到液压缸结构紧凑，加工、装配和维修方便。

2）尽量使活塞杆在承受最大负载时处于受拉状态，若受压则应具有良好的纵向稳定状态。长行程的活塞杆伸出时，还应加辅助支承，避免活塞杆下垂。

3）液压缸热胀冷缩时应不受阻碍，所以液压缸在安装、固定时只能一端定位。

4）根据液压缸具体工作条件，考虑是否有缓冲、排气和防尘等装置。

2. 主要尺寸的确定

（1）缸筒内径 D　根据液压缸推力 F 和选定工作压力 p，或者运动速度和输入流量，按本节有关公式确定缸筒内径 D 后，然后再从 GB/T 2348—1993 标准中选取相近的尺寸加以圆整。

（2）活塞杆直径 d　通常先满足液压缸速度或速度比 λ_v 来确定活塞杆的直径 d，按 GB/T 2348—1993 标准进行圆整，然后再按其结构强度和稳定性进行校核。

（3）液压缸缸筒长度 s　液压缸的缸筒长度 s 由最大工作行程长度决定，缸筒的长度一般最好不超过其内径的 20 倍。

（4）液压缸最小导向长度 H　当活塞杆全部外伸时，从活塞支承面中点到导向套滑动面中点的距离称为最小导向长度 H，如图 4-19 所示。若导向长度 H 太小，当活塞杆全部伸出时，液压缸的稳定性将变差；反之，又势必增加液压缸的长度。因此对一般液压缸必须有一个合适的导向长度，根据经验，当液压缸最大行程为 L，缸筒直径为 D 时，最小导向长度为

$$H = \frac{L}{20} + \frac{D}{2} \tag{4-28}$$

The width of the spacing sleeve $C = H - \frac{1}{2}(A + B)$.

3. Intensity checking

(1) Wall thickness δ of the cylinder tube As to low-pressure cylinder systems, the wall has enough strength, so intensity checking is not necessary. δ usually depends upon the structure technology. But for high pressure systems and especially when D is large, then intensity checking is needed.

When $D/\delta \geqslant 10$, δ is checked according to the thin-tube formula

$$\delta \geqslant \frac{p_y D}{2[\sigma]} \tag{4-29}$$

Where δ is the thickness of the thinnest position of the cylinder tube; D is the inner diameter of the cylinder tube; p_y the experimental pressure (when the rated working pressure of the cylinder $p_n \leqslant$ 16MPa, $p_y = 1.5p$; when $p_n > 16$MPa, $p_y = 1.25p$); $[\sigma]$ is the allowance of cylinder tube material, $[\sigma] = \sigma_b / n$, σ_b is the tensile strength and n is the safety coefficient, normally, $n = 5$.

When $D/\delta < 10$, δ is calculated according to the formula for the thick wall of the cylinder tube

$$\delta \geqslant \frac{D}{2}\left(\sqrt{\frac{[\sigma] + 0.4p_y}{[\sigma] - 1.3p_y}} - 1\right) \tag{4-30}$$

Most cylinder barrels are seamless steel tubes, and precise machining of the outer diameter is not necessary. After calculating the thickness of the wall, the outer diameter of the cylinder tube should be rounded to the nearest and larger standard diameter of the seamless steel tube.

(2) Diameter d of the piston rod checking The piston rod is mainly in tension or compression and its checking formula is

$$d \geqslant \sqrt{\frac{4F}{\pi[\sigma]} + d_1^{\,2}} \tag{4-31}$$

Where F is the pressure applied on the piston rod; d_1 is the hole diameter for the hollow piston rods. For solid piston rods, $d_1 = 0$; $[\sigma]$ is the allowance of the material of the piston rod, $[\sigma] = \sigma_b / n$, and generally $n \geqslant 1.4$.

If the length of the piston calculated $l \geqslant 10d$ and suppose the piston is in compression and the load on it surpasses the critical value, the piston will lose its stability. So stability checking according to the related formulae in material mechanics is necessary.

(3) Checking of the diameter of the connecting bolts in the cylinder When the cylinder tube and the cylinder cap are connected by a bolt, the bolt is both in tension and under torsion stress. The outer force on the bolt should be amplified by 1.3 times. The bolt must be checked according to the related formulae in the material mechanics.

4.2.4 Swing motors

The shaft of a swing motor can swing to-and-fro within an angle less than 360°. Swing motors can put out torque directly. There are two types of construction: single vane type and double vane type.

Fig. 4-20 shows a single vane swing motor, which consists of a stator, a cylinder, a swing

一般导向套滑动面长度 A，在 $D<80\text{mm}$ 时，可取 $A=(0.6\sim1.0)D$；在 $D>80\text{mm}$ 时，可取 $A=(0.6\sim1.0)d$。活塞宽度 $B=(0.6\sim1.0)D$。当导向长度 H 不够时，可在活塞杆上增加一个导向隔套 K（图 4-19）来增加 H 值。隔套 K 的宽度 $C=H-\frac{1}{2}(A+B)$。

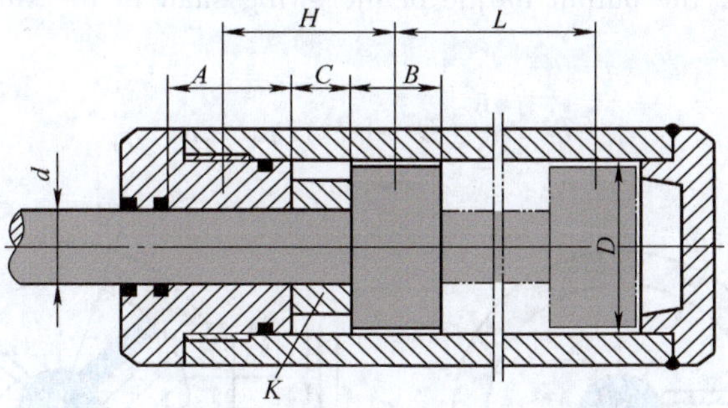

Fig. 4-19　Shortest guide length of cylinder（液压缸最小导向长度）

3. 强度校核

（1）缸筒壁厚 δ 的校核　对中低压系统，由于缸筒的壁厚 δ 往往根据结构工艺的要求来确定，它的强度足够，通常可以不必校核。但在高压系统并且缸筒内径 D 较大时，则必须对壁厚进行校核。

当 $D/\delta\geqslant10$ 时，可按薄壁筒公式来校核，即

$$\delta\geqslant\frac{p_y D}{2[\sigma]} \tag{4-29}$$

式中，δ 为缸筒壁厚；D 为缸筒内径；p_y 为试验压力，当缸的额定工作压力 $p_n\leqslant16\text{MPa}$ 时，取 $p_y=1.5p$；当 $p_n>16\text{MPa}$ 时，取 $p_y=1.25p$；$[\sigma]$ 为缸筒材料的许用应力；$[\sigma]=\sigma_b/n$，σ_b 为材料的抗拉强度，n 为安全系数，一般取 $n=5$。

当 $D/\delta<10$ 时，应按厚壁筒公式进行校核，即

$$\delta\geqslant\frac{D}{2}\left(\sqrt{\frac{[\sigma]+0.4p_y}{[\sigma]-1.3p_y}}-1\right) \tag{4-30}$$

因缸筒材料大多选用无缝钢管，外径不需精加工，计算出壁厚以后，缸筒外径应向大尺寸方向圆整成标准无缝钢管的外径。

（2）活塞杆直径 d 的校核　活塞杆主要承受拉、压作用力，其校核公式为

$$d\geqslant\sqrt{\frac{4F}{\pi[\sigma]}+d_1^{\,2}} \tag{4-31}$$

式中，F 为活塞杆上作用力；d_1 为空心活塞杆孔径，实心杆 $d_1=0$；$[\sigma]$ 为活塞杆材料的许用应力，$[\sigma]=\sigma_b/n$，通常取 $n\geqslant1.4$。

当活塞杆计算长度 $l\geqslant10d$ 时，受到轴向压缩负载超过某一临界值时，会失去稳定性，所以要按材料力学有关公式进行稳定性校核。

（3）液压缸联接螺栓的直径校核　当缸筒与缸盖用螺栓联接时，螺栓在工作中既承受拉应力又承受扭应力。计算时取螺栓所受外力的 1.3 倍，按材料力学有关公式进行校核。

4.2.4　摆动液压马达

摆动液压马达能实现小于 360°角度的往复摆动运动，它可直接输出转矩，主要有单叶

shaft, a vane, left and right supporting plate, press plate, etc. The airproof of the two working chamber is guaranteed by the vane and the frame shape seal kip around the clapboard edge. The stator is fixed on the motor, and the vane connects with the swing shaft. When the pressurized oil enters into the two oil ports, the vane drives the swing shaft to swing reciprocatingly. If the mechanical efficiency is considered, the output torque of the swing shaft in the single vane motor can be expressed as

$$T = \frac{b}{8}(D^2 - d^2)(p_1 - p_2)\eta_m \tag{4-32}$$

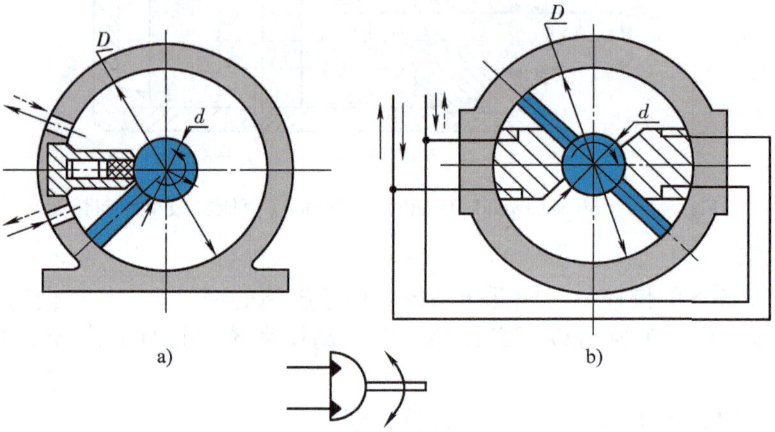

Fig. 4-20 Swing motors (摆动液压马达)

a) Single vane swing motor(单叶片式摆动液压马达)

b) Balanced vane rotary motor(双叶片式摆动液压马达) c) Graph symbol(图形符号)

From the above formula and the energy conservative principle, the output angular speed is

$$\omega = \frac{8q\eta_V}{b(D^2 - d^2)} \tag{4-33}$$

Where p_1 and p_2 are the pressures of the inlet and outlet ports; q is the input flow rate; b is the vane width; D is the diameter of the cylinder inner bore; d is the diameter of the swing shaft; η_V and η_m are the volumetric and the mechanical efficiencies respectively.

The swing angle of a single vane type is generally less than 300°, while that of a double vanes tgpe is usually less than 150°. Given the equal input pressure and flow rate as that for a single vane type, the output torque of the swing shaft in a double vanes type is double the torque of that in the single vane type with same parameters, while the swing angle only half that of the single vane type.

Swing motor is compact in design, and has a large output torque, but difficult to seal. They are only used in medium or low pressure systems for swing, position changing or intermission behavior.

片式和双叶片式两种结构形式。

图 4-20a 所示为单叶片式摆动液压马达。两个工作腔之间的密封靠叶片和隔板外缘所嵌的框形密封件来保证，定子块固定在缸体上，叶片和摆动轴固连在一起，当两油口相继通以压力油时，叶片即带动摆动轴作往复摆动，当考虑到机械效率时，单叶片式摆动液压马达的摆动轴输出转矩为

$$T = \frac{b}{8}(D^2 - d^2)(p_1 - p_2)\eta_m \tag{4-32}$$

根据能量守恒定律，结合式（4-32）得输出角速度为

$$\omega = \frac{8q\eta_V}{b(D^2 - d^2)} \tag{4-33}$$

式中，p_1、p_2 分别为单叶片式摆动液压马达进、出口压力；q 为流入流量；b 为叶片宽度；D 为缸体内孔直径；d 为摆动轴直径；η_V、η_m 分别为摆动缸的容积效率和机械效率。

单叶片式摆动液压马达的摆角一般不超过 300°，双叶片式摆动液压马达（图 4-20b）的摆角一般不超过 150°。当输入压力和流量不变时，双叶片式摆动液压马达摆动轴输出转矩是相同参数单叶片式摆动马达的两倍，而摆动角速度则是单叶片的一半。图 4-20c 所示为摆动液压马达的图形符号。

摆动液压马达结构紧凑，输出转矩大，但密封困难，一般只用于中、低压系统中往复摆动、转位或间歇运动的地方。

Chapter 5 Hydraulic Control Valves

5.1 Introduction

The types of hydraulic control valves used today are too numerous to be mentioned. Most valves, however, perform three primary functions in the hydraulic system: to control the oil pressure (pressure control valves), the flow rate (flow control valves) and the direction of oil flow (directional control valves) in the system to ensure the actuators in proper working on orders of the load requested.

5.1.1 Basic characteristics

Although there are many types of hydraulic control valves and they might be quite different from each other in either the functions or configurations, there exist basic characteristics.

1) In structure, each hydraulic control valve consists of a valve body, a valve core (spool, cone or ball) and the operation elements used to control the valve core. The three basic types of cores are shown in Fig. 5-1.

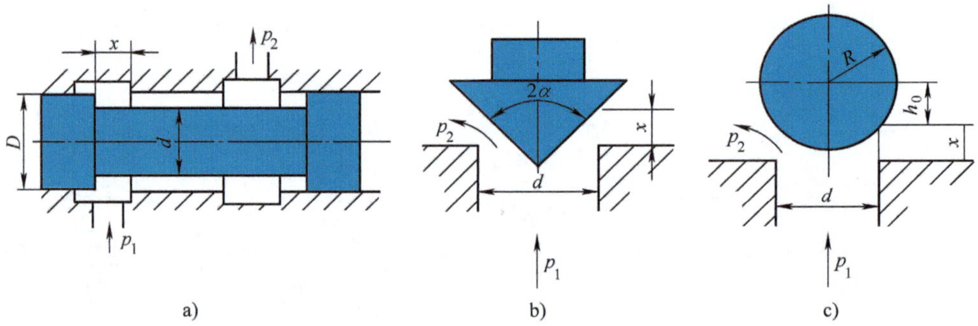

Fig. 5-1 Configuration types (阀的类型)

a) Spool valve(滑阀) b) Cone valve(锥阀) c) Ball valve(球阀)

2) In operating principle, the orifice flow formula should be kept true for the orifice size, the pressure differences between the inlet and outlet, and the flow rate through the valves.

5.1.2 Classification

The hydraulic control valves can be categorized according to different classification attributes, as shown in Tab. 5-1.

第 5 章 液压控制阀

5.1 概述

液压控制阀的种类繁多，但它们在液压系统的作用主要有三个方面：控制液压油的压力（压力控制阀）、流量（流量控制阀）和流动方向（方向控制阀），保证执行元件按照负载的需求进行工作。

5.1.1 液压阀的基本共同点

尽管液压阀的种类繁多，且各种阀的功能和结构形式也有较大的差异，但都具有基本共同点：

1）在结构上，所有液压阀均由阀体、阀芯（锥阀、滑阀或球阀）和驱动阀芯动作的元、部件组成。其中阀芯的结构如图 5-1 所示。

2）在工作原理上，所有液压阀的开口大小、进出口间的压差以及通过阀的流量之间的关系都符合孔口流量公式。

5.1.2 液压阀的分类

液压阀可按不同的特征进行分类，见表 5-1。

Tab. 5-1 Classification of hydraulic control valves（液压控制阀的分类）

Classification attributes（分类方法）	Types（种类）	Contents（类型）
The functions（按用途分）	The pressure control（压力控制阀）	Relief, pressure reducing, sequence, proportional pressure control, pressure switch, etc.（溢流阀、减压阀、顺序阀、比例压力控制阀、压力继电器等）
	The flow control（流量控制阀）	Throttle, speed control, overflow throttle, dividing, combining, proportional flow control, etc.（节流阀、调速阀、溢流节流阀、分流阀、集流阀、比例流量控制阀等）
	The directional control（方向控制阀）	Check, pilot-operated check, directional, proportional direction change, etc.（单向阀、液控单向阀、换向阀、比例方向控制阀等）
The manipulation（按操纵方式分）	Manually（人力操纵阀）	Handle or handwheel, pedal, lever（手把及手轮、踏板、杠杆）
	Mechanically（机械操纵阀）	Skid, spring, hydraulic, pneumatic（挡块、弹簧、液压、气动）
	Electrically（电动操纵阀）	The solenoid-operated, electro-hydraulic control（电磁铁控制、电-液联合控制螺纹式联接、法兰式连接）

(续)

Classification attributes (分类方法)	Types(种类)	Contents(类型)
Connection (按连接方式分)	The tubing(管式连接)	The tubing the flange(螺纹式联接、法兰式连接)
	Sub-plate mounting or modular (板式或叠加式连接)	Single/double-plate mounting, integration and sandwich valves (单层连接板式、双侧连接板式、集成块连接、叠加阀)
	The cartridge(插装式连接)	The screw, the flange(螺纹式插装、法兰式插装)
Control (按控制方式分)	The constant control valves or switches(定值或开关控制阀)	The pressure control, the flow control, the directional control(压力控制阀、流量控制阀、方向控制阀)
	The electro-hydraulic proportional control(电-液比例控制阀)	The electro-hydraulic, pressure/flow/directional/combined valves (电液比例压力阀、电液比例流量阀、电液比例换向阀、电液比例复合阀)
	The servo-control (伺服控制阀)	The electro-hydraulic, air-hydraulic, mechanical-hydraulic(电液伺服阀、气液伺服阀、机液伺服阀)
	The digital control (数字控制阀)	The digital control, pressure/flow/directional control valves(数字控制压力阀、数字控制流量阀、数字控制方向阀)

5.1.3 Basic requirements for hydraulic valves

1. Sensitive, credible, and low impact and vibration.

2. Low resistance when the orifice is opened completely, and good sealing ability when the orifice is closed.

3. Controllable parameters (such as the pressure and the flow rate) should be stable and insensitive under disturbance.

4. Compact in structure and good in installation, adjustment and applications.

5.2 Directional Control Valves

There are two types of general directional control valves: in-line check valves and direction changing valves.

5.2.1 In-line check valves

The in-line check valves include: the general type checks and hydraulic operated check valves.

1. The in-line check valves of general type

Functions: This valve allows free flow in one direction and prevents flow in the reverse direction.

Requirements for the configuration are low pressure losses in the flowing direction and good sealing performance in the reverse direction.

Fig. 5-2 shows a general type check valve in a threaded pipe connection. After pressurized oil flows through the inlet port P_1 to overcome the force of spring 3 and open valve core 2, it flows out from the outlet port P_2 via the radial orifice a and the axial orifice b on the valve core. As the flow

5.1.3　对液压阀的基本要求

1）动作灵敏、使用可靠，工作时冲击和振动要小。

2）阀口全开时，液流压力损失小；阀口关闭时，密封性能好。

3）所控制的参量（压力或流量）稳定，受外界干扰时变化量要小。

4）结构紧凑，安装、调试、维护方便，通用性好。

5.2　方向控制阀

方向控制阀主要有单向阀和换向阀两类。

5.2.1　单向阀

常用的单向阀有普通单向阀和液控单向阀两种。

1. 普通单向阀（单向阀）

普通单向阀的作用是只允许液流沿一个方向通过，不能反向流动。

结构要求：液流正向通过其时压力损失小，反向截止时密封性能好。

普通单向阀如图 5-2 所示。图中阀为螺纹管式联接。当压力油从进油口 P_1 流入时，克服弹簧 3 的作用力顶开阀芯 2，经阀芯上的径向孔 a 和轴向孔 b 从出口 P_2 流出。若液流反向，则压力油在弹簧作用下，使阀芯锥面紧压在阀座孔上，油液被截止而不能通过。在这里，弹簧力很小，仅起复位作用，因此正向开启压力只需 0.03～0.05MPa；反向截止时，因锥阀阀芯与阀座孔为线密封，且密封力随压力增高而增大，因此密封性能良好。

单向阀常被安装在泵的出口处，一方面可防止系统的压力冲击影响泵的正常工作，另一方面在泵不工作时可防止系统的油液倒流并经泵回油箱。单向阀还被用来分隔油路以防止干扰，或与其他阀并联组成复合阀，如单向减压阀、单向节流阀等。当安装在系统的回油路使回油具有一定背压或安装在泵的卸荷回路使泵维持一定的控制压力时，应更换刚度较大的弹簧，其正向开启压力 $p_{1k}=0.3\sim0.5\text{MPa}$。

2. 液控单向阀

液控单向阀除进、出油口 P_1、P_2 外，还有一个控制油口 P_c（图 5-3）。当控制油口不通压力油而通回油箱时，液控单向阀的作用与普通单向阀一样，油液只能从 P_1 到 P_2，不能反向流动。当控制油口 P_c 通压力油时，就有一个向上的液压力作用在控制活塞的下端，推动控制活塞克服单向阀阀芯上端的弹簧力顶开单向阀阀芯使阀口开启，正、反向的液流均可自由通过。液控单向阀既可以对反向液流起截止作用且密封性好，又可以在一定条件下允许正反向液流自由通过，因此多用在液压系统的保压或锁紧回路。

在高压系统中，液控单向阀反向开启前 P_2 口的压力很高，所以要使单向阀反向开启的控制压力也很高。为减小控制压力，可采用图 5-3b 所示的复式结构，在阀芯内装有卸荷小阀芯。控制活塞上行时先顶开小阀芯使主油路卸压，然后再顶开单向阀阀芯，其控制压力仅为工作压力的 4.5%。没有卸荷小阀芯的液控单向阀的控制压力为工作压力的 40%～50%。

需要指出的是，控制油口不工作时，应使其通回油箱，否则控制活塞不能复位。

direction reverses, the valve core returns against its valve seat under pressurized oil and the spring, shutting off the flow path. Note that the spring force is so low that it only performs the function of returning the valve core, so only a low open pressure of 0.03-0.05MPa is needed. The cone valve core and the seat hole form a line sealing that prevents backflow. The sealing force increases with the oil pressure, providing a reliable sealing in the reverse flow.

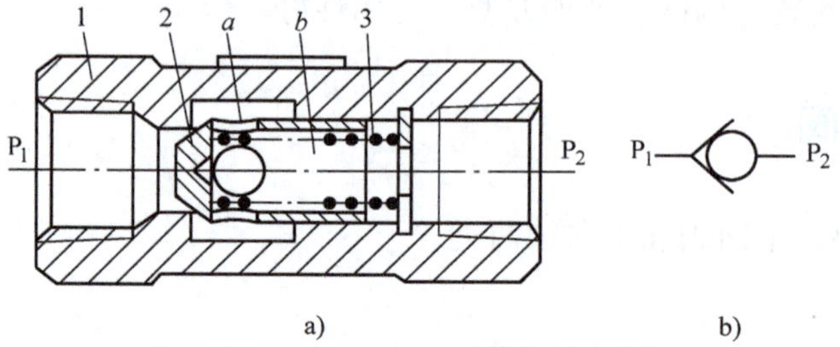

Fig. 5-2 Check valve (普通单向阀)

a) Configuration (结构图) b) Symbol(图形符号)

1—Valve body (阀体) 2—Valve core (阀芯) 3—Spring (弹簧)

A check valve is usually installed on the exit port of a pump. On one hand it is used for preventing the hydraulic impact from its system, and on the other hand it is used for preventing the oil flow from reversing back into the reservoir. Sometimes the check valve is used to separate the oil passage for preventing system turbulence. The check valve is connected with hydraulic elements to build multiple valves, such as check-pressure reducing valve or check-throttling valve. When it is installed in the oil circuit as a back check or control pressure, a high spring rate should be selected and the first open pressure is estimated by $p_{1k} = 0.3 \sim 0.5$MPa.

2. Hydraulic operated check valves

Besides inlet port P_1 and outlet port P_2, there is a pilot (remote) port P_c (see Fig. 5-3). If this port P_c is not connected to the pressure oil but to the reservoir, it is the same as the one-way (general type) check. However if it is connected with the pressure oil, and a force is given to the poppet (core) to open the poppet (core) by overcoming the spring force, and then letting oil flow from inlet P_1 to outlet P_2 or from P_2 to P_1. The check valves operated by a remote port is usually used for pressure held or locked circuits because of its good sealing performances.

Check valves that include external (as shown in Fig. 5-3a) or inner (as shown in Fig. 5-3b) discharge oil with pilot-controlled are applied widely in engineering. The former discharge port is connected to the inlet port, and the latter is discharged to the reservoir.

For a high pressure system, the pressure of P_2 on the port is quite high before the hydraulic operated check valve is inversely direction opened, which requires a high inversely direction pressure operated. To reduce the operated pressure, a compound structure shown in Fig. 5-3b is usually adopted, here there is a small discharged poppet (core) installed into the check's poppet (core). Usually the small discharged poppet (core) is open to allow the main circuit to discharge the pressure, and then open check's poppet (core). In this way the control pressure is only designed at 4.5% of the working pressure. Otherwise it is 40%-50% for a configuration without small discharged poppets (core).

Note that the port should be connected to the reservoir when it is not working, otherwise it cannot be returned.

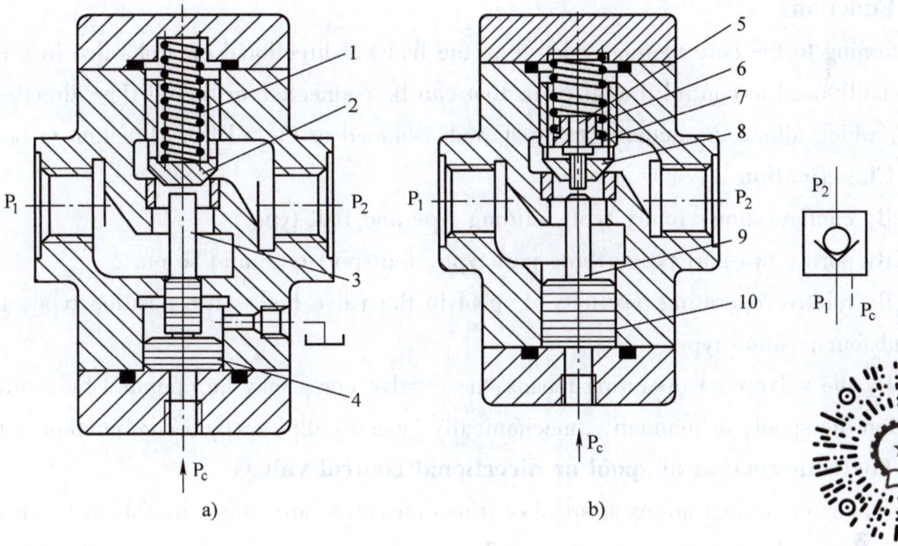

Fig. 5-3　Hydraulic-operated check valve（液控单向阀）

a）Hydraulic - operated general type check valve（普通液控单向阀）

b）Hydraulic-operated check valve with a compound structure（带有复合式结构液控单向阀）

1—Spring（弹簧）　2—Valve core（阀芯）　3—Handspike（推杆）　4—Control piston（控制活塞）

5—Small discharged core（卸载小阀芯）　6—Spring（弹簧）　7—Spring seat（弹簧座）

8—Valve core（阀芯）　9—Handspike（推杆）　10—Control piston（控制活塞）

5.2.2　换向阀

1. 功能

换向阀是利用阀芯在阀体中作相对运动，使油路接通、切断或改变流动方向，从而使执行元件起动、停止或变换运动方向的。

2. 分类

1）按结构类型可分为滑阀式、转阀式和球阀式。

2）按阀体连通的主油路数可分为二通、三通和四通等。

3）按阀芯在阀体内的工作位置可分为二位、三位和四位等。

4）按操作阀芯运动的方式可分为手动、机动、电磁动、液动和电液动等。

3. 滑阀式换向阀的结构

滑阀式换向阀最常见的结构形式见表 5-2。由表可见，不同的通数和位数构成了不同类型的换向阀，所谓二位阀、三位阀是指换向阀的阀芯有两个或三个不同的工作位置；二通阀、三通阀、四通阀是指其阀体上有两个、三个、四个各不相通且与系统中不同油管连接的油路接口。换向阀的功能主要是由其控制的通路数和工作位置所决定的。以表 5-2 中的三位五通阀为例，阀体上有 P、A、B、T_1、T_2 五个通口，阀芯有左、中、右三个工作位置。当阀芯处在图示中间位置时，五个通口都关闭；当阀芯移向左端时，通口 T_2 关闭，通口 P 和 B 相通，通口 A 和 T_1 相通；当阀芯移向右端时，通口 T_1 关闭，通口 P 和 A 相通，通口 B 和

5.2.2 Directional control valves

1. Functions

According to the core motions relative to the body, a directional control valve in a hydraulic system is usually used to control the oil paths that can be connected, or the oil flow direction that can be changed, which allows the actuator to be started, stopped or the oil flow direction to be shifted.

2. Classification

1) By configuration: rotary type, sliding type and ball type.

2) By ports: two-port type, three-port type, four-port type and so on.

3) By relative operating positions of spool in the valve body: two-position type, three-position type, and four-position type.

4) By the valve core operating means: these valve cores may be operated by a difference of oil pressure on the spool, or manually, mechanically, electrically, or by a combination of these means.

3. The construction of spool or directional control valve

The common configurations of spool or direction valves are shown in Tab. 5-2. Their differences lie in the ways and positions they possess. Two or three-position valves mean that their valve cores have two or three different working positions; similarly, two-, three- or four-way valves indicate that there are two, three, or four oil paths on the valve body which are separated from each other and connected with their independent oil lines. The functions of the direction valves depend upon their ways and working positions. Take the five-way, three-position valve shown in Tab. 5-2 for example, there are five ports (P, A, B, T_1, T_2) on the valve body and the valve core has three working positions (left, neutral, right). All the five ports are closed at neutral position. When the valve core moves to its left position, port T_2 is close, two flow paths are connected, one is from port P to port B, and the other is from port A to port T_1. Similarly, when the valve core works at its right position, port T_1 is close, two flows paths are connected, one is from port P to port A, and the other is from port B to port T_2. A valve in this design has a special neutral position where all the five ports are close, so it can be used to control the actuators to stop at any position. In addition, different oil return manners can be obtained because of the two ports for oil return.

Notes for the graphic symbols shown in the above table:

1) The square-pane indicates the working positions of a valve; the number of the square-pane equals the working-position number in the valve.

2) The arrow in a square-pane expresses the oil flow state.

3) The symbol of ⊥ or ⊤ means that the flow path is close.

4) The number of the connecting ports in a square-pane equals the number of flow ways. The inlet port is denoted by p, which is usually connected to the supplying oil line; The outlet port is denoted by T, which is usually connected to the reservoir. The other two ports are denoted by port A and port B, which are usually connected to the system working oil lines.

The directional control valves have two or more working positions, one of which is at normal condition (i. e., the valve core has not been operated by loading). The oil paths are usually setting at their normal condition when we draw a hydraulic system diagram.

T_2 相通。这种结构形式由于具有使五个通口都关闭的工作状态，故可使受它控制的执行元件在任意位置上停止运动，且具有两个回油口，可得到不同的回油方式。

Tab. 5-2 Configurations of spool type directional control valves
（滑阀式换向阀的结构形式）

<table>
<tr><th>Name
（名称）</th><th>Construction
（结构原理图）</th><th>Graphic symbol
（图形符号）</th><th colspan="3">Application
（使用场合）</th></tr>
<tr><td>Two-position, two-way
（二位二通）</td><td>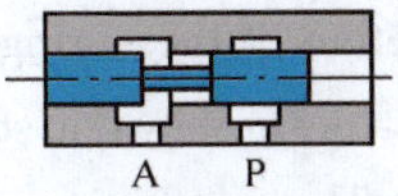</td><td>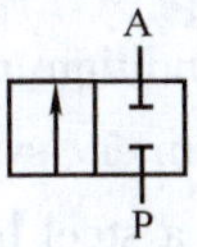</td><td colspan="3">To connect or cut off oil paths（similar to a switch in function）[控制油路的接通与切断（相当于一个开关）]</td></tr>
<tr><td>Two-position, three-way
（二位三通）</td><td>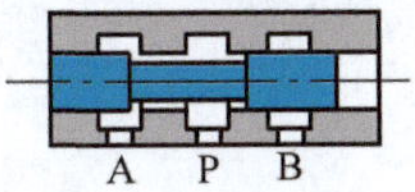</td><td>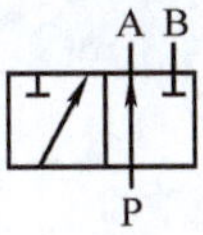</td><td colspan="3">To control flow directions（change from one direction to the other）[控制液流方向（从一个方向变换成另一个方向）]</td></tr>
<tr><td>Two-position, four-way
（二位四通）</td><td>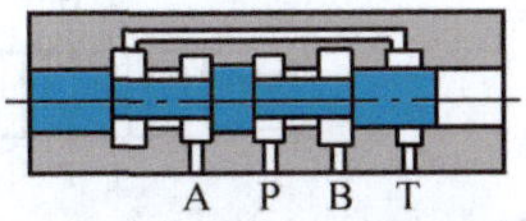</td><td>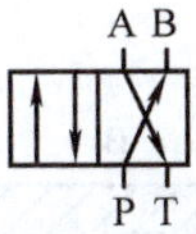</td><td rowspan="4">To shift the actuators（控制执行元件换向）</td><td>Stopping at any point is unavailable（不能使执行元件在任一位置上停止运动）</td><td rowspan="2">Identical oil return manner when the actuator moves in the two opposite directions（执行元件正、反向运动时回油方式相同）</td></tr>
<tr><td>Three-position, four-way
（三位四通）</td><td>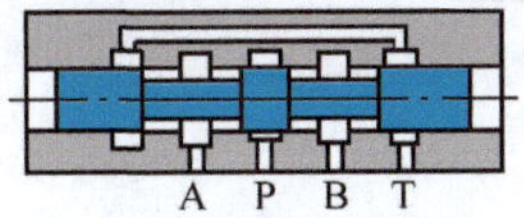</td><td>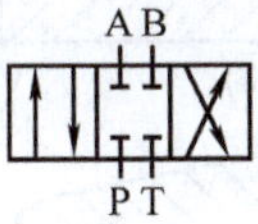</td><td>Stopping at any point is available（能使执行元件在任一位置上停止运动）</td></tr>
<tr><td>Two-position, five-way
（二位五通）</td><td>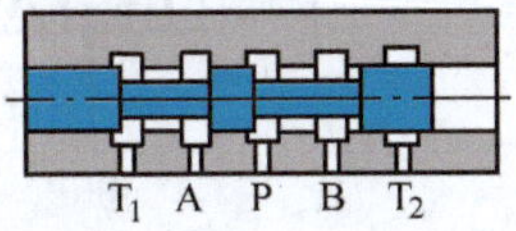</td><td></td><td>Stopping at any point is unavailable（不能使执行元件在任一位置上停止运动）</td><td rowspan="2">To realize different oil return manner when the actuator moves in the two opposite directions（执行元件正、反向运动时可以得到不同的回油方式）</td></tr>
<tr><td>Three-position, Five-way
（三位五通）</td><td>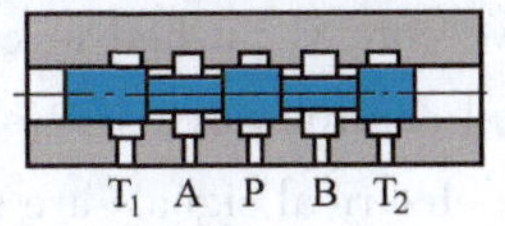</td><td>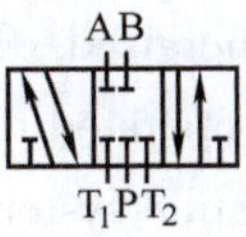</td><td>Stopping at any point is available（能使执行元件在任一位置上停止运动）</td></tr>
</table>

4. The operating manners for spool type directional control valves

The operating manners for spool type directional control valves include: manually operated (or mechanically operated), solenoid operated, pilot operated, electro-hydraulic-operated, etc. See Fig. 5-4, 5-5 and 5-6.

For a manually operated (or a mechanically operated) directional control valve, the valve core movement is accomplished by the external mechanical force. The main types include spring-centered and steel ball-centered (Fig. 5-4). Fig. 5-4a shows a spring-centered type. When the handle is released, the valve core returns to its normal condition under the action of the spring force. This type of valves are usually used in engineering machinery systems. Fig. 5-4b shows a steel ball-centered type. The three positions are accomplished by a steel ball setting. The graphic symbols of the above two types are shown in Fig. 5-4c and Fig. 5-4d respectively.

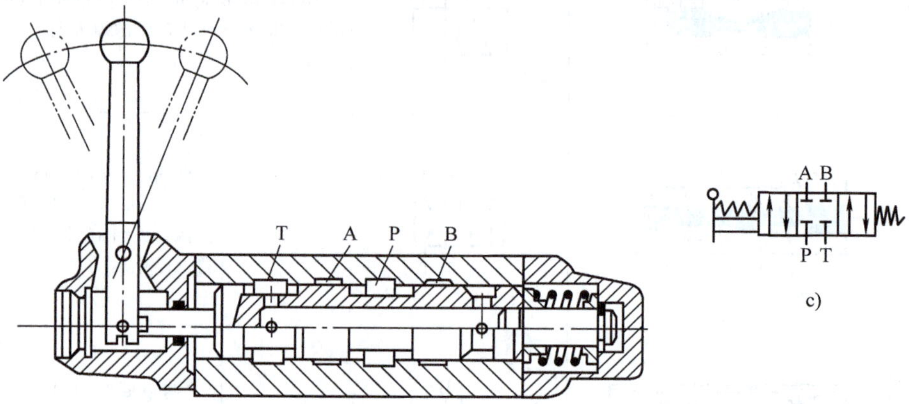

a) c)

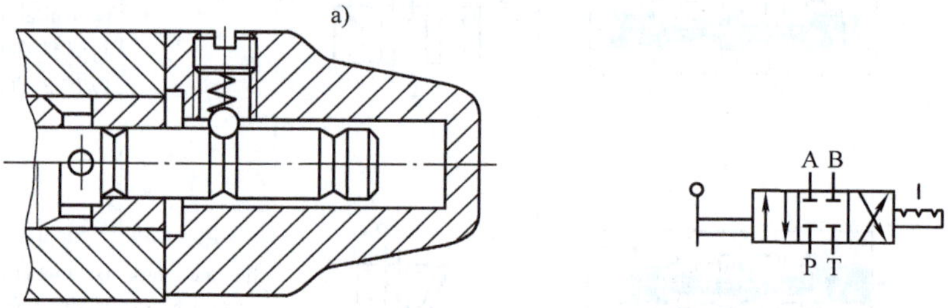

b) d)

Fig. 5-4 Three-position, four-way manual directional control valve(三位四通手动换向阀)

a) Spring-centered spool(弹簧自动复位式) b) Steel ball-centered spool(钢球定位式)

c) Graphics symbol for spring-centered spool(弹簧自动复位机构图形符号)

d) Graphics symbol for steel ball-centered spool(钢球定位结构图形符号)

A solenoid-operated directional control valve is shown in Fig. 5-5. The solenoid is attracted when energized, and released when it is de-energized. The valve core is pushed to control the flow direction based on this principle. A solenoid-operated directional control valve serves as an element to transfer signals between electrical and hydraulic systems. The electrical signals are sent by electrical components such as button switches, limit switches, and distance switches in the hydraulic system. With them, each operation and automatic sequential action can be accomplished. The solenoid-operated directional control valve shown in Fig. 5-5a is a two-position, three-way type. At the working state shown in the figure, the flow path is from port P to port A, while port B is cut off.

表 5-2 所列图形符号的含义如下：

1）用方框表示阀的工作位置，有几个方框就表示有几“位”。

2）方框内的箭头表示该位置上油路的接通状态。

3）方框内符号“⊥”或“⊤”表示该油路不通。

4）一个方框上、下边与外部连接的接口数即表示其通数。通常与系统供油路连接的进油口用 P 表示，与系统回油路相通的回油口用 T 表示，而与执行元件连接的油口用 A、B 表示。

换向阀都有两个或两个以上的工作位置，其中一个是常位，即阀芯未受外部操纵时所处的位置，绘制液压系统图时，油路一般应连接在常位上。

4. 滑阀式换向阀的操纵方式

滑阀式换向阀的操纵方式包括手动（机动）、电磁动、液动和电液联合驱动等，如图 5-4、图 5-5 和图 5-6 所示。

手动（机动）换向阀的阀芯运动是借助于机械外力实现的。如图 5-4 所示，主要有弹簧自动复位式和钢球定位式两种形式，图 5-4a 所示为弹簧自动复位式，放开手柄，阀芯在弹簧的作用下，自动回复中位，常用于工程机械系统中。图 5-4b 为钢球定位式，通过钢球滑阀可在三个位置定位。图 5-4c、d 所示为其图形符号。

二位三通电磁换向阀如图 5-5 所示。电磁换向阀是利用电磁铁的通电吸合与断电释放而直接推动阀芯来控制液流方向的。它是电气系统与液压系统之间的信号转换元件，它的电气信号由液压设备中的按钮开关、限位开关、行程开关等电气元件发出，从而可以使液压系统方便地实现各种操作及自动顺序动作。图 5-5a 所示为二位三通电磁阀的结构图。在图示位置，油口 P 和 A 相通，油口 B 断开；当电磁铁 1 通电吸合时，推杆 2 将阀芯 3 推向右端，这时油口 P 和 A 断开，而与 B 相通。当电磁铁断电释放时，弹簧 4 推动阀芯复位。图 5-5b 所示为其图形符号。

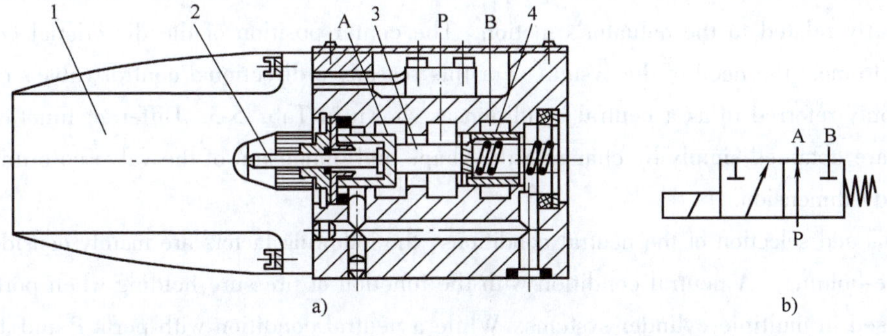

Fig. 5-5　Two-position, three-way solenoid-operated directional control valve（二位三通电磁换向阀）

a）Configuration（结构图）　b）Graphics symbol（图形符号）

1—Solenoid（电磁铁）　2—Handspike（推杆）　3—Valve core（阀芯）　4—Spring（弹簧）

电磁铁通常分为交流电磁铁和直流电磁铁两种。按电磁铁的铁心是否能够泡在油里，电磁铁又可分为干式和湿式。

图 5-6 所示为弹簧对中型三位四通电液换向阀，由图可见当两个电磁铁都不通电时，电磁阀阀芯 5 处于中位，液动阀（主阀）阀芯 1 因其两端都接通油箱，也处于中位。电磁铁 6 通电时，电磁阀阀芯 5 向右端移动，压力油 p' 经单向阀 2 接通主阀的左端，其右端的油则经节流阀 7 和电磁阀而接通油箱，于是主阀阀芯右移，移动的速度由节流阀 7 的开口大小决

When solenoid 1 is energized, the handspike 2 pushes the valve core 3 to move to the right position, changing the flow path from port P to port B and port A is cut off. When the solenoid is de-energized, spring 4 pushes the valve core return to its initial position. Fig. 5-5c shows its graphic symbol.

Fig. 5-6 shows a construction of middle-sized four-position, three-way electro-hydraulic directional control valve and its graphic symbol. When both of the solenoids are de-energized, valve core 5 of the solenoid valve is at its normal condition. Valve core 1 of the hydraulic operated (main) valve is also at its normal condition because two ends of the hydraulic operated valve are also connected to the reservoir. When solenoid 6 is energized, valve core 5 moves on to the right. Control pressurized oil p' flows to the left end of the main valve through check valve 2 and the oil from the right end is connected to reservoir via throttle valve 7 and the solenoid valve. All of these would cause the valve core of the main valve to move on to the right, whose speed can be adjusted by the orifice of throttle valve 7. Similarly, when solenoid 4 is actuated, the valve core of the solenoid valve moves on to the left and thus the main valve core, whose speed depends on the orifice of throttle valve 3.

For this type of valve, the main valve core operation is accomplished by the hydraulic pressurized oil fluid, which is controlled by a solenoid rather than by a solenoid directly. Thus the thrust force would be large and convenient to manipulate. The moving speeds in the left and right directions are controlled by throttle valves 3 and 7 respectively to reduce the hydraulic impact when actuator does the direction-shifting. In a large-sized hydraulic system, large flow rate via valves and thus large friction and flow force often occur. In this case, an electro-hydraulically operated directional control valve is used to replace the solenoid operated directional control valve due to the relative small thrust provided by the solenoid.

5. Functions of neutral positions

Three-position valve consists of two extreme positions and a center position. The two extreme positions are directly related to the actuator's motion. The center position of the directional control valve is designed to meet the need of the system. For this reason, a directional control valve's center position is commonly referred to as a neutral condition as shown in Tab. 5-3. Different functions of neutral positions are obtained simply by changing the shape and dimension of the valve core under a constant valve body dimension.

In the analysis and selection of the neutral conditions, the following factors are mainly considered:

(1) Pressure-holding A neutral condition with the function of pressure-holding when port P is blocked can be used in multiple-cylinder systems. While a neutral condition with ports P and T half open-type connected (such as X type) can be adopted for controlling the lines under a certain pressure in the system.

(2) Unloading condition When ports P and T are connected, the system is unloaded.

(3) The stability and precision during the directional shift When ports A and B are blocked, poor shift stability may occur due to hydraulic impact, but high precision can be obtained. However, when ports A and B are connected with port T, the low hydraulic impact could be reached, but a low precision occurs due to hard stop of the operating parts.

(4) Startup stability If one port is connected with the reservoir at neutral condition, startup instability may occur for there is no pressurized oil to perform the cushioning effect.

定。同理，当电磁铁 4 通电，电磁阀阀芯移向左位时，主阀阀芯也移向左位，其速度由节流阀 3 的开口大小决定。

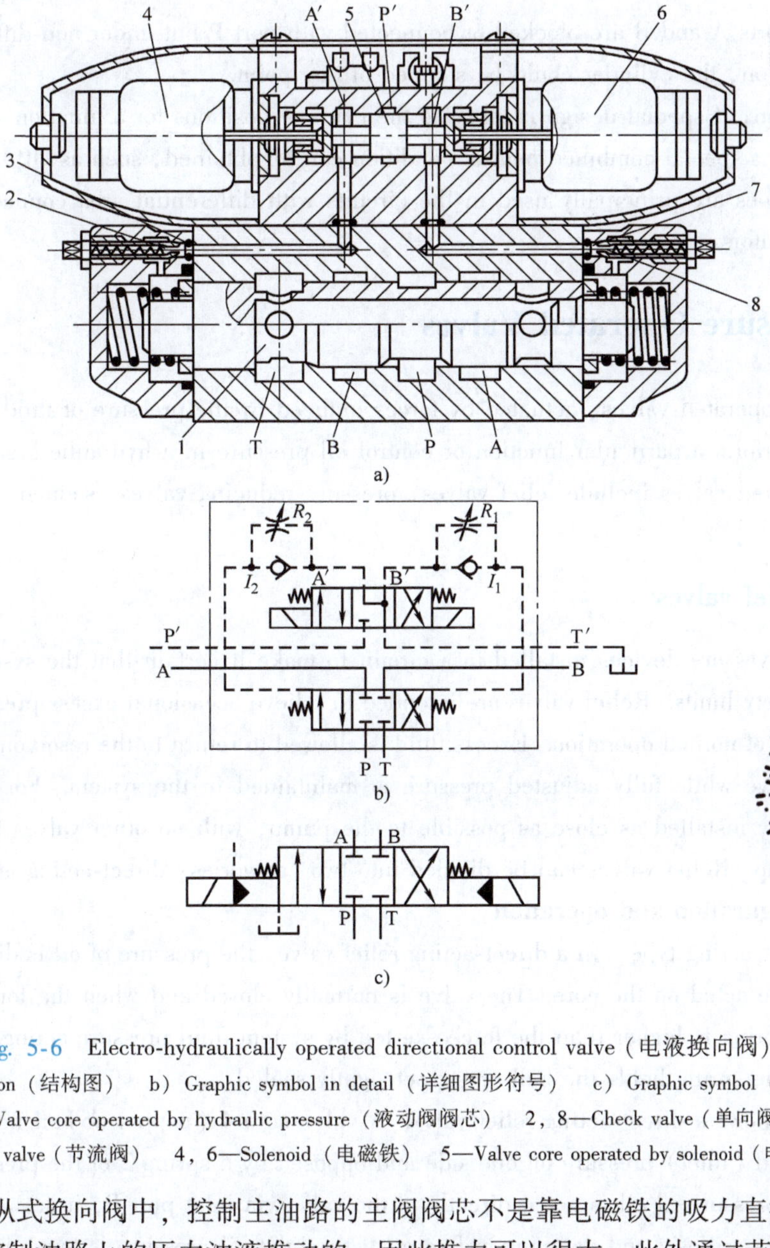

Fig. 5-6 Electro-hydraulically operated directional control valve（电液换向阀）

a）Configuration（结构图） b）Graphic symbol in detail（详细图形符号） c）Graphic symbol（图形符号）

1—Valve core operated by hydraulic pressure（液动阀阀芯） 2，8—Check valve（单向阀）

3，7—Throttle valve（节流阀） 4，6—Solenoid（电磁铁） 5—Valve core operated by solenoid（电磁阀阀芯）

在电-液操纵式换向阀中，控制主油路的主阀阀芯不是靠电磁铁的吸力直接推动，而是靠电磁铁操纵控制油路上的压力油液推动的，因此推力可以很大。此外通过节流阀 3 和 7 可以分别控制主阀阀芯的向左或向右的移动速度，这就使系统中的执行元件能够得到平稳无冲击的换向。所以，在大型液压设备中，当通过阀的流量较大时，作用在滑阀上的摩擦力和液动力较大，此时电磁换向阀的电磁铁推力相对太小，需要用电液换向阀来代替电磁换向阀。

5. 滑阀的中位机能

如前所述，三位阀有三个工作位置，根据需要，执行元件可在左位或右位工作。三位换向阀的阀芯在中间位置时，各油口间有不同的连通方式，可满足不同的使用要求，这种连通方式称为换向阀的中位机能。不同的中位机能是在阀体的尺寸不变的情况下，通过改变阀芯

(5) The cylinder needs at a "float" condition or can stop at any position For a neutral condition of horizontal cylinder , when ports A and B are connected, it should be at a "float" state; and in this case we can use the other mechanism to move the workbench and to change the valve condition. When ports A and B are blocked or connected with port P but under non-differential area connection condition, the cylinder could be stopped at any point.

In addition, if special design is taken to the extreme positions for a function similar to the neutral condition, a special combined neutral condition can be obtained, such as OP type and MP type. OP and MP types are principally used in the circuits with differential area connection to allow fast motion of actuators.

5.3 Pressure Operated Valves

Pressure operated valves, actuated by forces induced by the pressure of fluid and control sign, are used to perform a particular function or control oil pressure in a hydraulic system. The common pressure operated valves include relief valves, pressure-reducing valves, sequence valves and pressure switches.

5.3.1 Relief valves

Relief valves are devices installed in a circuit to make it certain that the system pressure does not exceed safety limits. Relief valves are intended to relieve occasional excess pressures arising during the course of normal operation. Excess fluid is allowed to return to the reservoir through an outlet port in the valve while fully adjusted pressure is maintained in the system. For safety, the relief valve is usually installed as close as possible to the pump, with no other valves between the relief valve and pump. Relief valves can be divided into two categories: direct-acting and pilot-operated.

1. Configuration and operation

(1) Direct-acting type In a direct-acting relief valve, the pressure of oil is directly balanced to the spring force acted on the core. The valve is normally closed and when the force exerted by the compression spring is higher than the force exerted by system fluid pressure acting on the core (ball or poppet), the spring holds the ball or poppet tightly sealed.

Fig. 5-7 shows a direct-acting relief valve. A valve core 4 (poppet or ball held) is exposed to the system fluid (inlet) pressure on one side and opposed by a spring 2 of the preset force F_t on the other. When pressurized oil is acted directly to the valve via inlet port P, it acts on the bottom c of valve core 4 via orifice f and damping orifice g. If the inlet oil pressure is low, the valve core stays at the lowest position (limited by the bottom nut) acted by the compression spring 2. Ports P and T are blocked by valve core 4 with a certain length of l, which is also called sealed length. In this case, the valve is closed. When the displacement equals the sealed length l, the valve is at the critical state for open. When the system pressure (p) exceeds the setting of the spring force F_t, the fluid unseats (x) the ball or poppet, allowing a controlled amount of fluid (q) to bypass to the reservoir, keeping the system pressure at the setting value. The spring reseats again the ball or poppet when enough fluid is released (bypassed) to drop the system pressure below the setting of the

的形状和尺寸得到的。三位四通换向阀、三位五通换向阀常见的中位机能见表 5-3。

Tab. 5-3 Neutral conditions of four/five-position, three-way directional control valves
（三位四通换向阀、三位五通换向阀的中位机能）

Type（滑阀机能）	Neutral condition（滑阀状态）	Symbol（中位符号） Four-way（四通）	Five-way（五通）	Features and application（特点及应用）
O	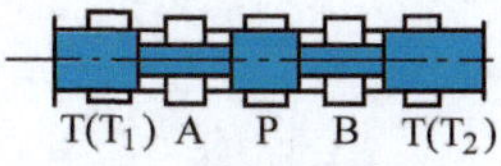		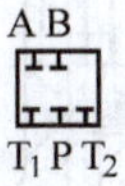	All ports are blocked and the two ends of the cylinder are sealed. Pump flow isn't returned to the reservoir. Can be mounted in parallel to perform complex operation（各油口全封闭，液压缸两腔闭锁，泵不卸荷，可用于多个换向阀并联工作）
H	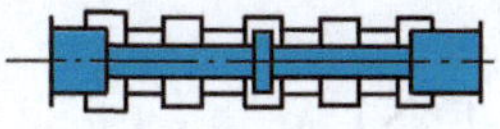			All ports are connected to each other. Allows free movement of an actuator while pump flow is returned to the reservoir（各油口互通，液压缸活塞浮动，泵卸荷）
Y	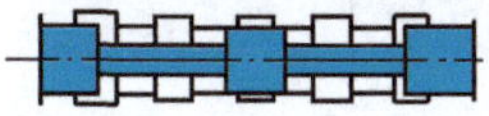			A float-center spool. Port P is blocked and ports A, B, and T are connected. Allows free movement of each actuator. Pump flow isn't returned to the reservoir（油口 A、B 通回油口 T，油口 P 封闭，活塞浮动，泵不卸荷）
J	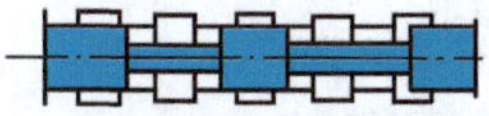			One end of the cylinder is sealed; the other is connected to the returned oil. The system isn't unloaded（系统不卸荷，缸一腔封闭，另一腔与回油连通）
C	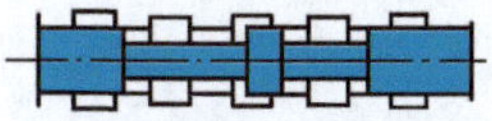			Ports P and A are connected, while B and T are blocked. One end of the cylinder is sealed. Pump flow isn't returned to the reservoir（油口 P 与 A 通，B 和 T 封闭，液压泵不卸荷，液压缸一腔闭锁）
P	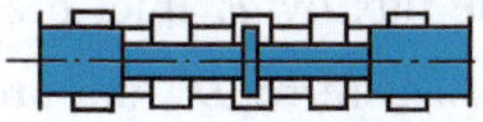			Port T is blocked and ports A, B, and P are connected. Can be used as an element of differential circuit for hydraulic cylinder（油口 P 与 A、B 油口连通，T 口封闭，可组成液压缸的差动回路）
K	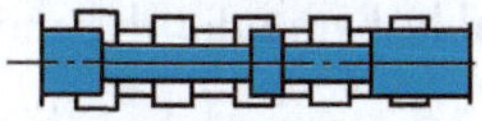		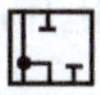	Ports P, A, T are connected, while port B is blocked. One end of the cylinder is sealed. Pump flow is returned to the reservoir（油口 P、A、T 互通，油口 B 封闭，液压缸一腔闭锁，泵卸荷）
X	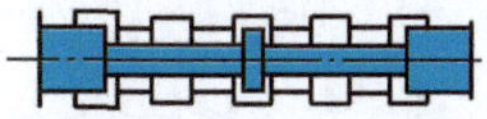			All ports are connected to each other with a half open throttle orifice. Allows free movement of an actuator while pump flow is returned to the reservoir under a given pressure drop（各油口半开启接通，液压泵压力油在一定压力下回油箱）
M	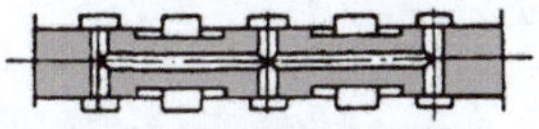			Ports P and T are connected and A and B are blocked. A tandem-center condition stops the motion of an actuator, but allows pump flow to return to reservior（油口 P 与 T 通，油口 A、B 封闭，液压泵卸荷，液压缸两腔闭锁）
U	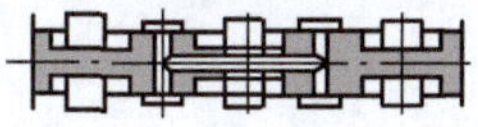			The two ends of the cylinder are connected to each other. The system isn't unloaded. Returned oil port is sealed（系统不卸荷，缸两腔连通，回油封闭）
N	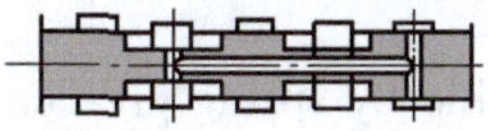			One end of the cylinder is connected to the returned oil; the other is sealed. The system isn't unloaded（系统不卸荷，缸一腔与回油连通，另一腔封闭）

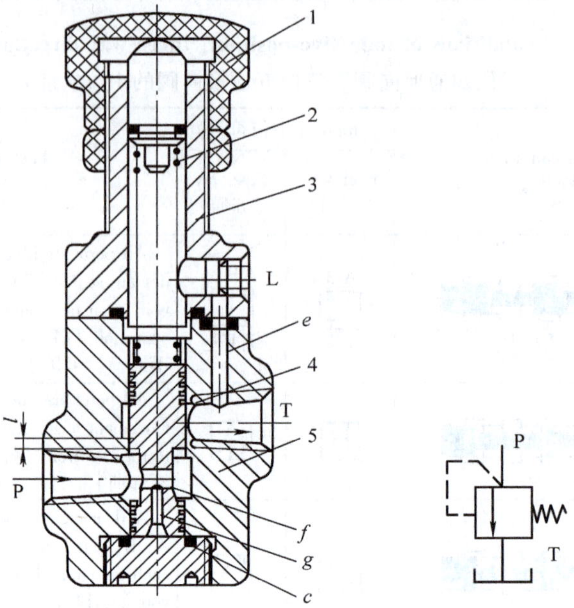

a)　　b)

Fig. 5-7　Direct-acting relief valve（直动式溢流阀）

a）Configuration（结构图）　b）Graphic symbol（图形符号）

1—Reguating screw（调节螺母）　2—Spring（弹簧）　3—Upper cover（上盖）

4—Valve core（阀芯）　5—Valve body（阀体）

spring. In this case the inlet pressure in the valve holds at a constant value and keeps a balanceable state due to the action of relief. If the orifice is denoted by x, the flux by q, inlet pressure by p, the rigidity of spring by K, pre-compression by x_0 and the diameter of core by D, the pressure at which a relief valve starts to open to allow fluid to flow through is known as the cracking pressure p_k, and the inlet pressure is denoted by p_s at the full-flow q_s and the stable hydraulic dynamic force by F_s, then the force balance equations can be obtained below:

1）The force balance equation to begin opening

$$p_k \frac{\pi D^2}{4} = K(x_0 + l) \tag{5-1}$$

2）The relief force balance equation

$$p \frac{\pi D^2}{4} = K(x_0 + l + x) + F_s \tag{5-2}$$

3）The pressure-flow rate equation

$$q = C_d \pi D x \sqrt{\frac{2}{\rho} p} \tag{5-3}$$

By combining equations（5-2）and（5-3）, inlet pressure p can be calculated under different flow rates.

Note here two points: ①The opening pressure p_k is adjustable by spring pre-compression x_0 and by controlling the inlet pressure p, which is called adjustment spring; ②The direct-acting valve is

在分析和选择中位机能时，通常需要考虑以下几点：

（1）系统保压 当P口被封闭，泵保压，可用于多缸系统。当P口与T口处于半开启状态时（如X型），系统仍保持一定压力供控制油路使用。

（2）系统卸荷 P口与T口相通，系统卸荷。

（3）换向平稳性与精度 当液压缸的A、B口均封闭时，换向过程中会产生液压冲击，换向平稳性较差，但换向精度高。反之，当A、B口均与T口相通时，换向过程液压冲击小，但工作部件不易制动，换向精度较低。

（4）起动平稳性 阀处于中位时，液压缸某腔若通油箱，则起动时该腔因无油液起缓冲作用，起动不太平稳。

（5）液压缸“浮动”和在任意位置上的停止 阀处于中位，当A、B口互通时，卧式液压缸呈“浮动”状态，可利用其他机构移动工作台，以调整其位置。当A、B口均被封闭或与P口连通时（在非差动情况下），则可使液压缸在任意位置停止下来。

三位换向阀除了中位有各种机能外，有时也把阀芯在某一端位置时的油口连通情况设计成特殊的机能，常用的有OP型和MP型等。OP型和MP型机能主要用于差动连接回路，以实现快速运动。

5.3 压力控制阀

压力控制阀有溢流阀、减压阀、顺序阀和压力继电器等，它们用来控制液压系统中的油液压力或通过压力信号实现控制。

5.3.1 溢流阀

溢流阀按结构形式不同分为直动式与先导式。它旁接在液压泵的出口以保证系统压力恒定或限制其最高压力，有时也旁接在执行元件的进口，对执行元件起安全保护作用。

1. 结构及工作原理

（1）直动式 直动式溢流阀的结构及图形符号如图5-7所示。压力油从进口P进入阀后，经孔f和阻尼孔g后作用在阀芯4的底面c上。当进口压力较低时，阀芯在弹簧2预调力作用下处于最下端，由底端螺母限位。由阀芯4与阀体5构成的节流口有重叠量，称为封油口长度l，将P与T口隔断，阀处于关闭状态。当进口P处压力升高至等于或大于弹簧力F_t时，阀芯开始向上移动。当阀芯上移量等于封油口长度l时，阀口处于开启的临界状态。若压力继续升高至阀口打开时，油液从P口经T口溢流回油箱。由于溢流阀的作用，在流量变化时，进口压力能基本保持恒定。此时阀芯处于受力平衡状态，阀口开度为x，通流量为q，进口压力为p。若弹簧刚度为K，预压缩量为x_0，阀芯直径为D，阀口刚开启时的进口压力为p_k，通过额定流量q_s时的进口压力为p_s，作用在阀芯上的稳态液动力为F_s，则得：

1）阀口刚开启时的阀芯受力平衡关系式为

$$p_k \frac{\pi D^2}{4} = K(x_0 + l) \tag{5-1}$$

2）阀口开启溢流时阀芯受力平衡关系式为

$$p \frac{\pi D^2}{4} = K(x_0 + l + x) + F_s \tag{5-2}$$

designed by a core (poppet or ball held) exposed to the system fluid pressure on one side and opposed by a spring of preset force on the other. So it needs a bigger spring force for a higher pressure or flow rate system. If so, the ability of spring will be lower and configuration is difficult. In practice, there are few direct-acting valves.

(2) Pilot-operated type Pilot-operated relief valves are used for the systems which require high pressure and large flow rate. Their common constructions are shown in Fig. 5-8. A pilot-operated relief valve operates in two stages: a pilot stage and a main stage. The main valve displacement depends upon the oil pressure differential on the two sides of the main valve.

Pressurized oil is directed to the down port of the main valve core through the inlet port of the main valve. The oil in the down port is induced to the upper port and the front of valve core 2 via damping orifice 1 to develop a hydraulic pressure of F_x on the pilot valve core. If F_x is less than the setting force spring F_{t2} on the other side, the pilot valve is closed and there exists no flow in the main valve. In this case the pressures in the down and upper ports of valve core 5 are equal. However, the upper port has a larger action area than the down port. The sum of this force differential and that of the main spring keep the poppet seated down. The main valve is still closed. As the system pressure rises, the pressure in the passage rises as well and when it reaches the setting of the pilot valve, i. e., $F_x > F_{t2}$, the pilot valve begins to open, the inlet pressurized oil flows back to the reservoir via damping orifice 1 and pilot orifice. Note that for the main valve, up port pressure p_1 is less than down port pressure p due to the pressure loss caused by the damping orifice. When the differential pressure $(p - p_1)$ is large enough to make the upper hydraulic pressure to overcome the force of spring 4 on the main valve, the main valve core 5 moves up and opens the orifice, allowing pressurized oil to bypass to the reservoir via the main valve. Under a given main valve orifice, both the pilot valve core and the main valve core are balanced out and the pressure-flow rate equation is fit for the orifice. The inlet pressure of the main valve can be determined. The pilot valve acts as a master adjustment and the piloted main relief valve poppet spool follows in a mirror image pattern to that of the pilot to provide the desired pressure value. The valve closes again when the inlet oil pressure drops below the setting of the pilot valve. The pilot operated relief valves have less pressure override than that of the direct-acting relief valves.

The static characteristics of a pilot-operated valve can be described by the following five equations.

1) Main valve force balance equation

$$pA = p_1A_1 + K_1(y_0 + y) + C_1\pi Dy\sin 2\alpha p \tag{5-4}$$

2) Pressure-flow rate equation for the main valve orifice

$$q = C_1\pi Dy\sqrt{\frac{2}{\rho}p} \tag{5-5}$$

3) Force balance equation for the pilot valve core

$$p_1A_x = p_1\frac{\pi d^2}{4} = K_2(x_0 + x) \tag{5-6}$$

4) Pressure-flow rate equation on pilot valve

$$q_x = C_2\pi dx\sin\varphi\sqrt{\frac{2}{\rho}p_1} \tag{5-7}$$

3）阀口开启溢流的压力流量方程为

$$q = C_d \pi D x \sqrt{\frac{2}{\rho} p} \tag{5-3}$$

联立求解式（5-2）和式（5-3）可求得不同流量下的进口压力。

注意两点：①调节弹簧的预压缩量 x_0，可以改变阀口的开启压力 p_k，进而调节控制阀的进口压力 p，此处弹簧称为调压弹簧；②直动式溢流阀因液压力直接与弹簧力相平衡而工作，若压力较高、流量较大，则要求调压弹簧具有很大的弹簧力，这不仅使调节性能变差，而且结构上也难以实现。所以滑阀直动式溢流阀一般只用于低压小流量处。

（2）先导式　当系统压力和流量较大时，通常使用先导式溢流阀。其常见的结构如图 5-8a 所示，它由先导阀和主阀两部分组成。这种阀的工作原理是利用主阀上、下两端油液压差来使主阀阀芯移动。图 5-8b 所示为其图形符号。

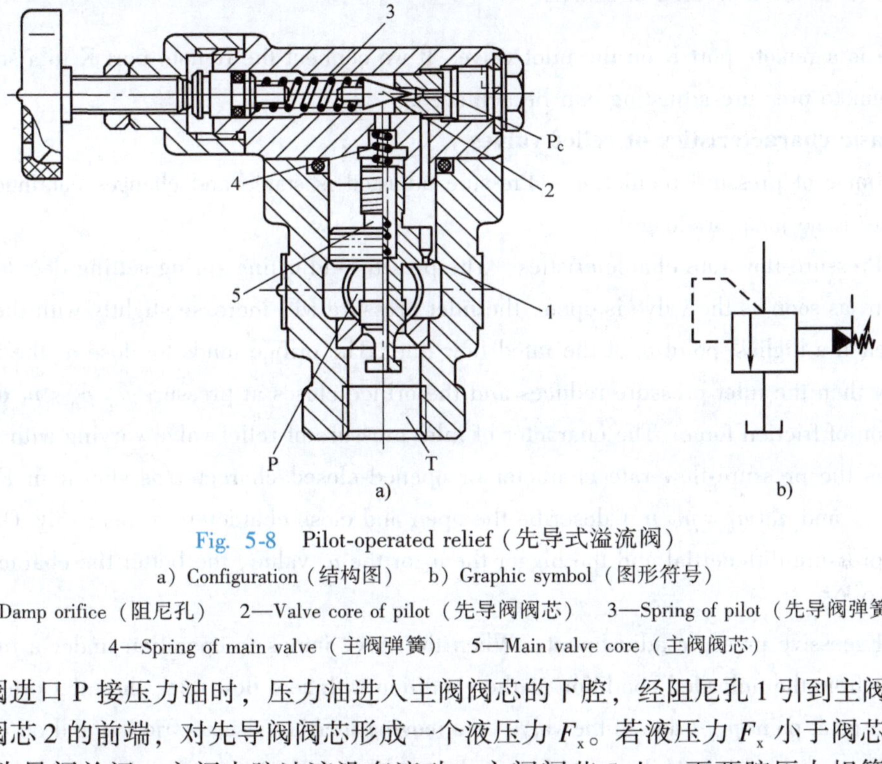

Fig. 5-8　Pilot-operated relief（先导式溢流阀）

a）Configuration（结构图）　b）Graphic symbol（图形符号）

1—Damp orifice（阻尼孔）　2—Valve core of pilot（先导阀阀芯）　3—Spring of pilot（先导阀弹簧）

4—Spring of main valve（主阀弹簧）　5—Main valve core（主阀阀芯）

当主阀进口 P 接压力油时，压力油进入主阀阀芯的下腔，经阻尼孔 1 引到主阀阀芯上腔及先导阀阀芯 2 的前端，对先导阀阀芯形成一个液压力 F_x。若液压力 F_x 小于阀芯另一端弹簧力 F_{t2}，先导阀关闭，主阀内腔油液没有流动，主阀阀芯 5 上、下两腔压力相等而上腔作用面积大于下腔作用面积。在主阀弹簧 4 和主阀阀芯上、下腔液压力共同作用下将主阀阀芯紧压在阀座孔上，主阀阀口关闭，没有溢流。若进口压力增大，则作用在先导阀阀芯上的液压力 F_x 也增大，当 $F_x > F_{t2}$ 时，先导阀阀口开启，溢流阀的进口压力油经阻尼孔 1、先导阀阀口溢流回油箱。由于阻尼孔的作用产生压力损失，主阀上腔压力 p_1 小于主阀下腔压力 p。当压差（$p - p_1$）足够大时，因压差形成的向上液压力克服主阀弹簧力 4 推动主阀阀芯 5 上移，主阀阀口开启，压力油经主阀阀口溢流回油箱。主阀阀口开度一定时，先导阀阀芯和主阀阀芯分别处于受力平衡状态，阀口满足压力流量方程，主阀进口压力为一确定值。

先导式溢流阀的静特性可用下列五个方程描述。

1）主阀阀芯受力平衡方程为

$$pA = p_1 A_1 + K_1 (y_0 + y) + C_1 \pi D y \sin 2\alpha p \tag{5-4}$$

5) Pressure-flow rate equation for the damping orifice

$$q_1 = q_x = \frac{\pi\phi^4}{128\mu l}(p - p_1) \tag{5-8}$$

Where K_1 and K_2 are the main valve spring rate and the pilot valve spring rate respectively; y_0 and x_0 are the pre-compression on main spring and pilot spring respectively; y and x are the open length on main valve and on pilot valve respectively; q and q_x are the passage flow in main valve and in pilot valve respectively; q_1 is the passage flow in damp orifice, $q_1 = q_x$; A_1 and A are the upper and end action areas on main valve respectively; D and d are the orifice diameters on main and pilot valves respectively; α and φ are the half-cone angles on the main and pilot valves respectively; ϕ and l are the diameter and the length of damp orifice respectively; μ is the viscosity of oil; ρ is the density of oil and the A_x is the hole area of seat, $A_x = \frac{\pi d^2}{4}$.

There is a remote port K on the pilot valve. If we connect the remote port K to a solenoid valve then the remote pressure-adjusting can be achieved.

2. Basic characteristics of relief valves

(1) Range of pressure regulation Pressure adjusted is stable and changes continuously up and down. There is no jump or sluggish.

(2) Pressure-flow rate characteristics The pressure-adjusting spring setting decides the cracking pressure as soon as the valve is open, the inlet pressure will increase slightly with the rise of flow rate and reach a highest point p_s at the rated flow rate. The orifice tends to close as the flow rate decreases. By then the inlet pressure reduces and the orifice closes at pressure p_b. $p_b < p_k$ due to different direction of friction force. The character of inlet pressure of relief valve varying with the flow rate is known as the pressure-flow rate character or opened-closed character as shown in Fig. 5-10. n_k ($n_k = p_k/p_s$) and n_b ($n_b = p_b/p_s$) describe the open and close characters respectively. Obviously, the lower the pressure diffenertial and the bigger the n_k or the n_b value, the better the character. Usually $n_k = 0.9 \sim 0.95$.

(3) Excessive pressure adjustment When the valve jumps to operation under a rated pressure or flow rate from the unloaded condition or the condition with zero flow rate, the inlet pressure reaches rapidly to a peak point p_{max} (due to the valve core moving inertia, viscous friction and oil compressibility). Finally the pressure stabilizes at the rated pressure p_s. The pressure difference Δp between the peak pressure and the rated full-flow pressure is named excessive pressure adjustment, generally, Δp is less than 30% that of the rated pressure. Fig. 5-10 shows the dynamic characteristic curves when the working conditions change from the zero pressure and the zero flow rate condition to the rated pressure and the rated flow rate.

5.3.2 Pressure-reducing valve

Pressure-reducing valves are used to drop the normal operating pressure of a main circuit as it is directed into the branch circuit to the required pressure in the branch circuit by flow resistance pressure differential. The desired lower pressure can be obtained by adjusting the regulating device on the valve. As soon as the desired pressure is reached in the secondary system, the valve partially

2）主阀阀口压力流量方程为

$$q = C_1 \pi D y \sqrt{\frac{2}{\rho} p} \tag{5-5}$$

3）先导阀阀芯受力平衡方程为

$$p_1 A_x = p_1 \frac{\pi d^2}{4} = K_2 (x_0 + x) \tag{5-6}$$

4）先导阀阀口压力流量方程为

$$q_x = C_2 \pi d x \sin\varphi \sqrt{\frac{2}{\rho} p_1} \tag{5-7}$$

5）流经阻尼孔的压力流量方程为

$$q_1 = q_x = \frac{\pi \phi^4}{128 \mu l} (p - p_1) \tag{5-8}$$

式中，K_1、K_2 分别为主阀弹簧、先导阀弹簧的刚度；y_0、x_0 分别为主阀弹簧、先导阀弹簧的预压缩量；y、x 分别为主阀和先导阀的阀口开度；q、q_x 分别为流经主阀阀口和先导阀口的流量；q_1 为流经阻尼孔的流量，$q_1 = q_x$；A_1、A 分别为主阀上、下腔作用面积；D、d 分别为主阀和先导阀阀座孔直径；α、φ 为主阀芯和先导阀芯半锥角；ϕ、l 分别为阻尼孔直径和长度；μ 为油液粘度；ρ 为油液密度；A_x 为先导阀座孔的面积，$A_x = \pi d^2/4$。

先导式溢流阀在先导阀前腔有一远控口 P_c，若接上电磁阀可以实现远控或多级调压。

2. 溢流阀的基本性能

（1）调压范围　在规定的范围内调节时，阀的输出压力能平稳地升降，无压力突跳或迟滞现象。

（2）压力-流量特性　在溢流阀调压弹簧的预压缩量调定之后，溢流阀的开启压力 p_k 即已确定，阀口开启后溢流阀的进口压力随溢流量的增加而略为升高，流量为额定值时的压力 p_s 最高，随着流量的减少阀口则反向趋于关闭；阀的进口压力降低，阀口关闭时的压力为 p_b。因摩擦力的方向不同，$p_b < p_k$。溢流阀的进口压力随流量变化而波动的性能称为压力-流量特性或启闭特性，如图 5-9 所示。压力流量特性的好坏用开启压力比 $n_k = p_k/p_s$、闭合压力比 $n_b = p_b/p_s$ 评价。显然调压偏差小好，n_k、n_b 大好，一般先导式溢流阀的 $n_k = 0.9 \sim 0.95$。

（3）压力超调量　当溢流阀由卸荷状态突然向额定压力工况转变或由零流量状态向额定压力、额定流量工况转变时，由于阀芯的运动惯性、粘性摩擦以及油液压缩性的影响，阀的进口压力将先迅速升高到某一峰值 p_{max} 然后逐渐衰减波动，最后稳定为额定压力 p_s。压力峰值与额定压力之差 Δp 称为压力超调量，一般限制超调量不得大于额定值的 30%。图 5-10

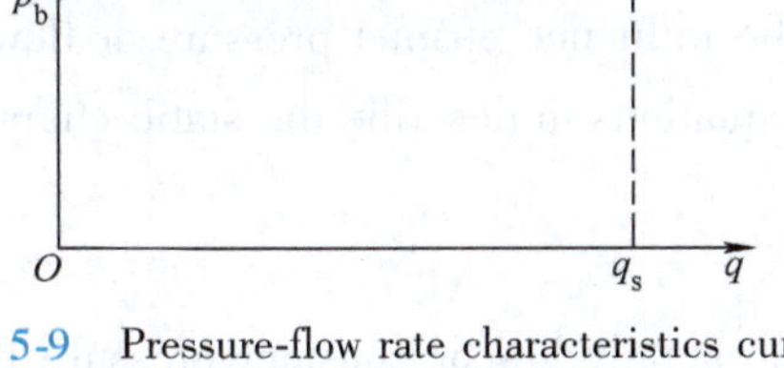

Fig. 5-9　Pressure-flow rate characteristics curves
（溢流阀的压力-流量特性曲线）

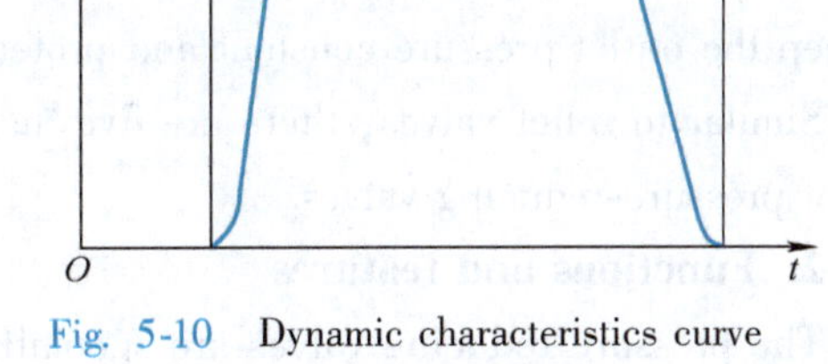

Fig. 5-10　Dynamic characteristics curve
（溢流阀的动态过程曲线）

closes so that just enough fluid flows through to maintain this desired pressure. The valve holds the desired reduced pressure at the outlet regardless of any fluctuation of the pressure at the inlet.

There are three kinds of pressure-reducing valves on request: ① to maintain outlet pressure invariableness, called fixed value pressure-reducing valves; ② to maintain a constant pressure differential between inlet and outlet ports, called fixed pressure differential pressure-reducing valves; ③to ensure that the outlet pressure is proportional to that of the inlet, called proportion pressure-reducing valves. The fixed values are applied widely. They will be introduced here.

1. Configuration and operating principle

Fig. 5-11 illuminates a pilot-operated pressure-reducing valve. It is the main valve orifice that reduces the pressure from inlet port p_1 to the outlet port p_2. At the same time, the pressure p_2 is introduced to the down port a_2 of the main valve via the passages on the valve body and the cover. The pressure p_2 then reaches to the upper port of the main valve core and the front port a_1 of the pilot through the damping orifices on the main valve core. The pilot valve is close when the pressure p_2 is lower than the spring setting under a small load. There is no oil flow through the damping orifice 3 on the main valve core and the pressures in the upper and down ports are equal. The main valve core stays at the lowest point under the action of spring, allowing fluid to flow unrestrictedly through the full-open orifice. The pressure p_2 increases with the outside load. When the pressure p_2 exceeds the setting of spring, the pilot valve is open, allowing the pressure p_2 to enter the upper port of the main valve core and also the pilot orifice via the damping orifice on the main valve core. The pressure p_2 then returns to the reservoir through the discharged ports. There is a pressure difference between the upper and down ports of the main valve due to damping orifice 3. Under this pressure difference the main valve core moves up by overcoming the upper spring force, the orifice is diminished and thus the outlet pressure-reducing is realized. When the valve is working, its outlet pressure is kept stable in a given value. For example, if the outlet pressure is decreased, the valve core moves down to enlarge the orifice and make the outlet pressure build up to a given value by weakening the action of pressure-reducing; whereas, if the outlet pressure is increased, the valve core moves up to diminish the orifice and make the outlet pressure fall down to a given value by enhancing the action of pressure-reducing. The outlet pressure p_2 can be changed simply by adjusting the preset of the spring.

The pilot valve of the pressure-reducing valve is similar to that of a relief valve. But the leaking oil from the spring port should be induced back to the reservoir due to the pressure existence in the inlet and outlet ports. Unlike a relief valve for the main valve parts, a pressure-reducing valve is normally open and the main valve core stays at the bottom under the spring force. The orifice is the largest in size and performs no flow-restricting, so no pressure-reducing function occurs in this case. It is the valve outlet pressurized oil that is induced to the front port of the pilot valve. The purpose is to keep the outlet pressure constant and protects it from the influence of inlet pressure or flow rate.

Similar to relief valves, there are five mathematical equations to describe the static characteristics of pressure-reducing valves.

2. Functions and features

The pressure reducing valves are specially designed to supply low or constant pressure fluid to some other part of a hydraulic secondary circuit from a high pressure source. The low pressure cir-

所示为溢流阀由零压、零流量过渡为额定压力、额定流量的动态过程曲线。

5.3.2　减压阀

减压阀是一种利用液流流过缝隙产生压力损失，使其出口压力低于进口压力的压力控制阀。按调节要求不同有：①用于保证出口压力为定值的定值减压阀；②用于保证进、出口压差不变的定差减压阀；③用于保证进、出口压力成比例的定比减压阀。其中定值减压阀应用最广，又简称为减压阀。这里只讨论出口定值减压阀。

1. 结构及工作原理

图 5-11 所示为先导式减压阀。进口压力油（压力为 p_1）经主阀阀口（减压口）流至出口，压力为 p_2。与此同时，出口压力油（压力为 p_2）经阀体、端盖上的通道进入主阀阀芯下腔 a_2，然后经主阀阀芯上的阻尼孔 3 到主阀阀芯上腔和先导阀的前腔 a_1。在负载较小、出口压力 p_2 低于调压弹簧所调定压力时，先导阀关闭，主阀阀芯阻尼孔 3 无液流通过，主阀阀芯上、下两腔压力相等，主阀阀芯在弹簧作用下处于最下端，阀口全开不起减压作用。若出口压力 p_2 随负载增大超过调压弹簧调定的压力时，先导阀阀口开启，主阀出口压力油（压力为 p_2）经主阀阀芯阻尼孔到主阀阀芯上腔与先导阀口，再经泄油口回油箱。因阻尼孔 3 的作用，主阀上、下两腔出现压差，主阀阀芯在压差作用下克服上端弹簧力向上运动，主阀阀口减小起减压作用。当阀口处于工作状态时，其出口压力始终维持在调定值。若出口压力减小，则阀芯下移，开大阀口，减压作用减弱，使出口压力回升到调定值；反之，若出口压力增大，则阀芯上移，关闭阀口，减压作用增强，使出口压力稳定不变。调节调压弹簧的预压缩量即可调节阀的出口压力。

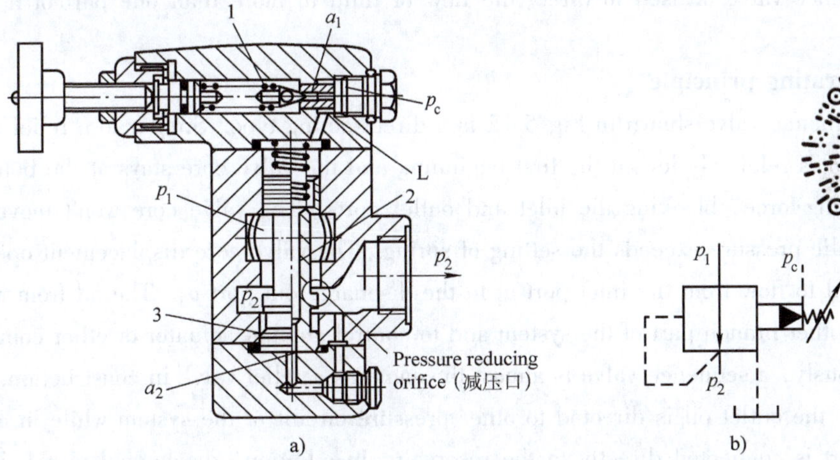

a)　　　　　　　　b)

Fig. 5-11　Pilot-operated pressure-reducing valve（先导式减压阀）

a）Configuration（结构图）　b）Graphic symbol（图形符号）

1—Valve core of pilot（先导阀阀芯）　2—Valve core of main valve（主阀阀芯）　3—Damping orifice（阻尼孔）

减压阀的先导阀与溢流阀的先导阀相似，由于其进、出油口均有压力，故弹簧腔的泄漏油需单独引回油箱。而主阀部分与溢流阀不同的是：在常位阀口常开，主阀阀芯在弹簧力作用下位于最下端，阀的开口最大，不起减压作用；引到先导阀前腔的是阀的出口压力油，以保证出口压力维持恒定，不受进口压力和通流流量的影响。

与溢流阀相似，同样可以用五个数学方程来描述它的静态特性。

cuits are sometimes used in clamping, oil lubricating circuits, or another form of mechanical device that may be damaged if the full-line pressure is applied. They are used quite widely in transfer machine equipment. It is necessary to note that the outlet pressure is related to external load: if the load-dependent pressure is lower than the preset pressure, then the outlet pressure is decided by the load. In this case, the pressure-reducing valve does not perform the reducing function, i. e, the pressure at the outlet is equal to that of the inlet. Thus the pressure-reducing valve cannot be operated when the outlet pressure is lower than the inlet pressure. Therefore the pilot-operated valve openning is the key to hold the desired reduced pressure at the outlet.

Compare with relief valves, their differences lie in: ① The pressure-reducing valve holds the outlet pressure constant, whereas the relief valve holds the desired reduced pressure at the inlet; ② The pressure-reducing valve is an normally open valve, while relief valve normally a close valve; ③ The pressure exists in both the inlet and outlet ports of pressure-reducing valves, so the leaking oil from the spring port of the pilot valve must be induced separately to the reservoir. But in relief valves, outlet pressure oil flows directly into the reservoir; the leaking oil from the spring port of the pilot valve discharges out through the inner passages of the valve body.

Similar to a relief valve, there is a remote port in a pressure-reducing valve which can be used for remote control or multiple-stage control by connecting it with a remote pressure-reducing valve.

5. 3. 3 Sequence valves

1. Functions

A sequence valve is used to direct the flow of fluid to more than one part of a circuit in sequence.

2. Operating principle

The sequence valve shown in Fig. 5-12 is a direct-acting type. Similar to a relief valve, the inlet pressure p_1 is relalively low at the first beginning and the valve core stays at the bottom under the action of spring force, blocking the inlet and outlet ports. The valve core won't move up until the down hydraulic pressure exceeds the setting of spring. The valve core displacement opens the orifice and allows oil to flow from the inlet port p_1 to the discharge out port p_2. The oil from port p_2 is connected to another branch part of the system and to control another actuator or other components operation. Obviously, a sequence valve is almost the same as a relief valve in construction. Their differences lie in: the outlet oil is directed to other pressure circuit of the system while in a relief valve, the outlet port is connected directly to the reservoir. In addition, discharged port L in a sequence valve must be connected to reservoir separately for the pressure existence in both the inlet and outlet ports. Fig. 5-13 shows a sequence valve of pilot operated type.

3. Control manners

According to different control pressures needed, sequence valve can be classified by the internal control and external control types. Internal control indicates that the open-close action of the valve core is controlled by inlet pressure, but by outside pressure for external control type. According to different arrangements of the upper cover or the bottom cover, four control manners can be obtained: internal control and external discharged, internal control and internal discharged, external

2. 功用和特点

减压阀在液压系统中主要用于降低与稳定某一支路的压力，如夹紧油路、润滑油路和控制油路。必须说明的是，减压阀的出口压力还与出口的负载有关，若因负载建立的压力低于调定压力，则出口压力由负载决定，此时减压阀不起减压作用，即减压阀保证出口压力恒定的条件是先导阀开启。

比较减压阀与溢流阀的工作原理和结构，可以将二者的差别归纳为以下三点：

1）减压阀保持出口压力为定值，而溢流阀则保持进口压力恒定。

2）阀不工作时，减压阀进、出油口相通；而溢流阀则进、出油口不通。

3）减压阀进、出油口都有压力，先导阀弹簧腔的泄漏油需单独引回油箱；而溢流阀的出口直接接回油箱，因此先导阀弹簧腔的泄漏油经阀体内流道内泄至出口。

与溢流阀相同的是，减压阀也可以在先导阀的远程调压口 p_c 处接远程调压阀以实现远控或多级调压。

5.3.3 顺序阀

1. 顺序阀的功能

顺序阀是一种利用压力控制阀口通断的压力阀，因用于控制多个执行元件的动作顺序而得名。

2. 工作原理

图 5-12 为直动式顺序阀。当进油口 P_1 压力较低时，阀芯在弹簧力的作用下处于下端位置，进油口和出油口不相通。当作用在控制小柱塞下端的油液压力大于弹簧的预压紧力时，顶起阀芯向上移动，阀口打开，油液便经阀口 P_1 从出油口 P_2 流出，从而操纵另一执行元件或其他元件动作。由此可见，顺序阀和溢流阀的结构基本相似，不同的是顺序阀的出油口通向系统的另一压力油路，而溢流阀的出油口通油箱，此外，由于顺序阀的进、出油口均为压力油，所以它的泄油口 L 必须单独外接油箱。图 5-13 所示为先导式顺序阀。

Fig. 5-12 Sequence valve operated directly（直动式顺序阀）

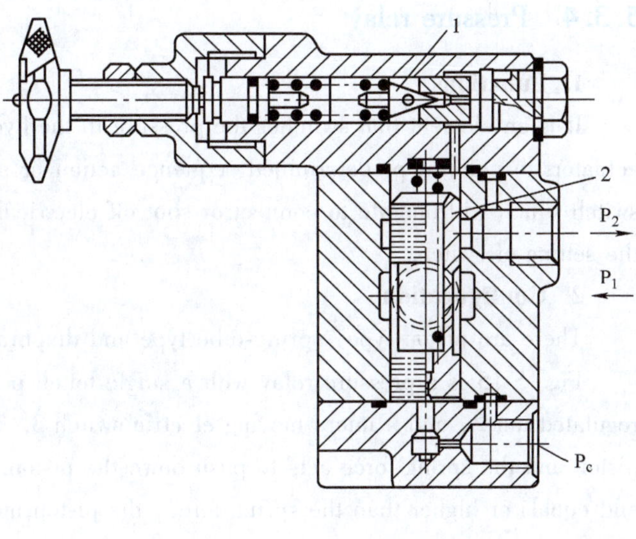

Fig. 5-13 Sequence valve operated by pilot（先导式顺序阀）

1—Pilot valve core（先导阀阀芯） 2—Main valve core（主阀阀芯）

control and external discharged, and external control and internal discharged, as shown in Fig. 5-14.

4. Features

Sequence valve with an internal control and internal discharged type can be used as a counter-balance valve or a back pressure valve; sequence valve with external control and internal discharged type is as an unloading valve; sequence valve with external control and external discharged type is similar to a pilot-operated two-way, two-position valve.

For a sequence valve with internal control and external discharged type or a same type of relief valve, the common point is that they are normally close and the core open action depends mainly upon the inlet pressure.

Let's compare the sequence valve with internal control and external discharged type with the relief valve, their commons and differences can be concluded as follows:

The common points: A sequence valve of the same control type is normally close. Whether the core in the valve will be open or not depends on the inlet pressure.

The difference points: The outlet pressure oil is used to supply pressure for some other part of a hydraulic secondary circuit. If when the outlet pressure built up by the load, or by secondary part of the circuit, exceeds the setting of valve, the inlet pressure of the valve equals the outlet pressure. If when the hydraulic pressure acting on the valve core exceeds the sum of the spring force and the flow dynamic force, the valve is fully open. If when the load-dependent outlet pressure is lower than the setting of valve, the inlet pressure of the valve equals the setting of valve. The valve core is at a balanced position under the three action forces of the hydraulic pressure, the spring force and the flow dynamic force. The orifice size is certain and meets the pressure-flow rate equation. For the outlet pressure is beyond zero, the leaking oil from the spring port is separately induced to the reservoir (external discharged).

5.3.4 Pressure relay

1. Functions

It is an element that switches the pressure in the hydraulic system to an electric sign and allows actuators to perform predetermined sequence action or security. This is accomplished by its jiggle switch which will operate to connect or shut off electrical paths when the hydraulic pressure reaches the setting of switch.

2. Configuration

There are piston type, spring-tube type and diaphragm type, etc.

Fig. 5-15 is a pressure relay with a single touch point piston, which consists of the piston 1, a regulated nut 2, and a micro-moving electric switch 3. The pressure oil acts on the end area of the piston and the spring force acts to push down the piston. When the pressure of the oil on the piston end equals or higher than the spring force, the piston moves up and presses the micro-moving electric switch in turn, the switch is on. If the pressure of oil on the piston end is lower than the spring force, the piston returns to its original position and the switch is off.

3. 顺序阀的控制形式

依控制压力的不同，顺序阀可分为内控式和外控式两种。内控式是利用阀的进口压力来控制阀芯的启闭，外控式是利用外来的控制压力油控制阀芯的启闭。通过改变上盖或底盖的装配位置可以实现顺序动作的内控外泄、内控内泄、外控外泄、外控内泄等 4 种类型，如图 5-14 所示。

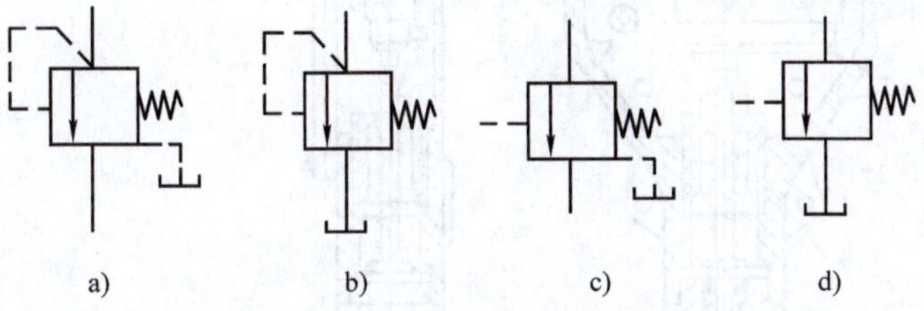

Fig. 5-14　Four control types of graphics symbols（顺序阀的四种控制形式）

a）Internal control external discharged（内控外泄）　b）Internal control internal discharged（内控内泄）

c）External control external discharged（外控外泄）　d）External control internal discharged（外控内泄）

4. 特点

顺序阀的四种控制形式如图 5-14 所示，其中内控内泄式用在系统中作平衡阀或背压阀；外控内泄式用做卸载阀；外控外泄式相当于一个液控二位二通阀。将其特点归纳如下：

内控外泄式顺序阀与溢流阀的相同之点是阀口常闭，由进口压力控制阀口的开启。区别是内控外泄式顺序阀调整压力油去工作，当因负载建立的出口压力高于阀的调定压力时，阀的进口压力等于出口压力，作用在阀芯上的液压力大于弹簧力和液动力，阀口全开；当负载所建立的出口压力低于阀的调定压力时，阀的进口压力等于调定压力，作用在阀芯上的液压力、弹簧力、液动力平衡，阀的开口一定，满足压力流量方程。因阀的出口压力不等于零，因此弹簧腔的泄漏油需单独引回油箱，即外泄。

5.3.4　压力继电器

1. 功能

压力继电器是一种将液压系统的压力信号转换为电信号输出的元件。其作用是：当液压系统压力升高到压力继电器的调整值时，通过压力继电器内的微动开关动作，接通或断开电气线路，实现执行元件的顺序控制或安全保护。

2. 结构特点

压力继电器按结构特点可分为柱塞式、弹簧管式和膜片式等。

图 5-15 所示为单触点柱塞式压力继电器，压力油作用在柱塞的下端，当系统压力升高达到或超过调定的压力值时，柱塞上移压微动开关触头，接通或断开电气线路。当系统压力小于调定值时，在弹簧力作用下，微动开关触头复位。其设定压力值靠螺帽调节。

5.4　流量控制阀

流量控制阀是依靠改变阀口通流截面积大小，即改变液阻实现流量调节的阀。流量控制阀有节流阀、调速阀、溢流节流阀和分流集流阀等。

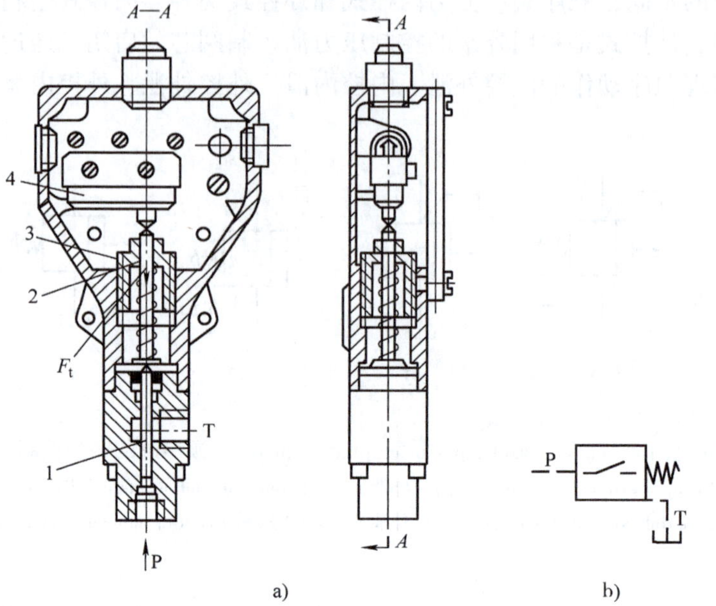

Fig. 5-15 Pressure relay with single touch point piston(单触点柱塞式压力继电器)

a) Configuration(结构图) b) Graphics symbol (图形符号)

1—Piston(柱塞) 2—Mandril (顶杆) 3—Regulating screw(调节螺帽) 4—Micro electrical switch(微动开关)

5.4 Flow Control Valves

The flow control valves are used to control the flow rate by changing the orifice flow area (hydraulic resistance). Types of flow control valves incorporate throttle, speed-regulating, relief-throttle, dividing-collecting, etc.

5.4.1 Operation

The pressure-flow rate equation discussed in Chapter 2 can be rewritten as follows

$$q = K_L A \Delta p^m \tag{5-9}$$

Where K_L is the throttle coefficient, usually a constant; A is the across-sectional area of the orifice or clearance; Δp is the pressure drop of the orifice or clearance; m is an exponent decided by the orifice shape, $0.5 \leqslant m \leqslant 1$, m is close to 0.5 for thin wall orifice and 1 for a slot.

The common configurations of throttling elements are shown in Fig. 5-16. The section area of poppet (core) - shaped $A = \pi dx \sin\beta$; triangle - notch $A = nx^2 \sin^2\alpha \tan\varphi$; rectangle $A = nb(x - x_d)$ and triangle cone - shaped $A = nx^2 \tan\beta$, where n is the number of orifices.

According to equation (5-9), under a given K_L, Δp and m, the flow rate can be adjusted by changing the flow area A (the hydraulic resistance), which is the operation principle of a flow control valve.

5.4.1　原理

在第 2 章中讨论过液流通过不同孔口的阀口流量公式，它可以写成通用表达式

$$q = K_L A \Delta p^m \tag{5-9}$$

式中，K_L 为节流系数，一般可视为常数；A 为孔口或缝隙的通流截面积；Δp 为孔口或缝隙的前后压差；m 为由节流口形状决定的指数，$0.5 \leqslant m \leqslant 1$，近似薄壁时 m 接近 0.5，近似细长孔时接近于 1。

几种节流口结构形式如图 5-16 所示。图中锥形结构的 $A = \pi dx\sin\beta$，三角槽形结构的 $A = nx^2\sin^2\alpha\tan\varphi$，矩形结构的 $A = nb(x - x_d)$，斜三角形结构的 $A = nx^2\tan\beta$（上列式中 n 为节流槽的个数）。

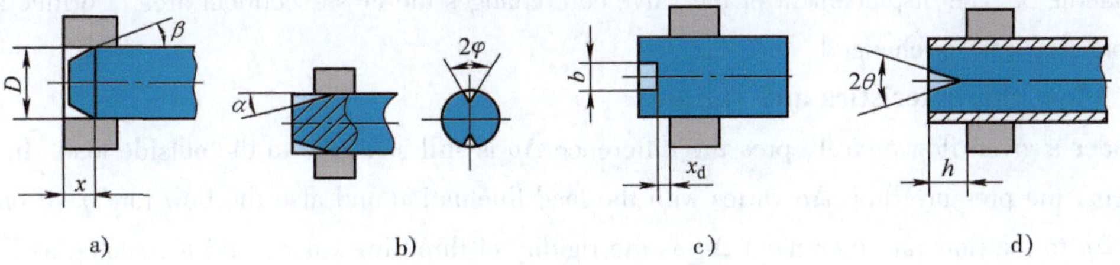

Fig. 5-16　Orifices types used in throttle valves（几种节流口的结构型式）

a) Poppet（core）-shaped（锥形）　b) Triangle-notch（三角槽形）

c) Rectangle（矩形）　d) Triangle cone-shaped（三角形）

由式（5-9）可知：在 K_L、Δp、m 一定时，改变通流截面积 A，即改变液阻的大小，实现流量调节，这就是流量控制阀的控制原理。

5.4.2　节流阀

节流阀是一种最简单的流量控制阀，其实质相当于一个可变节流口，即一种借助于控制机构使阀芯相对于阀体孔运动改变阀口过流面积的阀。常用在定量泵节流调速回路中实现调速。

1. 结构与原理

图 5-17 所示为一种普通节流阀。这种节流阀的通道呈轴向三角槽式。压力油从进油口 P_1 流入孔道 a 和阀芯 1 左端的三角槽进入孔道 b，再从出油口 P_2 流出。调节手柄 3，可通过

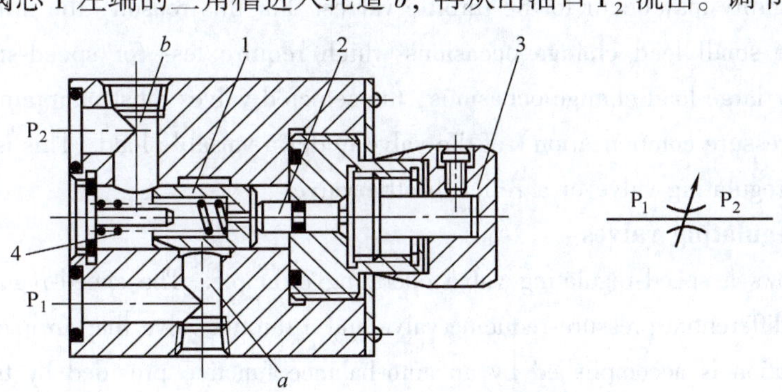

Fig. 5-17　Throttling valve（节流阀）

1—Valve core（阀芯）　2—Handspike（推杆）　3—Handle（手柄）　4—Spring（弹簧）

5.4.2 Throttle valves

A throttle valve is a simple flow control valve. A throttle valve is virtually equal to a variable orifice. The orifice size is changed by the valve core movement, which is controlled by a certain mechanism. A throttle valve is quite widely applied in the ration pump and the throttling regulating speed circuit.

1. Configuration and operating principle

Fig. 5-17 shows the construction and graphic symbol of a throttle valve. An axial triangle-notch is used as the throttle channel. The pressure oil flows following: P_1 (inlet port)→a (orifice channel) and the triangle-notch in the left of the valve core 1→*b* (orifice channel)→P_2 (outlet port) and then flow out from outlet P_2. The valve core 1 can be move in the axial direction under the action of handspike 2 by adjusted handle 3. The displacement of the valve core changes the cross-sectional area of orifice a and thus the output flow rate is changed.

2. Flow characteristics and rigidity

Under a given flow area A, pressure difference Δp is still sensitive to the outside load. In a hydraulic system, the pressure drop Δp varies with the load fluctuation and also the flow rate q. If one sets the ratio of Δp to the flow rate increment Δq as the rigidity of throttling valve, and it is noted as T, then the T can be obtained by

$$T = \frac{\partial \Delta p}{\partial q} = \frac{\Delta p^{1-m}}{K_L A m} \tag{5-10}$$

The larger the T value is, the better ability of the valve is. A thin-wall orifice is usually adopted due to its higher rigidity ($m = 0.5$) than that of a slot ($m = 1$). In addition, a higher Δp contributes to a higher rigidity, but Δp will also increase the pressure loss and result in blocked due to the smaller area. Generally $\Delta p = 0.15 \sim 0.4\text{MPa}$.

3. Application

Throttle valves are usually combined with fixed delivery pumps, relief valves to constitute a throttle speed-regulating system. The speed can be adjusted by changing the throttle orifice size.

5.4.3 Speed-regulating valves and relief-throttle valves

As described above, the flow rate through the orifice is sensitive to the pressure drop of the valve, so poor rigidity may occur to the throttle valves. For this reason, the throttle valves are only applicable for the small load change occasions which require less for speed-stability. To achieve speed-stability for large load change occasions, the selected valves must maintain a constant pressure drop Δp (i. e., pressure compensation) of the valve under a variable load. This is the operating principle of a speed-regulating valve or a relief-throttle valve.

1. Speed-regulating valves

Fig. 5-18 shows a speed-regulating valve operating principle. The speed-regulating valve is composed by a fixed differential pressure-reducing valve and a throttle valve that are mounted in series. The pressure compensation is accomplished by an auto-balance function provided by the pressure-reducing valve. This allows the valve pressure drop Δp to keep constant and thus the flow rate. When pressurized oil enters the speed-regulating valve, it passes through the orifice x of the pressure-reducing valve first to

推杆 2 使阀芯作轴向移动。改变节流口的通流截面积即可调节流量。

2. 流量特性与刚性

当节流阀的通流截面积 A 调定后，由于外负载的变化，引起阀前后压差 Δp 变化，也会导致流经阀口的流量 q 变化，即流量不稳定。一般定义节流阀通流截面积 A 一定时，节流阀前后压差 Δp 的变化量与流经阀的流量变化量之比为节流阀的刚性 T，即

$$T = \frac{\partial \Delta p}{\partial q} = \frac{\Delta p^{1-m}}{K_L A m} \tag{5-10}$$

刚性 T 越大，节流阀的性能越好。因此节流口制成薄壁孔型的（$m=0.5$）比制成细长孔型的（$m=1$）刚性大，故多作节流阀的阀口。另外，Δp 大有利于提高节流阀的刚性，但 Δp 过大，不仅造成压力损失的增大，而且可能导致阀口因面积太小而堵塞，因此一般取 $\Delta p = 0.15 \sim 0.4\text{MPa}$。

3. 应用

节流阀在液压系统中，主要与定量泵、溢流阀组成节流调速系统。调节节流阀的开口，便可以实现调速。

5.4.3　调速阀与溢流节流阀

通过节流阀阀口的流量因阀口前后压差变化而变化，刚性差，因此仅适用于执行元件工作负载变化不大且对速度稳定性要求不高的场合。为解决负载变化大的执行元件的速度稳定性问题，应采取措施保证负载变化时，使节流阀前后的压差 Δp 为常量。这就是调速阀和溢流节流阀的基本原理。

1. 调速阀

图 5-18 所示的调速阀由节流阀前面串联一个定差减压阀组合而成。由减压阀的自动平衡作用来进行压力补偿，使节流口前后压差 Δp 保持不变，从而使所通过的流量稳定。当压力油进入调速阀后，先经过定差减压阀的阀口（开度为 x），压力由 p_1 减至 p_2，然后经过节流阀阀口（开度为 y）流出，出口压力为 p_3。从图中可以看到，节流阀进、出口压力 p_2、p_3 经过阀体上的流道分别被引到定差减压阀阀芯的上、下两端（p_3 引到阀芯上弹簧端，p_2 引到阀芯下无弹簧端），作用在定差减压阀芯上的力有液压力、弹簧力和液动力。

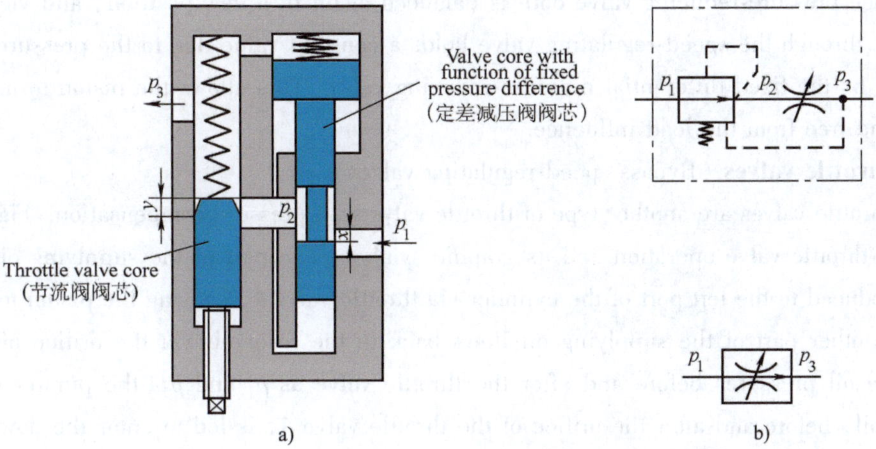

Fig. 5-18　Regulating speed valve（调速阀）

a）Operating principle（工作原理）　b）Graphics symbol（图形符号）

reduce the pressure from p_1 to p_2. The pressure p_2 then flows out via throttle orifice y, the outlet pressure of the throttle valve is p_3. As can be seen from the figure, the inlet and outlet pressures p_2 and p_3 are introduced to the two ends of the pressure-reducing valve core via the passages in the valve body (p_3 to the spring chamber, and p_2 to the other end). The forces acting on the fixed differential pressure-reducing valve core include: hydraulic pressure, spring force and flow dynamic force.

Static equations of a speed-regulating valve in operation are as follows:

$$p_2A = p_3A + F_t - F_s \tag{5-11}$$

$$q_1 = C_{d1}\pi dx\sqrt{\frac{2}{\rho}(p_1 - p_2)} \tag{5-12}$$

$$q_2 = C_{d2}A(y)\sqrt{\frac{2}{\rho}(p_2 - p_3)} \tag{5-13}$$

$$q_1 = q_2 = q \tag{5-14}$$

Where A is the action area of the pressure reducing valve; d is the diameter of pressure reducing valve orifice; q_1 is the flow rate through the pressure reducing valve orifice; q_2 is the flow rate through the throttle valve orifice; $A(y)$ is the across-sectional area of the throttle orifice; F_t is the spring force; F_s is the flow dynamic force.

For a given $A(y)$ and q, the inlet and outlet pressure drop $(p_2 - p_3)$ of the throttle is constant, according to the above equations for a fixed pressure-reducing valve core, i. e., $p_2 - p_3 = (F_t + F_s)/A$ = constant.

Action principle: Suppose the inlet pressure p_1 is a constant, the outlet pressure p_3 increases with the load. The increase of p_3 might lead to a sudden drop of the inlet and outlet pressure drop $(p_2 - p_3)$ in the speed-regulating valve and threaten to the balance state of the fixed differential pressure-reducing valve core. The down hydraulic pressure acting on the throttle valve core increases, causing the valve core to move downwards and enlarging the throttle orifice. The pressure drop of the throttle orifice falls down but p_2 is also increased. The result is that the pressure difference of the throttle valve $(p_2 - p_3)$ returns back to its setting. When the pressure drop is $p_2 - p_3 = (F_t + F_s)/A$, the fixed difference pressure-reducing valve core is balanced again in a new position, and vice versa. The flow rate through the speed-regulating valve holds a constant value due to the pressure compensation offered by the fixed differential pressure-reducing valve. This allows the piston to move at a stable speed and free from the load influence.

2. Relief-throttle valves (Bypass speed-regulating valves)

The relief-throttle valves are another type of throttle valve with pressure compensation. Fig. 5-19 indicates a relief-throttle valve operation and its graphic symbol. The part of the supplying oil from the pump is introduced to the left port of the cylinder via throttle valve 4, causing the piston to move in the right. The other part of the supplying oil flows back to the reservoir via the orifice of relief valve 3. Note the oil pressures before and after the throttle valve as p_1 and p_2 (the pump's output pressure). The oil, before and after the orifice of the throttle valve 4, is led to enter the down port c and the upper port a of the valve core of relief valve 3 respectively. The pressure p_2 and thus the pressure in port a increase with the load F on the cylinder piston, causing valve core 3 to move down

调速阀工作时的静态方程如下：

1）定差减压阀阀芯受力平衡方程为

$$p_2A = p_3A + F_t - F_s \tag{5-11}$$

2）流经定差减压阀阀口流量为

$$q_1 = C_{d1}\pi dx\sqrt{\frac{2}{\rho}(p_1 - p_2)} \tag{5-12}$$

3）流经节流阀阀口的流量为

$$q_2 = C_{d2}A(y)\sqrt{\frac{2}{\rho}(p_2 - p_3)} \tag{5-13}$$

4）流量连续性方程为

$$q_1 = q_2 = q \tag{5-14}$$

式中，A 为定差减压阀阀芯作用面积；d 为定差减压阀阀口处台肩直径；q_1 为流经定差减压阀阀口的流量；q_2 为流经节流阀阀口的流量；$A(y)$ 为节流阀开口截面积；F_t 为弹簧力；F_s 为液动力。

当上列方程成立时，对应于一定的节流阀开口面积 $A(y)$、流经阀的流量 q，因为节流阀的进、出口压差 (p_2-p_3) 由定差减压阀阀芯受力方程确定为一定值，即 $p_2-p_3=(F_t+F_s)/A=$ 常量。其作用原理：假定调速阀的进口压力 p_1 为定值，当出口压力 p_3 因负载增大而增加，导致调速阀的进出口压力差 (p_2-p_3) 突然减小的同时，因 p_3 的增大势必破坏定差减压阀阀芯原有的受力平衡状态，于是作用在减压阀阀芯上端的液压力增加，阀芯下移，减压阀的开度 x 加大，减压作用减弱，压降减小，因而使 p_2 也随之升高，结果使节流阀前后的压差 (p_2-p_3) 保持不变，当 $p_2-p_3=(F_t+F_s)/A$ 时，定差减压阀阀芯在新的位置平衡，反之亦然。由此可知，因定差减压阀的压力补偿作用，使通过调速阀的流量恒定不变，活塞的运动速度稳定，不受负载变化的影响。

2. 旁通型节流阀（溢流节流阀）

旁通型节流阀也是一种压力补偿型节流阀，图 5-19a 所示为其工作原理图。从液压泵输出的油液一部分经节流阀 4 进入液压缸左腔推动活塞向右运动，另一部分经溢流阀 3 的溢流口流回油箱，溢流阀 3 阀芯的上端 a 腔同节流阀 4 的出油口油液相通，其压力为 p_2；b 腔和下端 c 腔同节流阀 4 的进油口油液相通，其压力即为泵的输出压力 p_1，当液压缸活塞上的负载力 F 增大时，压力 p_2 升高，a 腔的压力也升高，使溢流阀 3 的阀芯下移，关小溢流口，这样就使液压泵的供油压力 p_1 增加，从而使节流阀 4 的进、出口压差 (p_1-p_2) 基本保持不变；同理，当负载力减小时，压力 p_2 下降，由于溢流阀 3 的阀芯相应动作，也就是 (p_1-p_2) 基本保持不变，这种溢流节流阀一般附带一个安全阀 2，以避免系统过载，图 5-19b、c 所示为该阀的图形符号。

溢流节流阀是通过 p_1 随 p_2 的变化来使流量基本上保持恒定的，它与调速阀虽都具有压力补偿的作用，但其组成调速系统时是有区别的，调速阀无论装在执行元件的进油路上或回油路上，执行元件上负载变化时，液压泵出口处压力都由定差减压阀保持不变，而溢流节流阀是通过 p_1 随 p_2（负载的压力）的变化来使流量基本上保持恒定的。因而使用溢流节流阀具有功率损耗低。发热量小的优点。但是，溢流节流阀中通过的流量比调速阀大（一般是系统的全部流量），阀芯运动时的阻力较大，弹簧较硬，其结果使节流阀前后压差 Δp 加大（需达 0.3～0.5MPa），因此它的稳定性稍差。

and reduce the relief orifice. In this case supplying pressure p_1 increases and the pressure drop $(p_1 - p_2)$ via the throttle valve 4 maintains constant. Similarly, when the load reduces, the pressure p_2 falls down but $(p_1 - p_2)$ still keeps constant due to the corresponding action of the valve core of relief valve 3. This type of relief-throttle valve usually incorporates a safety valve 2 to avoid overload in the system. The graphic symbol is shown in Fig. 5-19b and Fig. 5-19c.

In a relief-throttle valve, the inlet pressure p_1 increases with p_2; it is this operation that keeps the flow rate constant. Similar to a speed-regulating valve, a relief-throttle also performs the function of pressure compensation. Their systems, however, are not exactly the same in mechanics. Whether the speed-regulating valve is mounted in the inlet or the returned lines of a cylinder, a constant output pressure, regardless of the load changing, is guaranteed by the pressure-reducing valve. In a system with a relief-throttle valve, the pressure p_1 varies with p_2 (the load pressure). A relief-throttle valve has the advantages of low power loss and low heat generated. But the flow rate through a relief-throttle valve is larger (usually equals to the total flow rate of the system) than that through a speed-regulating valve. The resistance against the valve core movement is larger and the spring has a higher rigidity, which increases the pressure difference Δp over the throttle valve (it is usually up to 0.3-0.5MPa). So a system with a relief-throttle valve is less stable than that with a speed-regulating valve.

5.5 Manifold Valves and Cartridge Valves

5.5.1 Manifold valves

The manifold valves are designed on the basis of plate valves. The upper and down surfaces of each hydraulic valve are made the same as the underside of a plate valve as shown in Fig. 5-20a. A single manifold valve operates the same as the common valve in operation. But the difference is that each manifold valve has four ports P, A, B and T. These four ports are not only connected to each other but also connected with the corresponding ports on its upper and down valves. The manifold valves of the same specification are identical in ports position, connection or installation dimension. For a system installation, various manifold valves with the same specification but different functions are spliced together on a requested sequence on the hydraulic system chart, see Fig. 5-20b. In a system composed by manifold valves, the oil ports connected with the actuators are set on the bottom board, while the directional control valves are mounted at the highest position; and the others are mounted between them by bolts. Usually, a group of manifold valves only control one actuator. At many actuators occasion, many groups of manifold valves are mounted in parallel vertically and connected to the bottom board. In a hydraulic system with manifold valves, other connection devices are not needed for each unique valve connection. It is compact in design, small in size, leakage or pressure loss, particularly convenient and flexible to restructure for another hydraulic system. The manifold valves have been standardized. A hydraulic system chart is enough for design or install. It takes short time for system design and manufacture. Because of the advantages mentioned above, the manifold valves system has found a wide application in industries.

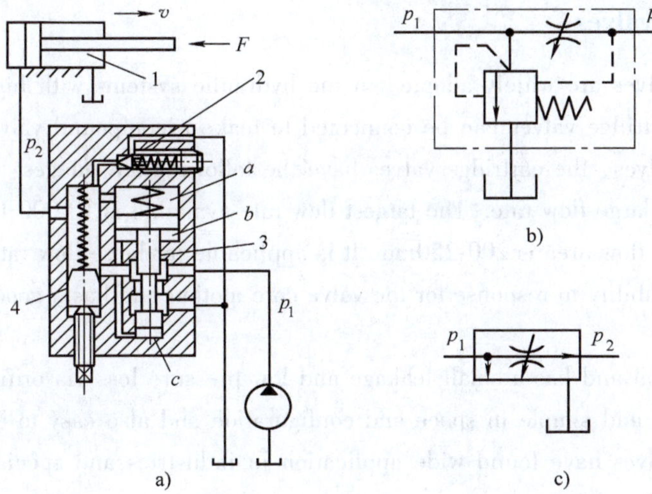

Fig. 5-19 Bypass speed-regulating valve（旁通型节流阀）

a) Operating princple（工作原理） b) Graphics symbol（图形符号） c) Simplified graphics（简化图形）

1—Actuator（液压缸） 2—Sefety valve（安全阀） 3—Relief valve（溢流阀） 4—Throttle valve（节流阀）

5.5 叠加阀和插装阀

5.5.1 叠加阀

叠加阀是以板式阀为基础，将各种液压阀的上、下面都做成像板式阀底面那样的连接面，做成叠装式结构，如图 5-20a 所示。单个叠加阀的工作原理与普通阀完全相同，所不同的是每个叠加阀都有四个油口 P、A、B、T，它除了具有液压阀的功能外，而且还起阀与阀之间油路通道作用。相同规格的各种叠加阀的油口位置、连接安装尺寸都相同。组成系统时，将相同规格的各种功能的叠加阀，按液压系统图的一定顺序叠加起来，即可组成叠加阀系统图，如图 5-20b 所示。在叠加阀组成的系统中，与执行元件连接的油口开在最下边的底板上，换向阀安装在最上面位置，其他的阀通过螺栓均安装在它们之间。通常一组叠加阀只控制一个执行元件，若系统有多个执行元件，可将多个叠加阀组竖立并排安装在串联底板上。用叠加阀组成的液压系统，阀与阀之间不需要其他连接体，因而结构紧凑，体积小，系统的泄漏及压力损失较小，尤其是液压系统更改较方便、灵活。叠加阀为标准的元件，设计中仅需绘出叠加阀系统图，即可进行安装，系统设计及制造周期短，应用广泛。

5.5.2 插装阀

插装阀在高压大流量的液压系统中应用很广。其元件已标准化，将几个插装式元件组合在一起便可组成复合阀。与普通液压阀相比，它有如下优点：

1）通流能力大，特别适用于大流量场合，它的最大通径可达 250mm，通过的最大流量可达 10 000L/min。

2）阀芯动作灵敏、抗堵塞能力强。

3）密封性好，泄漏小，油液流经阀口的压力损失小。

5.5.2 Cartridge valves

The cartridge valves are widely adopted in the hydraulic systems with high pressure and large flow rate. Several cartridge valves can be connected to make a compound valve. Compared with the common hydraulic valves, the cartridge valves have the following advantages:

1) It provides a large flow rate. The largest flow rate available is 10 000 L/min, and the maximum diameter for the flow area is 200-250mm. It is applicable for large flow rate occasions required.

2) It has acute ability to response for the valve core motion and has a good ability for anti-blocking.

3) It perfects seal and has a small leakage and low pressure loss via orifices.

4) It is compact and simple in space and configuration and also easy to be standardized.

The cartridge valves have found wide application in industries and specially in large flow and non-mineral oil industries.

1. Operating principle and their basic units

The basic units of a cartridge valve consist of a valve core, a bushing, a spring, the seals and so on. There are three different configurations, directional control units (Fig. 5-21a), pressure control units (Fig. 5-21b), and flow control units (Fig. 5-21c) to meet the different control needs. All include three ports: two are main ports A and B, and the third is control port X.

If we note the diameter of the valve core as D, the diameter of the valve seat hole as d, and the acting areas of ports A, B and X as A_A, A_B, and A_X respectively, then

$$A_A = \frac{\pi d^2}{4}, A_B = \frac{\pi(D^2 - d^2)}{4}, A_X = \frac{\pi D^2}{4}$$

The non-dimensional areas (area ratios)

$$a_{AX} = A_X/A_A, a_{BX} = A_X/A_B$$

For the directional control valve units (Fig. 5-21a), the half cone angle $\alpha = 45°$ and the nondimensioned area is $a_{AX} = a_{BX} = 2$. The area A equals area B, which completes double flowing.

For the pressure valve units, $a_{AX} = 1$ because of spool structure in reducing pressure valve, B is the inlet port and A the outlet; For relief and sequence valves, $\alpha = 15°$, $a_{AX} = 1.1$, port A is the inlet port and B the outlet.

For the flow valve units, $a_{AX} = 1$or 1.1, usually port A is the inlet port and B the outlet.

The two-way cartridge valve gets its name from the fact that the cartridge units have two ports used as the inlet and the outlet respectively. The open-close action of the valve depends upon the forces acting on the valve core. If we note the pressures at port A, B and X as p_A, p_B and p_X, the resilience on the spring as F_t, then

$$p_X A_X + F_t > p_A A_A + p_B A_B, \text{ valve closed}$$
$$p_X A_X + F_t \leqslant p_A A_A + p_B A_B, \text{valve opened}$$

From the above two equations, the hydraulic pressure p_X in port X can be controlled to connect or block ports A or B. If X port is connected to the reservoir (unloaded) $p_X = 0$ and the valve is opened; if X port is connected to inlet, $p_X = p_A$ or $p_X = p_B$ and the valve is closed. A controllable X port is called as the pilot operated port.

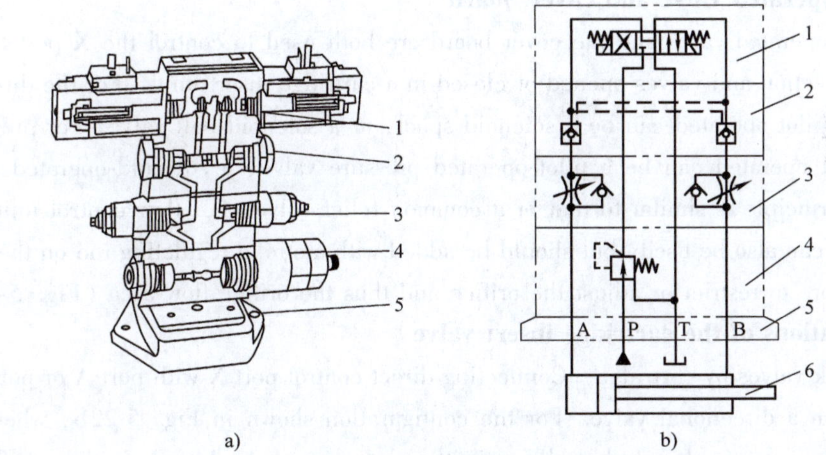

Fig. 5-20 Manifold valve（叠加阀）

a) Configuration（结构图） b) System（系统图）

1—Three-position，four-way solenoid operated valve（三位四通电磁换向阀）

2—Bidirectional hydraulic lock（双向液压锁） 3—Two-inlet-port check-throttle valve（双口进油路单向节流阀）

4—Pressure-reducing valve（叠加式减压阀） 5—Soleplate（底板） 6—Hydraulic actuator（液压缸）

4）结构紧凑、简单，易于实现标准化。特别是在一些大流量及介质为非矿物油的场合，优越性更为突出。

1. 工作原理和基本组件

插装阀基本组件由阀芯、阀套、弹簧和密封圈组成。根据其用途不同分为方向阀组件（图5-21a）、压力阀组件（图5-21b）和流量阀组件（图5-21c）三种。三种组件均有两个主油口A和B及一个控制油口X。

设阀芯直径为D、阀座孔直径为d，则油口A、B、X的作用面积A_A、A_B、A_X分别为

$$A_A=\frac{\pi d^2}{4},A_B=\frac{\pi(D^2-d^2)}{4},A_X=\frac{\pi D^2}{4}$$

面积比为

$$a_{AX}=A_X/A_A,a_{BX}=A_X/A_B$$

方向阀组件的阀芯半锥角$\alpha=45°$，面积比$a_{AX}=a_{BX}=2$，即油口A和B的作用面积相等，油口A、B可双向流动。

压力阀组件中减压阀阀芯为滑阀，即$a_{AX}=1$，油口B进油，油口A出油；溢流阀和顺序阀的阀芯半锥角$\alpha=15°$，面积比$a_{AX}=1.1$，油口A为进油口，油口B为出油口。

流量阀组件面积比$a_{AX}=1$或1.1，一般A口为进油口，B口为出油口。

因插装阀组件有两个进出油口，因此又称之为二通插装阀。若油口A、B、X的压力分别为p_A、p_B和p_X、阀芯上端的复位弹簧力为F_t，则工作时阀口开启或者关闭取决于阀芯的受力状况，即有：当$p_XA_X+F_t>p_AA_A+p_BA_B$时，阀口关闭；当$p_XA_X+F_t\leqslant p_AA_A+p_BA_B$时，阀口开启。

从此可以看出，改变控制口X的油液压力p_X，可以控制A、B油口的通断。如油口X通油箱，则$p_x=0$，阀口开启；如油口X与进口相通，则$p_X=p_A$或$p_X=p_B$，阀口关闭。改变油口X通油方式的阀称为先导阀。

2. Pilot-operated valve and cover board

The pilot-operated valve and the cover board are both used to control the X port to realize the oil circuit connection and valves opened or closed in a cartridge insert units. For the directional control units, the pilot operated can be a solenoid spool, or a solenoid ball valve. For pressure control units, the pilot operated can be a pilot-operated pressure valve, a solenoid-operated spool valve. Their control principle is similar to that of a common relief valve. For flow control units, solenoid-operated valve can also be used, but should be added with a stroke regulating rod on the cover board for the valve core to restrict or adjust the orifice and thus the orifice flow area (Fig. 5-21c).

3. Applications of the cartridge insert valve

(1) Check valves by cartridge Connecting direct control port X with port A or port B can buid a check valve in a directional valve. For the configuration shown in Fig. 5-22b, when fluid flows from A to B, the valve is closed, but also permitting a leakage from A to B via the annular clearance between the top of the valve core and the bushing hole. The sealing performance is not so good as that shown in Fig. 5-22a.

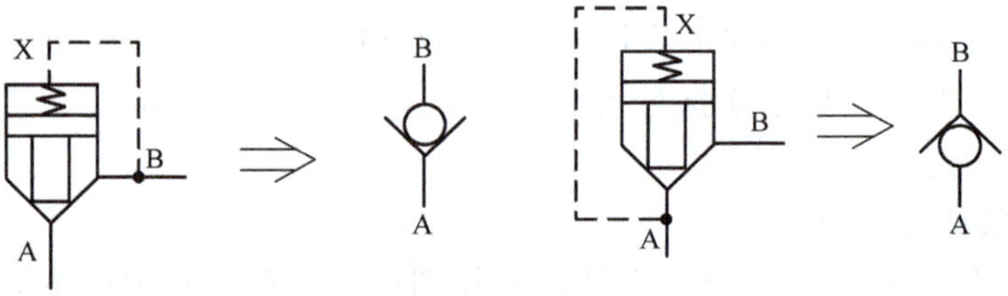

a) b)

Fig. 5-22 Cartridge insert check valve (插装式单向阀)

(2) Two-position, two-way valves by cartridge Fig. 5-23 shows a two-position, three-way directional control valve unit used to control oil direction. The pilot operated is a solenoid spool valve. As is shown in Fig. 5-23a, when the solenoid is not energized, port X is connected to the reservoir via the left port (normal condition) of the two-position, two-way valve. $p_x = 0$, and the valve can be opened to allow fluid to flow whether the oil comes from port A or port B. When the solenoid is energized, the valve works at its right position and port X is connected to port A. The hydraulic oil from port B opens the valve core and allows fluid flow. The valve cannot be opened by the oil from port A. In this concept, the valve shown in Fig. 5-23a is functionally the same as a check valve which only allows flow from port B to port A. Unlike the design shown in Fig. 5-23a, when the two-position, three-way valve shown in Fig. 5-23b works at its right position, the pressure in port X is always equal to the higher pressure due to the action of 'shuttle valve'. In this case, whether the oil passes through port A or port B, the valve is closed and ports A and B are blocked.

(3) Two-position, three-way valves by cartridge Fig. 5-24 shows a three-way insert cartridge valve consisting of two directional units in parallel connection with one pressure port P, one operating port A and one return port T. The flow directions through these two directional control valves are controlled by a two-position, four-way solenoid spool (pilot-operated valve). When the solenoid Y is

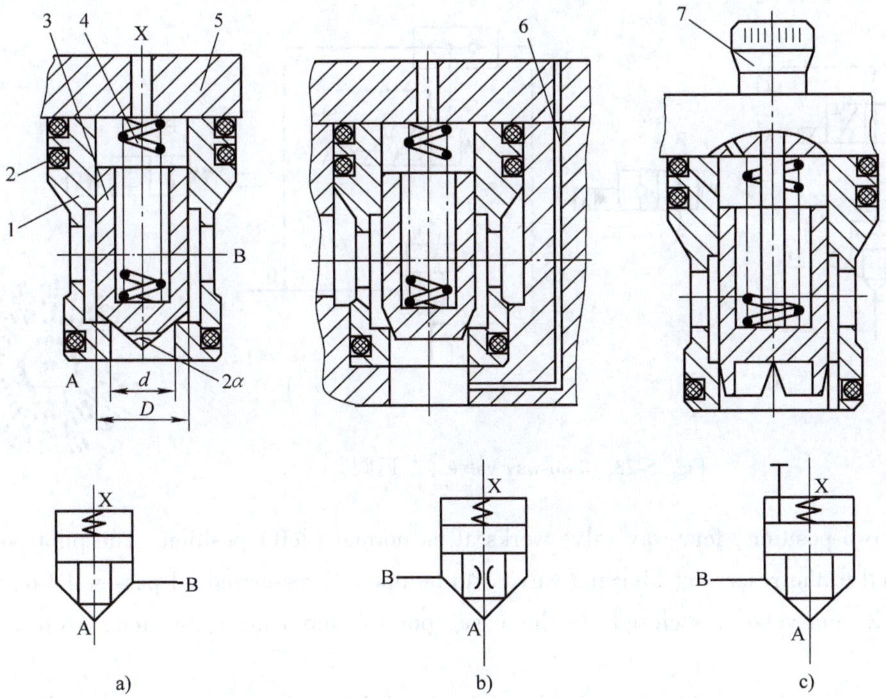

Fig. 5-21　Basic cartridge inserts units（插装阀基本组件）

a）Directional control valve units（方向阀组件）

b）Pressure valve units（压力阀组件）　c）Flow rate units（流量阀组件）

1—Bushing（阀套）　2—Seal ring（密封圈）　3—Poppet（core）（阀芯）　4—Spring（弹簧）

5—Cover（盖板）　6—Resistance hole（阻尼孔）　7—Poppet（core）regulating rod（阀芯行程调节杆）

2. 先导阀与盖板

先导阀通过盖板安装在阀块上，并经盖板上的油道来控制插装阀组件控制油口 X 的通油方式，从而控制阀口的开启和关闭。其中方向阀组件的先导阀可以是电磁滑阀，也可以是电磁球阀。压力阀组件的先导阀包括压力先导阀、电磁滑阀等，其控制原理与普通溢流阀完全相同。流量阀组件的先导阀除电磁滑阀外，还需在盖板上装阀芯行程调节杆，以限制、调节阀口开度大小，即改变阀口通流面积（图 5-21c）。

3. 插装阀的应用举例

（1）插装阀作单向阀　将方向阀组件的控制油口 X 通过阀块和盖板上的通道与油口 A 或 B 直接连通，可组成单向阀。其中图 5-22b 所示结构，反向（A→B）关闭时，控制腔的压力油可经过阀芯上端与阀套孔之间的环形间隙，向油口 B 泄漏，密封性能不及图 5-22a 所示的连接形式。

（2）插装阀作二位二通阀　如图 5-23 所示，由二位三通先导电磁滑阀控制方向阀组件控制腔的通油方式。如图 5-23a 所示，电磁铁 YA 失电时，控制油口 X 通过二位三通阀的常位通油箱，$p_x=0$，因此，无论 A 口来油，还是 B 口来油均可将阀口开启通油。电磁铁得电，二位三通阀右位工作，控制油口 X 与油口 A 接通，从 B 口来油可顶开阀芯通油，而 A 口来油则阀口关闭，相当于 B→A 的单向阀。与图 5-23a 不同，图 5-23b 所示结构在二位三通阀处于右位工作时，因梭阀的作用，控制油口 X 的压力始终为 A、B 两油口中压力较高

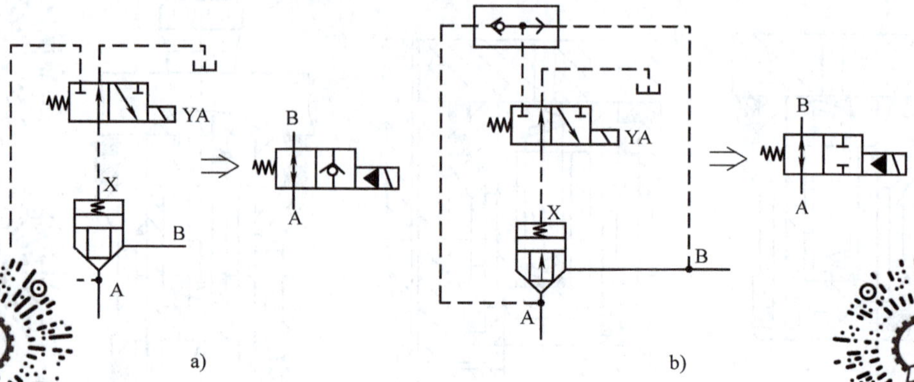

Fig. 5-23 Two-way valve (二通阀)

de-energized, the two-position, four-way valve works at its normal (left) position. The pilot port of valve 1 is connected to the reservoir, allowing valve 1 to be open. Pressurized oil passes through the pilot port of valve 2, but valve 2 is closed. In this case, ports A and T are connected, while port P is blocked.

(4) Four-way valves by cartridge As is shown in Fig. 5-25, a four-way insert cartridge valve is composited by two three-way insert cartridges with parallel connection. The four solenoid-operated directional valves are used for controlling the four directional units respectively and realize twelve different directional positions. However, the most widely used in practice is a three-position, four-way valve. The open-close actions of valves 1, 2, 3, and 4 are controlled as a group by a three-position, four-way solenoid spool.

5.6 Electro-hydraulic Servo Valves

An Electro-hydraulic servo valve functions as an electro-hydraulic transfer equipment and also a power amplifier, which is put into a lower electrical signal to get a powerful hydraulic pressure energy and realize the purpose of a displacement, speed, acceleration and force controls for actuators. Remote control, computer control and auto-control can be achieved for high-power, quick response hydraulic systems. All these lead to its wide uses in the industrial production.

Fig. 5-26 illustrates a nozzle flapper electro-hydraulic servo valve, which consists of an torque motor, a nozzle-flapper pilot stage, and a sliding spool main stage. An electric-hydraulic servo valve usually consists of three parts.

(1) The electro-hydraulic transfer equipment It can transfer the input electrical signal into a torque output (motor) or linear displacement output (hydraulic cylinder). A motor incorporates a permanent magnet 6, a magnetizer 4, a gag bit 3, a field winding (field coil) and a spring tube 5. The gag bit bears against the spring tube upside. The permanent magnet and the magnetizer constitute a permanent magnetic field. Four clearances (a, b, c, d) are enclosed by the magnetizer and the iron core.

者。因此，无论是 A 口来油，还是 B 口来油，阀口均处于关闭状态，油口 A 与 B 不通。

(3) 插装阀作二位三通阀　图 5-24 中三通插装阀由两个方向阀组件并联而成，对外形成一个压力油口 P，一个工作油口 A 和一个回油口 T。两组件的控制腔的通油方式由一个二位四通电磁滑阀（先导阀）控制。在电磁铁 YA 失电时，二位四通阀左位（常位）工作，阀 1 的控制腔接回油箱，阀口开启；阀 2 的控制腔接压力油，阀口关闭。于是油口 A 与 T 通，油口 P 不通。

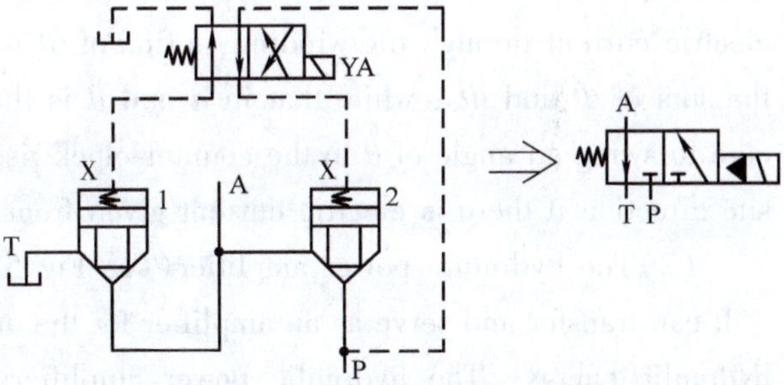

Fig. 5-24　Three-way valve（三通阀）

(4) 插装阀作四通阀　四通插装阀由两个三通阀并联而成。如图 5-25 所示，用四个二位三通电磁阀分别控制四个方向阀组件的开启和关闭，可以得到图示十二种机能。实际应用最多的是一个三位四通电磁阀成组控制阀 1、阀 2、阀 3 和阀 4 的开启和关闭的三位四通阀。

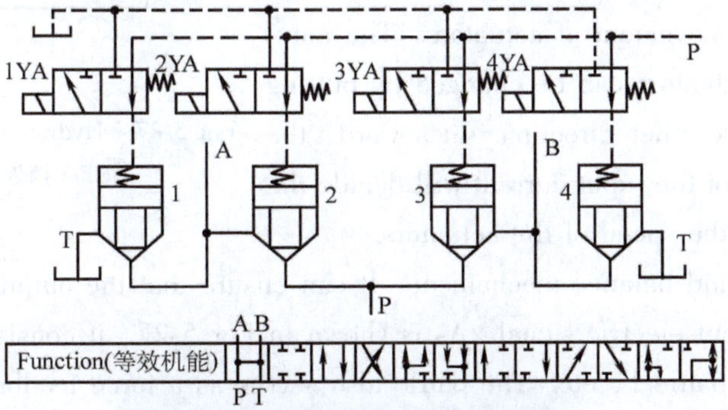

	1	2	3	4	5	6	7	8	9	10	11	12
1YA	+	−	−	+	−	+	−	−	−	+	+	−
2YA	+	−	+	−	+	−	+	−	−	−	+	−
3YA	+	−	−	+	+	−	−	−	+	−	−	+
4YA	+	−	+	−	−	+	−	+	−	−	−	+

Fig. 5-25　Four-way valve（四通阀）

5.6　电液伺服阀

电液伺服阀将电信号传递处理的灵活性和大功率液压系统控制相结合，可对大功率、快速响应的液压系统实现远距离控制、计算机控制和自动控制。同时它也是将小功率的电信号输入转换为大功率的液压能（压力和流量）输出，实现执行元件的位移、速度、加速度及力控制的一种装置。因而在现代工业生产中被广泛应用。

图 5-26 所示为喷嘴挡板式电液伺服阀，电液伺服阀通常由三部分组成。

(1) 电气-机械转换装置　用来将输入的电信号转换为转角或直线位移输出，输出转角的装置称为力矩马达，输出直线位移的装置称为力马达。力矩马达由永久磁铁 6、导磁体 4、衔铁 3、激磁线圈和弹簧管 5 组成。衔铁支承在弹簧管上，永久磁铁和导磁体形成一个永久

If there is no electric current through the field coil, the magnetic flux in the four clearances equals Φ_y and acts in the same direction. Φ_y makes the gag bit stay in its middle position. If there is electric current through the winding, a flux of Φ_s is produced. Thus the flux in clearances b and c is the sum of Φ_y and Φ_s, while that in a and d is the differential of Φ_y and Φ_s. This causes the iron core to swing an angle of θ in the counter-clockwise direction. The iron core will swing in the opposite direction if there is electric current given from the other direction.

(2) The hydraulic power amplifier (see Fig. 5-27)

It can transfer and serve as an amplifier for the output hydraulic energy. The hydraulic power amplifier is a four-sided spool valve. This amplifier converts the spool valve displacement x_v to the flow rate q_{vL} and pressure p_L on load request to realize actuators' operation. The higher the input current is, the bigger offset from the spool valve is, and the larger the output flow rate is, which allows a faster movement of actuators. The moving direction of the actuators can be changed by putting on the current from the other direction. In a word, the direction and amount of the input current will decide the moving direction and the speed of the actuators.

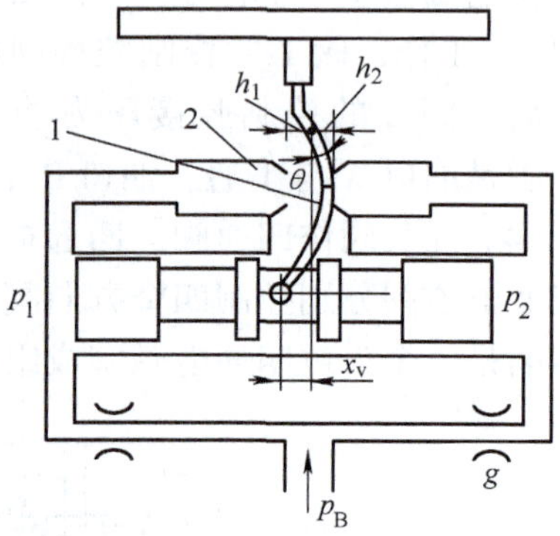

Fig. 5-27 Hydraulic amplifier (液压放大器)
1—Baffle(挡板) 2—Nozzles(喷嘴)

(3) A feedback and balance mechanism It can ensure that the output pressure or flow rate is proportional to the input electric signal. As is shown in Fig. 5-27, it consists of a fixed throttle orifice g, a nozzle 2, a baffle 1, etc. The baffle also serves as a force feedback spring for the amplifier. The spool valve is the actuator in this configuration. If there is no angle displacement put out from the motor, the baffle stays in its middle position. The clearances between the two nozzles meet this equation, $h_1 = h_2 = h_0$. For the nozzles and the fixed throttle orifice are identical in parameters, the hydraulic resistances of the two nozzles are equal, i. e., $p_1 = p_2$. The spool valve remains in its middle position due to the feedback spring. Here the feedback operation is that when and if the motor sets an angle displacement of θ, then $h_1 > h_2$. The hydraulic resistances of the two nozzles are no longer the same, and $p_2 > p_1$. The spool valve and also baffle down ball head move in the left, producing a clockwise torque to the iron core-baffle assembly and shortening the offset of the baffle between the two nozzles. As a result, the pressure differential of the spool valve $(p_2 - p_1)$ is reduced and the spool valve stops under a certain orifice x_v. This can be achieved because the hydraulic pressure on the spool valve is balanced with an elastic reaction force produced by the movement of the baffle down head.

Obviously, the electromagnetic torque is proportional to the input current. Different size of orifice can be obtained by changing the amount of input current. Putting the current from the other direction can make the valve core move in the other direction and realize an inverse control for flow.

As described above, the spool valve is balanced by the elastic reaction force of the feedback spring piece (baffle), so it is a force feedback type.

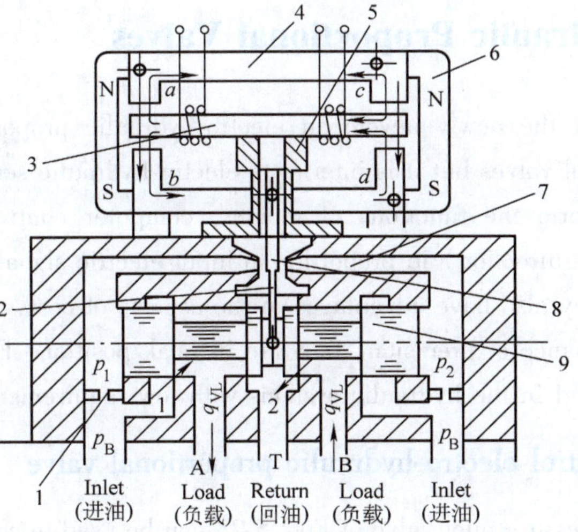

Fig. 5-26　Nozzle flapper electro-hydraulic servo valve（喷嘴挡板式电液伺服阀）

1—Fixed throttle orifice（节流孔）　2—Feed back pole（反馈杆）

3—Gag bit（衔铁）　4—Magnetizer（导磁体）　5—Spring pipe（弹簧管）

6—Permanent magnet（永久磁铁）　7—Nozzles（喷嘴）　8—Baffle（挡板）　9—Main valve（主阀）

磁场。激磁线圈中没有电流通过时，导磁体和铁心间的四个气隙 a、b、c、d 中的磁通都是 Φ_y，且方向相同，因此衔铁处于中间位置。当有控制电流通入线圈时，产生磁通 Φ_s，则在气隙 b、c 中，Φ_y 和 Φ_s 相加；在 a、d 中两者相减，于是铁心逆时针方向偏转 θ 角。如果控制电流反向，则偏转方向也相反。

（2）液压放大器　实现液压油控制功率的转换和放大。

液压放大器（图5-27）是一个滑阀，它将输入的滑阀位移 x_v 转换成负载流量 q_{vL} 和负载压力 p_L，以推动执行机构动作。输入的控制电流越大，滑阀的偏移量也越大，输出的流量越多，因而执行机构的运动速度也越高。如果改变控制电流的方向，就会使执行机构反向运动。因此，输入控制电流的方向和大小决定了执行机构的运动方向和速度。

（3）反馈和平衡机构　使电液伺服阀输出的流量或压力获得与输入电信号成比例的特性。如图5-27所示，它是由固定节流孔 g、喷嘴2、挡板1（兼作放大器的力反馈弹簧）组成。在这里，滑阀是它的执行元件。当力矩马达没有角位移输出时，挡板处于中位，两个喷嘴至挡板的缝隙 $h_1=h_2=h_0$，由于喷嘴孔和固定节流孔的参数一样，两喷嘴处的液阻相等，从而 $p_1=p_2$，滑阀由于反馈弹簧的作用而停在中位。当力矩马达输入一角位移 θ 时，$h_1>h_2$，两喷嘴处的液阻不等，$p_2>p_1$，滑阀向左移动。与此同时，挡板下端球头也随滑阀左移，在铁心挡板组件上产生一个顺时针方向的转矩，同时使挡板在两喷嘴间的偏移量减少，这就是反馈作用。反馈作用的结果，使滑阀两端的压差（p_2-p_1）减小。当滑阀上液压作用力与挡板下端球头因移动而产生的弹性反作用力平衡时，滑阀便停止移动，并保持在某一个开度 x_v 上。

显然，改变输入电流大小，可成比例地调节电磁力矩，从而得到不同的主阀开口大小。若改变输入电流的方向，则主滑阀阀芯产生反向位移，从而实现液流的反向控制。

从上述工作原理可知，滑阀的位置通过反馈弹簧片（挡板）的弹性力反馈而达到平衡位置，所以它属于力反馈式电液伺服阀。

5.7 Electro-hydraulic Proportional Valves

The performance of the newly developed electro-hydraulic proportional valves is better than common hydraulic control valves but it is inferior to electro-hydraulic servo valves. Electro-hydraulic proportional valves perform the functions of remote, computer controls for hydraulic parameters (pressure, flow rate and direction) in proportion to input electric signal. Compared with electro-hydraulic servo valves, they also have advantages in the aspects of cost, anti-contamination, etc. But as to the control performance or precision, they are in weak position. Electro-hydraulic proportional valves are usually adopted in the hydraulic systems with low requirements.

5.7.1 Pressure control electro-hydraulic proportional valve

The proportional pressure pilot valve of Fig. 5-28 can be used to connect with a common relief, pressure reducing or a sequence valve poppet (core) to become virtually any pilot controlled pressure device. Unlike a common pressure control valve, here the hydraulic pressure force acted on the valve is balanced out with electromagnetic force, rather than spring force. Adjusting the input current we can change the electromagnetic force, thus pressure in the front chamber of the pilot valve or in the upper port of the main valve can be adjusted. The inlet and outlet pressures of the main valve are controlled in this way.

Fig. 5-29 shows a proportional relief servo valve by direct detection. It has a spool pilot-operated valve. The inlet oil pressure (p) is directed to the left (action area A_0) of feedback rod 2 and then into the left (action area A_1) of pilot-operated spool 3 via fixed damping orifice R_1. The oil in the right end of the pilot-operated spool is introduced to the spool valve orifice and the upper port of the main valve. The oil in the upper port is admitted to the right (action area A_2) of the pilot-operated spool. When the main valve core is balanced out, R_3 (a dynamic pressure feedback damping orifice located between the pilot-operated spool orifice and the upper port of the main valve) performs no restricting function and the equal forces act on two ends of it. Only $A_1 - A_0 = A_2$ is adopted for design can we keep a hydraulic pressure of $F = pA_0$. When hydraulic pressure F is equal to the electromagnetic force F_E, the pilot-operated spool valve core is balanced out and remains in a certain position with a certain size of orifice. The pressure p_1 in the front chamber of the pilot-operated valve or in the upper port of the main valve is a constant, ($p_1 < p$). The main valve core is balanced out under its upper and down pressures p_1, p, the spring force and the flow force. In this case the main valve is open with a certain orifice to ensure that the inlet pressure p of the relief valve is proportional to the electromagnetic force. A variety of inlet pressures can be obtained by adjusting the input current.

If the inlet pressure p increases suddenly due to outside disturbance load, the balance state of the pilot-operated spool valve core is damaged and the valve core moves in the right. The orifice is enlarged, causing the pressure p_1 in the front chamber of the pilot valve or in the upper port of the main valve to fall down. In this case, the main valve core is no longer at a balance state. The main valve core moves up to enlarge its orifice and reduces the increased inlet pressure. When the inlet

5.7　电液比例阀

电液比例阀是一种性能介于普通液压控制阀和电液伺服阀之间的新阀种，它既可以根据输入电信号的大小连续地、成比例地对液压系统的参量（压力、流量及方向）实现远距离控制和计算机控制，又在制造成本、抗污染等方面优于电液伺服阀。但其控制性能和精度不如电液伺服阀，故广泛应用于要求不是很高的液压系统中。

5.7.1　电液比例压力阀

图 5-28 所示为电液比例压力先导阀，它与普通溢流阀、减压阀、顺序阀的主阀组合可构成电液比例溢流阀、电液比例减压阀和电液比例顺序阀。与普通压力先导阀不同，与阀芯上的液压力进行比较的是比例电磁铁的电磁吸力，不是弹簧力。改变输入电磁铁的电流大小，即可改变电磁吸力，从而改变先导阀的前腔压力，即主阀上腔压力，对主阀的进口或出口压力实现控制。

图 5-29 所示为直接检测式电液比例溢流阀。它的先导阀为滑阀结构，溢流阀的进口压力油（压力为 p）被直接引到先导滑阀反馈推杆 2 的左端（作用面积为 A_0），然后经过固定阻尼 R_1 到先导滑阀阀芯 3 的左端（作用面积为 A_1），进入先导滑阀阀口和主阀上腔，主阀上腔的压力油再引到先导滑阀的右端（作用面积为 A_2）。在主阀阀芯 1 处于稳定受力平衡状态时，先导滑阀阀口与主阀上腔之间的动压反馈阻尼 R_3 不起作用，因此作用在阀芯两端的压力相等。设计时取 $A_1 - A_0 = A_2$，于是作用在先导滑阀上的液压力 $F = pA_0$。当液压力 F 与比例电磁铁吸力 F_E 相等时，先导滑阀阀芯受力平衡，阀芯稳定在某一位置，先导滑阀开口一定，先导滑阀前腔压力即主阀上腔压力 p_1 为一定值（$p_1 < p$），主阀阀芯在上下两腔压力 p_1 和 p 及弹簧力、液动力的共同作用下处于受力平衡状态，主阀开口一定，保证溢流阀的进口压力 p 与电磁吸力成正比，调节输入的电流大小，即可调节阀的进口压力。

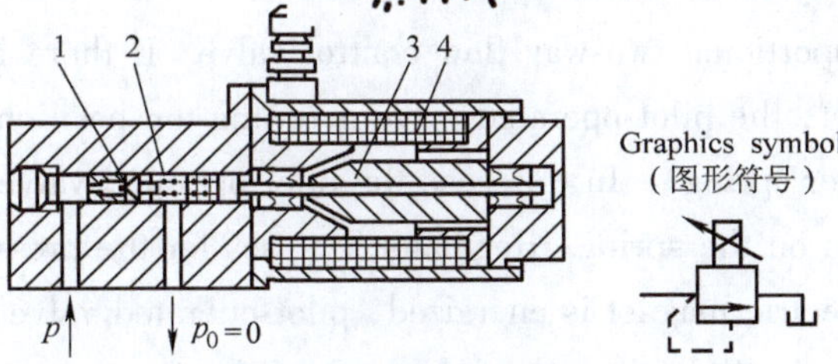

Fig. 5-28　Electro-hydraulic proportional pressure pilot
（电液比例压力先导阀）
1—Valve poppet（core）（阀芯）
2—Loaded spring（传力弹簧）　3—Handspike（推杆）
4—Proportional valve with solenoid operated（比例电磁铁）

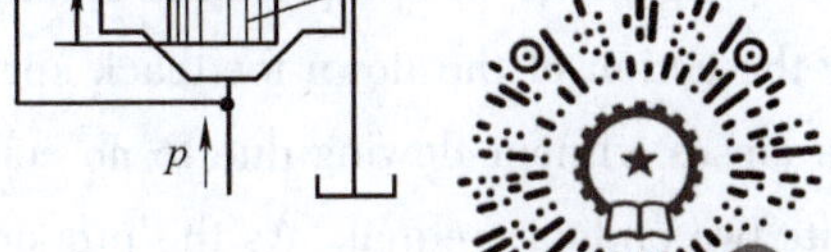

Fig. 5-29　Proportional relief by direct detecting
（直接检测式比例溢流阀）
1—Main valve poppet（core）（主阀阀芯）
2—Feed back Handspike（反馈推杆）
3—Pilot valve poppet（core）（先导滑阀阀芯）
4—Proportional valve with solenoid operated（比例电磁铁）

pressure p returns to its setting value, the valve cores of the pilot-operated spool valve and the main valve are again under balance states. The valve works in a new stable position.

5.7.2 Flow control electro-hydraulic proportional valves

A flow control electro-hydraulic proportional servo valve is made by simply replacing the manual control parts of flow control valves by a proportional electric magnet. Its construction and operation are introduced as follows.

1. Electro-hydraulic proportional throttle valves

Fig. 5-30 illuminates a two-way proportional throttle valve with a position feed back by the spring force. The main valve core 1 is a cartridge design. When the solenoid is energized, we input a given current and cause the electric magnet force to push the sliding spool 4 moving downward and open, so the pressure oil p_A flows from the inlet of the main valve 1 to the outlet of the main valve passage by the orifices R_1, R_2, and pilot-operated valve orifice. The main valve is moved up and opened, which permits the oil flow to the outlet of the valve under action of the pressure differential between the two end areas on the main valve by overcoming the spring force. This pressure differential is due to orifice R_1. At the same time, the main valve is moved up and results in the feedback spring reverse being compressed reversely until it is equal to the electricmagnet force. The pilot operated poppet (core) and main valve core are under balance position, and the forming open orifice on the main vale is in proportion to the electric input signal. Shifting the electric input signal, we can change the open orifice of the valve, which functions for regulating speed in a hydraulic system.

Different from the common flow control electro-hydraulic proportional servo valve, Fig. 5-30 illuminates another one with a proportional electric magnet in which the pressure of the upper chamber in the main valve is changed to regulate orifice by controlling the pilot operated orifice. At the same time, the displacement force of the main valve acts on the proportional electric magnet via a feed back spring, forming comparing force between the spring force and the proportional electric magnet. In this way, the control precision for the main valve displacement (orifice size) can be guaranteed. The main valve's displacement isn't limited by the stroke of the proportional electric magnet. The orifice can be designed large and thus allows more fluid to flow through.

2. Electro-hydraulic proportional two-way flow valves

Fig. 5-31 indicates an electro-hydraulic proportional two-way flow control valve. If there is no current signal for the proportional electric magnet, the pilot-operated valve is on the top position under the action of the down feedback spring (inner spring). In this case the pilot-operated valve orifice closes without flowing due to no compression on the spring. Regulator 3 closes for the pressures in its two ends are equal. As the proportional electric magnet is energized, pilot-operated valve core 1 moves down and opens the valve. Oil from pump flows into the inlet port of the flow-sensor via damping orifice R_1, orifice R_2 and the pilot valve orifice. The control pressure of regulator 3 p_2 is less than p_1 because of pressure loss caused by the flow resistance. When the pressure drop ($p_1 - p_2$) reaches a given value, the core of the regulator moves in the right and open, allowing the oil from the pump to flow into flow sensor 2 via the regulator orifice to prop up the core and the valve opens. At the same time of the core in the flow sensor moving up, the core displacement is transferred into the

若溢流阀的进口压力 p 因外界干扰而突然升高，则先导滑阀阀芯受力平衡被破坏，阀芯右移，阀口增大使先导阀前腔压力 p_1 减小，即主阀上腔压力减小，于是主阀阀芯受力平衡也被破坏，阀芯上移开大阀口使升高了的进口压力下降。当进口压力 p 恢复到原来值时，先导滑阀阀芯和主阀阀芯重新回到受力平衡位置，阀处在新的稳态位置工作。

5.7.2 电液比例流量阀

电液比例流量阀是将流量阀的手调部分改换为比例电磁铁而成。下面介绍它们的结构和工作原理。

1. 电液比例节流阀

图 5-30 所示为一种位移-弹簧力反馈型电液比例节流阀，主阀阀芯 1 为插装阀结构。当比例电磁铁输入一定的电流时，所产生的电磁吸力推动先导滑阀阀芯 4 下移，先导滑阀阀口开启，于是主阀进口的压力油 p_A 经阻尼 R_1 和 R_2、先导滑阀阀口流至主阀出口。因阻尼 R_1 的作用，R_1 前后出现压差，主阀阀芯在两端压差的作用下，克服弹簧力向上位移，主阀阀口开启，进、出油口连通。主阀阀芯向上位移导致反馈弹簧 2 反向受压缩，当反馈弹簧力与先导滑阀上端的电磁吸力相等时，先导滑阀阀芯和主阀阀芯同时处于受力平衡状态，主阀阀口大小与输入电流大小成比例。改变输入电流大小，即可改变阀口大小，在系统中起节流调速作用。

与普通电液比例流量阀不同，图 5-30 所示的电液比例节流阀的比例电磁铁是通过控制先导滑阀的开口改变主阀上腔压力来调节主阀开口大小的。在这里主阀的位移又经反馈弹簧作用到比例电磁铁上，由反馈弹簧力与比例电磁铁吸力进行比较。因此，不仅可以保证主阀位移量（开口量）的控制精度，而且主阀的位移量不受比例电磁铁行程的限制，阀口开度可以设计得较大，即阀的通流能力较大。

2. 电液比例流量阀

如图 5-31 所示，当比例电磁铁无电流信号输入时，先导滑阀由下端反馈弹簧（内弹簧）支承在最上端位置，此时弹簧无压缩量，先导滑阀阀口关闭，于是调节器 3 阀芯两端压力相等，调节器阀口关闭，无流量通过。当比例电磁铁输入一定电流信号产生一定的电磁吸力时，先导滑阀阀芯 1 向下移动，阀口开启，于是液压泵来油经阻尼 R_1、R_2、先导滑阀阀口到流量传感器的进油口。由于油液流动的压力损失，调节器 3 控制腔的压力 $p_2 < p_1$。当压差（$p_1 - p_2$）达到一定值时，调节器阀芯位移，阀口开启，液压泵来油经调节器阀口到流量传感器 2 进口，顶开阀芯，流量传感器阀口开启。在流量传感器阀芯上移的同时，阀芯的位移转换为反馈弹簧的弹簧力通过先导滑阀阀芯与电磁吸力相比较，当弹簧力与电磁吸力相等时，先导滑阀阀芯受力平衡。与此同时，调节器阀芯、流量传感器阀芯也受力平衡，所有阀口满足压力流量方程。压力油（压力为 p_1）经调节器阀口后降为 p_4，并作为流量传感器的进口压力，流量传感器的出口压力 p_5 由负载决定。

如负载压力 p_5 增大，则流量传感器受力平衡被破坏，阀芯下移，阀口有关小的趋势，这将使反馈弹簧力减小，先导阀阀芯下移、先导阀阀口增大，调节器控制腔压力 p_2 降低，调节器阀口增大使其减压作用减小，于是流量传感器进口压力 p_4 增大，导致流量传感器阀芯上移，阀口重新开大，当流量传感器阀口恢复到原来的开口大小时，先导阀阀芯受力重新平衡，流量阀在新的稳态位置下工作。

feedback spring force, forming a compared force between the spring force and the electromagnetic attracted force. When the spring force and the electromagnetic force are equal, the pilot-operated valve core, the regulator valve core and the valve core of the flow sensor are all balanced out and meet the pressure-flow rate equation. The oil pressure $p_1$from pump reduces to $p_4$after flowing through the regulator orifice. $P_4$is used as the input pressure for flow sensor. The output pressure of the flow sensor depends on outside load.

If the load-dependent pressure $p_5$increases, the balance state of the flow sensor is damaged and the valve core tends to close the orifice. This reduces the feedback spring force, forcing the pilot valve core to move down and enlarge the pilot valve orifice. The pressure in the pilot chamber of the regulator p_2 falls down, enlarging the orifice of the regulator to weaken its restricting function. This allows the inlet pressure of the flow sensor p_4 to increase, causing the valve core of the flow sensor to move up and enlarge its valve orifice again. When the size of orifice is equal to its previous size, the valve core of the pilot valve core is under a balance state again, allowing the flow valve to work at a new stable position.

Here the regulator functions as pressure compensation to ensure that the inlet and outlet pressure differential through the flow sensor is fixed and flow is stable. The flow sensor controls the valve core displacement of the regulator by flow detecting. So the stability of the electro-hydraulic proportional two-way flow valve is higher than the common electro-hydraulic proportional two-way regulating speed valve.

5.7.3 Electro-hydraulic proportional directional control valves

Fig. 5-32 illuminates an electro-hydraulic proportional directional control valve. It consists of two stages with prepositive stage (electro-hydraulic proportional bidirectional pressure-reducing valve) and amplifier stage (hydraulic-operated proportional bidirectional throttle valve).

In the prepositive stage, the displacement of valve core 1 in the pressure-reducing valve is bidirectionally controlled by the two proportion electric magnets (4 and 8) on both ends. If the proportional electric magnet 8 (left) is energized, an electromagnetic force is produced to drive valve core 1 in the right. The right valve opens, allowing supplying oil pressure p to fall down to p_c (control pressure) via the valve orifice. When p_c reacts back on the right end of the valve core (the left end is connected the reservoir p_0) via passage 3, a hydraulic pressure $F_1$is formed against the electromagnetic force F_{E1}. If $F_1 = F_{E1}$, the valve core 1 stops moving in the right and stays at a certain position with a certain size of valve orifice to hold the pressure p_c. It is clear that the control pressure p_C is independent from the supply oil pressure p, but varies in proportion to the electromagnetic force. Similarly, if current is put into the right end of the electric magnet, the pressure-reducing valve spool will move in the left and we can get a stable control oil pressure via the left valve orifice.

The amplifier stage consists of a valve body, a main valve core, left and right end covers, springs, damping screws 6 and 7, etc. When control pressure p_c from the prepositive stage acts on the right end area of the main valve core 5 after passing the damping orifice, the hydraulic pressure pushes the main spool 5 to move in the left by overcoming the left spring force. The left spring chamber of the valve core opens to the reservoir p_0, allowing the valve to open. The oil flows from port P

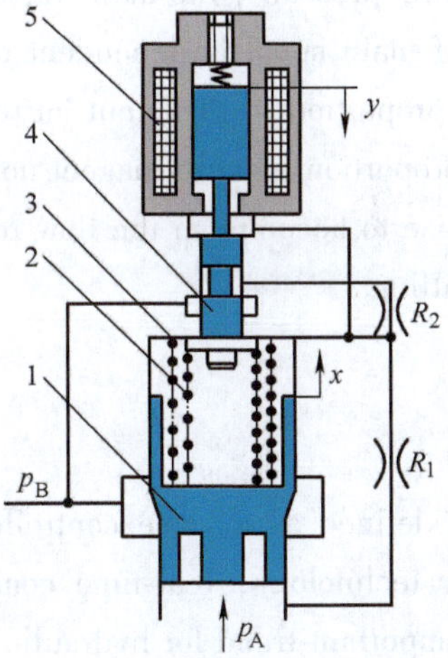

Fig. 5-30 Electro-hydraulic proportional two-way throttle valve（电液比例节流阀）

1—Main valve poppet（core）（主阀阀芯）

2—Feed back spring（反馈弹簧）

3—Reposition spring（复位弹簧）

4—Pilot valve poppet （core）（先导滑阀阀芯）

5—Proportional valve with solenoid operated（比例电磁阀）

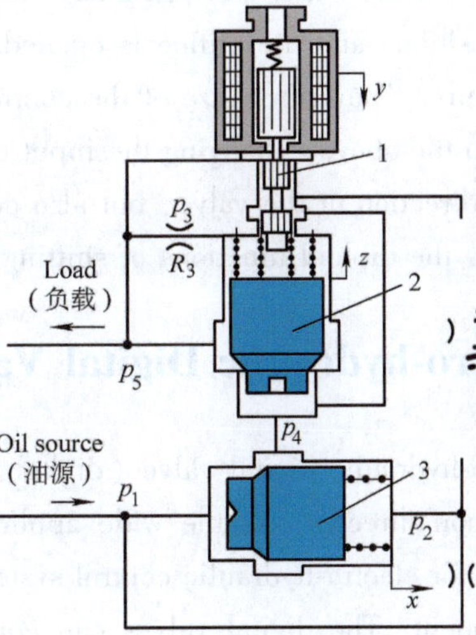

Fig. 5-31 Electro-hydraulic proportional two-way flow valve（电液比例流量阀）

1—Pilot valve（先导滑阀）

2—Flow sensor（流量传感器）

3—Regulator（调节器）

这里，调节器起压力补偿作用，保证流量传感器进、出口压差为定值与流经阀的流量稳定不变。由于调节器阀芯的位移是由流量传感器检测流量信号控制的，因此流量稳定性跟普通调速阀相比有很大的提高。

5.7.3 电液比例换向阀

如图 5-32 所示，电液比例换向阀是由前置级（电液比例双向减压阀）和放大级（液动比例双向节流阀）两部分组成。前置级由两端比例电磁铁 4、8 分别控制减压阀阀芯 1 的位移。如果左端比例电磁铁 8 输入电流，则产生一电磁吸力使减压阀阀芯 1 右移，右边阀口开启，供油压力油 p 经阀口后减压为 p_c（控制压力）。因 p_c 经流道 3 反馈作用到阀芯右端面（阀芯左端通回油 p_0），形成一个与电磁吸力方向相反的液压力 F_1，当 $F_1=F_{E1}$ 时，阀芯停止右移，稳定在某一位置上，减压阀右边阀口开度一定，压力 p_c 保持一个稳定值。显然压力 p_c 与供油压力 p 无关，仅与比例电磁铁的电磁吸力即输入电流大小成比例。同理，当右端比例电磁铁输入电流时，减压阀阀芯将左移，经左阀口减压后得到稳定的控制压力 p'_c。

放大级由阀体，主阀阀芯，左、右端盖、弹簧和螺钉 6、7 等组成。当前置级输出的控制压力 p_c 经节流孔缓冲后作用在主阀阀芯 5 右端时，液压力克服左端弹簧力使阀芯左移（阀芯左端弹簧腔通回油 p_0）开启阀口，油口 P 与 B 通、A 与 T 通。当弹簧力与液压力相等时，主阀阀芯停止左移，稳定在某一位置，阀开口大小一定。因此，主阀开口大小取决于输入的液流大小。控制阀芯的开口大小与输入电流大小成比例。

综上所述，改变比例电磁铁的输入电流，不仅可以改变阀的工作液流方向，而且可以控制阀口大小实现流量调节，即具有换向和节流的复合功能。

to B, and port A to T. When the spring force is equal to the hydraulic pressure, the main valve core is fixed in a position, and the orifice is opened. So the opening of main spool is dependent on the hydraulic pressure. The orifice size of the control valve core is in proportion to the input current.

To sum up the above, changing the input electric current to proportion electric magnet not only shifts the flow direction of the valve, but also controls the orifice size to accomplish the flow regulation, which has the diploid functions of shifting direction and throttling.

5.8 Electro-hydraulic Digital Valves

An electro-hydraulic digital valve (digital valve for short) is defined as a valve controlled by digital information directly. As the wide applications of computer technology, real-time computer control manner for electro-hydraulic control system has become an important trend for hydraulic technology development. The digital valves can connect with the computers directly without any D/A transfer unit. Compared with servo valves and proportional valves, digital valves have the advantages of compact design, good technics, low cost, anti-contamination, high repeat precision, reliability and low power loss. In the electro-hydraulic computer control systems, part of the proportional and servo valves are replaced with the digital valves, which creates a new field in the computer application to the hydraulic pressure systems.

5.8.1 Operation and makeups

For a computer, the ordinary signal is the binary signals, i. e, "on" and "off". There are many digital approaches used to control systems, but the mostly used is the incremental control approach caused by pulse number modulated (PNM) and the pulse width modulated (PWM) control strategies.

The incremental control approach uses a stepping motor-mechanical converter. Each electrical pulse (pulse numbers in per sampling cycle is increased or decreased based on the pulses before cycle) causes the stepping motor to rotate through a fixed amount and causes the hydraulic motor shaft to follow and the mechanical converter to put out displacement to control the orifice opening or closing. Fig. 5-33 is an increment type digital control system used in a hydraulic system.

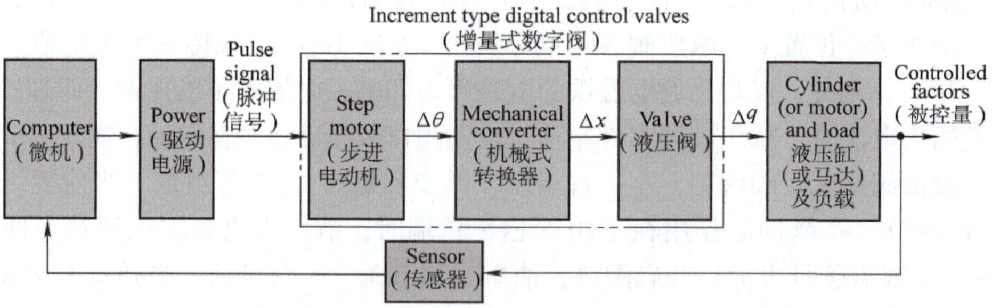

Fig. 5-33 Increment type digital control system(增量式数字阀控制系统框图)

The pulse width modulated digital valve is an approach with modulating and enlarging a sequential signal to a pulse signal via the pulse width modulated amplifier and then sending the amplified

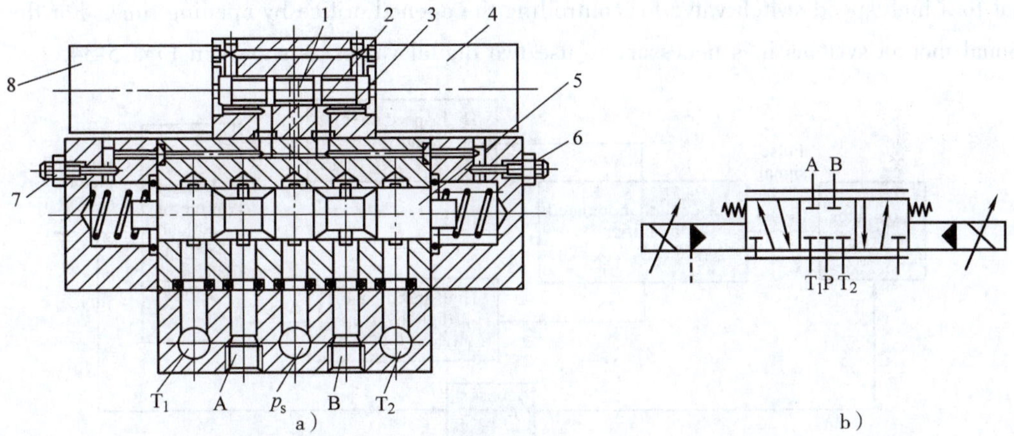

Fig. 5-32　Electro-hydraulic proportional directional control valve（电液比例换向阀）
a）Configuration（结构图）　b）Graphics symbol（图形符号）
1—Valve poppet（core）of pressure reducing valve（减压阀阀芯）
2，3—Flow channel（流道）　4，8—Proportional solenoid（比例电磁铁）
5—Main valve poppet（core）（主阀阀芯）　6，7—Screws（螺钉）

5.8　电液数字阀

用数字信号直接控制的阀称为电液数字阀，简称数字阀，由于计算机技术已获得广泛的应用，用计算机对电液控制系统进行实时控制已成为液压技术发展的一个重要趋势。数字阀可直接与计算机接口连接，不需要 D/A 转换。与伺服阀、比例阀相比，其结构简单、工艺性好、价格低廉、抗污染能力强、重复精度高、工作稳定可靠、能耗小。在计算机实时控制的电液系统中，已部分取代比例阀或伺服阀，为计算机在液压领域的应用开拓了一个新途径。

5.8.1　工作原理与组成

对计算机而言，最普通的信号是量化为两个量级的信号，即“开”和“关”。用数字量控制阀的方法很多，常用的是由脉数调制（PNM）演变而来的增量控制法以及脉宽调制（PWM）控制法。

增量控制数字阀采用步进电动机-机械转换器；通过步进电动机，在脉数（PNM）信号的基础上，使每个采样周期的步数在前一个采样周期步数上增加或减少步数，以达到需要的幅值；由机械转换器输出位移控制液压阀阀口的开启和关闭。图 5-33 所示为增量式数字阀用于液压系统的框图。

脉宽调制式数字阀通过脉宽调制放大器将连续信号调制为脉冲信号并放大，然后输送给高速开关数字阀，以开启时间的长短来控制阀的开口大小。在需要作两个方向运动的系统中，要用两个数字阀分别控制不同方向的运动，这种数字阀的控制系统框图如图 5-34 所示。

5.8.2　电-液数字阀的典型结构

图 5-35 所示为步进电动机直接驱动的数字式流量控制阀。当计算机给出脉冲信号后，

signal to a high speed switch valve for controlling the opened orifice by opening time. For the two directional motion systems it is necessary to use two digital valves as shown in Fig. 5-34.

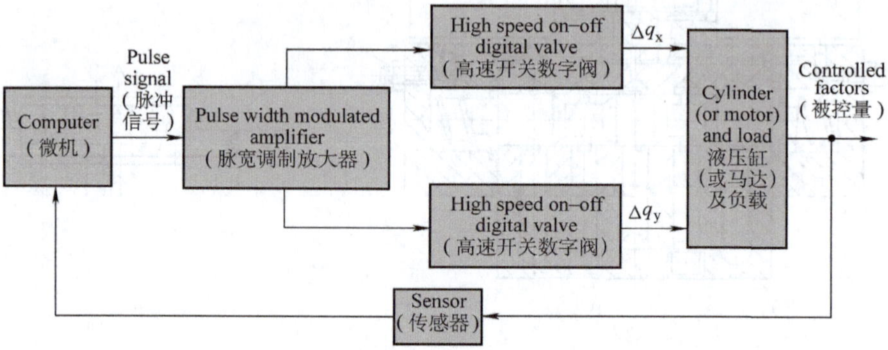

Fig. 5-34 PNM digital valve control system(脉宽调制式数字阀控制系统框图)

5.8.2 Typical construction of electro-hydraulic digital valves

Fig. 5-35 shows a digital flow control valve construction. It is driven directly by a stepping motor. When a pulse signal is given by a computer, the stepping motor 1 rotates an angle $\Delta\theta$. The screw thread rod 2 in the mechanical converter turns this angle $\Delta\theta$ into an axial displacement Δx to drive directly throttling valve core 3 and open this valve orifice. As long as the stepping motor rotates a step number it will control a fixed open orifice, which reaches the flow rate control.

Note that this is an open-loop control system, in which a zero position sensor 6 is mounted, see Fig. 5-35. At the end of each control cycle, the valve core returns to its zero position controlled by sensor 6. This ensures each working cycle to start at the same position and thus makes a higher repeat precision for the valve.

步进电动机 1 转过一个角度 $\Delta\theta$，作为机械转换装置的滚珠丝杠 2 将旋转角度 $\Delta\theta$ 转换为轴向位移 Δx 直接驱动阀芯 3，开启阀口。步进电动机转过一定的步数，相当于阀芯具有一定的开度，从而实现流量控制。

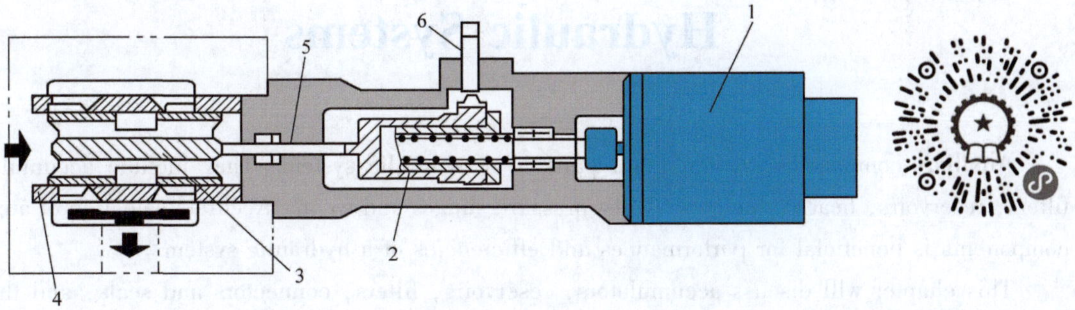

Fig. 5-35　Digital flow control valve（数字式流量控制阀）

1—Step motor（步进电动机）　2—Ball screw（滚珠丝杠）　3—Valve core（阀芯）
4—Seat（阀套）　5—Connecting rod（连杆）　6—Zero position sensor（零位移传感器）

这种阀是开环控制的，但装有零位移传感器 6（图 5-35），在每个控制周期终了时，阀芯可由零位移传感器控制回到零位，以保证每个工作周期都从相同的位置开始，使阀的重复精度比较高。

Chapter 6 Auxiliary Components for Hydraulic Systems

Auxiliary components are important parts in a hydraulic system. They include accumulators, filters, reservoirs, heat exchangers, seals, pressure gauges and so on. Accurate selection of auxiliary components is beneficial for performances and efficiencies of a hydraulic system.

This chapter will discuss accumulators, reservoirs, filters, connectors and seals; and the rest can be looked up from involved books or manuals.

6.1 Accumulators

6.1.1 Operating principles

There are three types of accumulators: spring loaded, weight loaded and charged gas. The common used is charged gas. Gas and oil fluid can be divided into piston type (Fig. 6-1a), bag type or flexible type (Fig. 6-1b).

Fig. 6-2 shows an accumulator, which basically consists of 4 parts: shell body 1, interlayer 3, air loaded 2 (weight loaded or spring loaded) on the upper of the interlayer and operating oil 4 located under the interlayer and connected with system.

It includes two processes: filled in oil fluid (accumulation) and displacement out oil fluid (release).

1. Filling in oil fluid (accumulation)

It is an equilibrium state of pressure on the two sides of upper and down on the interlayer with the operating oil volume as V_1 after discharging out oil fluid (release) shown in Fig. 6-2a. When operating oil pressure in the accumulator is increased with the oil pressure in hydraulic system, the interlayer losses equilibrium and moves up under the action of operating difference; at the same time oil is filled into the accumulator until a new pressure equilibrium on the two sides of the interlayer occurs and keeps with the operating oil volume as V_2 as shown in Fig. 6-2b.

2. Discharging out oil fluid (release)

When the system pressure is lower than the pressure of the operating oil, the interlay is moved down under the action of gas pressure (or weight loaded or spring loaded); at the same time oil is discharged out of the accumulator and flows into the system until a new pressure equilibrium on the two sides of the interlayer as shown in Fig. 6-2c.

Hydraulic accumulators store oil under pressure. They do this by letting pressurized oil lift weights, compress springs, or compress gas volumes. Then, if the system pressure lowers below the pressure potential of the energy storing device, the weight drops, the spring extends, or the gas ex-

第 6 章　液压辅件

液压辅件主要包括蓄能器、过滤器、油箱、热交换器、密封装置、压力表装置等；液压辅件的合理设计与选用，将在很大程度上影响液压系统的效率、噪声、温升、工作可靠性等技术性能，必须予以重视。

本章主要论述蓄能器、过滤器、油箱、管件和密封装置的结构和原理，其余辅件可查阅相关的液压设计手册。

6.1　蓄能器

6.1.1　工作原理

蓄能器有弹簧式、重锤式和充气式三类，常用的是充气式。在蓄能器中气体和油液被隔离。根据隔离的方式不同，充气式常用的有活塞式（图 6-1a）和气囊式（图 6-1b）。

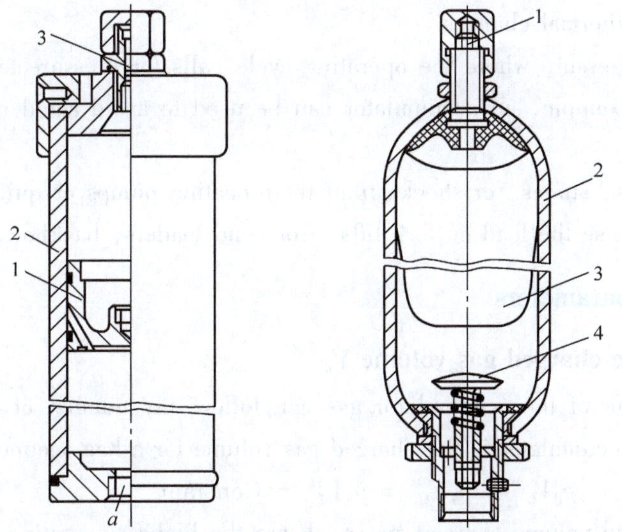

Fig. 6-1　Charged gas accumulator（充气式蓄能器）

a）Piston type（活塞式）

1—Piston（活塞）　2—Cylinder（缸筒）　3—Gas valve（充气阀）

b）Bag type（气囊式）

1—Gas valve（充气阀）　2—Steel shell（壳体）　3—Bladder（气囊）　4—Poppet valve（限位阀）

图 6-2 所示为蓄能器工作原理示意图，蓄能器基本上由四个部分组成：壳体 1，隔层 3，隔层上的可压缩气体 2（或重锤或弹簧），以及隔层下部与系统相连的工作液体 4。

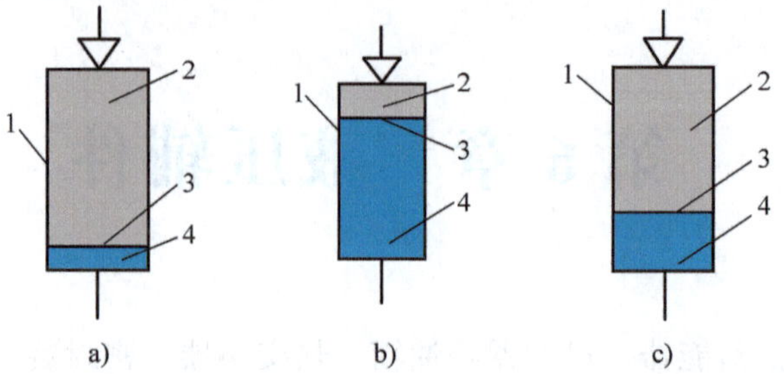

Fig. 6-2 Operating principle of accumulator (蓄能器工作原理示意图)
1—Shell (壳体) 2—Gas (or weight or spring) [可压缩气体(重锤或弹簧)]
3—Interlayer(隔层) 4—Operating fluid(工作液体)

pands. Oil will be expelled from the accumulator under pressure.

6.1.2 Applications

Such an accumulator can be used to advantage in a number of hydraulic applications:

1) To supply oil lost through leakage in a system held under pressure but not operating for long periods. To serve as an auxiliary or emergency source of hydraulic power where pump or feed line have failed. To operate a secondary circuit intermittently in connection with a constant duty primary circuit without affecting the speed or loading of the primary circuit. To compensate for expanding and contracting volumes due to thermal changes.

2) To supply holding pressure where the operating cycle calls for pressure to be applied for a long periods of time. For example, an accumulator can be used to avoid accident when the pump fails randomicity at starting.

3) To absorb pulsations, surges, or shocks from reciprocating pumps or quick-closing valves. To cushion and lower the noise the load on fork lifts, front end loaders, backhoes, etc.

6.1.3 Calculation of parameters

1. Calculation for the charged gas volume V_0

The changes in the rate of the accumulator gas can follow any number of polytropic change curves. The volume of an accumulator is the charged gas volume for a bag accumulator, i. e.

$$p_0 V_0^n = p_1 V_1^n = p_2 V_2^n = \text{Constant} \tag{6-1}$$

Where p_0, V_0 are pressure and volume charged gas; p_1, V_1 are the highest operating pressure and volume at this pressure; p_2, V_2 are the lowest operating pressure and volume at this pressure and n is the exponential.

(1) For the isothermal condition They involve no change in temperature, and no internal energy changes occur in the gas. Take $n=1$, then

$$V_0 = \frac{\Delta V}{p_0\left(\frac{1}{p_1} - \frac{1}{p_2}\right)} \tag{6-2}$$

(2) For the adiabatic condition The other bounding condition is a reversible adiabatic (isen-

其工作过程可分为充液、排液两个阶段。

1. 充液阶段（贮能阶段）

图 6-2a 所示为蓄能器排液后的状态，这时隔层上、下工作的液体和气体（重力或弹簧力）处于平衡状态，此时工作液体体积为 V_1。

当系统压力增高时，工作液的压力也随之增高，破坏了原来的平衡状态，在液压力的作用下，隔层向上移动，系统中的工作液进入蓄能器（这时的体积增加为 V_2），直至达到平衡状态，如图 6-2b 所示。这时系统中有体积为（V_2-V_1）的工作液进入蓄能器贮存起来，此阶段称为充液阶段。在此阶段蓄能器贮存了一定压力和体积为（V_2-V_1）的工作液。

2. 排液阶段（释放阶段）

当系统压力小于蓄能器的工作液压力时，在气体的压力（重力或弹簧力）的作用下，隔层下移，工作液向系统排放，直至平衡状态，如图 6-2c 所示。此阶段称为排液阶段。此阶段把在充液阶段贮存的工作液部分或全部排到系统中。

从上述可知，只要系统压力有变化，蓄能器中的工作液的压力就随之变化，根据力平衡原理，隔层就移动，工作液体积就随之改变，如此反复充液、排液，便达到贮存和释放液压能的目的。

6.1.2 用途

蓄能器是一种用来贮存和释放液压能的装置，合理利用蓄能器是节约能源的手段，其主要功用如下：

1）作辅助动力源。在液压系统工作循环中不同阶段需要的流量变化很大时，常采用蓄能器和一个流量较小的泵组成油源。当系统需要很小流量时，蓄能器将液压泵多余的流量贮存起来；当系统短时期需要较大流量时，蓄能器将贮存的压力油释放出来与泵一起向系统供油。在某些特殊的场合，如驱动泵的原动机发生故障时，蓄能器可作应急能源避免事故发生。

2）保压和补充泄漏。有的液压系统需要较长时间保压而液压泵卸荷，此时可利用蓄能器释放所贮存的压力油，补偿系统的泄漏，保持系统的压力。

3）吸收压力冲击与压力脉动。由于液压阀的突然关闭或换向，系统可能产生压力冲击，此时可在压力冲击处安装蓄能器起吸收作用，使压力冲击峰值降低。如在泵的出口处安装蓄能器，可吸收泵的压力脉动，降低噪声，提高系统工作的平稳性。

6.1.3 参数计算

1. 蓄能器的容积计算

蓄能器的容积计算指的是气腔中气体容积。由气体定律有

$$p_0V_0^{\ n}=p_1V_1^{\ n}=p_2V_2^{\ n}=\text{常数} \tag{6-1}$$

式中，p_0、V_0 分别为蓄能器贮油前的充气压力和气室容积；p_1、V_1 分别为蓄能器最高工作压力和最高工作压力下的气体体积；p_2、V_2 分别为蓄能器维持的最低工作压力和最低工作压力下的气体体积；n 为多变指数。

（1）蓄能器在等温条件下工作时　一般用于维持压力，补偿泄露的蓄能器，其释放能量的速度缓慢，可认为气体在等温条件下工作（$n=1$），可按下式计算

tropic) change which takes place so rapidly that no flow of heat into or out of the gas occurs. The defining relationship is

$$V_0 = \frac{\Delta V}{p_0^{\frac{1}{n}}\left[\left(\frac{1}{p_1}\right)^{\frac{1}{n}} - \left(\frac{1}{p_2}\right)^{\frac{1}{n}}\right]} \tag{6-3}$$

The adiabatic exponent n is the ratio of the specific heat at constant pressure over the specific heat at constant volume. For dry air and dry gaseous nitrogen it is 1.4 over the pressure ranges used for most hydraulic systems.

(3) For the polytropic condition Practically accumulator gas can follow any number of polytropic changed, which is isothermal during accumulating and adiabatic during discharging due to gas expanded. It can be so calculated by

$$V_0 = \frac{\Delta V}{p_0^{\frac{1}{n}}\left[\left(\frac{1}{p_1}\right)^{\frac{1}{n}} - \left(\frac{1}{p_2}\right)^{\frac{1}{n}}\right]} \tag{6-4}$$

Where n is a polytropic change exponent, usually $n = 1.25$.

2. Estimation of pressure in an accumulator

(1) Calculation for lowest operating pressure p_2 The lowest operating pressure p_2 include two parts, which can be calculated by

$$p_2 = p_{\mathrm{imax}} - \sum \Delta p_{\mathrm{max}} \tag{6-5}$$

Where p_{imax} is needed highest operating pressure by working load; $\sum \Delta p_{\mathrm{max}}$ is pressure loss.

(2) Estimation of highest pressure p_1 Estimation of highest pressure in an accumulator p_1 should consider the factors of lift of system and stability. Below an empirical formula is recommended.

$$p_1 = (1.18\text{-}1.25)p_2 \tag{6-6}$$

(3) Estimation of charged gas pressure p_0 For a bag accumulator

$$p_0 = (0.8\text{-}0.85)p_1 \tag{6-7}$$

6.1.4 Application examples

Loaded $F = 1\mathrm{MN}$ when the press machine is working. The moving distance of piston is $S = 3.6$ m with the moving velocity $v = 0.6\mathrm{m/s}$. The hydraulic system pressure $p = 21$ MPa, repetition time is 80s, and the total efficiency of the pump is 0.85. Under the both cases of using or no using the accumulator respectively:

(1) How large is the piston area used?

(2) What is the oil flow rate when piston is moved once?

(3) How high is the power of pump needed?

(4) And how much the volume of accumulator is? (Note: the highest pressure could be reduced 20% when using the accumulator.)

Solution:

Under no accumulator situation:

(1) Piston area needed

$$V_0 = \frac{\Delta V}{p_0\left(\frac{1}{p_1} - \frac{1}{p_2}\right)} \tag{6-2}$$

（2）蓄能器在绝热条件下工作时　当蓄能器用于大量供油时，其释放能量迅速，一般可认为气体在绝热条件下工作，可按下式计算

$$V_0 = \frac{\Delta V}{p_0^{\frac{1}{n}}\left[\left(\frac{1}{p_1}\right)^{\frac{1}{n}} - \left(\frac{1}{p_2}\right)^{\frac{1}{n}}\right]} \tag{6-3}$$

对于干空气和氮气，取多变指数 $n=1.4$。

（3）蓄能器在多变过程时　实际上蓄能器工作过程大多属于多变过程，在贮油时，气体压缩为等温过程，放油时气体膨胀为绝热过程，所以应按下式计算

$$V_0 = \frac{\Delta V}{p_0^{\frac{1}{n}}\left[\left(\frac{1}{p_1}\right)^{\frac{1}{n}} - \left(\frac{1}{p_2}\right)^{\frac{1}{n}}\right]} \tag{6-4}$$

其中，多变指数一般推荐 $n=1.25$。

2. 蓄能器的压力确定

（1）蓄能器的最低工作压力 p_2 的确定　蓄能器的最低工作压力 p_2 应能满足执行机构最大负载工作时所需压力。可按下式计算

$$p_2 = p_{\mathrm{imax}} - \sum \Delta p_{\max} \tag{6-5}$$

式中，p_{imax} 为执行机构所需最大工作压力；$\sum \Delta p_{\max}$ 为蓄能器到最远的执行机构的最大的压力损失之和。

（2）蓄能器最高压力 p_1 的确定　蓄能器的最高压力 p_1 的确定，既要考虑到蓄能器寿命，又要考虑到能适当增加有效排油量；系统压力又不至于过高，且相对稳定。常用的经验公式为

$$p_1 = (1.25 \sim 1.18)p_2 \tag{6-6}$$

（3）蓄能器的充气压力 p_0 的确定　对于气囊式蓄能器，一般取

$$p_0 = (0.8 \sim 0.85)p_1 \tag{6-7}$$

6.1.4 蓄能器应用举例

某液压机在压制时负载 $F=1\mathrm{MN}$，柱塞行程 $S=3.6\mathrm{m}$，运动速度为 $v=0.6\mathrm{m/s}$，系统压力 $p=21\mathrm{MPa}$，每次循环时间为80s，设液压泵的总效率为0.85。求分别不用蓄能器和使用蓄能器时：

（1）所需的柱塞面积；

（2）柱塞移动一次所需输入的液体体积；

（3）所需的传动功率；

（4）蓄能器的容量。（注：用蓄能器时压力允许下降20%）

解：不用蓄能器时：

（1）所需的柱塞面积

$$A_1 = \frac{F}{p} = \frac{1\times10^6}{21\times10^6}\mathrm{m}^2 = 0.0476\mathrm{m}^2 \tag{6-8}$$

$$A_1 = \frac{F}{p} = \frac{1 \times 10^6}{21 \times 10^6}\text{m}^3 = 0.0476\text{m}^2 \tag{6-8}$$

(2) Oil volume needed for piston moving once

$$V_1 = A_1 S = 0.0476 \times 3.6\text{m}^3/\text{s} = 0.171\text{m}^3 \tag{6-9}$$

As the moving speed of piston $v = 0.6\text{m/s}$, so the flow rate needed per second is

$$Q_1 = A_1 v = 0.0476 \times 0.6 = 0.0286\text{m}^3/\text{s} \tag{6-10}$$

(3) Transmission power P_1 (set total efficiency as 0.85), then

$$P_1 = \frac{pQ_1}{10^3 \eta} = \frac{21 \times 10^6 \times 0.0286}{10^3 \times 0.85}\text{W} = 706.6\text{kW} \tag{6-11}$$

Under using accumulator situation:

Since the highest pressure is reduced 20% and the system highest pressure $p_2 = 21\text{MPa}$, so the lowest pressure $p_1 = 21(1 - 20\%) = 16.8\text{MPa}$, i. e. , it should keep the action force of 1MN under the lowest pressure p_1.

(1) Piston area needed is

$$A_1 = \frac{F}{p_1} = \frac{1 \times 10^6}{16.8 \times 10^6}\text{m}^2 = 0.0595\text{m}^2 \tag{6-12}$$

(2) Oil volume needed for piston moving once

$$V_2 = A_2 S = 0.0595 \times 3.6\text{m}^3 = 0.214\text{m}^3 \tag{6-13}$$

As the moving speed of piston $v = 0.6\text{m/s}$, so the flow rate needed per second is

$$t = \frac{S}{v} = \frac{3.6}{0.6}\text{s} = 6\text{s} \tag{6-14}$$

Because of per circle 80s, so supply oil to system only needs 6s and rest 74s for filling oil into the accumulator. In this case, the flow rate by pumping is

$$Q_2 = \frac{0.214}{80}\text{m}^3/\text{s} = 0.002\,68\text{m}^3/\text{s} \tag{6-15}$$

(3) Transmission power is

$$P_2 = \frac{pQ_2}{10^3 \eta} = \frac{21 \times 10^6 \times 0.002\,68}{10^3 \times 0.85}\text{W} = 66.21\text{kW} \tag{6-16}$$

It is shown that $P_2 < P_1$. It saves energy by using the accumulator.

(4) Volume needed in an accumulator A output oil flow rate under 6s is $74 \times 0.002\,68 = 0.198\text{m}^3$. The pressure reduced from 21 MPa to 16.8 MPa. Set charged pressure as

$$p_p = 0.8 p_1 = 0.8 \times 16.8\text{MPa} = 13.4\text{MPa}$$

Under adiabatic (isentropic), for Nitrogen the radio $k = 1.4$. So, the volume of the accumulator is

$$V_0 = \frac{(p_1 p_2)^{1/k} \Delta V}{p_0^{1/k}(p_2^{1/k} - p_1^{1/k})} = \frac{(16.8 \times 21)^{0.71} \times 0.198}{13.4^{0.71}(21^{0.71} - 16.8^{0.71})}\text{m}^3 = 1.588\text{m}^3 \tag{6-17}$$

6.2 Filters

6.2.1 Functions of a filter

It is estimated that there are faults of 70%-80% due to the oil contaminated. The oil contami-

(2) 柱塞移动一次所需输入的液体体积

$$V_1 = A_1S = 0.0476 \times 3.6\text{m}^3 = 0.171\text{m}^3 \tag{6-9}$$

因柱塞运动速度 $v=0.6\text{m/s}$，故一秒内所需油量为：

$$Q_1 = A_1v = 0.0476 \times 0.6\text{m}^3/\text{s} = 0.0286\text{m}^3/\text{s} \tag{6-10}$$

(3) 所需的传动功率 P_1（设液压泵总效率为 0.85）

$$P_1 = \frac{pQ_1}{10^3\eta} = \frac{21 \times 10^6 \times 0.0286}{10^3 \times 0.85}\text{W} = 706.6\text{kW} \tag{6-11}$$

用蓄能器时：

压力允许下降 20%，系统最高工作压力 $p_2=21\text{MPa}$，最低压力 $p_1=21(1-20\%)\ \text{MPa}=16.8\text{MPa}$，在 p_1 时也应能保证 1MN 的力。

(1) 所需的柱塞面积

$$A_1 = \frac{F}{p_1} = \frac{1 \times 10^6}{16.8 \times 10^6}\text{m}^2 = 0.0595\text{m}^2 \tag{6-12}$$

(2) 柱塞移动一次所需输入的液体体积

$$V_2 = A_2S = 0.0595 \times 3.6\text{m}^3 = 0.214\text{m}^3 \tag{6-13}$$

因柱塞速度 $v=0.6\text{m/s}$，故移动 $S=3.6\text{m}$ 所需时间为

$$t = \frac{S}{v} = \frac{3.6}{0.6}\text{s} = 6\text{s} \tag{6-14}$$

循环的时间为 80s，因此，这 6s 内蓄能器与液压泵同时向系统供油，而 74s 内液压泵向蓄能器贮油。因此液压泵的流量为

$$Q_2 = \frac{0.214}{80}\text{m}^3/\text{s} = 0.00268\text{m}^3/\text{s} \tag{6-15}$$

(3) 传动功率

$$P_2 = \frac{pQ_2}{10^3\eta} = \frac{21 \times 10^6 \times 0.00268}{10^3 \times 0.85}\text{W} = 66.21\text{kW} \tag{6-16}$$

可见，比不用蓄能器时节省了许多能量。

(4) 蓄能器的容量　因蓄能器在 6s 内输出的油量为 $74 \times 0.00268\text{m}^3 = 0.198\text{m}^3$，而压力从 21MPa 降到 16.8MPa，取充气压力

$$p_p = 0.8p_1 = 0.8 \times 16.8\text{MPa} = 13.4\text{MPa}$$

绝热膨胀时，对于氮气取 $k=1.4$，故蓄能器的容量为

$$V_0 = \frac{(p_1p_2)^{1/k}\Delta V}{p_0^{1/k}(p_2^{1/k} - p_1^{1/k})} = \frac{(16.8 \times 21)^{0.71} \times 0.198}{13.4^{0.71}(21^{0.71} - 16.8^{0.71})}\text{m}^3 = 1.588\text{m}^3 \tag{6-17}$$

6.2　过滤器

6.2.1　过滤器的功用

据统计，液压系统中有 70% ~80% 的故障和液压油的污染有关。油液被污染所混入的颗粒状杂质或油液自身氧化生成的氧化物等，将加速液压元件的磨损、卡死运动件或堵塞阀口、工作间隙和小孔，使元件失效。油液的污染程度直接影响元件和系统的正常工作及可靠

nated with impurity or oxide will result in abrasion or block of elements quicken up. The filter is used to try to eliminate the impurity and make a clean oil pipe system. Thus it is important to choose filter for controlling the contamination and keeping a normal hydraulic system.

6.2.2 The types of filter elements

Filter elements are divided into two kinds of coarse and precise filters in terms of the precision. They can be divided into mesh, line-gap, paper-core and magnetism filters in terms of configuration. And they can be also divided into three classifies: surface-type, depth-type and magnetism filters in terms of the principle of filtration.

1. Surface-type elements

The surface of elements touches directly with hydraulic medium and consists of a single layer of material through which the fluid must pass and keep the impurity outside of it. The most common are wire-type filter and a line clearance-type filter.

Fig. 6-3a is a wire-type filter. The structure of the wire-type filter is around holes plastic or metal which covers with 1-2 layers brass wires on a framework. The precise is usually 0.08-0.18mm and determined by the dimension of hole and numbers of layer. The advantages of wire-type filters are simple configuration and convenient cleanliness. The pressure loss cannot be exceeded 0.01MPa and it is usually installed in the suction of pump. Fig. 6-3b is a line clearance-type filter. The material of core layer consists of copper or aluminium. It works by the tiny gap. The precision is 0.03-0.1mm and the pressure loss is 0.07-0.35MPa. It has the features with simple structure but lower intensity and usually is installed in a lower pressure pipe or suction of pump.

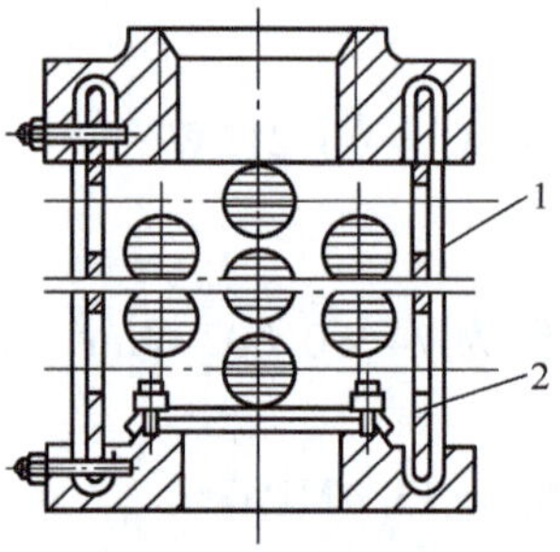

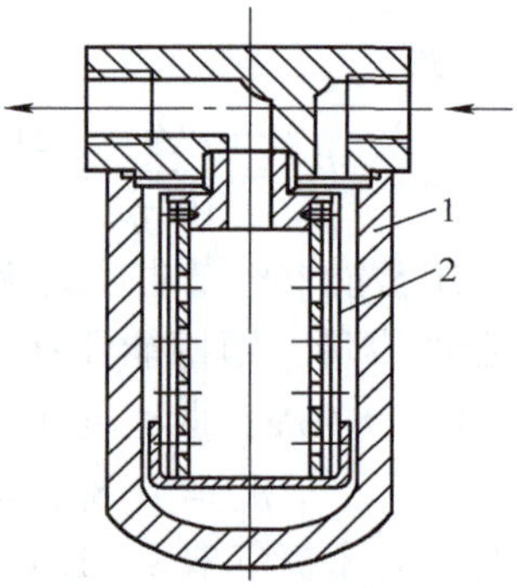

Fig. 6-3 Surface filter (表面型过滤器)

a) Wire filter(网式过滤器) b) Line filter (线隙式过滤器)

1—Copper net(铜丝网) 2—Framework(筒形骨架)

2. Depth-type elements

Because of its construction, a depth-type element has many pores of various sizes. The depth-type elements force the fluid to pass through multiple layers of material and the dirt is caught because of the intertwining path that the fluid takes. The core material consists of paper, metal, woolen or various fibres. Fig. 6-4a is a paper-type core filter and Fig. 6-4b is a sintered filter. They have the features of a high precision with 0.01-0.06mm. It has also a big pressure loss of 0.03-0.2MPa and usually used in a high precision situation.

性。过滤器的功用就是滤去油液中的杂质，维护油液的清洁，防止油液污染，保证系统正常工作。

6.2.2 过滤器的类型

过滤器按过滤精度可分为粗过滤器和精过滤器两大类；按滤芯的结构可分为网式、线隙式、烧结式、纸芯式等；按过滤材料的过滤原理可分为表面型、深度型、磁性三种。

1. 表面型过滤器

这种过滤器的滤芯表面与液压介质接触，就像筛网一样把污物阻留在其外表面。最常用的有网式和线隙式过滤器两种。

图6-3a所示为网式过滤器，其结构是在周围开有很多窗孔的塑料或金属筒形骨架2上包着1~2层铜丝网1。过滤精度由网孔大小和层数决定，过滤精度为0.08~0.18mm。网式过滤器结构简单，清洁方便，压力损失不超过0.01MPa，常安装在泵吸油口。图6-3b所示为线隙式过滤器，滤芯是用铜或铝线绕在筒形骨架上，依靠线间的微小缝隙进行过滤，过滤精度为0.03~0.1mm，压力损失为0.07~0.35MPa，其结构简单，但滤芯强度较低，常用于回油低压管路或泵吸油口。

2. 深度型过滤器

这种过滤器的滤芯由多孔可透性材料制成，材料内部具有曲折迂回的通道，大于表面孔径的粒子直接被拦截在靠油液上游的外表面，而较小污染粒子进入过滤材料内部，撞到通道壁上，滤芯的吸附及迂回曲折通道有利污染粒子的沉积和截留。这种滤芯材料有纸芯、烧结金属、毛毡和各种纤维类等。图6-4a所示为纸芯式过滤器，图6-4b所示为烧结式过滤器。这种过滤器的优点是过滤精度高，可达0.01~0.06mm，但压力损失也较大，一般为0.03~0.2MPa，常用于过滤精度要求较高的场合。

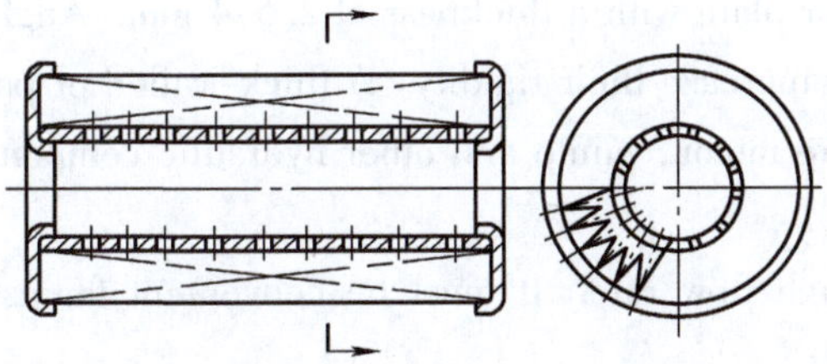
a)

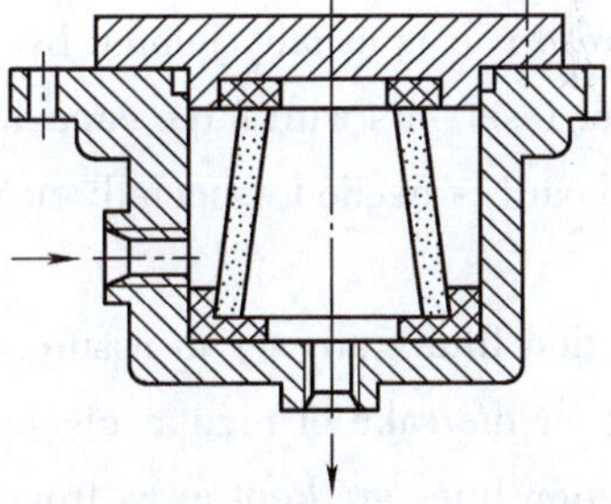
b)

Fig. 6-4 Depth filter（深度型过滤器）

a) Paper-type core filter（纸芯式过滤器） b) Sinter filter（烧结式过滤器）

3. 磁性过滤器

其原理是利用磁铁吸附油液中的铁质微粒，但对其他污染颗粒不起作用，常与其他形式滤芯一起制成复合式过滤器，适用于金属切削机床液压系统中。

3. Magnetism filters

The principle of magnetism filters is an absorbent-type element, which absorbs the particle of metal in the oil. However it cannot absorb the other non-metal particle. So it is usually combined with other type of filters and used in hydraulic system for metal-cutting machine tool.

6.3 Reservoirs

1. Functions and construction

A reservoir provides oil to make up for system leakage, allows space for the expansion or contraction of the oil with changes of temperature, enables gas bubbles to rise out of the oil and dirt particles to sink to the bottom, and allows room for heat dissipation or for the inclusion of thermal control devices. In some specialized systems a reservoir is also a low-level pressure control system.

There are two kinds of reservoirs: integer and separateness. Integer reservoir is installed with machine; it has the advantages of compact and easy recycles for leaked oil but also the disadvantages of difficult dissipation for heat and difficult maintenance. The separateness reservoir is separately installed; it has been used widely in engineering with many advantages of convenient maintenance, dissipation heat and reducing vibration. Fig. 6-5 shows a small separate reservoir.

2. Problems in design

1) The determination to the reservoir fluid capacity is the key to the whole reservoir design. As to the systems under heavy load and continuously run for long time, a reservoir fluid capacity is mainly designed in terms of thermal-balance. As to the other systems, a reservoir capacity is simply selected 3-8 times the flow rate of the pump: for low pressure systems, 2-4 times; for medium pressure, 5-7 times; for high pressure, 10-12 times. Furthermore, the oil lever must be kept under 80% the reservoir height to prevent the returned oil from overflow.

2) A reservoir is usually jointed by armor plate with a thickness of 2.5-4 mm. Angle iron is also needed for reservoirs with large capacity to increase their rigidity. A thick-walled or part-strengthened cover board is needed especially in where motor, pump and other hydraulic components are installed.

3) Suction filters are set to assure enough flow rate. It must be convenient for assembly and disassembly for the sake of regular cleanness.

4) Suction lines are kept away from return lines as far as possible. Usually a clap board with a height 3/4 the oil lever is set between them to increase the distance of the oil cycle and help heat-exchange, sedimentation, air bubble separation. The ends of the suction and return lines are chopped in 45° (the tilted incisions are faced to the reservoir wall) to enlarge through-flow area. The distances among the pipe ends, the reservoir bottom and the reservoir wall should be designed more than three times the diameter of the pipe. The distance between the filter and the reservoir bottom must be longer than 20mm.

5) The cover board of the reservoir and the ends of the pipes must be safely sealed to prevent oil contamination. An air filter is set on the cover board. It also serves as venthole to prevent negative pressure on the reservoir or in the process of adding oil.

6.3 油箱

1. 油箱的功用和结构

油箱的主要功用：贮存液压系统所需的足够油液，散发油液中的热量，分离油液中气体及沉淀污物等。

油箱有整体式和分离式两种。整体式利用本机的内腔作为油箱。其结构紧凑，漏油易于回收。但散热性差，易使主机产生热变形，影响精度，再则维修不便，机身复杂。分离式油箱单独设置，与主机分开，它布置灵活，维护方便，可减小油箱发热和液压振动对主机工作精度的影响，便于设计成通用化、系列化的产品，因而得到广泛的应用，特别是组合机床、自动线和精密设备等。图 6-5 所示为分离式油箱。

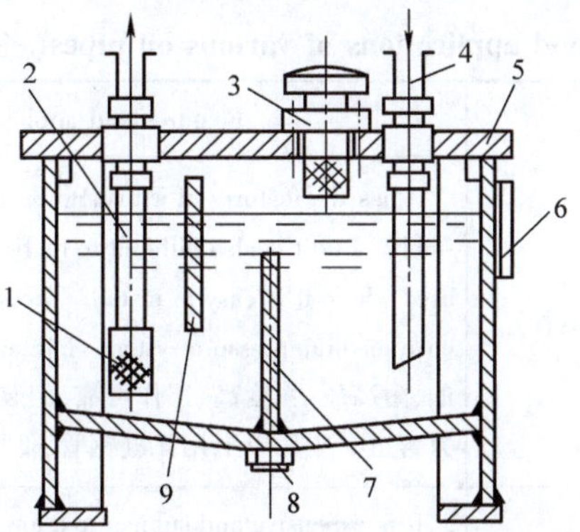

Fig. 6-5 Separate reservoir（分离式油箱）

1—Wire type filter（网式过滤器） 2—Suction oil pipe（吸油管） 3—Air filter（空气过滤器）
4—Oil charged pipe（加油管） 5—Top cover（顶盖） 6—Indicator of oil position（油面指示器）
7，9—Clapboard（隔板） 8—Oil escape valve（放油塞）

2. 油箱设计时应注意的问题

1）油箱容量的确定是油箱设计的关键。主要根据热平衡来确定，但这只是在系统负载较大，长期连续工作时才有必要进行。通常油箱的容量取液压泵每分钟流量的 3 ~ 8 倍进行估算，低压系统为额定流量的 2 ~ 4 倍，中压系统为 5 ~ 7 倍，高压系统为 10 ~ 12 倍。此外，还要考虑到液压系统回油到油箱不至溢出，油面高度一般不超过油箱高度的 80%。

2）油箱一般用 2.5 ~ 4mm 的钢板焊接而成的，对于大容量的油箱还要加焊角铁、加强筋以增强其刚性。特别是在油箱盖板上安装电动机、泵和液压元件时，盖板要适当加厚和局部加强。

3）油箱中应设吸油过滤器，要有足够的通流能力。因需经常清洗过滤器，所以在油箱结构上要考虑拆卸方便。

4）吸油管和回油管要尽量远离，它们之间要用隔板隔开，增加油液循环的距离，以利散热、沉淀杂质和分离气泡，隔板高度一般为液面高度的 3/4。吸油管和回油管管端切成 45°，以增大通流面积，斜切口应面向箱壁。管端与箱底、箱壁间距离应大于管径的三倍，

6) An oil level gauge and a thermometer are usually set on the side wall of the reservoir. The detailed information can be found in related books or manuals.

6.4 Piping/Tubing and Connectors

Piping connectors include oil pipes and connectors. The functions of piping connections are used to join hydraulic elements and transfer fluid connectors. Piping connections should have good rigidity, sealing, less leak, low pressure loss and convenient replacement and fixing.

1. Oil piping

Piping is made from a variety of materials, including steel, copper, brass, nylon, plastics, and rubber pipe. It is decided by system pressure and operating conditions. The features and applications are listed in Tab. 6-1.

Tab. 6-1 The features and applications of various oil pipes(各种油管的特点及适用场合)

Types(种类)		Features and applications(特征和适用场所)
Tubing(硬管)	Steel tube(钢管)	It has the features of withstand oil and high pressure, high rigidity and working stably. But it is also difficult to be bended in installation. High pressure is usually used where it is easy to install. The steel tubing seamless steel tube is used in above medium pressure system. Jointed tubing is used in a low pressure system(耐油、耐高压、强度高、工作可靠,但装配时不便弯曲,常在装拆方便处用作压力管道。中压以上用无缝钢管,低压用焊接钢管)
	Copper pipe(纯铜管)	It is expensive and subject to a low pressure (6.5-10MPa). It withstands with a low shock and vibration. It is easy to be bended and corrosive. It is commonly used where the system is difficult to install, such as instrument(价高、承载能力低(6.5~10MPa),抗冲击和振动能力差,易使油液氧化,但易弯曲成各种形状,常用在仪表和液压系统装配不方便)
Hose(软管)	Plastic pipe(塑料管)	It has the features of withstand oil, low price, convenient installation, and easy to age for a long use. It is used in low pressure (<0.5MPa), such as return oil pipe.(耐油、价低、装配方便,长期使用易老化,只适用于压力低于0.5MPa的回油管或泄油管)
	Nylon pipe(尼龙管)	It is an ivory-white and sight flowing. It is low in price and easy to be bended when heated. It is easy to be bended, enlarged and formed after cooling. It is easy to install. It tends to be used late.(乳白色透明,可观察流动情况,价低,加热后可随意弯曲,扩口、冷却后定形,安装方便,承压能力因材料而异(2.5~8MPa),今后有扩大使用的可能)
	Rubber pipe(橡胶管)	It is used where the hydraulic line is to be subjected to movement. It has two types of low and high pressure hose. The high pressure hose is made with several plies of braided steel wire reinforcement to withstand working pressures. It is expensive and used in a high pressure system. The low pressure type is made of resistant oil rubber with plies of sailcloth inner tube, which is used in low pressure system.(用于相对运动部件的连接,分高压和低压两种。高压软管有耐油橡胶夹有几层钢丝编织网(层数越多耐压越高)制成,价高,用于压力管路。低压软管有耐油橡胶夹帆布制成,用于回油管路)

过滤器距箱底不应小于20mm。

5）为防止油液污染，油箱盖板、管口处都要妥善密封。盖板上还应设置空气过滤器，它还兼作为加油和防止油箱出现负压而设置的通气孔。

6）油箱侧壁应设置油位计及温度计。

具体尺寸、结构可参看有关资料及设计手册。

6.4 管件

管件包括油管和管接头，其功用是连接液压元件和输送液压油。它应保证有足够强度，密封性好，无泄漏，压力损失小和装拆方便等。

1. 油管

液压系统常用油管有钢管、纯铜管、尼龙管、塑料管和橡胶软管等。应当根据液压装置工作条件和压力大小来选择油管。各种油管的特点及适用场合见表6-1。

油管尺寸的确定：油管尺寸主要指内径 d 和壁厚 δ，它是根据液压系统的流量和压力来计算，即

$$d = 2\sqrt{\frac{q}{\pi v}} \tag{6-18}$$

$$\delta = \frac{pd}{2[\sigma]} \tag{6-19}$$

式中，p、q 分别为管内的工作压力和最大流量；v 为允许流速，吸油管取0.5～1.5m/s，回油管取1.5～2m/s，压力油管取2.5～5m/s（压力高、流量大、管路短时取大值），控制油管取2～3m/s，橡胶软管取<4m/s，短管及局部收缩处取5～7m/s；$[\sigma]$ 为管材的许用应力，对钢管 $[\sigma]=\sigma_b/n$，σ_b 为抗拉强度，n 为安全系数，对钢管，$p<7\text{MPa}$ 时取 $n=8$，$7\text{MPa}<p<17.5\text{MPa}$ 时取 $n=6$，$p>17.5\text{MPa}$ 时取 $n=4$。

计算出 d 和 δ 后，可查有关手册按标准系列值选取。

2. 管接头

管接头是油管与液压元件、油管与油管之间可拆卸的连接件。管接头必须在强度足够的前提下，在压力冲击和振动下要保持管路密封性好、连接牢固、外形尺寸小、加工工艺性好、压力损失小等要求。管接头种类繁多，具体规格与品种可查阅有关手册。管接头的联接螺纹采用国家标准米制锥螺纹（ZM）和普通细牙螺纹（M）。在液压系统中常用的管接头有扩口式管接头（图6-6）、焊接式管接头（图6-7）、卡套式管接头（图6-8）、橡胶软管接头（图6-9）和快速管接头（图6-10）等。其中快速管接头是一种快速装拆的接头，适用于需经常接通和断开的管路系统。图示为油路接通的情况。当需油路断开时，用力将外套6向左推，钢球8即从接头体10的槽中退出，再拉出接头体10，两单向阀阀芯4和11分别在弹簧3和12的作用下外伸，顶在接头体2和10的阀座上而封闭油路。当需重新接通油路时，仍将外套6左推，并插入接头体10，使单向阀阀芯4和11互相顶紧，油路接通。

液压系统的泄漏问题大都出现在管路的接头上，所以对接头形式、材料与管路的设计、安装都要认真对待，否则将影响液压系统的工作性能。

Just as other tubular materials, tubing is measured by its inside diameter d, and thickness of wall δ, i. e.

$$d = 2\sqrt{\frac{q}{\pi v}} \tag{6-18}$$

$$\delta = \frac{pd}{2[\sigma]} \tag{6-19}$$

Where p, q are working pressure and the highest flow rate respectively; v is flow speed permitted, suction pipe is 0. 5-1. 5m/s, return pipe is 1. 5-2m/s, the pressure pipe is 2. 5-5m/s and control oil pipe is 2-3m/s, the rubber pipe is <4m/s, short pipe or local contraction is 5-7m/s. $[\sigma]$ is a permitted stress and $[\sigma] = \frac{\sigma_b}{n}$ for steel tubing; σ_b is a flex intensity; n is a safe coefficient, and $n = 8$ when its pressure is at $p < 7$MPa, $n = 6$ when its pressure is at 7MPa $< p < 17.5$MPa, and $n = 4$ when pressure is at $p > 17.5$MPa for steel tubing.

The values of d and δ after calculation should be chosen by looking up from references.

2. Connectors

Connectors-more frequently called fittings-are used to join pipe lines and to connect pipe lines to components. No matter what the configuration of the connector is used, it performs three tasks under enough intensity: the connector must join to the line in a firm, leakproof manner; the connector must carry any loads imposed on it by the hydraulic system or by the line; and the connector must provide a seal between the parts being joined. Tubing connector design by Chinese country's standard, metric international standard thread, is denoted as ZM and plain thread denoted as M. There are many types of connectors which can be looked up from references. They include flared shape (Fig. 6-6), joining style (Fig. 6-7), clip style (Fig. 6-8), hosepipe connector (Fig. 6-9) and fast connector (Fig. 6-10).

One should pay more attention to decide which style connector should be chosen because of the big leaking trouble between piping and connectors.

6. 5 Seals and Sealing Devices

The sealing devices are used to prevent oil from internal and external leakage, hold back pollutant and keep the needed system pressure. Abilities of sealing devices affect directly on operating efficiency and performances and are important for hydraulic system.

Types and characters:

There are many kinds of sealing devices; however according to the situation of application, there are two main kinds of sealing devices: dynamic and static.

1. Clearance sealing

Clearance sealing is the simplest type utilizing gap δ between two motions such as piston, cylinder or valve as shown in Fig. 6-11. The fitting clearance between cylinder surfaces is related to the diameter. The clearance is 0. 005- 0. 017mm between spool core and hole. It has an advantage of low friction force. However it cannot be compensated after abrasion.

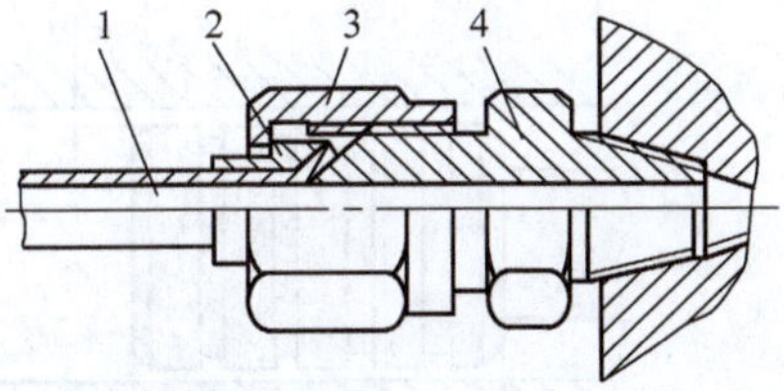

Fig. 6-6　Flared connector（扩口式管接头）

1—Pipe connected（接管）　2—Oriented sleeve（导套）

3—Nut（螺母）　4—Connection body（接头体）

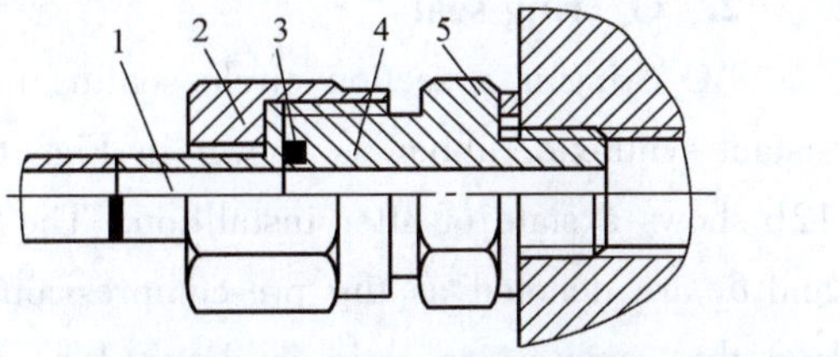

Fig. 6-7　Joining style connector（焊接式管接头）

1—Pipe connected（接管）　2—Nut（螺母）

3—“O” ring seal（O 形密封圈）

4—Connection body（接头体）

5—Compound Seals（组合密封圈）

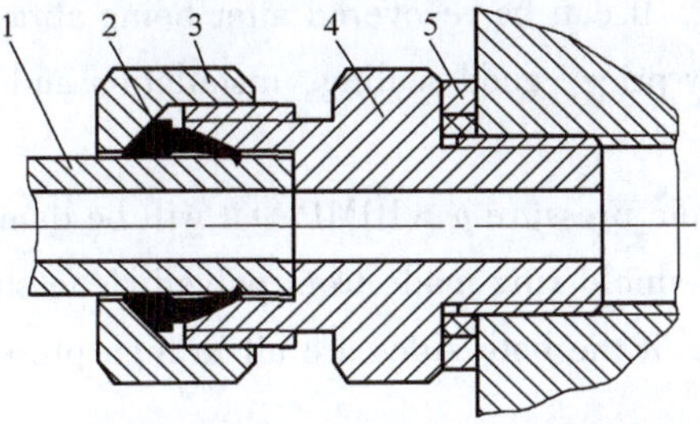

Fig. 6-8　Clip style connector（卡套式管接头）

1—Pipe connected（接管）　2—Clip（卡套）

3—Nut（螺母）　4—Connection body（接头体）

5—Compound Seals（组合密封圈）

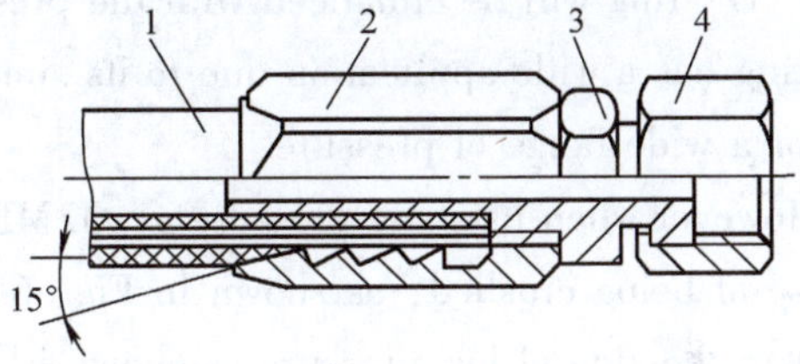

Fig. 6-9　Hosepipe connector（橡胶软管接头）

1—Rubber pipe（胶管）　2—Shell（外壳）

3—Connection body（接头体）　4—Nut（螺母）

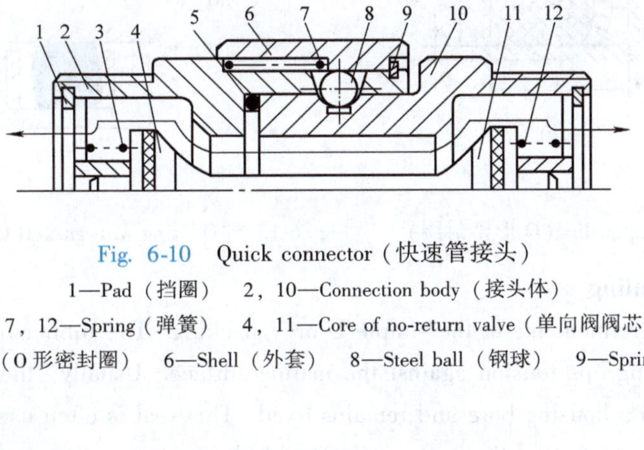

Fig. 6-10　Quick connector（快速管接头）

1—Pad（挡圈）　2，10—Connection body（接头体）

3，7，12—Spring（弹簧）　4，11—Core of no-return valve（单向阀阀芯）

5—“O” ring seal（O 形密封圈）　6—Shell（外套）　8—Steel ball（钢球）　9—Spring ring（弹簧圈）

6.5　密封装置

密封装置是用来防止液压系统油液的内、外泄漏以及外界灰尘和异物的侵入，保证系统建立所需的工作压力的。密封装置的性能直接影响液压系统的工作性能和效率，是衡量液压系统性能的一个重要指标。

密封装置的类型及特点：

密封装置的种类很多，按被密封部位运动情况，可分为动密封和静密封两大类。常用的密封有以下几种。

2. "O" ring seal

"O" ring is a section circle sealing made of oil resistant synthetic rubber as shown in Fig. 6-12a. Fig 6-12b shows a state of after installtion. The symbols of δ_1 and δ_2 are denoted as the pre-compression respectively, and the compression ratio is denoted as w, i. e, $w = [(d_0 - h)/d_0] \times 100\%$, the other symbols in this expression are shown in Fig. 6-12. The compression ratio w should reach 15%-20% for static sealing, 10%-20% for reciprocate sealing and 5%-10% for sealing of turn around motion respectively. The ability of sealing of "O" ring will be enhanced with the pressure rising. It can be recovered after being abrasion. "O" ring has a wide application due to its simplicity, low price, good sealing, installation and suitable for a wide range of pressure.

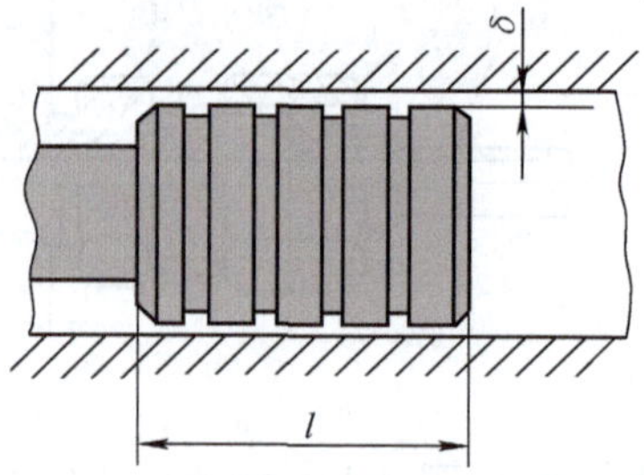

Fig. 6-11 Clearance sealing (间隙密封)

However when its static pressure $p > 32$MPa or dynamic pressure $p > 10$MPa, it will be damaged because of being crushed, as shown in Fig. 6-13a. So a shield ring made from polyethylene should be put in the side of low pressure as shown in Fig. 6-13b. if the both sides are all in high pressure, two shield rings are needed as shown in Fig. 6-13c.

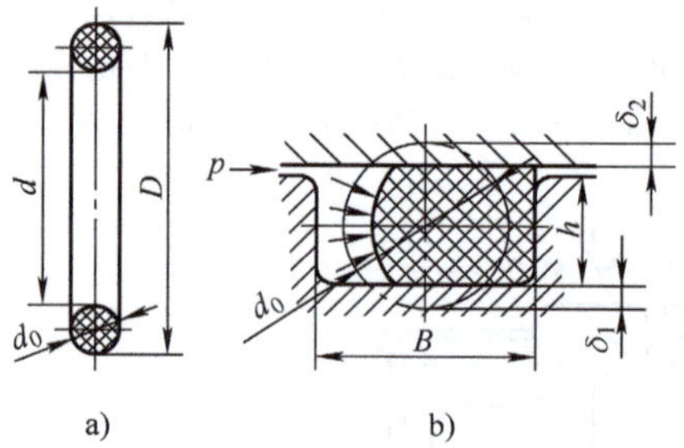

Fig. 6-12 "O" ring sealing(O 形密封圈)

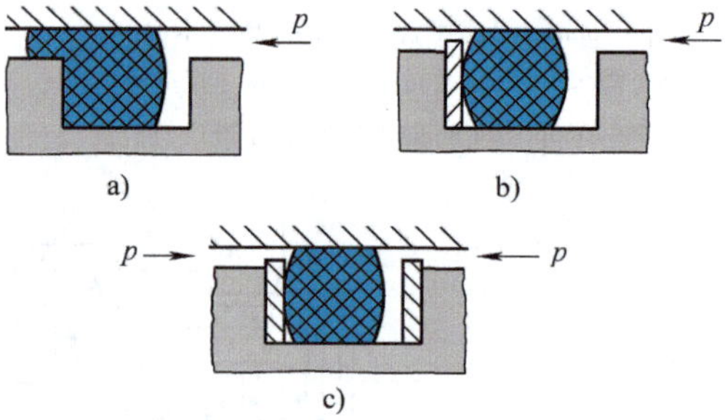

Fig. 6-13 "O" ring with gasket(O 形密封圈的挡圈安装)

3. Lip style sealing

These seals are refinements of the simple U or V packing. The rubber lip is ringed by a spring which gives the sealing lips tension against the mating surface. Usually, the seal has a metal case which is pressed into a housing bore and remains fixed. This seal is often used to seal rotary shafts. The lip normally faces in toward the system oil. Double-lip seals are sometimes used to seal in fluids on both sides of an area. The common styles are y-style (Fig. 6-14, 15), Y_x-style (Fig. 6-16) and V-style (Fig. 6-17).

(1) Y-style Fig. 6-14 and 6-15 show Y-style seals. It is made of rubber proof against oil. In the installation of Y-style the lip should bear against the side exposed to high pressure. When the working pressure reaches more than 14MPa or at a high change of pressure and at a high slid speed, a fixed ring is needed in order to prevent from the overturn of Y-style seal and have a good seal ability as shown in Fig. 6-15. Y-style seal is usually used in the situation with less than 32MPa of working pressure, −30-100℃ of operating temperature and 0.5m/s slid speed.

1. 间隙密封

间隙密封是利用相对运动之间微小间隙 δ 起密封作用。常用于柱塞、活塞或滑阀的圆柱面配合副中，如图 6-11 所示。其圆柱面配合间隙与直径大小有关，对于阀芯与阀孔一般取 $\delta = 0.005 \sim 0.017\text{mm}$。这种密封的优点是摩擦力小，缺点是磨损后不能自动补偿。

2. O 形密封圈

O 形密封圈一般用耐油橡胶制成，其截面为圆形，如图 6-12a 所示。图 6-12b 所示为装入密封沟槽后的情况，δ_1、δ_2 分别为其预压缩量，通常用压缩率 W 表示，即 $W = [(d_0 - h)/d_0] \times 100\%$，式中符号意义见图示；对于静密封、往复运动密封和回转运动密封，压缩率应分别达到 15% ~20%、10% ~20% 和 5% ~10%，才能取得满意的密封效果。随压力的增加能自动地提高密封性能，并在磨损后具有自动补偿的能力。由于 O 形密封圈结构简单、密封性好、成本低、装拆方便，高低压均可用。

但是当静密封压力 $p > 32\text{MPa}$ 或动密封压力 $p > 10\text{MPa}$ 时，O 形密封圈有可能被压力油挤入间隙而损坏，如图 6-13a 所示。为此在 O 形密封圈低压侧安置聚四氟乙烯挡圈，如图 6-13b 所示。当双向受压力油作用时，两侧都要加挡圈，如图 6-13c 所示。

3. 唇形密封圈

唇形密封圈是依靠密封圈的唇口受液压力作用变形，使唇边贴紧密封面而进行密封的，液压力越高，唇边贴得越紧，并且具有磨损后自动补偿的能力。这类密封一般用于往复运动密封。常见的有 Y 形（图 6-14、图 6-15）、Y_x 形（图 6-16）、V 形（图 6-17）等。

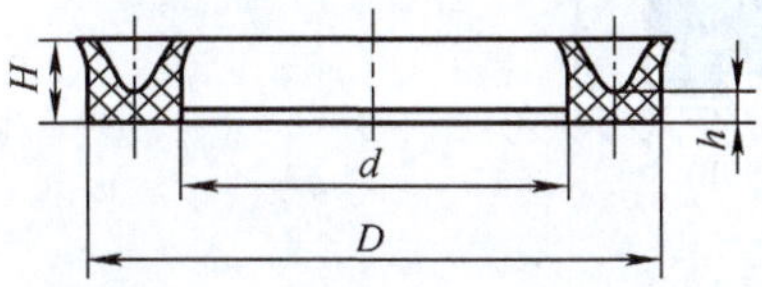

Fig. 6-14 Y-style lip seal

（Y 形密封圈）

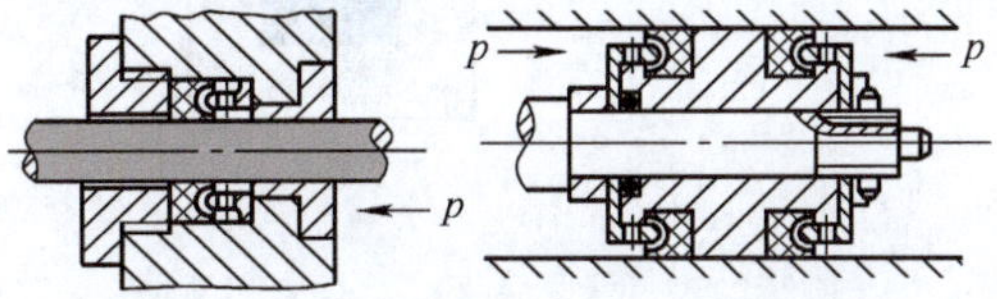

Fig. 6-15 Install for Y-style seal

（Y 形密封圈的安装及支承环结构）

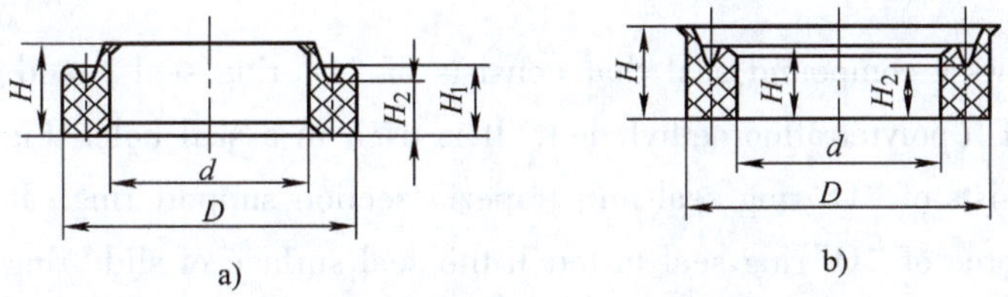

Fig. 6-16 Y_x-style seal（Y_x 形密封圈）

a) Hole used（孔用） b) Spindle used（轴用）

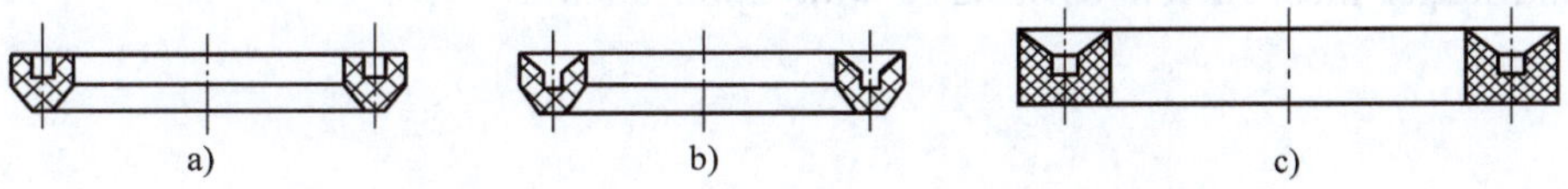

Fig. 6-17 V-style lip seal（V 形密封圈）

a) Support ring（支承环） b) Sealing ring（密封环） c) Press ring（压环）

(2) Y_x-style Y_x-style is an improvement of Y-style sealing and made of polyesteramide. As shown in Fig. 6-16, the ratio of the height of the section surface to its width is larger than 2, so it is hard to be overturned and has a good stability. Y_x-style has two lips with different height. The short one is used as seal side, it touches directly with the sealed surface with a low friction force and another long one touches non-slide surface and increases compression and friction force in order to be stable.

(3) V-style Fig. 6-17 shows a V-style sealing. It is made of layers textile and consists of support, seal and pressure rings. When the working pressure $p > 10$MPa, several seal rings are needed in order to meet the need of sealing. When it is installed the V-style orifice should bear against the side exposed to high pressure. V-style is usually used in the environment with a working pressure $p \leqslant 50$MPa and a temperature range −40-80℃.

4. Compound Seals

With the development of hydraulic technology, one requires high temperature-resistant, speed-resistant, lower friction and long life used to the reciprocator, so there are compound seals made of PTFE (polytetrafluoroethylene) with oil-resistant synthetic rubber and a new kinds of seal is shown in Fig. 6-18.

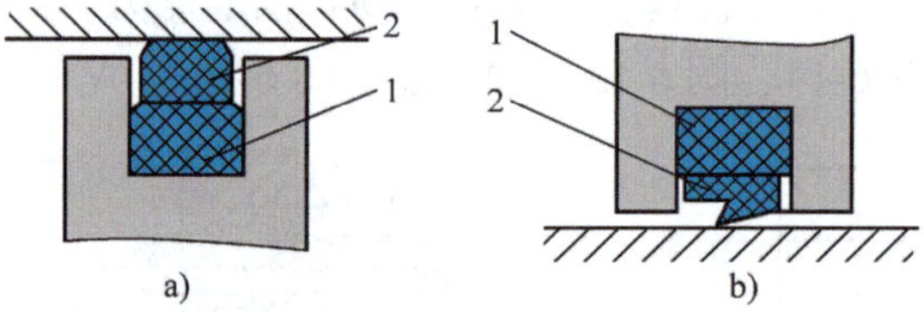

Fig. 6-18 Compound seals made of PTFE and oil-resistant synthetic rubber (橡胶组合密封装置)
a) Used in hole(孔密封) b) Used in axes(轴密封)
1—"O" ring seal(O 形密封圈) 2—Slid ring(滑环)

Fig. 6-18a shows a compound seal that consists of "O" ring seal 1 and rectangle section slid ring 2 made of PTFE (polytetrafluoroethylene). It is used in a seal hole. Fig. 6-18b shows a compound seal that consists of "O" ring seal and trapezia section support ring. It utilizes the pre-compression distortion force of "O" ring seal to touch the seal surface of slide ring (or support ring) so as to keep the sealing property. The "O" ring seal does not touch directly with surface sealed and so has no problems of friction, and torsion. It has the advantage of both materials of rubber and PTFE (polytetrafluoroethylene), and it has not only a good sealing property but also less friction force, stability and longer life. Therefore, it has a wide application.

(1) Y 形密封圈　图 6-14 所示为 Y 形密封圈，用耐油橡胶压制而成。如图 6-15 所示，安装 Y 形密封圈时，唇口一定要对着压力高的一侧。当工作压力大于 14MPa 或压力波动较大，滑动速度较高时，为了防止 Y 形密封圈的翻转，应加支承环固定密封圈，支承环上有小孔，使压力油经小孔作用到密封圈唇边上，以保证良好密封，如图 6-15 所示。Y 形密封一般适合工作压力不大于 32MPa、工作温度为 -30 ~ 100℃、滑动速度小于或等于 0.5m/s 的场合。

(2) Y_x 形密封圈　Y_x 形密封圈是由 Y 形密封圈改进设计，通常用聚氨酯材料压制而成的。如图 6-16 所示，其断面高度与宽度之比大于 2，因而不易翻转，稳定性好，分为轴用与孔用两种。Y_x 形密封圈的两个唇边高度不等，其短边为密封边，与密封面接触，滑动摩擦阻力小；长边与非滑动表面相接触，增加了压缩量，使摩擦阻力增大，工作时不易窜动。

(3) V 形密封圈　图 6-17 所示为 V 形密封圈，它是由多层涂胶织物压制而成的，由支承环、密封环和压环三部分组成一套使用。当工作压力 $p > 10\text{MPa}$ 时，可以根据压力大小，适当增加密封环的数量，以满足密封要求。安装时，V 形密封圈的 V 形口一定要面向压力高的一侧。V 形密封圈适宜在工作压力 $p \leqslant 50\text{MPa}$、工作温度为 -40 ~ 80℃ 条件下工作。

4. 组合密封装置

随着液压技术的发展，对密封装置提出了耐高压、高温、高速，低摩擦因数，长寿命和可靠性等方面的要求，于是出现了一些新型密封装置，如图 6-18 所示。

如图 6-18a 所示，橡胶组合密封装置由 O 形密封圈 1 与截面为矩形的聚四氟乙烯滑环 2 组合而成，用于孔密封；如图 6-18b 所示，橡胶组合密封装置由 O 形圈与截面为阶梯形的支持环组成，用于轴密封。利用 O 形密封圈的弹性变形所产生的预压力将滑环（或支持环）紧贴在密封面上起密封作用的。O 形密封圈不与密封面直接接触，不存在摩擦、扭转、啃伤等问题，它充分发挥了橡胶密封圈与聚四氟乙烯滑环（或支持环）的各自优点，不仅密封可靠，摩擦力小而稳定，而且使用寿命比普通橡胶密封圈提高近百倍，应用日益广泛。

Chapter 7 Basic Hydraulic Circuits

The materials presented in the previous chapters deal with fundamentals and system components. In this chapter we shall discuss basic hydraulic circuits. Any hydraulic system is constituted by several basic hydraulic circuits. A hydraulic basic circuit is a group of components such as pumps, actuators, control valves, and conductors so arranged that they will perform a useful task.

When analyzing or designing a hydraulic circuit, the following three important considerations must be taken into account: ①safety of operation; ②performance of desired function; ③efficiency of operation.

Let's classify the functions in hydraulic system by:

1) Pressure control circuits: control entire or partial operating pressure.

2) Speed control circuits: control and regulate motion speed of actuator elements.

3) Directional control circuits: control and shift motion direction of actuator.

4) Multi-actuator control circuits: control and shift operating cycle between several actuators.

The most basic hydraulic circuits will be detailed in this chapter. To be familiar with them is very important for analysis, design and use of the hydraulic system.

7.1 Pressure Control Circuits

Pressure control circuits are used to control and regulate the entire and partial pressure to realize a satisfied force and torque requirements, enhance working efficiency and improve the properties in a hydraulic system. Pressure control circuits include pressure regulated circuit, pressure-venting circuit, pressure-reducing circuit, pressure-increasing circuit, pressure counter-balance circuit, pressure-holding circuit and pressure-releasing circuit.

7.1.1 Pressure regulated circuit

Pressure regulated circuit is used to predetermine or limit the highest system pressure or fulfill multistage pressure shift to an actuator. Relief valve is usually used here.

1. Pressure regulated from remote port

Fig. 7-1a shows a simple pressure regulated circuit. Consider that one disconnects the remote control port and only the relief-valve pilot 1 is installed into the system. When we regulate the throttle orifice valve 2 and make the overflow oil piped into tank via the valve 1, the pressure of system can be determined by the relief-valve 1 setting. If we connect a valve 3 to the remote control port, and then pressure of the system can be determined by the valve 3, however the pressure of valve 3 setting should be lower than that of the valve 1.

第 7 章　液压基本回路

本书的前面几个章节已对液压流体力学基础知识和液压元件作了详细介绍。这一章节将讨论液压基本回路。任何一个液压系统都是由一些基本回路组成的，所谓液压基本回路是由一组能实现某种特定功能的液压元件的组合。

在分析和设计液压回路时，要重点考虑以下三方面问题：①系统工作的安全性；②所要完成的工作任务；③工作效率。

按其在液压系统中的功用不同，基本回路可分为：

1）压力控制回路——控制整个系统或局部油路的工作压力。

2）速度控制回路——控制和调节执行元件的速度。

3）方向控制回路——控制执行元件运动方向的变换和锁停。

4）多执行元件控制回路——控制几个执行元件相互间的工作循环。

本章讨论的是最常见的液压基本回路。熟悉和掌握它们的组成、功用、性能和应用，有助于更好地分析、设计和使用各种液压系统。

7.1　压力控制回路

压力控制回路的作用是利用压力控制阀来控制整个液压系统或局部油路的压力，达到调压、卸荷、减压、增压、平衡、保压、卸压等目的，以满足执行元件对力或力矩的要求，提高系统的工作效率，改善系统的工作性能等。

7.1.1　调压回路

调压回路的功用在于调定或限制液压系统的最高工作压力，或者使执行机构在工作过程的不同阶段实现多级压力变换。一般由溢流阀来实现这一功能。

1. 调压回路

图 7-1a 所示为最基本的调压回路。若只安装一先导式溢流阀 1（此时远控口需封住），当调节节流阀 2 时，系统多余的油通过先导式溢流阀 1 溢流回油箱，使系统压力保持在先导式溢流阀 1 调定值，对系统起限压保护作用。若在先导式溢流阀 1 的远控口串接一远程调压阀 3，且先导式溢流阀 1 的调定压力必须大于远程调压阀 3 的调定压力，则系统压力可由远程调压阀 3 远程调节。

2. 多级调压回路

图 7-1b 所示为多级调压回路。主溢流阀 1 的遥控口通过三位四通换向阀 4 分别接具有不同调定压力的远程调压阀 2 和 3，且主溢流阀 1 的调定压力必须大于远程调压阀 2 和 3。当换向阀处于左位时，压力由远程调压阀 2 调定；换向阀处于右位时，压力由远程调压阀 3 调定；换向阀处于中位时，由主溢流阀 1 来调定系统最大压力。

2. Multistage pressure regulated circuit

Fig. 7-1b is a three-stage pressure regulated circuit. The remote control port in the main relief valve 1 is connected to a three-way and four ports directional control valve 4. Two oil ports on the valve 4 are connected with different pressure setting remote pressure regulated valves 2 and 3 respectively, and the pressure setting of valves 2 and 3 should be both lower than that of valve 1. So system pressure can be controlled by the directional valve 4. When valve 4 is operating at left envelope, system pressure is determined by valve 2, while at the right system pressure is determined by valve 3; and if it is at neutral position the highest system pressure can be limited by the main valve 1.

3. Stepless pressure regulated circuit

Fig. 7-1c shows a scale pressure regulated circuit with an electro-hydraulic scale valve to fulfill pressure stepless regulated. The pressure stepless can be regulated by the input current.

7.1.2 Pressure-venting circuit

The function of a pressure-venting circuit is to unload the pumping and to keep the pump idling when the hydraulic system is during the regular off-working, which will save the energy, reduce the heating and get a long life of use. Because the output power of a pump equals to the pressure multiplying by flow, there are two methods of unloaded pump: one is venting by pressure (pump operating under zero or near zero pressure, is called the pressure-venting) and the other is venting by flow rate (pump operating under zero or near zero flow rates, is called flow-venting). Both of the two methods release directly the fluid to the tank.

1. Pressure-venting by the neutral of a directional valve

Venting for a fixed delivery pump can be done by the neutral of a directional valve, such as open-center directional valve 'H' type, 'M' type or 'K' type. Usually an against-pressure valve is installed in this circuit to maintain a lower control pressure for operating hydraulic elements as shown in Fig. 7-2a.

2. Pressure-venting by pilot-operated relief valve

We connect a two-way and two ports solenoid-actuated valve 2 to a pilot-actuated relief valve 1. When the solenoid is actuated, the remote port connects the tank directly to let the pump unload. The throttle orifice *b* between the remote port and solenoid valve is used to prevent the pressure shock, as shown in Fig. 7-2b.

3. Pressure-venting by a pump limited pressure

Fig. 7-2c is a venting system by zero flow rates, which uses a variable displacement pump limited. When the piston of hydraulic cylinder 3 extends to the end or at neutral of the directional valve 2, pressure of pump 1 is increased and flow rate is diminishing to quantity supplied for leaking of cylinder and directional valve. A relief valve 4 is used as a safe valve to prevent system from abnormal pressure.

4. Pressure-venting by an accumulator

Fig. 7-2d is a pressure-venting circuit with an accumulator. When the pressure of this circuit reaches the pressure setting of sequence valve 2, the invariable displacement pump vents by valve 2

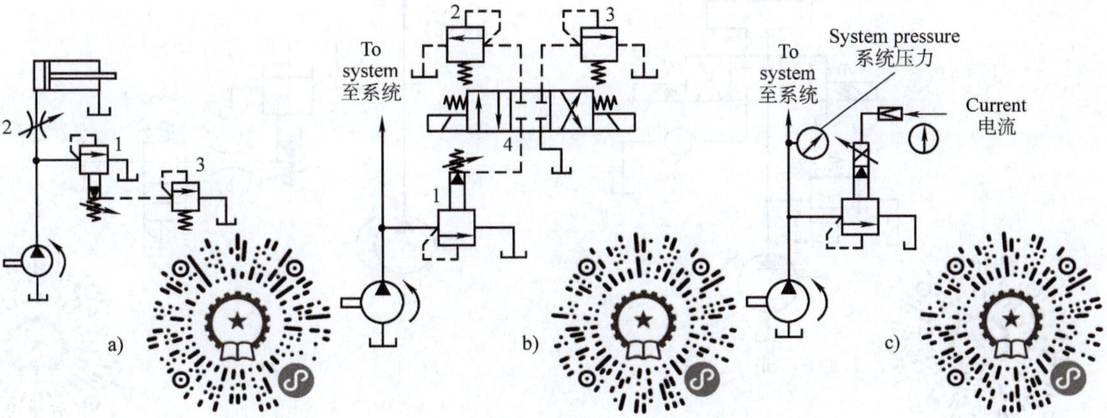

Fig. 7-1　Pressure regulated circuits（调压回路）

a) Pressure regulated from remote port（远程调压回路）

b) Multistage pressure regulated circuit（多级调压回路）

c) Stepless pressure regulated circuit（无级调压回路）

3. 无级调压回路

图 7-1c 所示为无级调压回路。该回路采用了电-液比例溢流阀，调节输入电流即可无级调压。

7.1.2　卸荷回路

卸荷回路的功用是在系统执行元件频繁短时间停止工作，一般都采用使泵卸荷，而不频繁起停电动机，以节省功率消耗，减少系统发热，延长泵和电动机的使用寿命。因为泵的输出功率等于压力和流量的乘积，因此卸荷的方法有两种，一是将泵的出口直接接回油箱，泵在零压或接近零压下工作；另一是使泵在零流量或接近零流量下工作。前者称为压力卸荷；后者称为流量卸荷，仅适用于变量泵。

1. 用换向阀中位机能的卸荷回路

定量泵可采用中位机能为 H、M 或 K 型的换向阀来实现泵卸荷，如图 7-2a 所示。一般这种阀本身装有预控压力阀，或者采用在回油路上安装背压阀 *a*，以保证主液动阀动作的控制压力。

2. 用先导式溢流阀的卸荷回路

如图 7-2b 所示，在先导式溢流阀 1 的远控口串接一个二位二通电磁阀 2。在远控口与电磁阀间可设置阻尼 *b*，以防止卸载或升压时产生液压冲击。

3. 用限压式变量泵的卸荷回路

如图 7-2c 所示，当液压缸 3 活塞运动碰到挡块或换向阀 2 处于中位时，泵 1 的压力升高，当压力达到调定压力时，泵的流量减小到只补充液压缸或换向阀的泄漏，回路实现保压卸荷。回路中可以不用安全阀 4，但为了防止泵的压力补偿失灵和换向阀换向过程中的压力冲击，加一安全阀较好。

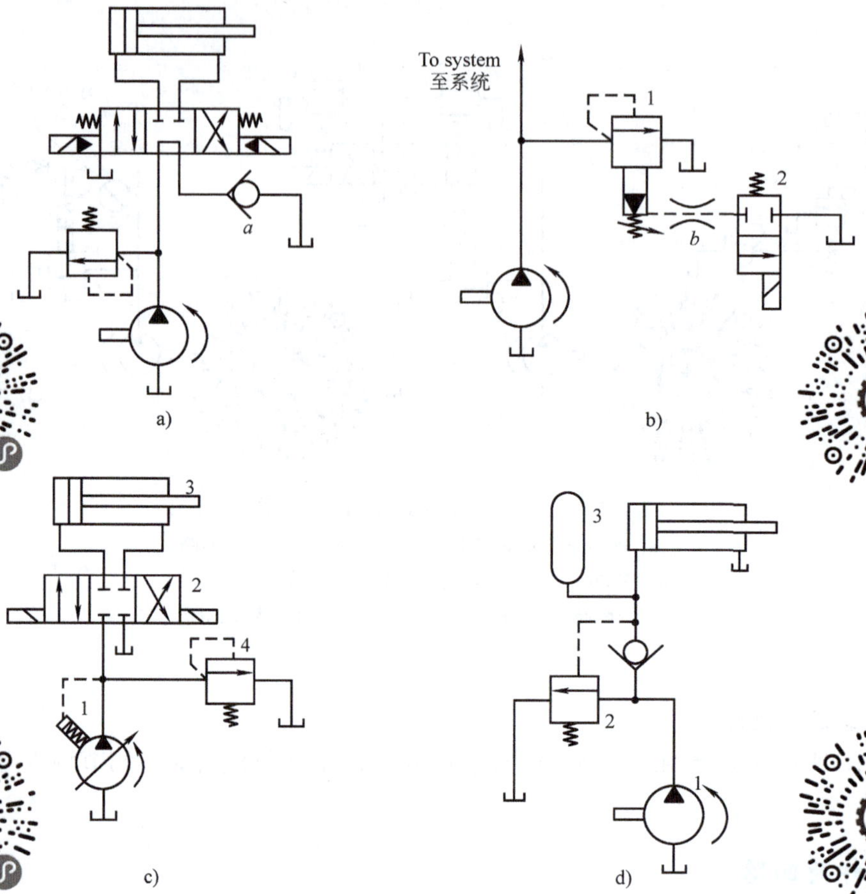

Fig. 7-2 Pressure-venting circuits (卸荷回路)

a) Pressure-venting by the neutral of a directional valve(用换向阀中位机能的卸荷回路)

b) Pressure-venting by a solenoid operated relief valve(用先导式溢流阀的卸荷回路)

c) Pressure-venting by a pump limited pressure(用限压式变量泵的卸荷回路)

d) Pressure-venting by an accumulator(用蓄能器的卸荷回路)

and system pressure is kept by accumulator 3. When and if the pressure of circuit is lower than the pressure rated by relief valve 2, the valve is closed to let the pump supply to system again. Thus the valve can regulate automatically the pump by venting or rising pressure.

7.1.3 Pressure-reducing circuit

The function of pressure-reducing circuits: they are used to drop a stable branch circuit pressure below the normal operating pressure of system. For example, pressure-reducing circuits are usually applied to clamping of a machine tool and the lubricating path.

Usually a fixed pressure-reducing valve is installed to a branch circuit such as Fig. 7-3a. Valve 3 in this circuit is a check valve in order to prevent the hydraulic cylinder 4 from any fluctuation of the pressure at its inlet when the main operating pressure is lower than the rated value of pressure-reducing valve 2.

Fig. 7-3b is a two-stage pressure-reducing circuit. Valve 6 is connected to the remote control

4. 用蓄能器保持系统压力的卸荷回路

如图 7-2d 所示，当回路压力达到顺序阀 2 调定压力时，泵通过阀 2 卸荷，由蓄能器 3 保持系统压力；当回路压力下降到低于顺序阀 2 的调定压力时，顺序阀 2 关闭，泵恢复向系统供油。

7.1.3　减压回路

减压回路的功用：使系统某一支路具有低于系统压力调定值的稳定工作压力，如机床的工件夹紧、导轨润滑等常需用减压回路。

最常见的减压回路是在所需低压的支路上串接一定值减压阀，如图 7-3a 所示。回路中的单向阀 3 用于当主油路压力低于减压阀 2 的调定值时，防止液压缸 4 的压力受其干扰，起短时间保压作用。

图 7-3b 所示为二级减压回路。在先导式减压阀 2 的遥控口接二位二通换向阀 5 和远程调压阀 3，当二位二通换向阀处于图示位置时，液压缸 4 的压力由减压阀 2 的调定压力决定；当二位二通换向阀处于右位时，液压缸 4 的压力由远程调压阀 6 的调定压力决定。阀 6 的调定压力必须低于阀 2。泵的最大工作压力由溢流阀 1 调定。

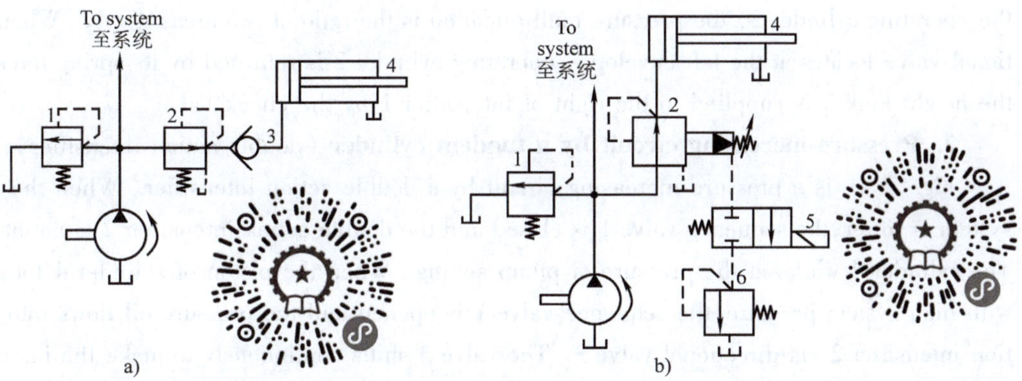

Fig. 7-3　Pressure-reducing circuits（减压回路）

a）One-stage pressure-reducing（单级减压回路）　b）Two-stage pressure-reducing（二级减压回路）

1—Direct-acting relief valve（直动式溢流阀）　2—Rated pressure-reducing valve（定值减压阀）

3—Check valve（单向阀）　4—Hydraulic cylinder（液压缸）

5—Two-position，two way directional valve（二位二通换向阀）

6—Remote control valve（远程调压阀）

要液压回路工作可靠，减压阀的最低调定压力应不小于 0.5MPa，最高调定压力至少应比系统压力小 0.5MPa。当减压回路的执行元件需调速时，调速元件应放在减压阀的后面，以免减压阀工作时其泄漏口泄漏对调整速度产生影响。

7.1.4　增压回路

增压回路的功用：用来使系统中某一支路获得比液压泵供油压力高的压力。

port on the pilot-operated pressure-reducing valve 2 which is operated by valve 5. When the two-way and two-port directional control valve is at the left envelope (shown in Fig. 7-3b), the pressure in cylinder 4 should be determined by valve 2; and when at the right envelope the pressure should be determined by remote control valve 6. Here the setting pressure for valve 6 should be lower than that of the valve 2. The highest pressure on this pump can be rated by relief valve 1.

To ensure a stable working speed, the lowest pressure should not be less than 0. 5MPa, and the highest setting pressure should be at least less than 0. 5 MPa of the system pressure. When the actuator elements speed in a pressure-reducing circuit needs to be regulated, the speed-regulating valve should be installed in the latter of the reducing-valve in order to avoid the affection of speed during the working.

7. 1. 4 Pressure-increasing circuit

Pressure-increasing circuits are used to enhance a stable branch circuit pressure above the normal operating pressure of system.

1. Pressure-increasing circuit by a single action booster

Pressure-increasing circuit by a single action intensifier is shown in Fig. 7-4a. When the directional valve locates at the right envelope, the oil pressure ($p_2 = p_1 A_1 / A_2$) from intensifier 1 flows into the operating cylinder 2, the pressure multiplication is the ratio of two areas A_1 / A_2. When the directional valve locates at the left envelope, operating cylinder 2 is returned by its spring force. Oil from the height tank 3 is supplied to the right of intensifier 1 by the check valve.

2. Pressure-increasing circuit by a tandem cylinder (double action intensifier)

Fig. 7-4b is a pressure-increasing circuit by a double action intensifier. When the pressure of system is lower, the sequence valve 1 is closed and the double action intensifier 2 is not at operating. The cylinder 4 works at the pressure of pump setting. When the piston of sylinder 4 touches a load with high system pressure, the sequence valve 1 is opened and the pressure oil flows into double action intensifier 2 via directional valve 3. The valve 3 shifts continuously to make the intensifier 2 always in motion, the pressure-increased oil will flow into the right of the operating cylinder 4 by check 7 or 8. Check valve 5 or 6 separates the higher pressure chamber from the lower pressure chamber. It can continuously supply the higher oil pressure, so it is suitable for longer pressure-increasing distance needed.

7. 1. 5 Pressure counter-balance circuit

A counter-balance circuit is suitable for use to a circuit having a vertical cylinder. The counter-balance valve will prevent the ram from dropping rapidly due to gravity when the system is not in operation and also improve the working stabilities during the load dropping. The hydraulic resistance should be set on the rod end of the vertical cylinder to produce a balance force against it.

The Fig. 7-5 shows three general pressure counter-balance circuits. Referring to Fig. 7-5a, it is a counter-balance circuit done by a combination of check-sequence valve. The setting pressure of sequence valve should be more than that of the total gravity and hold the total load at a position. If the total load is very high and then a higher setting pressure is needed for the sequence valve. So it will

1. 单作用增压器的增压回路

图7-4a所示，当换向阀电磁铁得电处于右位时，增压器1输出压力为$p_2=p_1A_1/A_2$的压力油进入工作缸2，其增压倍数是增压器两缸有效面积之比A_1/A_2。换向阀处于左位时，工作缸2靠弹簧力回程，高位油箱3经单向阀向增压器1右腔补油。

2. 双作用增压器的增压回路

如图7-4b所示，当系统压力较低时，顺序阀1关闭，双作用增压器2不起作用，液压缸4按泵供油压力工作。当活塞遇到较大负载时，系统压力升高，顺序阀1被打开，压力油经换向阀3进入双作用增压器，只要换向阀3不断切换，增压器2就不断往复运动，高压油就连续经单向阀7或8进入工作缸4右腔，此时单向阀5或6有效地隔开了增压器的高低压油路。它能连续输出高压油，适用于增压行程要求较长的情况。

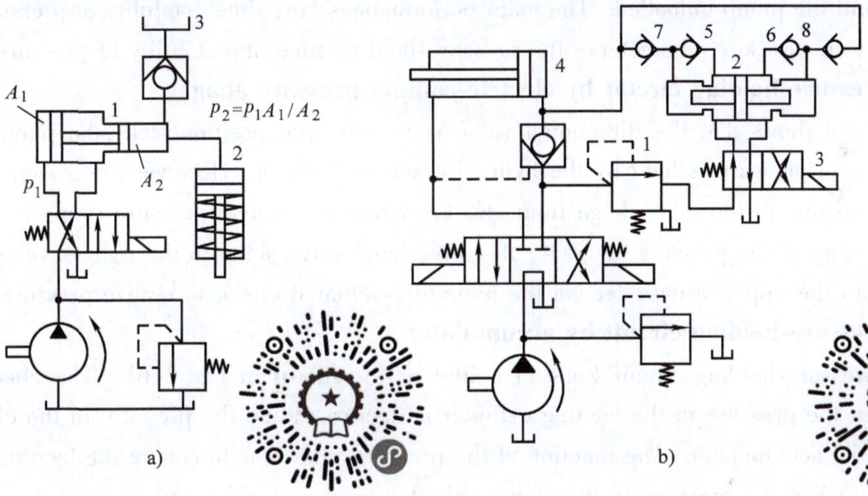

Fig. 7-4 Pressure-increasing circuits（增压回路）

a）Single action pressure-increasing circuit（单作用增压器的增压回路）

b）Bi-action pressure-increasing circuit（双作用增压器的增压回路）

7.1.5 平衡回路

平衡回路的功用：用于防止垂直运动的工作部件由于自重而自行下滑，同时还可改善工作下行时的运动平稳性。可在立式液压缸的下行回路上设置适当液阻，使其回油腔产生一定的背压与自重相平衡。

图7-5所示是三种常见的平衡回路。图7-5a所示为采用单向顺序阀的平衡回路。顺序阀的调整压力应稍大于工作部件所造成的缸下腔的背压，使工作部件不会因自重而下滑。但当自重较大时，顺序阀的压力需调得较高，消耗功率较大，故仅适用于工作部件重量不太大的场合，此处的顺序阀又被称作平衡阀。

图7-5b所示为采用液控单向阀的平衡回路。图7-5c所示为采用液控顺序阀的平衡回路。阀的启闭均靠缸上腔的压力来控制。在图示位置，缸下腔回路均无法排出，工作部件不致因自重而下滑。这两种回路在缸下腔回路上均设置单向节流阀，使工作部件下行运动平稳。因回路背压较小，功率消耗也较小。

consume more power and is suitable to use in the case of light weight. Here the sequence valve is called counter-balance valve.

Fig. 7-5b shows one by a hydraulic-operated check valve. Fig. 7-5c shows a pressure counter-balance circuit done by a hydraulic-operated sequence valve. The counter-balance valve is controlled by the upper pressure of cylinder. In the states of Fig. 7-5b or Fig. 7-5c hydraulic oil does not vent from the end of the cylinder so the workpiece does not drop and is held in a position. Two circuits set a check-throttle valve to ensure the workpiece stability. Since it has less against pressure so it is smaller in power consumption.

7.1.6 Pressure-holding circuit

A pressure-holding circuit is used to hold the pressure under the conditions of hydraulic cylinder fixed and the pump unloaded. The main performances are: time, stability of pressure and power loss. There are two key parameters: the pressure-holding time and stability of pressure-standing.

1. Pressure-holding circuit by electric-contact pressure guage

Fig. 7-6a shows that the directional valve is in a neutral position state, the pump is unloaded and the system pressure is held by the hydraulic-actuated check. However the system pressure will be downtrend due to the oil leakage from check. When the system pressure drops to the minimum pressure setting of the pressure gauge 4, the directional valve shifts to the right envelope. And then oil flows into the upper of cylinder via the hydraulic-actuated check to realize pressure holding.

2. Pressure-holding circuit by accumulator

The pressure-holding circuit (one of cylinders) is shown in Fig. 7-6b. The check valve 3 is closed when the pressure in the feeding cylinder is decreased and the pressure in the clamp cylinder is held by the accumulator. The function of the pressure relay 4 is to ensure the feeding cylinder operating only when the pressure in the clamp cylinder reaches a given pressure value.

3. Pressure-holding circuit by an auxiliary pump

An auxiliary pump 5 with low flow rate and high pressure is set in a common circuit, as shown in Fig. 7-6c. When the hydraulic cylinder has finished the press working and requires pressure-holding, we shift directional valve 2 at the neutral position via the pressure relay 4 and the main pump 1 is unloaded (hydraulic oil to tank). At the same time, the directional valve with two-position, two-way valve 8 is at the left envelope, which allows the auxiliary pump 5 to supply oil to point a in order to maintain the system pressure.

7.2 Speed Control Circuits

Speed control circuits, an important part in a hydraulic system, aim at regulating the motion speed of the actuator elements. The characteristics of speed-load and power will determine the performances, features and applications in a hydraulic system. Details will be discussed in this section.

7.2.1 The principles of speed regulating and control

The main types of actuators are hydraulic cylinder and motor in a hydraulic system. When oil

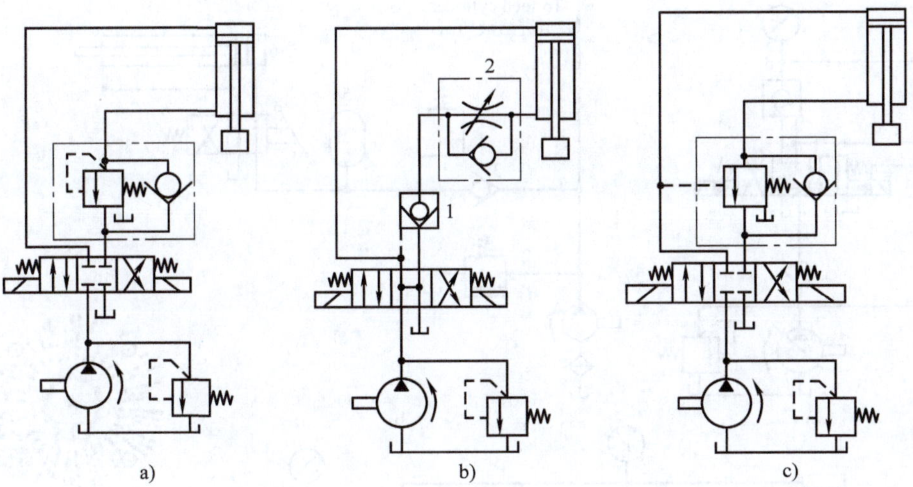

Fig. 7-5　Pressure counter-balance circuits（平衡回路）

a）Counter-balance circuit done by a combined of check-sequence valve（采用单向顺序阀的平

b）Counter-balance circuit done by a hydraulic-operated check valve（采用液控单向阀的平衡

c）Counter-balance circuit done by a hydranlic-operated sequence valve（采用液控顺序阀的平衡回路）

7.1.6　保压回路

保压回路的功能：使系统在液压缸停止不动、泵卸荷的工况下，仍保持稳定不变的压力。其性能的主要指标为保压时间、压力稳定性及功率损耗。

1. 采用电接触式压力表控制的保压回路

如图 7-6a 所示，换向阀 2 处中位，泵 1 卸荷，系统压力由液控单向阀 3 保压；因阀泄漏，液压缸上腔压力下降至压力表 4 下限触点调定压力时，触点接通，使换向阀 2 右位接入回路，使泵 1 给液压缸上腔补油，实现保压。

2. 采用蓄能器的保压回路

如图 7-6b 所示，该回路为多缸系统中的一缸保压回路，当进给缸压力降低时，单向阀 3 关闭，夹紧缸压力由蓄能器持续补油保压。压力继电器 4 的作用是：只有当夹紧缸的压力达到预定值时，进给缸才能动作。

3. 采用辅助泵的保压回路

如图 7-6c 所示，在回路中增设一台小流量高压泵即泵 5。当液压缸加压完毕要求保压时，由压力继电器 4 发信，换向阀 2 处于中位，泵 1 卸载；同时二位二通换向阀 8 处于左位，由辅助泵即泵 5 向封闭的保压系统 a 点供油，以维持系统压力稳定。

7.2　速度控制回路

调速回路的功用在于调节液压执行元件的运动速度，是液压系统的核心部分。调速回路的速度-负载特性和功率特性等基本决定了液压系统的性能、特点和应用，为此应予以分析和讨论。

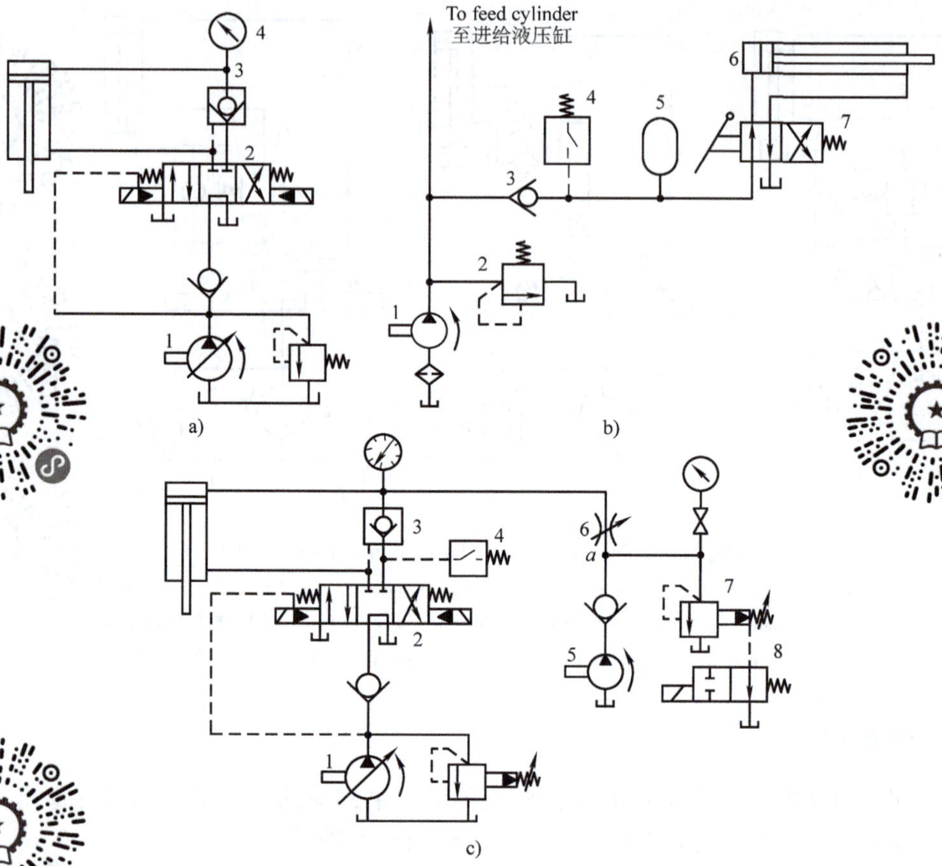

Fig. 7-6 Pressure-holding circuits(保压回路)

a) Pressure-holding circuit by electric-contact pressure guage (采用电接触式压力表的保压回路)

1—Pump(泵) 2—Directional valve(换向阀)

3—Hydraulic-actuated check valve (液控单向阀) 4—Pressure gauge(压力表开关,压力继电器)

b) Pressure-holding circuit by accumulator(采用蓄能器的保压回路)

1—Pump(泵) 2—Relief valve (溢流阀) 3—Check valve (单向阀) 4—Pressure relay(压力继电器)

5—Accumulator(蓄能器) 6—Cylinder(液压缸) 7—Directional valve(方向阀)

c) Pressure-holding circuit by auxiliary pump(采用辅助泵的保压回路)

1,5—Pump(泵) 2—Directional valve(换向阀) 3—Hydraulic actuated check valve (液控单向阀)

4—Pressure relay(压力继电器) 6—Throttle valve(节流阀) 7—Pilot relief valve (先导溢流阀)

8—Two-position, two-way unloading valve(二位二通阀)

compressibility and leakage are neglected, then

The speed of hydraulic cylinder is

$$v = q/A$$

The speed of hydraulic motor is

$$n = q/V_M$$

Where q is the input flow rate of hydraulic cylinder; A is the effective action area; V_M is oil displacement of the motor.

As two expressions mentioned above for a hydraulic cylinder, the speed regulating can be done only by regulating the input flow rate q of the cylinder. Similarly, for a hydraulic motor the speed

7.2.1 速度调节与控制原理

液压系统中的执行元件主要是液压缸和液压马达，在不考虑泄漏的情况下，液压缸的速度为

$$v = q/A$$

液压马达的转速为

$$n = q/V_M$$

式中，q 为输入液压缸或液压马达的流量；A 为液压缸的有效作用面积；V_M 为液压马达的排量。

由上面两式可知，对于液压缸来说，能用改变输入液压缸流量的办法来调速。对变量液压马达，既可用改变输入流量的办法来调速，也可用改变马达排量的办法来调速。改变输入执行元件的流量，根据液压泵是否变量分为定量泵节流调速回路和变量泵容积调速回路。

7.2.2 定量泵节流调速回路

定量泵节流调速回路根据节流阀在回路中安装位置的不同分为进油节流调速、回油节流调速和旁路节流调速三种基本形式。

1. 进油节流调速回路

进油节流调速回路是将节流阀串接在液压泵和液压缸之间，用它来控制进入液压缸的流量以达到调速的目的。如图 7-7a 所示，调速时，定量泵多余的油液通过溢流阀流回油箱，这是这种回路能够正常工作的必要条件。压力经溢流阀调定后，基本保持恒定不变。

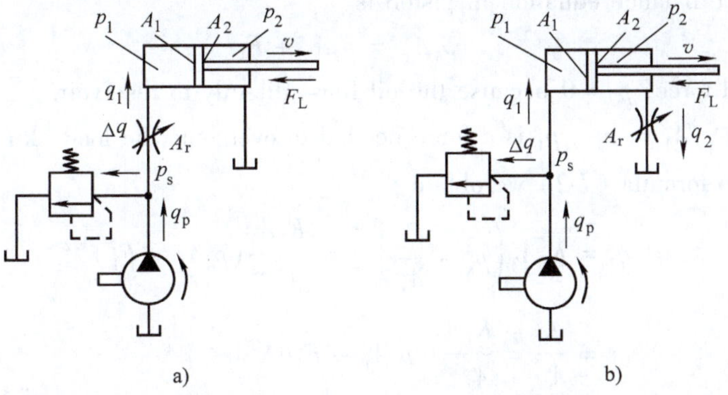

Fig. 7-7 Inlet and outlet throttle speed-regulating circuit（进油/回油调速回路）

a）Inlet throttle speed-regulating circuit（进口节流调速回路）

b）Outlet throttle speed-regulating circuit（回油节流调速回路）

（1）速度-负载特性　在不考虑回路中各处摩擦力的作用时，液压缸活塞运动速度为

$$v = q_1/A_1 \tag{7-1}$$

流经节流阀的流量为

$$q_1 = K_L A_T \sqrt{\Delta p} = K_L A_T \sqrt{p_s - p_1} \tag{7-2}$$

式中，A_T 为节流阀通流面积；K_L 为取决于节流阀阀口和油液特性的液阻系数；p_s 为液压泵出口压力；p_1 为液压缸进油腔压力；Δp 为节流阀两端压差。

regulating can be done by both of the two methods, i. e. , input flow rate q or displacement of motor V_m. The input flow rate changed can be realized by invariable displacement pump with throttle regulating speed valve or variable displacement pump capacity speed-regulating circuits.

7.2.2 Throttle speed-regulating circuit by invariable displacement pump

There are three basic types for the invariable displacement pump throttle speed-regulating circuits according to different installment position. They are inlet, exit and bypass circuits.

1. Inlet throttle speed-regulating circuit

A throttle orifice valve set between pump and cylinder is used for regulating the inlet flow rate of hydraulic and also the motion speed of it. It is called the inlet throttle orifice speed-regulating as shown in Fig. 7-7a. It is regulating the excess oil piping directly to tank via relief valve, which is a necessary condition for working of inlet throttle speed-regulating circuits.

(1) The features of speed-load When several frictions in a circuit are neglected the motion speed of a hydraulic cylinder is

$$v = q_1/A_1 \tag{7-1}$$

and flow rate into the throttle orifice valve

$$q_1 = K_L A_T \sqrt{\Delta p} = K_L A_T \sqrt{p_s - p_1} \tag{7-2}$$

Where A_T is the section area of throttle valve; K_L is hydraulic resistance coefficient depending on the features of throttle orifice and hydraulic oil; p_s is the output pressure of hydraulic pump; p_1 is inlet pressure of the hydraulic; Δp is the pressure difference of the throttle orifice.

Then the force balance equation in piston is

$$p_1 A_1 = p_2 A_2 + F_L \tag{7-3}$$

Where F_L is a load force; $p_2 = 0$ because the oil flows directly to reservoir.

Thus, $p_1 = F_L/A_1 = p_L$, p_L is a force needed to overcome the load, known as load pressure. Substituting p_1 into formula (7-2) we obtain

$$q_1 = K_L A_T \left(p_s - \frac{F_L}{A_1}\right)^{1/2} = \frac{K_L A_T}{A_1^{1/2}}(p_s A_1 - F_L)^{1/2} \tag{7-4}$$

$$v = \frac{q_1}{A_1} = \frac{K_L A_T}{A_1^{3/2}}(p_s A_1 - F_L)^{1/2} \tag{7-5}$$

The speed-load performance can be shown in equation (7-5) that expresses the relation between speed v and load F_L. If we take the v in the y-coordinate and F_L in the x-coordinate then the curves by different flux area A_T are called characteristic curves of speed-load as shown in Fig. 7-8.

From formula (7-5) and Fig. 7-8 it can be seen that the piston motion velocity v is proportional to flux area of throttle orifice A_T. Thus only adjusting the area A_T can realize the arbitrary speed-regulation. The range of speed regulation is very wide, $R_{cmax} = \dfrac{v_{max}}{v_{min}} \approx 100$. The character of speed variation with load can be described by speed rigidity k_v as below

$$k_v = -\frac{\partial F}{\partial v} = -\frac{1}{\tan\theta} \tag{7-6}$$

活塞的受力平衡方程为

$$p_1A_1 = p_2A_2 + F_L \tag{7-3}$$

式中，F_L 为负载力；p_2 为液压缸回油腔压力，由于回油腔通油箱，$p_2=0$。

所以 $p_1 = F_L/A_1 = p_L$，p_L 为克服负载所需的压力，称为负载压力。将 p_1 代入式（7-2）得

$$q_1 = K_LA_T\left(p_s - \frac{F_L}{A_1}\right)^{1/2} = \frac{K_LA_T}{A_1^{1/2}}(p_sA_1 - F_L)^{1/2} \tag{7-4}$$

$$v = \frac{q_1}{A_1} = \frac{K_LA_T}{A_1^{3/2}}(p_sA_1 - F_L)^{1/2} \tag{7-5}$$

式（7-5）即为进油节流调速回路的速度-负载特性方程，它反映了速度 v 与负载 F_L 的关系。若以 v 为纵坐标，F_L 为横坐标，将式（7-5）按不同节流阀通流面积 A_T 作图，可得一组抛物线，称为进油节流调速回路的速度-负载特性曲线，如图 7-8 所示。

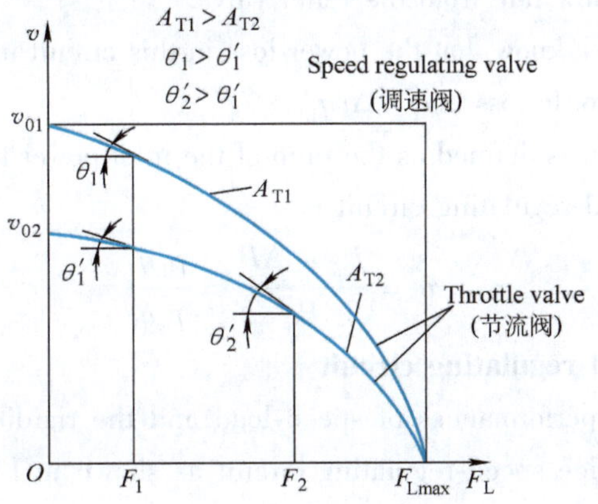

Fig. 7-8　Speed-load performances of inlet throttle（进油节流调速回路速度-负载特性）

从式（7-5）和图 7-8 看出，活塞的运动速度 v 与节流阀通流面积 A_T 成正比，调节 A_T 就能实现无级调速。这种回路的调速范围较大，$R_{cmax} = v_{max}/v_{min} \approx 100$。速度随负载变化而变化的程度，常用速度刚性 k_v 来评定。

$$k_v = -\frac{\partial F}{\partial v} = -\frac{1}{\tan\theta} \tag{7-6}$$

它表示负载变化时回路阻抗速度变化的能力，由式（7-5）和式（7-6）可得

$$k_v = -\frac{\partial F_L}{\partial v} = \frac{2A_1^{3/2}}{K_LA_T}(p_sA_1 - F_L)^{1/2} = \frac{2(p_sA_1 - F_L)}{v} \tag{7-7}$$

由式（7-7）可以看到，当节流阀通流面积 A_T 一定时，负载 F_L 越小，速度刚性 k_v 越大；当负载 F_L 一定时，活塞速度越低，速度刚性 k_v 越大。

（2）功率特性　液压泵输出功率为

$$P_p = p_sq_p = 常量$$

液压缸输出的有效功率为

$$P_1 = F_Lv = F_L\frac{q_L}{A_1} = p_Lq_L$$

Formula (7-6) tells us the ability of the circuit resistance to the change of speed when load is unstable. Meanwhile from formulae(7-5) and (7-6) we obtain

$$k_v = -\frac{\partial F_L}{\partial v} = \frac{2A_1^{3/2}}{K_L A_T}(p_s A_1 - F_L)^{1/2} = \frac{2(p_s A_1 - F_L)}{v} \tag{7-7}$$

Formula (7-7) shows that the lower load F_L has stronger speed rigidity k_v at a given area A_T. The lower the piston motion speed is, the stronger the speed rigidity k_v is at a given load F_L.

(2) Performance of power The outlet power of pump $P_p = p_s q_p$ = constant, and the effective outlet power of the hydraulic cylinder is $P_1 = F_L v = F_L \frac{q_L}{A_1} = p_L q_L$, where $q_L = q_1$ is the load flow rate. Then the circuit power loss is

$$\begin{aligned} \Delta P &= P_p - P_1 = p_s q_p - p_L q_L \\ &= p_s(q_L + \Delta q) - (p_s - \Delta p_1)q_L = p_s \Delta q + \Delta p q_L \end{aligned} \tag{7-8}$$

Where $\Delta q = q_p - q_L$ is the flow rate from the relief valve.

From formula (7-8), we know that the power loss in this circuit includes two parts: the overflow loss $\Delta p_1 = p_s \Delta q$ and the throttle loss $\Delta p_2 = \Delta p q_L$.

The efficiency of circuit is defined as the ratio of the inlet power to the outlet power, so the efficiency of inlet throttle speed-regulating circuit is

$$\eta = \frac{P_p - \Delta P}{P_p} = \frac{p_L q_L}{p_s q_p} \tag{7-9}$$

2. Exit throttle speed-regulating circuit

In the same way, the performances of speed-load and the rigidity of speed for the exit of hydraulic cylinder throttle orifice speed-regulating circuit as shown in Fig. 7-7b are described as follows

$$v = \frac{K_L A_T}{A_2^{3/2}}(p_s A_1 - F_L)^{1/2} \tag{7-10}$$

$$k_v = -\frac{\partial F_L}{\partial v} = \frac{2A_2^{3/2}}{K_L A_T}(p_s A_1 - F_L)^{1/2} = \frac{2(p_s A_1 - F_L)}{v} \tag{7-11}$$

3. Bypass throttle orifice speed-regulating circuit

This is done by installing a throttle orifice valve in a branch circuit where it is a parallel connection with the hydraulic cylinder as shown in Fig. 7-9a. The flow rate from the fixed displacement pump is q_p, and then one part (Δq) flows into the reservoir by the throttle valve and another part (q_1) flows into hydraulic cylinder allowing the piston motion. Speed-regulation valve is done by adjusting the area of the throttle orifice valve. Here the relief valve is usually closed since the overflow is acted by the throttle orifice valve and the relief valve is used as a safety valve for the system with the maximum rating pressure 1.1-1.2 times of pressure of load.

(1) Performance of speed-load It is similar to the situation mentioned above; the speed-load performance of bypass throttle orifice circuit can be obtained in terms of equations of mass continuity, pressure-flow rate and force balance on the piston. But here leaking flow rate Δq_p should be added into the main flow rate q_p from output of pump. Thus, the expression of speed is

式中，q_L 为负载流量，即进入液压缸的流量，这里 $q_L = q_1$。

回路的功率损失为

$$\Delta P = P_p - P_1 = p_s q_p - p_L q_L$$
$$= p_s(q_L + \Delta q) - (p_s - \Delta p_1) q_L = p_s \Delta q + \Delta p q_L \tag{7-8}$$

式中，Δq 为溢流阀的溢流量，$\Delta q = q_p - q_L$。

由式（7-8）可知，这种调速回路的功率损失由两部分组成：溢流损失 $\Delta p_1 = p_s \Delta q$ 和节流损失 $\Delta p_2 = \Delta p q_L$。

回路的输出功率与回路的输入功率之比定义为回路效率。进油节流调速的回路效率为

$$\eta = \frac{P_p - \Delta P}{P_p} = \frac{p_L q_L}{p_s q_p} \tag{7-9}$$

2. 回油节流调速回路

如图 7-7b 所示，对于回油节流调速回路，用同样的分析方法可得到与进油节流调速回路相似的速度-负载特性、速度刚性分别为

$$v = \frac{K_L A_T}{A_2^{3/2}} (p_s A_1 - F_L)^{1/2} \tag{7-10}$$

$$k_v = -\frac{\partial F_L}{\partial v} = \frac{2A_2^{3/2}}{K_L A_T} (p_s A_1 - F_L)^{1/2} = \frac{2(p_s A_1 - F_L)}{v} \tag{7-11}$$

3. 旁路节流调速回路

这种节流调速回路是将节流阀装在液压缸并联的支路上，如图 7-9a 所示。定量泵输出的流量 q_p 一部分 Δq 通过节流阀溢回油箱，一部分 q_1 进入液压缸，使活塞获得一定运动速度。调节节流阀的通流面积，即可调节进入液压缸的流量，从而实现调速。由于溢流功能由节流阀来完成，故正常工作时溢流阀处于关闭状态，溢流阀作安全阀用，其调定压力为最大负载压力的1.1～1.2倍。液压泵的供油压力 p_p 取决于负载。

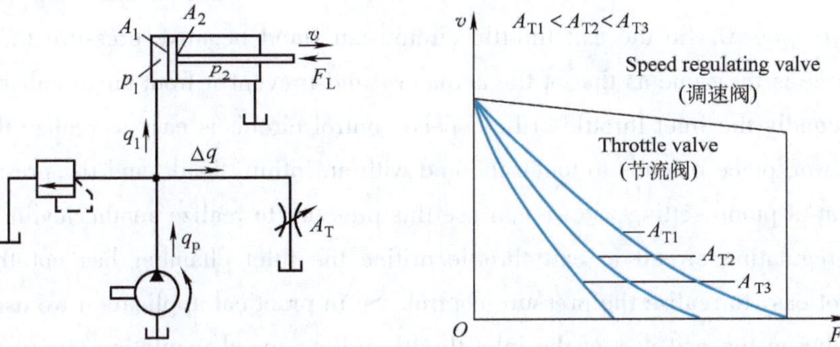

Fig. 7-9　Bypass throttle orifice speed-regulating（旁路节流调速）

a）Circuit（回路）

b）Speed-regulating performanles（速度-负载特性）

（1）速度负载特性　如同式（7-5）的推导过程，由流量连续性方程、节流阀的压力流量方程和活塞的受力平衡方程，可得旁路节流调速回路的速度-负载特性方程。需要指出，由于泵的工作压力随负载而变化，泵的输出流量 q_p 应计入，泵的泄漏量随压力而变化的变

$$v = \frac{q_1}{A_1} = \frac{q_{pt} - \Delta q_p - \Delta q}{A_1} = \frac{q_{pt} - \lambda_p \left(\frac{F_L}{A_1}\right) - K_L A_T \left(\frac{F_L}{A_1}\right)^{1/2}}{A_1} \tag{7-12}$$

Where q_{pt} is the theoretical flow rate of the pump; λ_p is the leaking coefficient.

The rigidity of speed is described as

$$k_v = -\frac{\partial F_L}{\partial v} = \frac{A_1^{\ 2}}{\lambda_p + \frac{1}{2} K_L A_T \left(\frac{F_L}{A_1}\right)^{-1/2}} = \frac{2A_1 F_L}{\lambda_p \left(\frac{F_L}{A_1}\right) + q_{pt} - A_1 v} \tag{7-13}$$

(2) Characters of power The output power from pump is

$$P_p = p_L q_p$$

Where p_L is the load pressure with $p_L = F_L/A_1$.

The output power from hydraulic cylinder

$$P_1 = F_L v = p_L A_1 v = p_L q_1$$

The power loss is

$$\Delta P = P_p - P_1 = p_L q_p - p_L q_1 = p_L \Delta q \tag{7-14}$$

The efficiency of the circuit is

$$\eta = \frac{P_p - \Delta P}{P_p} = \frac{p_L q_1}{p_L q_p} = \frac{q_1}{q_p} \tag{7-15}$$

4. The comparisons for three throttle speed-regulating circuits

1) Comparing Equation (7-11) with (7-7), we can easily find that they have the same speed-load characteristics and the rigidity of speed. Thus the analysis and conclusions for the inlet are all suitable for those of exit throttle speed-regulating circuits.

2) There are two application differences between inlet and exit throttle orifice speed-regulating: firstly the throttle orifice valve in the exit circuit is used for engendering an against-pressure in the return chamber, i. e., $p_2 \neq 0$. So the exit throttle circuit can stand negative pressure (i. e., the direction of action force is the same as that of the actuator) and prevent it from air to enhance the motion stabilities. Secondly the inlet throttle orifice speed control circuit is easy to realize the pressure control. When the workpiece extends to touch the end with an infinity load, and the pressure in cylinder will reach that of pump setting, so we can use this pressure to realize another cylinder control. And in the speed-regulating circuit of exit throttle orifice the inlet chamber has not the pressure change thus it is not easy to realize the pressure control. So in practical application we usually set an against-pressure valve in the exit port of the inlet throttle orifice speed-regulating circuit, which can enhance the general performance of the circuit and has advantages of both the inlet and exit circuits, But it will consume much more power and result in oil heating.

3) The comparison of three circuits in rigidity. According to the formula (7-12) we can draw a set of curves of speed-load by different throttle section areas A_T as shown in Fig. 7-9b. From the formula (7-12) and Fig. 7-9b, speed is decreased with the increasing of load at a given flux area A_T; the bigger the load is, the higher the speed rigidity is. However, the smaller flux area is (piston motion speed is higher), the stronger the speed rigidity is at a given load, which is different from the

化量为 Δq_p。因此，速度表达式为

$$v = \frac{q_1}{A_1} = \frac{q_{pt} - \Delta q_p - \Delta q}{A_1} = \frac{q_{pt} - \lambda_p\left(\frac{F_L}{A_1}\right) - K_L A_T\left(\frac{F_L}{A_1}\right)^{1/2}}{A_1} \tag{7-12}$$

式中，q_{pt} 为泵的理论流量；λ_p 为泵的泄漏系数；其他符号意义同前。

速度刚度为

$$k_v = -\frac{\partial F_L}{\partial v} = \frac{A_1{}^2}{\lambda_p + \frac{1}{2}K_L A_T\left(\frac{F_L}{A_1}\right)^{-1/2}} = \frac{2A_1 F_L}{\lambda_p\left(\frac{F_L}{A_1}\right) + q_{pt} - A_1 v} \tag{7-13}$$

（2）功率特性　液压泵的输出功率为

$$P_p = p_L q_p$$

式中，p_L 为负载压力，$p_L = F_L/A_1$。

液压缸的输出功率为

$$P_1 = F_L v = p_L A_1 v = p_L q_1$$

功率损失为

$$\Delta P = P_p - P_1 = p_L q_p - p_L q_1 = p_L \Delta q \tag{7-14}$$

回路效率为

$$\eta = \frac{P_p - \Delta P}{P_p} = \frac{p_L q_1}{p_L q_p} = \frac{q_1}{q_p} \tag{7-15}$$

4. 进油、回油和旁路三种节流阀调速回路性能的比较

1）比较式（7-11）与式（7-7）可以看出，回油节流阀调速与进油节流阀调速的速度-负载特性及速度刚度完全相同。因此，对进油节流阀调速回路的分析和结论都适用于回油节流调速回路。

2）进油节流与回油节流调速回路在使用中主要存在两点差异：一是回油节流阀调速回路的节流阀使缸的回油腔形成一定的背压（$p_2 \neq 0$），因而能够承受一定的负值负载（即作用力的方向与执行元件的运动方向相同的负载），并且可以防止空气从回油路吸入，以提高缸的运动平稳性。另一点是进油节流阀调速回路容易实现压力控制。因当工作部件在行程终点碰到死挡铁后，缸的进油腔油压会上升到等于泵的压力，利用这个压力变化，可使并联于此处的压力继电器发出电气信号，对系统的下步动作实现控制。而在回油节流阀调速时，进油腔压力没有变化，不易实现压力控制。因此为了提高回路的综合性能，一般常采用进油节流阀调速，并在回路上加背压阀，使其兼具二者的优点，但会增加一定的功率消耗，增大油液的发热量。

3）三种调速回路的刚度比较。根据式（7-12），选取不同的节流阀通流面积 A_T 可作出一组速度-负载特性曲线，如图 7-9b 所示。由式（7-12）和图 7-9b 可看出，当节流阀通流面积一定而负载增加时速度显著下降，旁通节流调速回路负载越大，速度刚性越大；当负载一定时，节流阀通流面积越小（活塞运动速度越高），速度刚性越大。这与进油和回油两种节流调速回路（图 7-7）的结论正好相反。由于负载变化引起泵的泄漏对速度产生附加影响，导致旁通回路的速度负载特性较进油和回油节流调速回路要差。

从图 7-9b 还可看出，旁通回路的最大承载能力随着节流阀通流面积 A_T 的增加而减小。当 $F_{Lmax} = (q_p/kA_T)^2 A_1$ 时，泵的全部流量经节流阀流回油箱，液压缸的速度为零，继续增大

situations of the inlet and outlet throttle speed-regulating circuits (as shown in Fig. 7-7). Because of the leakage due to the change load it will affect the speed stability and result in a poorer speed-load performance, while the inlet and exit circuits do not have such troubles.

From Fig. 7-9b, the maximum ability of bearing the weight of load is decreased with the increasing of the flux area A_T. If we put on a maximum load force $F_{Lmax} = \left(\frac{q_p}{kA_T}\right)^2 A_1$, all the oil from pump flow into the tank via throttle valve and the hydraulic cylinder will stop with a zero speed. So bypass circuit has the disadvantages of lower loads and also narrow range of regulating speed.

Because of the poor speed-loading performance, application of bypass throttle orifice speed regulating circuts is less. Generally it is applied in the situations with a high speed, heavy load with lower operation stability required, such as the main cutting motion system and machine feeding system.

4) The comparisons of three circuits in power loss. From formulae(7-14) and (7-15), there is only the throttle orifice loss without the overflow loss. So the bypass circuit has the advantages of lower power loss and higher efficiency than that of the inlet and outlet circuits. The bypass circuit is usually used in the positions that do not require high power and stability.

5) Start-up performance after shut down. In outlet throttle orifice speed regulating circuits, the oil in the return chamber flows directly into the tank to make the work-machine impact ahead instantly due to no back pressure at start-up. However, the start-up impact for the inlet throttle orifice speed regulating circuits can be avoided by regulating the inlet throttle orifice valve.

5. Improvement of the throttle orifice speed-regulating circuit

The reason of poor speed rigidity in throttle orifice speed-regulating is the unstable pressure difference on two ends of the valve, which results in the flow rate instability. So we should try to use the regulating speed valve (pressure-compensated) instead of the throttle orifice valve in order to improve circuit performances.

Similar to the throttle orifice valve, the circuits with regulating speed valve are (pressure-compensated) classified into three positions of inlet, outlet and bypass as shown in Fig. 7-10 (a, b, and c). But now the performances of speed rigidity are getting better than those of throttle types since pressure compensates to keep the pressure differential constant. The comparison curves of performance are shown in Fig. 7-8 and 7-9(b). Here we need to pay attention to the pressure differential on two ends of the regulating speed valve which should be bigger than 0.5MPa for middle pressure systems and 1MPa for high pressure systems to ensure the action of pressure-compensation.

7.2.3 Volume speed-regulating circuit

The volume speed-regulating circuits are performed by regulating the volume of pump or motor. Compared with the throttle speed-regulating circuit, the volume speed-regulating circuit has the advantages of higher efficiency of circuit and is suitable for application in high speed and powerful situations due to no throttle orifice loss and overflow loss. The popular circuits will be detailed.

A_T 已不起调速作用，即旁通调速回路在低速时承载能力低，调速范围也小。

由于旁通节流调速回路的速度-负载特性很软，所以应用比进、出口节流调速回路少，一般只用于高速、重载、对速度平稳性要求不高的较大功率系统，如牛头刨床主运动系统、输送机械液压系统等。

4）三种调速回路功率损失的比较。由式（7-14）和式（7-15）看出，旁路节流调速回路只有节流损失，而无溢流损失，因而功率损失比进油和回油两种节流阀调速回路小，效率高。这种调速回路一般用于功率较大且对速度稳定性要求不很高的场合。

5）停机后的起动性能。长期停机后，回油节流阀调速回路中液压缸回油腔的油液会泄漏流回油箱。当液压泵重新起动时，背压不能立即建立会引起瞬间工作机构的前冲现象。而在进油节流调速回路中，因为进油路上有节流阀控制流量，只要在开车时关小节流阀即可避免起动冲击。

5. 改善节流调速负载特性的回路

采用节流阀的节流调速回路速度刚性差，主要是由于负载变化引起节流阀前后压差变化，使通过节流阀的流量发生了变化的缘故。在负载变化较大而又要求速度稳定时，这种调速回路远不能满足要求。如果用调速阀代替节流阀，回路的负载特性将大为提高。

根据调速阀在回路中安放的位置不同，有进油调速、回油调速和旁通型调速等多种方式，如图7-10a、b、c所示。它们的回路构成、工作原理同它们各自对应的节流阀调速回路基本一样。由于调速阀本身能在负载变化的条件下保证节流阀两端压差基本不变，因而回路的速度刚性大为提高，如图7-8和图7-9b所示，旁路节流调速回路的最大承载能力也不因活塞速度的降低而减小。需要指出，为了保证调速阀中定差减压阀起到压力补偿作用，调速阀两端压差必须大于一定数值，中低压调速阀为0.5MPa，高压调速阀为1MPa，否则调速阀和节流阀调速回路的负载特性将没有区别。由于调速阀的最小压差比节流阀的压差大，所以其调速回路的功率损失比节流阀调速回路要大一些。

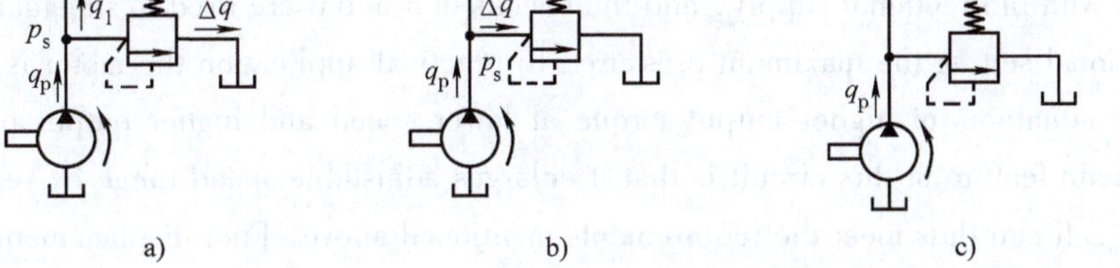

Fig. 7-10 Speed-regulating circuits with regulating speed valve（pressure-compensated）（采用调速阀的调速回路）

a）Inlet（进油调速回路） b）Outlet（回油调速回路） c）Bypass（旁通型调速回路）

7.2.3 容积调速回路

容积调速回路是通过改变泵或马达的排量来实现调速的。与节流调速回路相比，这种调

1. Regulating speed circuit by variable displacement pump and invariable displacement motor

Fig. 7-11a gives a circuit of a closed-circuit. A safety valve 3 (relief valve) is used for preventing overload in the hydraulic system. The low-rate flow pump 4 is used for replenishing leakage from pump 1 and motor 2 and replacing partial heat from the closed-circuit to decrease the temperature.

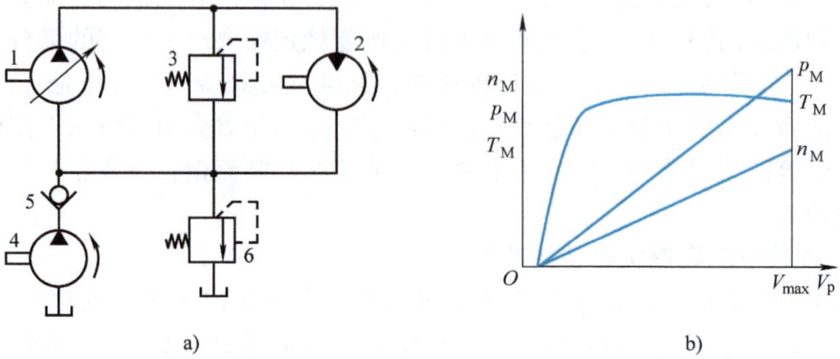

Fig. 7-11 Variable displacement pump-fixed displacement motor
(变量泵-定量马达调速)
a) Circuit (回路) b) Characteristics curves (回路特性曲线)

In this circuit the rotational speed n_p of pump and the displacement V_M of motor are constant. The motor speed n_M and output power P_M are varied proportionally by changing the pump displacement V_p. The torque capacity T_M of the motor and operating pressure Δp of the circuit depend on the load and are independent of the speed, so it is called invariableness torque circuit. The characteristic curves of the circuit are shown in Fig. 7-11b. The range of regulating speed is 1:40.

2. Regulating speed circuit by variable displacement pump and variable displacement motor

Fig. 7-12a is a speed-regulating circuit composed of a bidirectional variable displacement pump and a bidirectional variable motor. Shifting direction of supply oil of the pump and the motor can get bidirectional rotating clockwise or anticlockwise. The checks of 3 and 4 are used to transmit oil from the auxiliary pump 8 with bidirectional supply, and the checks of 5 and 6 are used to safeguard relief valve 7 with bidirectional setting the maximum pressure. In practical application the motor is usually operating under both situations of higher output torque at lower speed and higher output power at higher speed. The main feature of this circuit is that it enlarges adjustable speed range by regulating V_p and V_M at fixed n_p. It can thus meet the requirements mentioned above. First displacement of motor is set to the maximum at low speeds and regulates the displacement of pump until maximum value and the output power is linearly increased. In this case the motor can obtain the maximum and maintain a constant torque because of the maximum motor displacement. Second at high speeds the pump is kept at maximal displacement and regulate the motor displacement from maximal to lower. Motor rpm is going up continuously and output torque is decreased. In this way the pump is at the maximal and constant output power, so the motor is at constant power. The performance is shown in Fig. 7-

速回路既无溢流损失，又无节流损失，回路效率较高，适用于高速、大功率场合。常用的回路有：

1. 变量泵-定量马达调速回路

图 7-11a 所示回路为闭式回路。回路中设置安全阀 3，用以防止过载；小流量的补油泵 4 用于补充泵 1 和马达 2 的泄漏，并可置换闭式回路中部分发热油液，降低其温升。

在这种回路中，液压泵的转速 n_p 和液压马达的排量 V_M 为常量，改变泵的排量 V_p 可使马达转速 n_M 和输出功率 P_M 随之成比例地变化，马达的输出转矩 T_M 和回路的工作压力 Δp 取决于负载转矩，不会因调速而发生变化，所以这种回路常被称为恒转矩调速回路。回路特性曲线如图 7-11b 所示。该回路调速范围较大，可达到 1:40。

2. 变量泵-变量马达调速回路

图 7-12a 为双向变量泵-双向变量马达调速回路。回路中各元件对称布置，改变变换泵的供油方向，即可实现马达正反向旋转。单向阀 3 和 4 用于补油泵 8 能双向补油，单向阀 5 和 6 使安全阀 7 在两个方向都起过载保护作用。该回路 n_p 为常数，V_p 和 V_M 都可调，所以扩大了马达的调速范围。一般机械要求低速时有较大的输出转矩，高速时能提供较大的输出功率。采用这种回路恰好可以达到这个要求。在低速段，先将马达排量调至最大，用变量泵调速，当泵的排量由小变大时，马达转速随之升高，输出功率也随之线性增加。此时因马达排量最大，马达能获得最大输出转矩。且处于恒转矩状态。在高速段，泵为最大排量，用变量马达调速，将马达排量由大调小，马达转速继续升高，输出转矩随之降低。此时因泵处于最大输出功率状态不变，故马达处于恒功率状态。回路特性曲线如图 7-12b 所示。由于泵和马达的排量都可改变，扩大了回路的调速范围，可达 1:100。

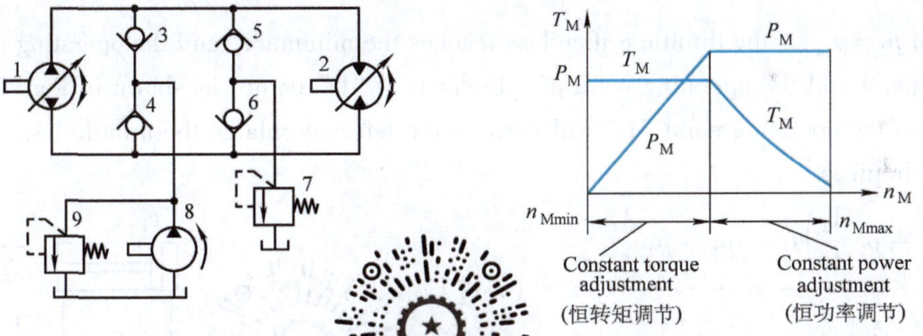

Fig. 7-12　Variable displacement pump-variable displacement motor
（变量泵-变量马达调速）
a) Circuit（回路）　b) Characteristic curves（回路特性曲线）

3. 容积节流调速回路

容积节流调速回路采用压力补偿型变量泵供油，用流量控制阀调节进入或流出液压缸的流量来调节其运动速度，并使变量泵的输油量自动地与液压缸所需流量相适应。这种调速回路没有溢流损失，效率较高，速度稳定性比容积调速回路好，适用于中小功率与调速范围大的场合。

（1）限压式变量泵和调速阀的容积节流调速回路　如图 7-13a 所示，限压式变量泵 1 输

12b. Since both displacements of pump and motor can be changed, the range of speed-regulating in this circuit can be extended to 1:100.

3. Volume-throttle speed-regulating circuit

(1) Volume-throttle speed-regulating circuit by pressure-limiting variable displacement pump and speed-regulating valve. As is shown in Fig. 7-13a, the pressure oil from pressure-limiting variable pump 1 flows into the left inlet of cylinder and the oil from right outlet of cylinder is flowing back to the reservoir via against-pressure valve 4. We can adjust the flux area of throttle orifice A_T to adapt the output flow of pump q_p to flow into cylinder q_1 for speed-regulating of cylinder motion. At the moment of regulating valve closed, the outlet flow rate from pump is more than that of cylinder, i. e. , $q_p > q_1$ due to q_1 drop. And this results in the outlet pressure p_p increases and the outlet flow rate q_p from the pressure-limited variable pump decreases automatically. Whereas if the orifice of regulating-speed valve is enlarged to result in the $q_p < q_1$. And then the outlet pressure from the pump p_p is decreased and outlet flow rate q_p increased automatically until $q_p = q_1$. Thus for any given open-area A_T, the pump will produce flow adapted to the desired value for the cylinder to realize the speed-regulation.

Features of the circuit: the curves are shown in Fig. 7-13b. In Fig. 7-13b it can be seen that it has no overflow loss but has throttle orifice loss related to the operating pressure of cylinder p_1. If q_1 is described as the operating flow rate into the cylinder, the flow rate from pump should be $q_p = q_1$ and supply pressure is p_p, then working pressure in the cylinder would be

$$p_2 \frac{A_2}{A_1} \leqslant p_1 \leqslant (p_p - \Delta p) \tag{7-16}$$

When $p_1 = p_{1max}$, the throttle orifice loss reaches the minimum, and the operating point of pump is at "a" point and the operating point of cylinder is at "b" point (as shown in Fig. 7-13b). If p_1 is decreased the operating point "b" will move to the left and enlarge the throttle loss. The efficiency of the circuit is

$$\eta_c = \frac{\left(p_1 - p_2 \frac{A_2}{A_1}\right) q_1}{p_p q_p} = \frac{p_1 - p_2 \frac{A_2}{A_1}}{p_p} \tag{7-17}$$

From the formula (7-17), it can be shown that the less flow rate from pump is, the higher the outlet pressure p_p from pump is. And the less the load is, the less pressure p_1 is. Thus this circuit has low circuit efficiency when it is used in the position of low-speed and less load.

(2) Volume-throttle speed-regulating circuit by pressure-differential variable displacement pump and throttle valve It has been shown in Fig. 7-14. The throttle valve is installed in the inlet of the cylinder. The pressure-differential is built and exerts feedback on the two control pistons of

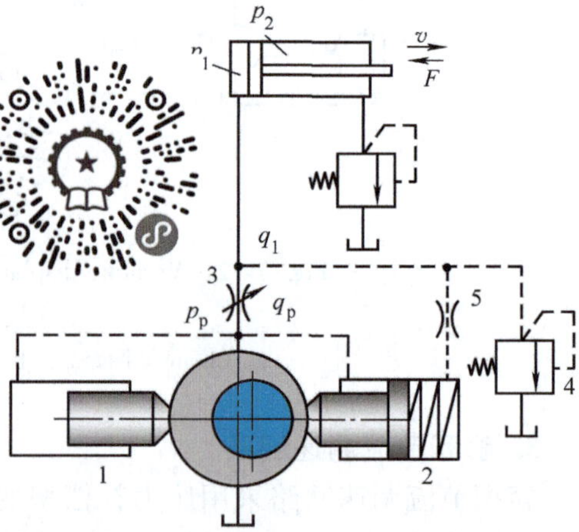

Fig. 7-14 Volume speed-regulating circuit by pressure-differential variable displacement pump and throttle valve
(差压式变量泵和节流阀的容积节流调速回路)

出的压力油经调速阀 2 进入液压缸左工作腔，回油经背压阀 4 返回油箱。改变调速阀中节流阀的通流面积 A_T 的大小，就可以调节液压缸的运动速度，泵的输出流量 q_p 和通过调速阀进入液压缸的流量 q_1 自相适应。其过程是：当关小调速阀的瞬间，q_1 减小，泵的输出流量还未来得及改变，出现了 $q_p > q_1$，导致泵的出口压力增大，从而使限压式变量泵输出流量自动减小，直至 $q_p = q_1$；反之，开大调速阀的瞬间，将出现 $q_p < q_1$，使泵出口压力减小，输出流量自动增加，直至 $q_p = q_1$。对应于调速阀一定的开口 A_T，泵输出相应的流量。

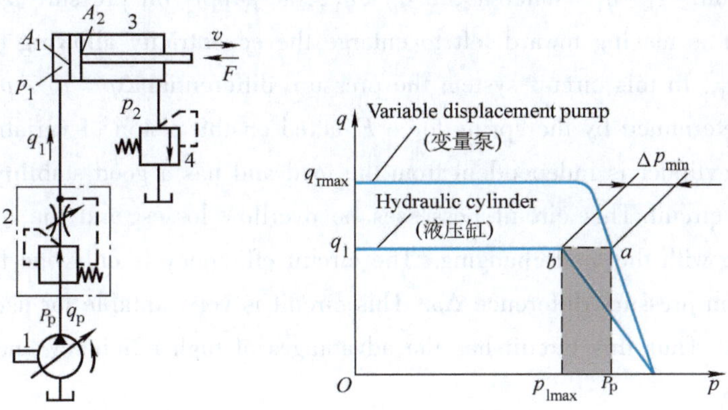

a)　　　　　　　　　　　　　　b)

Fig. 7-13 Volume-throttle speed-regulating circuit by pressure-limiting variable displacement pump and speed-regulating valve

（限压式变量泵和节流阀的容积节流调速）

a) Circuit（回路） b) Charateristic curves（调速特性曲线）

回路的调速特性曲线如图 7-13b 所示，由图可见，这种回路无溢流损失，但有节流损失，其大小与液压缸工作压力 p_1 有关。当进入液压缸的工作流量为 q_1 时，泵的供油流量应为 $q_p = q_1$，供油压力为 p_p，此时液压缸工作腔压力的正常工作范围是

$$p_2 \frac{A_2}{A_1} \leqslant p_1 \leqslant (p_p - \Delta p) \tag{7-16}$$

如图 7-13b 所示，当 $p_1 = p_{1max}$ 时，回路的节流损失最小，此时液压泵的工作点为 a，液压缸工作点为 b，若 p_1 减小，则 b 点向左移动，节流损失加大。回路的效率为

$$\eta_c = \frac{\left(p_1 - p_2 \dfrac{A_2}{A_1}\right) q_1}{p_p q_p} = \frac{p_1 - p_2 \dfrac{A_2}{A_1}}{p_p} \tag{7-17}$$

由上式看出：泵的输出流量越小，泵的压力 p_p 就越高；负载越小，p_1 就越小。因而在速度低、负载小的场合，回路效率不高。

（2）差压式变量泵和节流阀的容积节流调速回路　如图 7-14 所示，在液压缸的进油路上接一节流阀，节流阀两端的压差反馈作用在变量泵的两个控制活塞（柱塞）上。其中柱塞 1 的面积和活塞 2 的活塞杆面积相等。因此变量泵定子的偏心距大小，也就是泵的流量受到节流阀两端压差的控制。溢流阀 4 作安全阀用，固定阻尼 5 用于防止定子移动过快引起振荡。改变节流阀开口，就可以控制进入液压缸的流量 q_1，并使泵的输出流量 q_p 自动与 q_1 相适应。若 $q_p > q_1$，泵的供油压力 p_p 上升，则泵的定子在控制活塞作用下右移，减小偏心距、

variable pump. The area of piston 1 is equal to that of piston 2. Thus the flow rate on the eccentricity of stator in the pump is regulated by the pressure-differential. Relief valve 4 is used as a safety valve and a fixed hole 5 is used to prevent the stator from oscillation. The flow rate into cylinder q_1 can be adjusted by change of throttle orifice and adapted automatically the outlet flow rate from pump q_p to flow rate into cylinder q_1. If and when the flow rate $q_p > q_1$ the supply oil pressure p_p will be increased. In this case the stator is moving toward right to reduce the eccentricity allowing the flow rate q_p to reduce until $q_p \approx q_1$. Whereas, if $q_p < q_1$, the supply oil pressure p_p will be reduced. In this case the stator is moving toward left to enlarge the eccentricity allowing the flow rate q_p to increase until $q_p \approx q_1$. In this circuit system the pressure differential $\Delta p = p_p - p_1$ on the two sides of throttle valve is determined by the spring force F_t acted on the piston of variable pump, and so the flow rate into the cylinder is independent from the load and has a good stability.

Efficiency of circuit. This circuit possesses no overflow losses, and the oil supply pressure of pump is increasing with the load changing. The circuit efficiency is only one term of throttle losses that is resulted from pressure difference Δp. This circuit is very suitable for use in the situation of a larger change load. Thus this circuit has the advantages of high efficiency and low heat. The efficiency of circuit is

$$\eta_c = \frac{p_1 q_1}{p_p q_p} = \frac{p_1}{p_1 + \Delta p} \tag{7-18}$$

7.2.4 Fast-speed movement circuit

A fast-speed circuit is used to allow the hydraulic cylinder to obtain a fast and desired movement speed in order to shorten the movement time when the rod of cylinder extends. It will enhance the production efficiency.

1. Fast-speed movement circuit by a differential area actuator

It is shown in the Fig. 7-15. When the directional valve is working at the right position, the oil from outlet of pump and right outlet port of the cylinder flows together into the left inlet port of the cylinder. It allows the piston with rod to move at a high speed. This circuit is called the differential area connection circuit.

2. Fast speed movement circuit by double pumps oil supply

A fast-speed circuit by double pumps oil supply is shown in Fig. 7-16. The hydraulic energy unit consists of pump 1 (with low pressure and high displacement) and pump 2 (with high pressure and low displacement). The unloading valve 3 and relief valve 5 will be connected respectively with the desired pressure value when the cylinder is at fast speed movement (unloading) and desired maximum pressure of low speed value when the cylinder is at low speed movement (loading). When the directional valve 6 is at the right envelope and the circuit system pressure is lower than the rating pressure of valve 3, two pumps together supply the oil to the system, allowing the piston of actuator and rod to move toward right with a high speed. At the right envelope, the circuit system pressure is higher than the rating pressure of valve 3. In this case pump 1 is unloaded through valve 3 directly to tank, and only pump 2 supplies oil to the system, allowing the piston and rod to move toward right at a lower speed. The rating pressure in valve 3 is at least lower than 10%-20% of valve 5. This

使 q_p 减小至 $q_p \approx q_1$。反之，若 $q_p < q_1$，泵的供油压力 p_p 将下降，则引起定子左移，加大偏心距，使 q_p 增大至 $q_p \approx q_1$。回路中，节流阀两端的压差 $\Delta p = p_p - p_1$ 基本上由作用在变量泵控制活塞上的弹簧力 F_t 来确定，因此输入液压缸的流量不受负载变化的影响，故回路具有良好的稳速特性。

这种调速回路不但没有溢流损失，而且泵的供油压力随负载变化，回路中的功率损失也只有节流阀处压降 Δp 所造成的节流损失一项，因此回路效率高，发热小。特别适用于负载变化较大、速度较低的中、小功率，如组合机床的进给系统中。其回路效率为

$$\eta_c = \frac{p_1 q_1}{p_p q_p} = \frac{p_1}{p_1 + \Delta p} \tag{7-18}$$

7.2.4　快速运动回路

快速运动回路的功用在于使执行元件获得所需高速，缩短机械空程运动时间，以提高生产率或充分利用功率。

1. 液压缸差动连接快速运动回路

如图 7-15 所示，换向阀处于右位时，液压缸为差动连接，则有杆腔的回油和液压泵供油合在一起进入液压缸无杆腔，使活塞快速向右运动。

2. 双泵供油快速运动回路

如图 7-16 所示，低压大流量泵 1 和高压小流量泵 2 组成的双联泵作动力源。卸荷阀 3 和溢流阀 5 的压力分别按快速（空程）所需的压力与慢速（工作行程）所需的最大工作压力调定阀。换向阀 6 处于图示位置，系统压力低于卸荷阀 3 调定压力时，两个泵同时向系统供油，活塞快速向右运动；当换向阀 6 处于右位，系统压力达到或超过卸荷阀 3 的调定压力时，大流量泵 1 通过卸荷阀 3 卸荷，单向阀 4 自动关闭，只有小流量泵向系统供油，活塞慢速向右运动。卸荷阀 3 的调定压力至少应比溢流阀 5 的调定压力低 10% ~20%，大流量泵 1 卸荷减少了功率消耗，回路效率较高。该回路常用在执行元件快进和工进速度相差较大的场合。

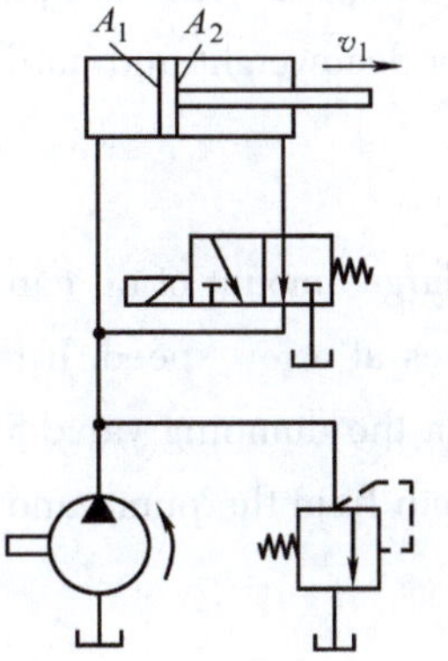

Fig. 7-15　Fast-speed circuits by a differential area actuator
（液压缸差动连接快速回路）

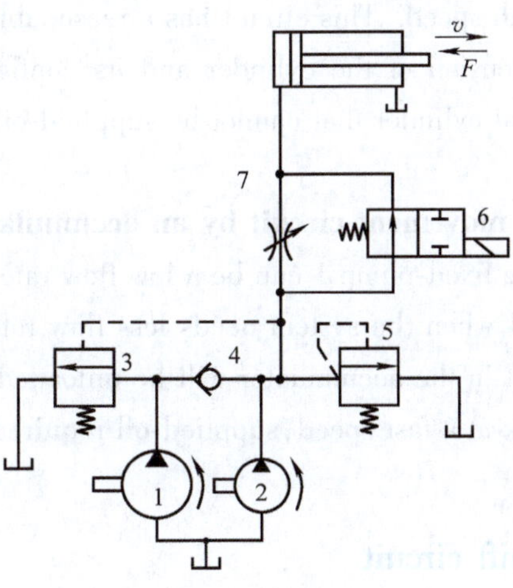

Fig. 7-16　Fast-speed circuits by double pumps oil supply
（双泵供油快速运动回路）

circuit takes advantage of high efficiency because the pump is unloaded to tank with decreasing loss of power and is widely used in differential speed observably needed.

3. Fast-speed movement circuit by increased-speed cylinder

Fast-speed movement circuit by increased-speed cylinder is shown in Fig. 7-17. It is a combination of piston ring cylinder and plunger cylinder. When the directional valve is at the left envelope end, the pressure oil flows into the small chamber 2 via the hole of the plunger and pushes the piston to move towards the right end and the desired oil in the big chamber 5 can be obtained from the check valve 3 drawn from the tank. The oil in the right of increased-speed cylinder 1 can be returned to tank via directional valve. When the cylinder touches with the workpiece, it will increase the load and result in a high return oil pressure. Then the sequence valve 4 is opened and at the same time the high pressure flows into the big chamber 5 and the check 3 is closed to allow the piston to slow its movement at a high push force. When the directional valve is at the right envelope end the pressure oil flows into the right of the cylinder 1 and at the same time check 3 is opened, the return oil in the big chamber 5 flows into the tank and the piston retractes at a high speed. This circuit has a reasonable power use but the ratio of increased-speed is restricted by dimension of the cylinder and its configuration is complex. This circuit is usually used in the horizontal cylinder that cannot be supplied-oil needed by deadweight and hard to move at a fast speed.

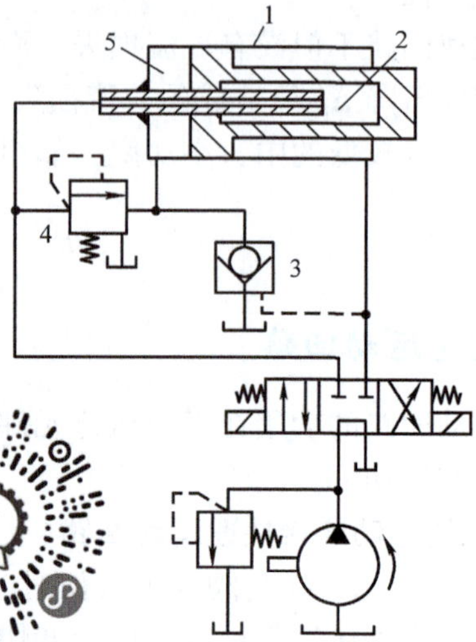

Fig. 7-17 Fast-speed circuits by increased-speed cylinder (采用增速缸的快速运动回路)

1—Increased-speed cylinder(增速缸)

2—Small chamber in the cylinder (小腔)

3—Check (单向阀) 4—Sequence valve (卸荷阀)

5—Big chamber in the cylinder (大腔)

4. Fast-speed movement circuit by an accumulator

In Fig. 7-18 the fixed-pump 1 can be a low flow rate pump. A large amount of oil can be deposited in accumulator 4 when the system needs less flow rate or operates at a low speed. If the pressure rises the pressure oil in the accumulator will be unloaded to tank via the dumping valve 5. When the system needs to work at a fast speed, supplied-oil required will be both from the pump and accumulator together.

7.2.5 Speed shift circuit

Speed shift circuit is used to realize changes between two different speed. This type of circuit usually requires a high shift stability and precision.

3. 用增速缸的快速运动回路

图 7-17 所示为采用增速缸的快速运动回路。增速缸是由活塞缸与柱塞缸复合而成的。当换向阀左位接入回路时，压力油经柱塞孔进入增速缸小腔 2，推动活塞快速向右移动，大腔 5 所需油液经液控单向阀 3 从油箱吸取，活塞缸右腔的油液经换向阀回油箱。当执行元件接触工件使负载增加时，回路压力升高，使顺序阀 4 开启，高压油关闭液控单向阀 3，并进入增速缸大腔 5，活塞转换成慢速运动，且推力增大。当换向阀右位接入回路时，压力油进入活塞缸右腔，同时打开液控单向阀 3，大腔 5 的回油排回油箱，活塞快速向左退回。这种回路功率利用比较合理，但增速比受增速缸尺寸的限制，结构比较复杂。该回路常应用于卧式液压缸不能利用运动部件的自重充液作快速运动的场合。

4. 采用蓄能器的快速运动回路

如图 7-18 所示，泵 1 可选较小的流量规格，当系统不需要流量或工作速度很低时，泵的全部流量或大部分流量进入蓄能器 4 贮存，蓄能器压力升高后，通过卸荷阀 5 使泵卸荷；当系统要求快速运动时，由泵和蓄能器同时向系统供油。

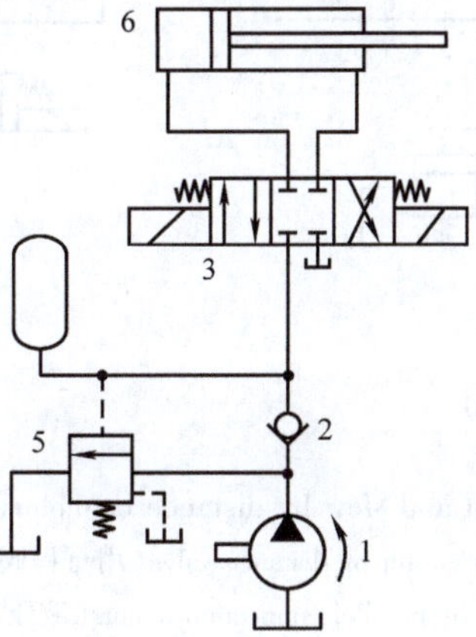

Fig. 7-18　Fast-speed circuit by accumulator
（采用蓄能器的快速运动回路）
1—Pump（泵）　2—Check（单向阀）
3—Directional valve（换向阀）　4—Accumulator（蓄能器）
5—Sequence valve（卸荷阀）　6—Hydraulic cylinder（液压缸）

7.2.5　速度换接回路

速度换接回路用于执行元件实现不同速度之间的切换，有快速—慢速、慢速—慢速的换接。这种回路要求具有较高的换接平稳性和换接精度。

1. 快、慢速换接回路

（1）采用行程阀（电磁阀）的速度换接回路　如图 7-19a 所示，换向阀处于图示位置，液压缸活塞快进到预定位置，活塞杆上挡块压下行程阀 1，行程阀关闭，液压缸右腔油液必须通过节流阀 2 才能流回油箱，活塞运动转为慢速工进。

（2）液压马达速度换接回路　采用两个液压马达串联、或并联，以达到两档速度换接的目的。

1. Speed shift between fast and slow

(1) Speed shift by distance valve (or solenoid pilot-actuated valve) When and if the circuit is at the position of Fig. 7-19a, hydraulic oil from pump flows into the left of actuator and pushes the piston rod to extend at a fast speed until the baffle on the end of piston gives a press-force down on the distance valve 1, allowing the right oil of the actuator to slowly flow into tank only via the throttle valve 2, so the piston is turned to slow motion.

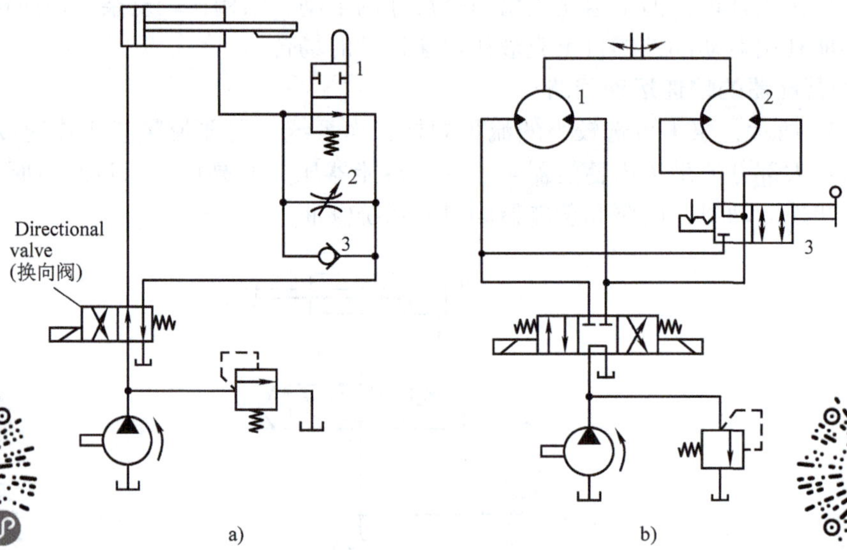

Fig. 7-19 Speed shift between fast and slow by distance or motor(采用行程阀或马达的速度换接回路)

a) Speed shift circuit by distance valve(用行程阀的速度换接回路)

b) Speed shift circuit by parallel connection motors(液压马达并联双速换接回路)

(2) Speed shift between two speeds by motors Speed shift between two speeds can be done by two motors. Fig. 7-19b is a circuit by parallel connection of two motors. When the directional valve 3 is at the left envelope end, the pressure oil is used to drive motor 1 and the motor 2 is rotating without load. When the directional valve 3 is at the right envelope end, the two motors are connected by type of parallel connection and the motors 1 and 2 have got 1/2 of total oil flow-rate respectively. The directional valve 3 is used to realize the shift between two speeds.

2. Speed shift of two slow speeds

Fig. 7-20a is a speed shift circuit between two slow speeds. It is done by series connection of two speed-regulating valves. In this circuit the speed of the second is set lower than the first one, so the orifice area of valve B is less than the area of valve A. Fig. 7-20b is a two stages slow speed circuit and it is done by parallel connection of two speed-regulating valves. In this circuit two orifice areas of speed-regulating valves can be adjusted separately without interfering each other.

7.3 Directional Control Circuits

A directional control circuit is used to control the actuator on operating or stopping or shifting

图 7-19b 所示为液压马达并联双速换接回路，手动换向阀 3 处于左位时，压力油只驱动马达 1，马达 2 空转；手动换向阀 3 右位时，马达 1 和 2 并联。若两马达排量相等，则并联时进入每个马达的流量减少一半，转速相应降低一半，而转矩增加一倍。手动换向阀 3 实现马达速度的切换，不管阀处于何位，回路的输出功率相同。

2. 两种慢速的换接回路

图 7-20a 所示为两个调速阀串联的速度换接回路，它只能用于第二进给速度小于第一进给速度的场合，故调速阀 *B* 的开口小于调速阀 *A* 的开口。图 7-20b 所示为两个调速阀并联的速度换接回路，这里两个进给速度可以分别调整，互不影响。

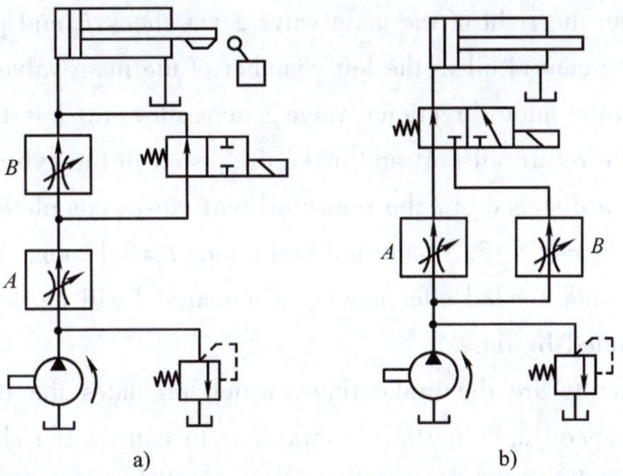

Fig. 7-20　Speed shift done by series or parallel connection of two speed-regulating valves
（调速阀串、并联速度换接回路）
a）Speed shift done by series connection of two speed-regulating valves（调速阀串联的速度换接回路）
b）Speed shift done by parallel connection of two speed-regulating valves（调速阀并联的速度换接回路）

7.3　方向控制回路

通过控制进入执行元件液流的通、断或变向来实现液压系统执行元件的起动、停止或换向的回路称为方向控制回路。

7.3.1　换向回路

对于普通的换向回路，只需在泵与执行元件之间采用标准的换向阀即可。当对换向精度平稳性要求较高时，则需要采用复杂的换向回路。本节主要介绍两种复杂的换向回路：时间控制制动式换向回路和行程控制制动式换向回路。

1. 时间控制制动式换向回路

图 7-21 所示为一种时间控制制动式换向回路。这种回路中的主油路只受主换向阀 2 控制。当先导阀 3 处于左位时，压力控制油经过单向阀 I_2 流进主换向阀 2 的右端，推动主换向阀的阀芯向左移动，同时主换向阀左端的控制油经过节流阀 J_1 流回油箱。在主换向阀阀

flow direction of oil fluid.

7. 3. 1 Directional circuit

A general directional valve can be connected between actuator and pump for a simple directional circuit. When a circuit needs to realize reciprocating motions continuously, the complex circuits are needed in this case. So two types of directional circuits, i. e. , directional shift controlled by time and by distance will be detailed in this section respectively.

1. The directional circuit controlled by time

The directional circuit controlled by time is shown in Fig. 7-21. In Fig. 7-21 the main oil line is only controlled by the main directional valve 2. If the pilot valve 3 moves towards left, the pressure control oil flows into the right of the main valve 2 via check I_2 and pushes the main valve move towards left, and then the control oil in the left chamber of the main valve 2 flows into tank via throttle valve J_1. Here when the main directional valve 2 moves towards left, the brake taper surface T on it will close gradually the return oil port and make the piston in the cylinder move slowly. And after the main valve 2 moves a distance of l the return oil port closes completely to make the piston stop. The half taper degree is $\alpha = 1.5°\text{-}3.5°$ and the brake long $l = 3\text{-}12$mm. When the throttle orifices of J_1 and J_2 are rated, the time needed after moving distance of l will be determined. Thus it is called directional circuit controlled by time.

The features of circuits are the brake time, which regulates the throttle orifices of $J_1$and J_2 valves according to the speed and inertia of workpiece to control the shift shock and enhance the work efficiency. On the other hand the main directional valve uses an "H" neutral which is useful for reducing the shift shock and enhancing the stability. But when the workpiece moves at a high speed, the shift shock will be very high and result in a low shift precision. Thus it is suitable for use to the position lower shift precision needed, such as plan grinding machine.

2. The directional circuit controlled by distance

Fig. 7-22indicates a directional circuit controlled by the distance. The main oil line is controlled not only by the main directional valve 2, but also the pilot valve 3. When the pilot valve 3 moves towards left, the right brake taper surface T on it will close gradually the right return oil port in the cylinder and make the piston move slowly. This will pre-brake the piston. And when the port is nearly closed (the axial opening still remains 0. 2-0. 5mm) and the movement speed of piston is very low, the main directional valve shifts. And when the main valve 2 moves towards the left and cuts the main oil path, the piston will be stopped and the valve 2 will be started in the opposite direction. It is called directional circuit controlled by distance. The half taper degree is $\alpha = 1.5°\text{-}3.5°$ and the brake length $l = 3\text{-}12$mm.

The features of the directional circuit controlled by distance are: whatever the original speed of hydraulic cylinder is, the pilot valve 3 will always first move a rated distance of l and then realize the two steps of pre-brake and end-brake. So this circuit has the advantages of shift stability and low shock. But the workpiece movement speed will affect the brake time and shift shock due to a invariable brake distance in the pilot valve. Thus this circuit is suitable for use in the position of low movement speed but high shift precision needed, such as the hydraulic system of inner or external circular

芯向左移动的同时，其阀芯上的制动锥面 T 逐渐将回油的通道关小，从而使液压缸活塞速度逐渐减慢，并在主换向阀 2 的阀芯移过 l 距离后将通道封闭，使液压缸活塞停止运动。主换向阀阀芯上的制动锥的半锥角度一般取 $\alpha = 1.5° \sim 3.5°$，制动锥长度取 $l = 3 \sim 12\text{mm}$。当节流阀 J_1 和 J_2 的开口大小调定后，主换向阀阀芯移过距离 l 所需的时间就确定不变。因此这种制动方式被称为时间控制制动式换向回路。

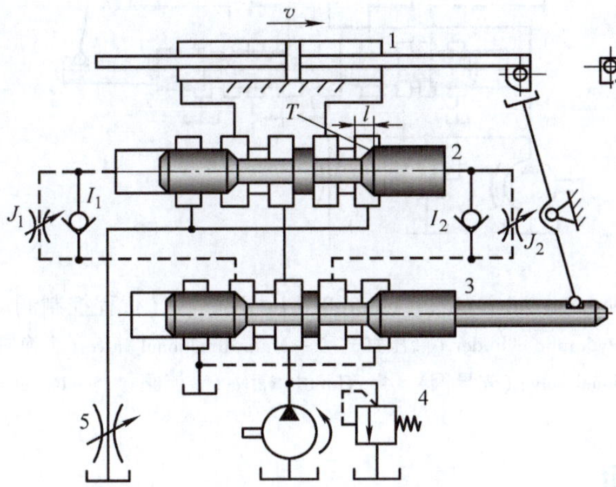

Fig. 7-21　Directional valve circuit controlled by time（时间控制制动式换向回路）

1—Hydraulic actuator（液压缸）　2—Main directional valve（主换向阀）

3—Pilot valve（先导阀）　4—Relief valve（溢流阀）　5—Throttle valve（节流阀）

时间制动控制式换向回路的优点在于：制动时间可根据主机部件运动速度的快慢和惯性的大小，通过节流阀 J_1 和 J_2 的开口量得到调节，如工作台速度高、质量大时，可把制动时间调得长一些，以利于消除换向冲击；反之，则可调得短一些，以使其换向平稳，又提高工作效率。此外主换向阀的中位机能采用“H”型，有利于减小液压冲击和提高换向平稳性。其缺点：工作部件速度大时，其制动过程中的冲出量就较大，换向精度较低。因此这种回路适用于工作部件运动速度大，换向精度要求不高的场合，如平面磨床液压系统。

2. 行程控制制动式换向回路

如图 7-22 所示，这种回路中的主油路受主换向阀 2 控制外，还受先导阀 3 的控制。当先导阀 3 在换向过程中向左位移动时，先导阀阀芯的右制动锥面 T 将逐渐关小液压缸右腔的回油通道而使液压缸活塞速度逐渐减慢，对液压缸活塞进行预制动。当回油通道被关得很小（轴向开口量尚留约 0.2 ~ 0.5mm）、液压缸活塞速度变得很慢时，主换向阀 2 的控制油路才开始切换。若主换向阀 2 向左移动，切断主油路通道，则液压缸活塞停止运动，并随即使它在相反的方向起动。由于先导阀总是要移动一段固定的行程 l，将工作部件先进行预制动后，再由换向阀来使它换向。这种回路称为行程控制制动式换向回路。

行程控制制动式换向回路换向精度较高，冲出量较小。但由于先导阀的制动行程恒定不变，制动时间的长短和换向冲击的大小将受到运动部件速度的影响。所以这种回路适用于主机工作部件运动速度不大，但换向精度要求较高的场合，如内、外圆磨床的液压系统中。

grinding machine.

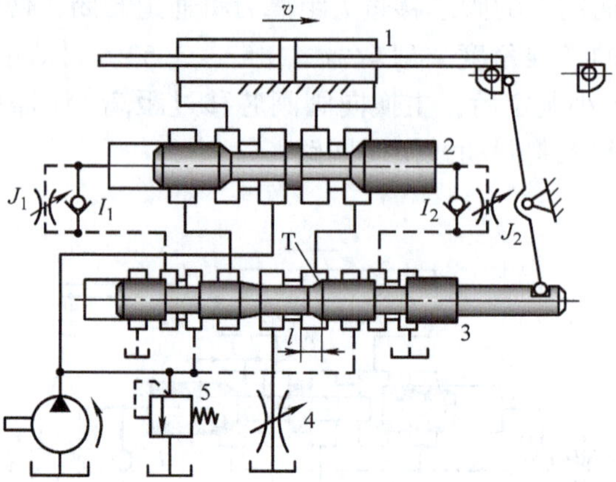

Fig. 7-22 Directional valve circuit controlled by distance (行程控制制动式换向回路)
1—Hydraulic cylinder (液压缸) 2—Main directional valve (主换向阀)
3—Pilot directional valve (先导阀) 4—Throttle valve (节流阀) 5—Relief valve(溢流阀)

7.3.2 Locked circuit

In many cylinder applications, it is necessary to lock the cylinder so that its piston cannot be moved due to external force acting on the piston rod. One common method for locking a cylinder is by using hydraulic pilot check valves, as shown in Fig. 7-23. The hydraulic-actuated checks are connected with both ports of inlet and exit of the cylinder respectively to lock the piston in either direction. An external force, acting on the piston rod, will not move the piston in either direction because reverse flow through either pilot check valve is not permitted under this condition. The leakage from the locked circuit by the pilot-operated check valve is less because of a good taper seal structure. The locked precision for the circuit depends mainly on the leakage from cylinder.

7.4 Multi-actuator Control Circuits

As one power pump supplies oil to multiple actuators, the scheduled task can be accomplished by controlling the pressure, flow rate and distance.

7.4.1 Sequence action circuit

Sequence action circuits allow several actuator motions to obey the setting scheduled actions in turn. There are two kinds of control methods: pressure and distance controls.

1. Sequence action circuit by pressure control

Utilizing the change of pressure during operating to realize actuator sequence actions is a typical control feature for hydraulic system. Fig. 7-24 is a sequence action circuit done by a sequence valve, for instance the action sequence of a drill press is: ①hold parts→② aiguille squash feed→③aiguille exit→④loosen parts. When the directional control valve 5 is shifted to its left envelope mode,

7.3.2 锁紧回路

锁紧回路的功用是使液压缸能在任意位置停留，且不会因外力作用而移动位置。常用的方法之一是采用液控单向阀作锁紧元件，如图 7-23 所示。在液压缸的进、出油路上都串接一液控单向阀（液压锁），使活塞双向锁紧。由于液控单向阀阀座为锥阀式结构，密封性好，所以回路泄漏极少。回路的锁紧的精度主要取决于液压缸的泄漏。

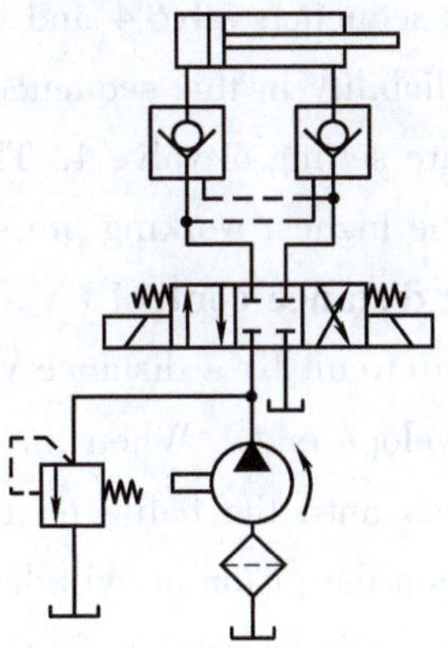

Fig. 7-23 Locked circuit using hydraulic pilot check valves（采用液控单向阀的锁紧回路）

7.4 多执行元件控制回路

当液压系统由一个油源向多个执行元件供油时，可以通过压力、流量、行程控制来实现多执行元件预定动作的要求。

7.4.1 顺序动作回路

顺序动作回路的功用在于使几个执行元件严格按照预定顺序依次动作。按控制方式不同，分为压力控制和行程控制两种。

1. 压力控制顺序动作回路

利用液压系统工作过程中的压力变化来使执行元件按顺序先后动作是液压系统独具的控制特性。图 7-24 所示为采用顺序阀控制的顺序动作回路。钻床液压系统的动作顺序为：①夹紧工件→②钻头进给→③钻头退出→④松开工件。当电磁换向阀 5 得电，阀左位接入回路时，夹紧缸 1 的活塞向右运动，夹紧工件后回路压力升高到顺序阀 3 的调定压力，顺序阀 3 开启，钻孔缸 2 活塞才向右运动进行钻孔。钻孔完毕，电磁换向阀 5 失电，阀右位接入回路，钻孔缸 2 活塞先退到左端点。钻孔缸 2 活塞退到左端点后回路压力升高，打开顺序阀 4，再使夹紧缸 1 活塞退回原位。回路顺序动作的可靠性取决于顺序阀性能及其压力调定值，即它的调定压力应比前一动作的最高工作压力大 10% ~15%，否则会造成误动作。

2. 行程控制顺序动作回路

图 7-25a 所示为采用行程阀控制的顺序动作回路。图示位置两液压缸活塞均退至左端点。电磁阀 3 得电，阀左位接入回路后，缸 1 活塞先向右运动，当活塞杆上挡块压下行程阀 4 后，缸 2 活塞才向右运动；电磁阀 3 失电，阀右位接入回路，缸 1 活塞先退回。其挡块离开行程阀 4 后，缸 2 活塞才退回。

the piston in the clamped cylinder 1 extends toward right. When the piston touches and holds the parts, the pressure of system will be increased until the pressure reaches the setting of the sequence valve 3, at the same time the sequence 3 is opened. And then the piston in the cylinder 2 is just extended to drill the hole of workpiece. When the process of drill hole is finished, the valve 5 is shifted into its right envelope mode and the piston in cylinder 2 retracts to the end of the left. When the piston in the cylinder is retracted to the left envelope mode, the system pressure will be increased until it reaches the pressure setting of the sequence valve 4 and opens the valve to let the cylinder 1 retract to the original position. The reliability in this sequence action circuit depends on the performance of sequence valve 4 and pressure setting of valve 4. That is, the pressure setting of the valve 4 should be 10%-15% heigher than the highest working pressure, otherwise error actions will occur.

2. Sequence action circuit by distance control

Fig. 7-25a is a sequence action circuit by a distance valve. The figure shows that the two pistons of cylinders are at their left envelope ends. When the solenoid valve 3 is at the left envelope mode, the piston of cylinder 1 extends until the baffle on the end of piston gives a pressure force down on the distance valve 4, and then the piston of cylinder 2 extends. When solenoid valve 3 is at the right envelope mode, the piston of cylinder retracts firstly and when the baffle leaves the distance valve 4, piston of cylinder 2 retracts.

Fig. 7-25b is a sequence action circuit controlled by the distance switches. When we turn on the start button firstly, solenoid valve 1YA is energized and the piston of cylinder 1 extends until the baffle on it touches and puts a pressure force down on the distance switch "2st". And then the solenoid 2YA is energized and allows the piston in cylinder 2 to extend. The retraction processing is followed when the piston in cylinder 2 extends until it puts a pressure force down on the distance switch "3st" and the 1YA loses electricity, which makes the piston in the cylinder 1 retract until it presses down distance switch "1st". And then it makes 2YA lose electricity to let piston in cylinder 2 retract. The advantage of this circuit is that we adjust only the baffle on the end of the piston and the action sequences of cylinder can be regulated randomly by electro control, so it can be applied.

7.4.2 Synchronization circuit

The synchronization circuits are used to overcome the loads, friction resistance, leakage, and fabrication or structure error in actuators and ensure the cylinder motion at a same speed or displacement.

1. Synchronization circuit by a speed-regulating valve

Fig. 7-26 is a synchronization circuit. Two speed-regulating valves are connected respectively in the inlet or exit port of the two parallel cylinders. We only adjust carefully the two speed-regulating valves until both cylinder motions are bidirectionally synchronized. This circuit has the advantage of simple structure and the disadvantage of poor regulating precision. So it is not suitable for use to the position of more load changed. We utilize the concentrate (or distributary) flow valves instead of the speed-regulating valve to obtain the high regulating precision, but it has the disadvantage of low efficiency and so it is suitable to be used in the position of low pressure systems.

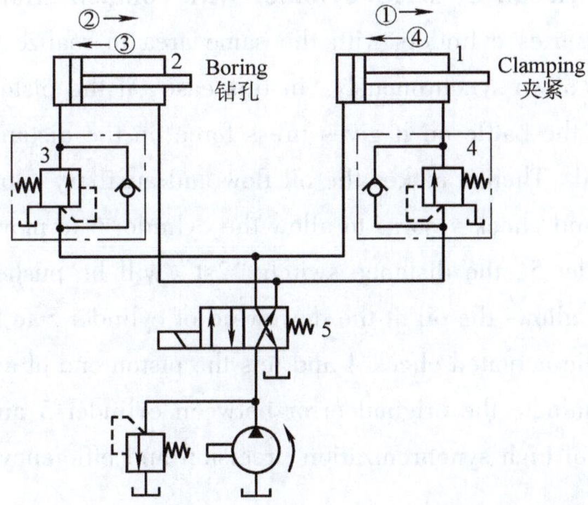

Fig. 7-24　Sequence action circuit controlled by sequence valve（采用顺序阀控制的顺序动作回路）

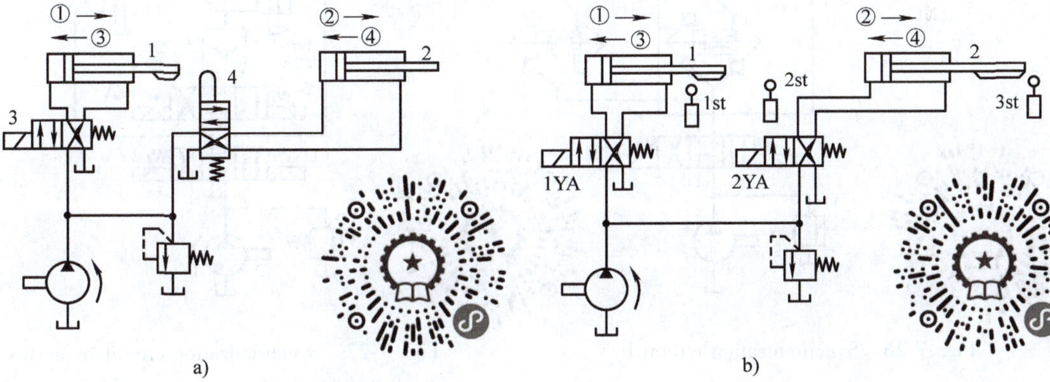

Fig. 7-25　Sequence action circuits（行程控制顺序动作回路）

a) Sequence action circuit by distance valve（采用行程阀控制的顺序动作回路）

b) Sequence action circuit by distance switch（采用行程开关控制的顺序动作回路）

图 7-25b 所示为采用行程开关控制电磁换向阀的顺序回路。按起动按钮，电磁铁 1YA 得电，缸 1 活塞先向右运动，当活塞杆上的挡块压下行程开关 2st 后，使电磁铁 2YA 得电，缸 2 活塞才向右运动，直到压下 3st，使 1YA 失电，缸 1 活塞向左退回，而后压下行程开关 1st，使 2YA 失电，缸 2 活塞再退回。回路优点是：调整行程比较方便，改变电控线路即可改变动作顺序，灵活方便，应用广泛。

7.4.2 同步回路

同步回路的功用是使系统中多个执行元件克服负载、摩擦阻力、泄漏、制造质量和结构变形上的差异，而保证在运动中的位移量相同或以相同的速度运动。

1. 采用调速阀的同步回路

图 7-26 所示，在两个并联液压缸的进（回）油路上分别串接一个调速阀，细调两个调速阀的开口大小，控制进入两液压缸或自两液压缸流出的流量，即可实现双向同步。这种回

2. Synchronization circuit by series cylinder with compensation unit

Fig. 7-27 shows two series cylinders with the same area to realize the synchronization action. When two cylinders move down synchronously, in this case, if the piston of cylinder 5 reaches the end point in advance and the baffle on it gives press force on the distance switch "1st", it lets the solenoid 3YA be energized. Then it makes the oil flow and auxiliary supply to the upper of cylinder 6 via directional valve 3 and check 4 so as to allow the cylinder 6 to move to end. If cylinder 6 reaches the end before cylinder 5, the distance switch "2st" will be pushed down to let the solenoid 4YA be energized, which allows the oil at the downward of cylinder 5 to flow into the tank via directional valve 3 and hydraulic-actuated check 4 and lets the piston end of cylinder 5 move downward to the end point so as to eliminate the original error between cylinder 5 and cylinder 6. This kind of circuit has the advantage of high synchronization precision and efficiency.

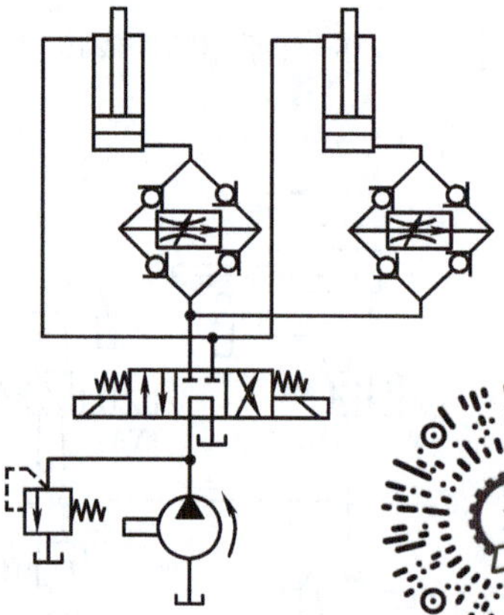

Fig. 7-26 Synchronization circuit by speed-regulating valve
(采用调速阀的同步回路)

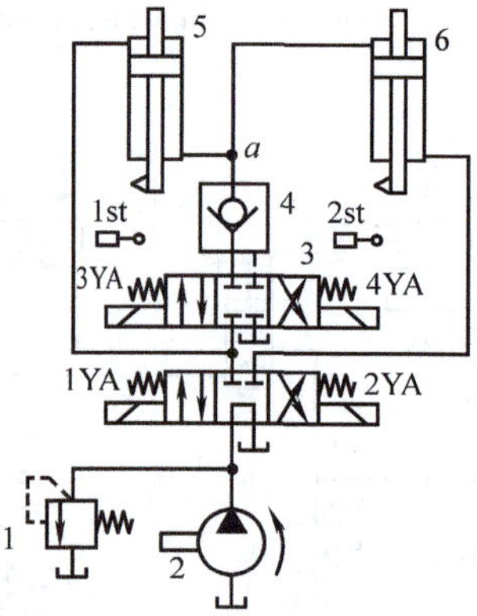

Fig. 7-27 Synchronization circuit by series cylinder with compensation unit
(带补偿装置的串联缸同步回路)

3. Synchronization circuits by synchronous cylinder or synchronous motor

Fig. 7-28a is a synchronization circuit by a pair of synchro-cylinders. The synchro-cylinder 3 is a hydraulic cylinder that consists of two pistons with same areas and one common rod. When the pistons move toward the right or left, the same volume of oil will be output or input, which is acted for distributing the flow, to let the two cylinders move for synchronous motion.

It is similar to the synchro-cylinder, we utilize two synchro-motors with the same displacement, which is acted for distributing flow, to realize that two cylinder move for synchronous motion as shown in Fig. 7-28b. The throttle valve 4 is used to eliminate the error between the two cylinders when they move to the end.

This kind of circuit has the advantage of high synchronous motion precision and efficiency but it has also the disadvantage of special element needed, complexity and high cost demanded.

4. Synchronization circuits by scale or servo valves

When the hydraulic system requires a high synchronous motion, a precision scale or servo valve should be used. Fig. 7-29 shows an example; according to the signal from sensors B and C, valve A

路结构简单，同步精度不高，不宜用于偏载或负载变化频繁的场合。采用分流集流阀（同步阀）代替调速阀来控制两液压缸的进入或流出的流量，可使两液压缸在承受不同负载时仍能实现速度同步。由于同步作用靠分流集流阀自动调整，使用较为方便，但效率低，压力损失大，不宜用于低压系统。

2. 采用串联液压缸的同步回路

如图 7-27 所示，有效工作面积相等的两个液压缸串联起来便可实现两缸同步。回路中设置有补偿装置，以消除两同步缸所积累的位置误差，其原理是：当两缸活塞同时下行时，若缸 5 活塞先到达行程端点，则挡块压下行程开关 1st，电磁铁 3YA 得电，换向阀 3 左位接入回路，压力油经换向阀 3 和液控单向阀 4 进入缸 6 上腔，进行补油，使其活塞继续下行到达行程端点。如果缸 6 活塞先到达端点，行程开关 2st 使电磁铁 4YA 得电、换向阀 3 右位接入回路，压力油进入液控单向阀 4 的控制腔，打开阀 4，缸 5 下腔与油箱接通，使其活塞继续下行到达行程端点，从而消除积累误差。这种回路同步精度较高，回路效率也较高。

3. 采用同步缸或同步马达的同步回路

图 7-28a 所示是一种采用同步缸或同步马达的同步回路。同步缸 3 是两个尺寸相同的缸体和两个活塞共用一个活塞杆的液压缸，活塞向左或向右运动时输出（或输入）相等容积的油液，在回路中起着配流的作用，使有效面积相等的两个液压缸实现双向同步运动。

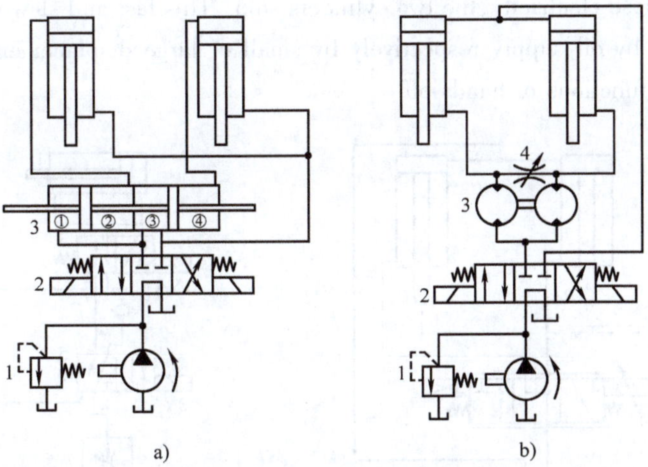

Fig. 7-28　Synchronization circuit by synchronous cylinder or synchronous motor
（采用同步缸或同步马达的同步回路）
a）Synchronization circuit by synchronous cylinder（同步缸同步回路）
b）Synchronization circuit by synchronous motor（同步马达同步回路）

和同步缸一样，用两个同轴等排量双向液压马达 3 作为等流量分流装置，输出相同流量的油液也可实现两缸双向同步。如图 7-28b 所示，节流阀 4 用于在行程端点消除两缸位置误差。

这种回路的同步精度比采用调速阀的同步回路高，但需专用装置，费用较高。

4. 采用比例阀或伺服阀的同步回路

当液压系统有很高的同步精度要求时，必须采用比例阀或伺服阀的同步回路。图 7-29 所示为一例，伺服阀 *A* 根据两个位移传感器 *B*、*C* 的反馈信号，持续不断地调整阀口开度，

regulates continuously the orifice area to control the flow rate at flow inlet or outlet and to realize synchronous motion. A cheap proportional valve can be used to replace a servo valve, but its precision will be worse.

7.4.3 Hands-off circuit

This circuit is used to realize hands-off cylinders motion during carrying their own operating tasks. Fig. 7-30 is a hands-off circuit by two pumps supplying oil to the system. Hydraulic cylinder 1 and 2 finish automatically in a cycle motion of "fast extends→slow extends→fast retracts" respectively. When the solenoid 1YA, 2YA are energized, two cylinders obtain the oil from the big displacement pump 10 to realize fast extends with the different areas. If the cylinder 1 finishes the fast extends before the other cylinder and the baffle presses down on the distance switch to make the solenoid 3YA energized and 1YA lose energy, the oil of the system is supplied by the small displacement pump 9 with speed-regulating 7 to get slow speed motion. So in this case the cylinder 1 is moving with the hands-off cylinder 2. When the two cylinders are in slow extend and all oil is supplied by pump 9, and if cylinder 1 finishes the task before the cylinder 2, the baffle touches down on distance switch to allow the solenoid 1YA and 3YA to be energized. Cylinder 1 gets oil from pump 10 and retracts fast, but by then cylinder 2 still gets oil from pump 9 and is hands-off to cylinder 1. When all solenoids lose electricity, the two cylinders stop. This fast and slow motion speed of the circuit will be realized by oil supply respectively by small or large displacement pumps. It uses solenoids to control the functions of hands-off.

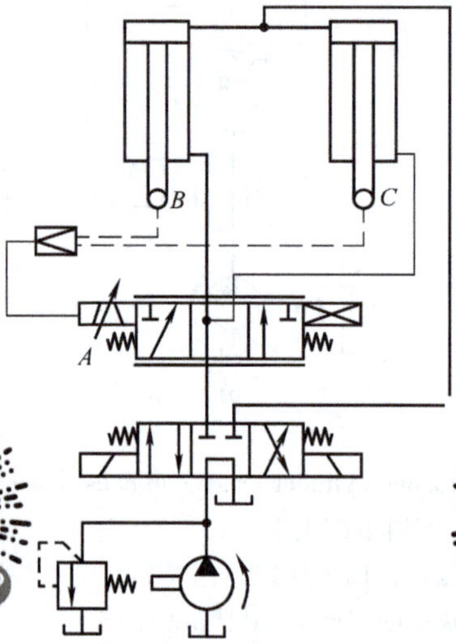

Fig. 7-29 Synchronization circuits by scale or servo valves
(采用比例阀或伺服阀的同步回路)

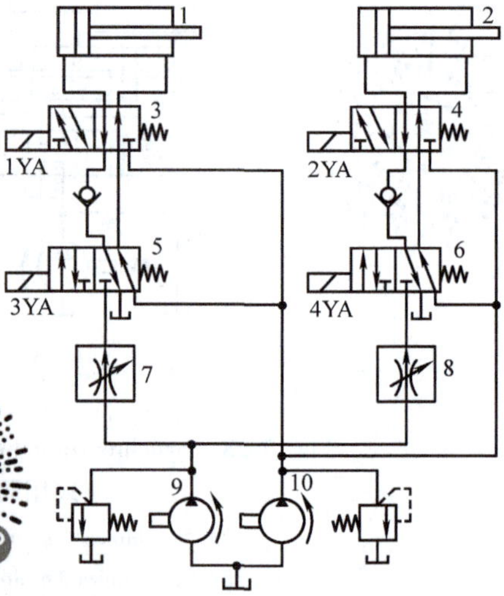

Fig. 7-30 Multi-cylinder hands-off circuits with fast and slow speed
(多缸快、慢速互不干扰回路)

控制两个液压缸的输入或输出流量，使它们获得双向同步运动。也可采用价廉的比例阀来代替伺服阀，但精度要差些。

7.4.3 互不干扰回路

这种回路的功用是使系统中几个执行元件在完成各自工作循环时彼此互不影响。图 7-30 所示为通过双泵供油来实现多缸快、慢速互不干扰的回路。缸 1 和缸 2 各自要完成“快进—工进—快退”的自动工作循环。当电磁铁 1YA、2YA 得电时，两缸均由大流量泵 10 供油，并作差动连接实现快进。如果缸 1 先完成快进动作，挡块和行程开关使电磁铁 3YA 得电，1YA 失电，大流量泵进入缸 1 的油路被切断，而改为小流量泵 9 供油，由调速阀 7 获得慢速工进，不受缸 2 快进的影响。当两缸均转为工进、都由小流量泵 9 供油后，若缸 1 先完成了工进，则挡块和行程开关使电磁铁 1YA、3YA 都得电，缸 1 改由大流量泵10 供油，使活塞快速返回，这时缸 2 仍由泵 9 供油继续完成工进，不受缸 1 影响。当所有电磁铁都失电时，两缸都停止运动。此回路采用快、慢速运动由大、小流量泵分别供油，并由相应的电磁阀进行控制的方案来保证两缸快、慢速运动互不干扰。

Chapter 8 Examples of Hydraulic System

The material presented in the previous chapters dealt with fundamentals of system components and basic hydraulic circuits. In this chapter we shall discuss the hydraulic transmission system.

A hydraulic transmission system is built by several basic hydraulic circuits and designed to satisfy a given set of machinery or apparatus requirements. When analyzing or designing a hydraulic circuit, the following important considerations must be taken into account:

1. Find out the working principle for a given set of machine or apparatus and understand the performance of the desired function for speeds and pressures and directions during the machine working periods.

2. Read the sketch of the hydraulic system in advance and understand the elements included, classified into several basic hydraulic circuit units by actuators.

3. Analyze the principles of several basic circuits one by one and find out the relationship between basic circuits to tidy up the lines.

4. Safety of operation.

5. Efficiency of operation.

6. Sum up and comprehend the features of the hydraulic system based on the functions of several basic hydraulic circuits mentioned above.

How to analyze a hydraulic system is a troublesome problem for beginners. Here we shall show the methods and steps by four practical examples of hydraulic transmission systems from industry. We aim to give the students further understanding of the functions of hydraulic elements and the composition of hydraulic circuits, which enable them to read and analyze complex hydraulic system circuits.

8.1 Hydraulic System of Power-Slipway for Combined Machine Tools

8.1.1 Introduction

Combined machine tools are one type of machine tools with high efficiency, high degree of automation, which are usually used for special purpose. They consist of standard units and some special units. They can not only accomplish several working procedures such as hole drilling, expansion, reaming, boring, milling, tapping, etc, but also provide assistant operations such as workbench rotation, position, clamping, feeding, etc. Thus they can also build automatic machine line.

第 8 章　液压系统实例

前面章节已经对液压元件和液压传动的基本回路进行详细的论述。这一章将讨论液压传动系统。

液压传动系统是根据机械设备的工作要求，选用适当的液压基本回路经过有机组合而成的。阅读一个较复杂的液压系统图，可分为以下几个步骤：

1）了解机器设备的工作原理，从而了解机械设备的工况对液压系统的要求；了解在工作循环中的各个工况对力、速度和方向这三个参数的要求。

2）初读液压系统图，了解系统中包含哪些元件，并且以执行元件为中心，将系统分解为若干个子系统。

3）逐个分析每一个子系统，了解其执行元件与相应的阀、泵之间的关系和由哪些基本回路组成。参照电磁铁动作表和执行元件的动作要求，理清其液流路线。

4）工作系统的安全性。

5）工作效率。

6）在全面读懂液压系统的基础上，根据系统所使用的基本回路的性能，对系统作综合分析，归纳总结整个液压系统的特点，以加深对液压系统的理解。

如何分析液压传动系统对初学者来说比较难。本章介绍来自工业生产中的四种液压传动系统，其目的是让学生进一步掌握分析复杂的液压传动系统的方法和步骤，更好地理解和掌握液压元件的结构与原理、液压基本回路以及它们在液压传动系统中的作用。

8.1　组合机床动力滑台液压系统

8.1.1　概述

组合机床是由通用部件和部分专用部件组成的高效、专用、自动化程度较高的机床。它能完成钻、扩、铰、镗、铣、攻螺纹等工序和工作台转位、定位、夹紧、输送等辅助动作，可用来组成自动线。

8.1.2　YT4543 型动力滑台液压系统工作原理

图 8-1 所示为 YT4543 型动力滑台液压系统原理图，该系统的动作循环见表 8-1。这个系统能够实现的工作循环是：快进→一工进→二工进→死挡铁停留→快退→原位停止。实现该工作循环的工作原理是：

1. 快进

按下起动按钮，电磁铁 1YA 得电，先导电磁阀 4 处于左位，使液动换向阀 3 阀芯右移，左位接入系统，其主油路为：

8.1.2 The principle of hydraulic system for YT4543 type power-slipway

The principle chart and operation cycle table of hydraulic system for YT4543 type power-slipway are shown in Fig. 8-1 and Tab. 8-1 respectively. In this system, a pressure-limiting variable vane pump is used to supply oil. Eectro-hydraulic directional valves are used to change oil direction and a distance valve is used to accomplish the shift between quick-working feed and slow-working feed. A solenoid valve is used to realize the replacement of the two different speeds of working feed, and a speed adjusting valve is used to ensure feeding speed stability. Mechanically or electrically, these components can accomplish a semi-automation cycle, i. e., quick feed→working feed→end block dwell→quick retract→stop (at the start position). The working operations are detailed as follows:

Tab. 8-1 YT4543 type power-slipway hydraulic system operation cycle
(YT4543 型动力滑台液压系统的动作循环表)

Component(元件) / Operation (动作)	1YA	2YA	3YA	Pressure switch (压力继电器)6	Distance valve (行程阀)9
Quick feed (differential-area connection) 快进(差动)	+	−	−	−	Actuated (导通)
Working feed 1(一工进)	+	−	−	−	Released(切断)
Working feed 2(二工进)	+	−	+	−	Released(切断)
End block dwell(死挡铁停留)	+	−	+	+	Released(切断)
Quick retract(快退)	−	+	−	−	Released(切断)
Stop(原位停止)	−	−	−	−	Actuated(导通)

1. Quick feed

When the button is started, solenoid 1YA is energized and the solenoid pilot valve 4 shifts at the left position to move the hydraulic-operator direction valve core 3 toward right, which makes the left position of valve 3 enter the system. The main oil line is:

Inlet oil line is pump 1→ check 2→ left position of directional control valve (hydraulic actuated)3→ distance valve 13 (normal position) → left chamber of the hydraulic actuator.

Return oil line is the right chamber of hydraulic actuator → left position of directional control valve (hydraulic actuated) 3→ check 7→distance valve13 (normal position) → left chamber of the hydraulic actuator.

The system pressure is very low due to power slipway unloaded. The sequence valve 6 is closed and the oil line is in area differential connection to realize the slipway quick feed. The variable displacement pump 1 has the maximal flow rate in this case.

2. Working feed 1

When the distance valve 13 is actuated by block on the slipway, the oil line to the enveloped end of the cylinder via valve 13 is cut off. However, the solenoid on valve 10 is still de-energized, causing the oil to pass through speed-regulating valve 8. The system pressure increases to open sequence valve 6 and reduce the output flow rate of the pump until a new equilibrium of flow rate after

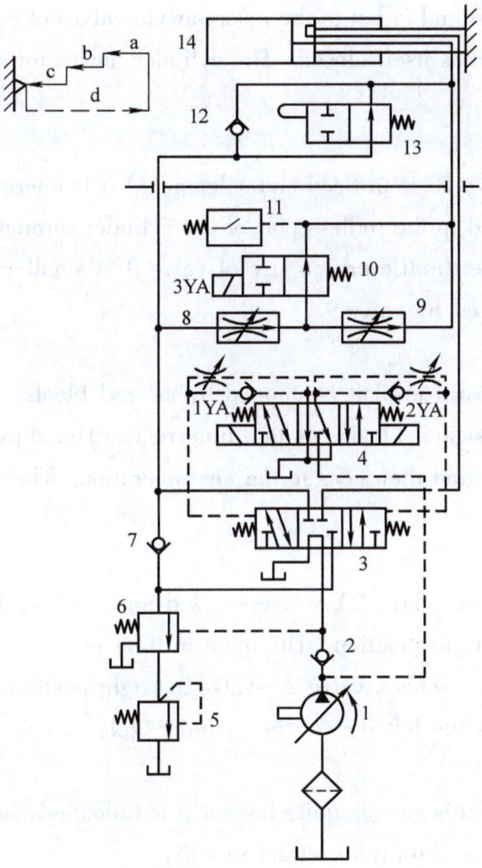

Fig. 8-1 YT4543 type power-slipway hydraulic system（YT4543 型动力滑台液压系统图）

1—Hydraulic pump（液压泵） 2，7，12—Check valve（单向阀）

3—Directional control valve by hydraulic actuated（液动换向阀） 4—Pilot solenoid valve（先导电磁阀）

5—Against pressure valve（背压阀） 6—Sequence valve（顺序阀） 8，9—Speed adjusting valve（调速阀）

10—Solenoid valve（电磁阀） 11—Pressure switch（压力继电器） 13—Distance valve（行程阀）

14—Hydraulic cylinder（液压缸）

a—Quick feed（快进） b—Working feed 1（一工进） c—Working feed 2（二工进） d—Retraction and stop（返回停止）

对于进油路，液压泵 1→单向阀 2→液动换向阀 3 左位→行程阀 13（常态位）→液压缸左腔。

对于回油路，液压缸右腔→液动换向阀 3 左位→单向阀 7→行程阀 13（常态位）→液压缸左腔。

由于动力滑台空载，系统压力低，液控顺序阀 6 关闭，液压缸成差动连接，且液压泵 1 有最大输出流量，滑台向左快进。

2. 一工进

快进到预定位置，滑台上的行程挡块压下行程阀 13，使原来通过行程阀 13 进入液压缸无杆腔的油路切断。此时电磁阀 10 的电磁铁 3YA 处于失电状态，调速阀 8 接入系统进油路，系统压

passing through valve 8.

The flow rate through the enveloped end of the cylinder depends upon the throttle orifice size of valve 8. The oil from the rod end is led to the reservoir via valves of 3, 6 and 5. The pressure difference from check valve 7 makes itself closed. The cylinder moves toward the left at a speed of working feed 1.

3. Working feed 2

When the distance valve 10 is pressed and solenoid 3YA is energized, the line through valve 10 is cut off. The oil is directed to the rodless port of the cylinder through valves 8 and 9, reducing the slipway speed. Note that the throttle orifice size of valve 9 is smaller than that of valve 8, and the slipway speed can be adjusted by valve 9.

4. End block dwell

The slipway moves forward until it is stopped by an end block. The pressure in the enveloped end increases and pressure switch 11 signals to a time-relay. The slipway holds and touches with the end block for a certain time and then goes to the next operation. The slipway hold time can be regulated by time-relay.

5. Quick retract

Given a sign by the time-relay, 2YA is energized and 1YA and 3YA are de-energized. Both valves 4 and 3 work in the right position. The main oil line is:

Inlet oil line is pump 1 →check valve 2→valve 3 (right position) →right port of the cylinder;

The returned oil line is the left port of the cylinder→check valve 12→valve 3 (right position) →reservoir.

The system pressure in this case is quite low for it is unloaded, and the output flow rate reaches the largest point, causing the slipway to retract rapidly.

6. Stop at the start position

When the slipway returns to its start position, the distance switch at the start position is pressed by a block and 1YA, 2YA, 3YA are released. Valves 4 and 3 return to their original positions and the slipway stops. Pump 1 is under an unloaded condition via valve 3.

8.1.3 Characteristics

This hydraulic system consists of the following basic circuits: a volume throttling speed control circuit is composed by a pressure-limiting variable vane pump and a speed adjusting valve mounted in the oil inlet line, a quick moving circuit with differential-area connection, a direction shift circuit controlled by electro-hydraulic directional valves, a speed regulating circuit accomplished by distance valves, solenoid valves, sequence valves, etc, and a relief circuit released by an electro-hydraulic directional valve with M type neutral function.

1) A volume throttling speed control circuit is composed by a pressure-limiting variable vane pump and a speed adjusting valve. This circuit not only meets the requirements of large speed regulating range and low-speed stability, but also allows for a high efficiency. An against-pressure valve is set in the returned oil line for working feed. The purpose is to improve the speed stability (i. e., prevent air from going into the system, increase the rigidity) and to enable the slipway to undertake

力升高。压力的升高，一方面使顺序阀 6 打开，另一方面使限压式变量泵的流量减小，直到与经过调速阀 8 后的流量相同为止。

这时进入液压缸无杆腔的流量由调速阀 8 的开口大小决定。液压缸有杆腔的油液则通过液动换向阀 3 后经顺序阀 6 和背压阀 5 回油箱（两侧的压差使单向阀 7 关闭）。滑台以第一种工进速度向左运动。

3. 二工进

一工进结束时，挡块压下行程开关，使电磁铁 3YA 得电，经电磁阀 10 的通路被切断。此时油液需经调速阀 8 与 9 才能进入液压缸无杆腔。由于调速阀 9 的开口比调速阀 8 小，滑台的速度减小，速度大小由调速阀 9 的开口大小决定。

4. 死挡铁停留

当滑台二工进终了碰上死挡铁后，滑台停止运动。液压缸无杆腔压力升高，达到压力继电器 11 的调定值，压力继电器动作，经过时间继电器的延时，再发出电信号，使滑台退回，滑台的停留时间可由时间继电器调定。

5. 快退

当时间继电器经延时发出信号，2YA 得电，1YA、3YA 失电，先导电磁阀 4 和液动换向阀 3 处于右位。主油路为：

对于进油路，液压泵 1→单向阀 2→液动换向阀 3 右位→液压缸右腔。

对于回油路，液压缸左腔→单向阀 12→液动换向阀 3 右位→油箱。

由于此时为空载，系统压力低，液压泵 1 输出的流量最大，滑台向右快退。

6. 原位停止

当滑台快退到原位时，挡块压下原位行程开关，使电磁铁 1YA、2YA 和 3YA 都失电，先导电磁阀 4 和液动换向阀 3 处于中位，滑台停止运动，液压泵 1 通过液动换向阀 3 中位（M 型）卸荷。

8.1.3　YT4543 型动力滑台液压系统特点

YT4543 型动力滑台液压系统包括以下一些基本回路：由限压式变量叶片泵和进油路调速阀组成的容积节流调速回路，差动连接快速运动回路，电液换向阀的换向回路，由行程阀、电磁阀和液控顺序阀等联合控制的速度切换回路以及用中位为 M 型机能的电液换向阀的卸荷回路等。液压系统的性能就由这些基本回路所决定。该系统有以下几个特点：

1）采用了由限压式变量泵和调速阀组成的容积节流调速回路。它既满足系统调速范围大，低速稳定性好的要求，又提高了系统的效率。进给时，在回油路上增加了一个背压阀，这样做一方面是为了改善速度稳定性（避免空气渗入系统，提高传动刚度），另一方面是为了使滑台能承受一定的与运动方向一致的切削力。

2）采用限压式变量泵和差动连接两个措施实现快进，这样既能得到较高的快进速度，又不致使系统效率过低。动力滑台快进和快退速度均为最大进给速度的 10 倍，泵的流量自动变化，即在快速行程时输出最大流量，工进时只输出与液压缸需要相适应的流量，死挡铁停留时只输出补偿系统泄漏所需的流量。系统无溢流损失，效率高。

3）采用行程阀和液控顺序阀使快进转换为工进，动作平稳可靠，转换的位置精度比较高。至于两个工进之间的换接则由于两者速度都较低，采用电磁阀完全能保证换接精度。

a certain cutting force in the moving direction.

2) A pressure-limiting variable pump and differential-area connection method are used in this system to accomplish quick feed. By this way, we can obtain higher speed for quick feed and prevent the system's excessively low efficiency. Both the speeds for the quick feed and quick retract are 10 times that of the largest speed for working feed. The flow rate of the pump is adjusted automatically: providing largest output for quick feed, required flow rate by the cylinder for working feed, leakage compensation for compensating system for the dwell in front of the end block. There is no relief loss in this system, but a high efficiency.

3) Using distance valves and sequence valves for quick-working feeds shift. This can achieve a stable and reliable operation and a higher precision positioning. For the shift from working feed 1 to working feed 2, both the required speeds are relatively low, so solenoid valves can ensure the shift precision.

8.2 Hydraulic System of Plastic Injection Moulding Machines

8.2.1 Introduction

A plastic injection moulding machine has the function of heating or melting the plastic particle resin to flowing stage and then injecting the fluid into a mode chamber and forming a plastic product via an injection nozzle holding a given set of pressure and time.

The injection moulding process is shown below:

Closing mode chamber → injection seat moving onwards → injecting → holding pressure → $\left\{\begin{matrix} \textit{cooling} \\ \textit{preplastic} \end{matrix}\right\}$ → injection seat going back → opening mode chamber → pushing out the product → put out cylinder going back → closing mode chamber again

Functions above are done by hydraulic actuators which are called as the close mode chamber cylinder, injection seat moving cylinder, pre-plastic hydraulic motor, injecting cylinder and push out cylinder.

Desired hydraulic systems should have a force enough for closing mode chamber, adjustable speeds for opening/closing mode chamber, adjustable pressure and speed for injecting processing and holding hydraulic pressure. The system should have safety units.

8.2.2 Operating principle of hydraulic system of SZ-250A type plastic injection moulding machine

Injection moulding machine SZ-250A belongs to a middle and small mould with max. 250cm^3 capacity of injection. Fig. 8-2 shows the outline of the machine and its hydraulic system. The action sequences of solenoid energized are shown in Tab. 8-2.

1. Safety gate close

The safety gate is used to ensure security during operation. The close mould cylinder is stopped until the sequence valve 6 is at the normal position, which happpens only when the safety gate is closed and then the system action procedure begins to operate.

8.2　塑料注射成型机液压系统

8.2.1　概述

塑料注射成型机简称注塑机。它将颗粒状的塑料加热熔化到流动状态，用注射装置快速高压注入模腔，保压一定时间，冷却后成型为塑料制品。

注塑机的工作循环如下：

合模→注射座前移→注射→保压→{冷却 / 预塑}→注射座后退→开模→顶出制品→顶出缸后退→合模

以上动作分别由合模缸、注射座移动缸、预塑液压马达、注射缸和顶出缸完成。

注塑机液压系统要求有足够的合模力，可调节的合模、开模速度，可调节的注射压力和注射速度及可调节的保压压力，系统还应设有安全联锁装置。

8.2.2　SZ-250A 型注射机液压系统工作原理

SZ-250A 型注塑机属中小型注塑机，每次最大注射容量为 250cm³。图 8-2 为其外形及液压系统图。各执行元件的动作循环主要依靠行程开关切换电磁换向阀来实现。电磁铁动作顺序见表 8-2。

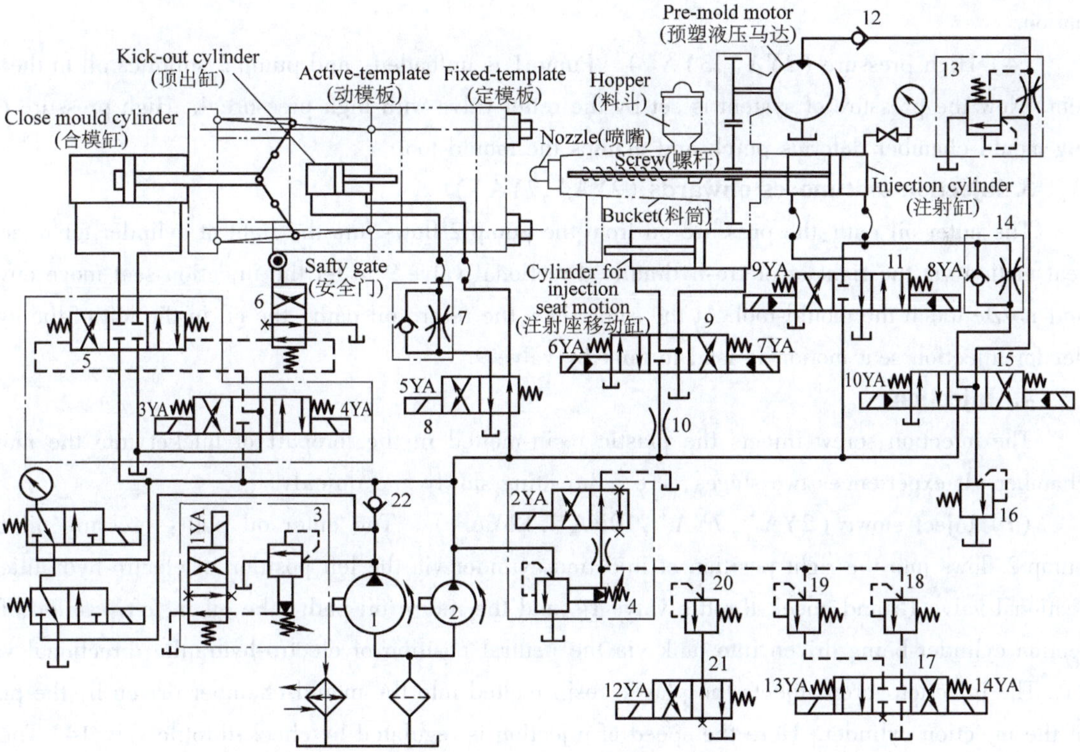

Fig. 8-2　The outline and hydraulic system of SZ-250A type injection moulding machine
（SZ-250A 型注塑机组件外形及液压系统图）

2. Close the mould-chamber

When the active-template moves quickly and closes to the fixed-template, the hydraulic system pressure turns into a low pressure and slow speed stage. Everything is affirmed normally in the mould-chamber and the system turns into high pressure to close the mould-chamber. Utilizing the hydraulic-machine for closing the mould-chamber and the close mould cylinder is used to push the template for opening or closing the mould-chamber by symmetry 5-perch structure which has functions of forcing enhancement and self clamping. The process of mould-chamber has 4 stages:

(1) Close the mould-chamber slowly ($2YA^+$, $3YA^+$) Pump 1 with a high flow rate is unloaded by solenoid valve 3 and the pressure of pump 2 with a low flow rate is set by relief valve 4; the main oil path: the pressure oil from pump 2 is going into the left-chamber of the close mould sylinder via the right position of the electro-hydraulic directional valve 5 to push the piston and the 5-perch structure to close mould-chamber slowly. At the same time, the return oil path: the oil in the right of close mould cylinder flows to the tank via valve 5 and cooler.

(2) Close the mould-chamber quickly ($1YA^+$, $2YA^+$, $3YA^+$) The distance switch gives an electro-signal to let the 1YA be energized and pump 1 under load again. The oil from pump 1 by the check valve 22 combines with the oil from pump 2 supply together to close the mould-chamber quickly.

(3) Low pressure and slow speed ($2YA^+$, $3YA^+$, $13YA^+$) Pump 1 is unloaded again; the pressure of pump 2 is set by the remote pressure-regulating valve 18. Because of the low pressure setting in valve 18 to make a lower closing force, the mould surface cannot be damaged by contamination.

(4) High pressure ($2YA^+$, $3YA^+$) Pump 1 is unloaded, and pump 2 supplies oil to the system. Now the pressure of system is set by the relief valve with high pressure 4. High pressure closing mould-chamber deforms perch and clamps the mould-tool.

3. Injection seat moves onwards ($2YA^+$, $7YA^+$)

The enter oil path: the pressure oil from the pump 2 flows into the right of cylinder for injection seat motion via the right of eletro-hydraulic directional valve 9 to let the injection seat move onward and nozzle touch the mould-tool. At the same time, the return oil path: the oil in the left of the cylinder for injection seat motion is lead to tank by valve 9.

4. Injection

The injection screw injects the plastic resin-melted in the forepart of bucket into the mould-chamber. It experiences two stages, i. e., injecting slowly and quickly.

(1) Inject slowly ($2YA^+$, $7YA^+$, $10YA^+$, $12YA^+$) The enter oil path: pressure oil from pump 2 flows into the right position of injection cylinder via the left position of electro-hydraulic directional valve 15 and check-throttle valve 14, and the return oil path: the oil in left position of injection cylinder being driven into tank via the neutral position of electro-hydraulic directional valve 11. The injection screw injects the plastic resin-melted into the mould-chamber driven by the piston of the injection cylinder. Here the speed of injection is regulated by check-throttle valve14. The remote valve 20 is used to rate the pressure.

(2) Inject quickly($1YA^+$, $2YA^+$, $7YA^+$, $8YA^+$, $10YA^+$, $12YA^+$) The enter oil path: the oil from pump 1 and pump 2 flows into the right position of injection cylinder via the right position of

Tab. 8-2　Action sequences of solenoid energized for SZ-250A type injection moulding machine

（SZ-250A 型注塑机电磁铁动作顺序表）

Action procedures（动作顺序）		1YA	2YA	3YA	4YA	5YA	6YA	7YA	8YA	9YA	10YA	11YA	12YA	13YA	14YA
		Action sequences of solenoid energized SZ-250A 型注塑机电磁铁动作顺序表													
Close mould-chamber（合模）	Slow speed（慢速）		+	+											
	Fast speed（快速）	+	+	+											
	Low pressure and slow speed（低压慢速）		+	+										+	
	High pressure（高压）		+	+											
Injection seat moving onwards（注射座前移）			+												
Injection（注射）	Slowly（慢速）		+					+			+		+		
	Fast（快速）	+	+					+	+		+		+		
Holding pressure（保压）			+					+			+				+
{*Cooling* / *Preplastic*} {冷却 / 预塑}		+	+					+				+			
Prevent flowing（防流涎）			+					+		+					
Injection seat withdraw（注射座后退）			+				+								
Open mode-chamber（开模）	Slow open（慢速）		+		+										
	Fast open（快速）	+	+		+										
Kick-out products（顶出制品）	Going onwards（前进）		+			+									
	Withdraw（后退）		+												
Screw withdrawing（螺杆后退）			+							+					

electro-hydraulic directional valve 11; the return oil path: the oil in the left position of injection cylinder is driven to tank by valve 11. Because the two pumps supply together oil to the system, rather than via the check valve 14, the fast injection has been realized. Now the remote valve 20 is used to ensure safety.

5. Holding pressure($2YA^+$, $7YA^+$, $10YA^+$, $14YA^+$)

A small quantity of oil is required in the process of holding pressure, so pump 1 is unloaded and the system oil is supplied by pump 2. And the rest oil is returned to the tank via relief valve 4. The holding pressure is set by remote pressure-regulating valve 19.

6. Pre-plastic($1YA^+$, $2YA^+$, $7YA^+$, $11YA^+$)

When the holding pressure process is finished, the plastic particle resin in the bucket is jammed into the forepart of the bucket with the rotating of screw for heating and the plastic is melted, which creates a pressure. When the pressure of the material melted overcomes the resistance force for piston of injection cylinder withdrawal, the screw also starts to withdraw. When the screw goes back to the position to the quantity of injection required, it stops and prepares for the next injection. By then the product in the mould-chamber has been formed.

In the transmission above, the screw is driven by the pre-mold motor and the gear machine. The pressure oil from pump 1 and pump 2 flows into motor via the right position of electro-hydraulic directional valve 15, speed-regulating valve 13 and check 12. The rotational speed of motor is controlled by speed-regulating valve 13; the relief valve 4 is a safety valve. When the pressure of melted plastic rises to propel the injection cylinder to withdraw, the oil in the right position of injection cylinder flows into the tank via check-throttle valve 14, the right position of electro-hydraulic valve 15 and back pressure valve 16. The back pressure is controlled by valve 16. At the same time the left chamber in the injection cylinder forms a partial vacuum (negative pressure) and the oil in the reservoir flows into the left chamber via the valve 11 under the atmospheric pressure.

7. Prevent flowing($2YA^+$, $7YA^+$, $9YA^+$)

This process is used to reduce the pressure to prevent the melted material from flowing out from the forepart of the nozzle by moving the screw to withdraw a small distance after the pre-plastic procedure. The enter oil path: the pump is unloaded and the oil from pump 2 on the one hand flows into the right position of the cylinder for injection seat motion via the right position of valve 9 to allow the nozzle to touch the mould tool, and on the other hand into the left chamber of injection cylinder via the left position of valve 11 to compel the screw withdraw. The return oil path: both the oil in the left chamber of the cylinder for injection seat motion and the oil in the right chamber of injection cylinder flow into the reservoir by valve 9 and valve 11 respectively.

8. Injection seat withdraws($2YA^+$, $6YA^+$)

After holding the pressure, the injection seat withdraws. Pump 1 is unloaded and the oil from pump 2 drives the injection seat withdrawing.

9. Open the mould-chamber

In the opening process, the mould-chamber experiences two stages:

(1) Open the mould-chamber slowly ($2YA^+$ or $1YA^+$, $4YA^+$) Pump 1 (or pump 2) is unloaded and the oil from pump 2 (or pump 1) flows into the right chamber of cylinder for closing

1. 关安全门

为保证操作安全，注塑机都装有安全门。关闭安全门，行程阀 6 恢复常位，合模缸才能动作，开始整个动作循环。

2. 合模

动模板慢速起动、快速前移，接近定模板时，液压系统转为低压、慢速控制。在确认模具内没有异物存在时，系统转为高压使模具闭合。这里采用了液压-机械式合模机构，合模缸通过对称五连杆结构推动模板进行开模和合模，连杆机构具有增力和自锁作用。合模过程经历 4 个阶段：

（1）慢速合模（$2YA^+$、$3YA^+$）　大流量泵 1 通过电磁溢流阀 3 卸荷，小流量泵 2 的压力由溢流阀 4 调定。主油路：小流量泵 2 压力油→电液换向阀 5 右位→合模缸左腔，推动活塞带动连杆慢速合模。回油路：合模缸右腔油液经电-液换向阀 5 和冷却器回油箱。

（2）快速合模（$1YA^+$、$2YA^+$、$3YA^+$）　慢速合模转快速合模时，由行程开关发令使 1YA 得电，大流量泵 1 不再卸荷，其压力油经单向阀 22 与小流量泵 2 的供油汇合，同时向合模缸供油，实现快速合模，最高压力由阀 3 限定。

（3）低压合模（$2YA^+$、$3YA^+$、$13YA^+$）　大流量泵 1 卸荷，小流量泵 2 的压力由远程调压阀 18 控制。因远程调压阀 18 所调压力较低，合模缸推力较小，即使两个模板间有硬质异物，也不致损坏模具表面。

（4）高压合模（$2YA^+$、$3YA^+$）　大流量泵 1 卸荷，小流量泵 2 供油，系统压力由高压溢流阀 4 控制。高压合模并使连杆产生弹性变形，牢固地锁紧模具。

3. 注射座前移（$2YA^+$、$7YA^+$）

主油路：小流量泵 2 的压力油→电磁换向阀 9 右位→注射座移动缸右腔，注射座前移使喷嘴与模具接触；回油路：注射座移动缸左腔油液经电磁换向阀 9 回油箱。

4. 注射

注射螺杆以一定的压力和速度将料筒前端的熔料经喷嘴注入模腔。分慢速注射和快速注射两种。

（1）慢速注射（$2YA^+$、$7YA^+$、$10YA^+$、$12YA^+$）　主油路：小流量泵 2 的压力油→电液换向阀 15 左位→单向节流阀 14→注射缸右腔。回油路：注射缸左腔油液经电液换向阀 11 中位回油箱，注射缸活塞带动注射螺杆慢速注射，注射速度由单向节流阀 14 调节。远程调压阀 20 起定压作用。

（2）快速注射（$1YA^+$、$2YA^+$、$7YA^+$、$8YA^+$、$10YA^+$、$12YA^+$）　主油路：大流量泵 1 和小流量泵 2 的压力油→电-液换向阀 11 右位→注射缸右腔。回油路：注射缸的左腔油液经电液换向阀 11 回油箱。由于两个泵同时供油，且不经过单向节流阀 14，注射速度加快。此时，远程调压阀 20 起安全阀的作用。

5. 保压（$2YA^+$、$7YA^+$、$10YA^+$、$14YA^+$）

由于注射缸对模腔内的熔料实行保压并补塑，只需少量油液，所以大流量泵 1 卸荷，小流量泵 2 单独供油，多余的油液经溢流阀 4 流回油箱，保压压力由远程调压阀 19 调节。

6. 预塑（$1YA^+$、$2YA^+$、$7YA^+$、$11YA^+$）

保压完毕，从料斗加入的物料随着螺杆的转动被带至料筒前端，进行加热塑化，并建立起一定压力。当螺杆头部熔料压力到达能克服注射缸活塞退回的阻力时，螺杆开始后退。后退到预定位置，即螺杆头部熔料达到所需注射量时，螺杆停止转动和后退，准备下一次注

mould-chamber via the left position of electro-hydraulic directional valve 5 and the oil in the left of the cylinder into tank by valve 5.

(2) Open the mould-chamber quickly ($1YA^+$, $2YA^+$, $4YA^+$) The oil from pump 1 and pump 2 supplies the system to speed up the open mould-chamber.

10. Kick-out the products

(1) Kick-out cylinder moves the products out ($2YA^+$, $5YA^+$) Here pump 1 is unloaded and the oil from pump 2 flows into the left chamber of the kick-out cylinder via the left position of electro-hydraulic directional valve 8 and check 7 to push the product out. The motion speed is regulated by check-throttle valve 7, and the relief valve 4 is used to restrict the pressure.

(2) Kick-out cylinder withdraws($2YA^+$) The oil from pump 2 drives the kick-out cylinder withdrawing via valve 8.

11. Screw withdraws and goes onwards

Sometimes the screw withdraws for being taken down. The solenoids 2YA and 9YA are energized to make pump 1 unloaded and the oil from pump 2 flows into the left chamber of injection cylinder, Via the left position of valve 11. Then the piston drives the screw back. When selenoids 2YA and 8YA are energized, the screw will move onwards.

8.2.3 Characteristics

1) The pressure of injection cylinder acts directly on the screw, thus the ratio of the pressure of injecting p_z to the pressure of injection cylinder p is D^2/d^2 (D indicates the diameter of injection cylinder, and d the screw). Usually the plastic injection moulding machine prepares with three different diameter screws to cater for different pressure requirements during machining. At system pressure $p = 14\text{MPa}$, an injection pressure $p_z = 40\text{-}150\text{MPa}$ can be reached.

2) This injection moulding machine utilizes the hydraulic-machine of enhancing force or intensified cylinder closing mould-chamber equipments to ensure the force enough for closing mould-chamber and prevent plastic from overflowing during high pressure injection.

3) According to the technology of injecting moulding the system is prepared with 2 sets of pumps to meet the different speeds desired during the processes of starting /closing and injecting moulding. The radio of the high speed to the low speed is 50-100. When it is working at the high speed both pumps supply to the system; and at a slow speed only pump 2 (the flux is 48 L/min) does and pump 1 (the flux is 194 L/min) is unloaded. It has a reasonable usage in system power.

4) The system utilizes several parallel remote pilot operated pressure-regulating valves to realize the multi-pressure controls. If we use an electro-hydraulic and scale pressure valve for multi-pressure controls and further an electro-hydraulic and scale flow-regulating valve for speed-regulating, it not only decreases the hydraulic components, reducing the shock and noise during shift action directions of valves but also creates a conditions for the computer control.

5) Utilization of the distance switches for action procedures of multi-actuator, which is flexible and convenient.

射。与此同时，在模腔内的制品冷却成型。

螺杆转动由预塑液压马达通过齿轮机构驱动。主油路：大流量泵1和小流量泵2的压力油→电液换向阀15右位→旁通型调速阀13→单向阀12→马达。马达的转速由旁通型调速阀13控制，溢流阀4为安全阀。螺杆头部熔料压力迫使注射缸后退。回油路：注射缸右腔油液→单向节流阀14→电液换向阀15右位→背压阀16回油箱。其背压力由背压阀16控制。同时注射缸左腔产生局部真空，油箱的油液在大气压作用下经电液换向阀11中位进入其内。

7. 防流涎（$2YA^+$、$7YA^+$、$9YA^+$）

采用直通开敞式喷嘴时，预塑加料结束，要使螺杆后退一小段距离，减小料筒前端压力，防止喷嘴端部物料流出。大流量泵1卸荷，主油路：小流量泵2压力油分成两路，一路经电磁换向阀9右位进入注射座移动缸右腔，使喷嘴与模具保持接触；一路经电液换向阀11左位进入注射缸左腔，使螺杆强制后退。注射座移动缸左腔和注射缸右腔油液分别经电磁换向阀9和电液换向阀11回油箱。

8. 注射座后退（$2YA^+$、$6YA^+$）

保压结束，注射座后退。大流量泵1卸荷，小流量泵2压力油经电磁换向阀9左位使注射座后退。

9. 开模

开模速度一般为慢—快。

（1）慢速开模（$2YA^+$或$1YA^+$、$4YA^+$）　大流量泵1（或小流量泵2）卸荷，小流量泵2（或大流量泵1）压力油经电液换向阀5左位进入合模缸右腔，左腔油液经电液换向阀5回油箱。

（2）快速开模（$1YA^+$、$2YA^+$、$4YA^+$）　大流量泵1和小流量泵2合流向合模缸右腔供油，开模速度加快。

10. 顶出

（1）顶出缸前进（$2YA^+$、$5YA^+$）　大流量泵1卸荷，小流量泵2压力油经电磁换向阀8左位、单向节流阀7进入顶出缸左腔，推动顶出杆顶出制品，其运动速度由单向节流阀7调节，溢流阀4为定压阀。

（2）顶出缸后退（$2YA^+$）　小流量泵2的压力油经电磁换向阀8常位使顶出缸后退。

11. 螺杆后退和前进

为了拆卸螺杆，有时需要螺杆后退。这时电磁铁2YA、9YA得电，大流量泵1卸荷，小流量泵2的压力油左位进入注射缸的左腔，注射缸活塞带动螺杆后退。当电磁铁2YA、8YA得电时，螺杆前进。

8.2.3　特点

1）注射缸液压力直接作用在螺杆上，因此注射压力p_z与注射缸的油压p的比值为D^2/d^2（D为注射缸活塞直径，d为螺杆直径）。一般注塑机都配备三种不同直径的螺杆以满足加工不同塑料对注射压力的要求。在系统压力$p=14\text{MPa}$时，可获得注射压力$p_z=40\sim150\text{MPa}$。

2）该注塑机采用了液压-机械增力合模机构，以保证足够的合模力和防止高压注射时模具离缝产生塑料溢边。另外，也可以采用增压器的合模装置。

8.3 Hydraulic System of YA32-200 Type Four-Column Universal Press Machine

8.3.1 Introduction

Hydraulic press machine is a mechanical equipment wildly used in the press working technology such as forging, stamping, cold-extrusion, alignment, bending, powder metallurgy, molding, etc. It is an earliest application of the hydraulic transmission. Based on the working medium (oil, water/emulsification liquid), hydraulic presses can be divided into oil hydraulic presses and water hydraulic presses. The purpose of this section is to introduce the hydraulic system of the YA32-200 type four-column universal press machine, with the working medium of oil. The max pressing force of the main cylinder is 2000kN. The main actions for the hydraulic system incorporate: the sliding block's rapidly moving downward in the main cylinder, pressure-adding at low speed, pressure-keeping, pressure-venting, rapidly return, stop at any point and the piston's extending and retracting in cylinder 1, etc. Sometimes a sheet is required to be clamped by cylinder 1 when is drawn. In this case pressure in the down position of cylinder 1 must be above zero when cylinder 1 moves downward with the main cylinder. During a working cycle, the pressure and the flow rate in the system may vary in a large scale, so attention must be paid to the reasonable utilization of the power.

8.3.2 Operating principle

Fig. 8-3 shows its operation. There are two pumps in the system: a main pump 1 is a constant-power (pressure compensation) variable pump which can supply high pressure (the highest pressure can reach 32MPa) and large flow rate and then an auxiliary invariable pump 2 is used to supply low pressure and small flow rate. The auxiliary pump is mainly used to supply the controlled oil for electro-hydraulic valves, and the output pressure can be adjusted by relief valve 5.

1. Main cylinder movement

(1) Moving downwards rapidly By starting the button, the solenoids of 1YA and 5YA are energized. Electro-hydraulic valve 6 is shifted to its right position due to the low pressure oil acting on its left position. The low pressure oil from the auxiliary pump also opens check valve 13 via valve 14. The oil supplied by pump 1 is directed to the upper port of main cylinder 10 through valve 6, check valve 8. The oil in the down port of the main cylinder returns to the reservoir through check vale 13, valve 6 and valve 15. Sliding block of the main cylinder moves downwards rapidly under deadweight. Note that even the largest flow rate from pump 1 cannot fully compensate for the upper empty space of the main cylinder. Partial vacuum is formed, causing the oil in filling reservoir (located at the top of the cylinder) to enter the upper port of the main cylinder via check valve (filling valve) 9 under atmosphere and oil level.

(2) Slowly close to the workpiece and pressurize it When the distance switch 2SQ is depressed by block on sliding block , solenoid 5YA is de-energized to return valve 14 and close valve 13. The returned oil from the main cylinder flows into the reservoir via against pressure (balancing) valve 12, valve 6 (right position) and valve 15 (neutral position). The sliding block cannot move

3）根据塑料注射成型工艺，模具的启闭过程和塑料注射的各阶段速度不一样，而且快、慢速之比可达 50～100，为此该注塑机采用了双泵供油系统，快速时双泵合流，慢速时小流量泵 2（流量为 48L/min）供油，大流量泵 1（流量为 194L/min）卸荷，系统功率利用比较合理。

4）通过多个并联的远程调压阀控制实现系统所需多级压力。如果系统采用电-液比例压力阀来实现对多级压力调节，再加上电-液比例流量阀实现调速，不仅可减少元件，减小压力及速度变换过程中的冲击和噪声，还为实现计算机控制创造条件。

5）注塑机的多执行元件的循环动作主要依靠行程开关按事先编程的顺序完成。这种方式灵活方便。

8.3 YA32-200 型四柱万能液压机液压系统

8.3.1 概述

液压机是锻压、冲压、冷挤、校直、弯曲、粉末冶金、成形等压力加工工艺中广泛应用的机械设备。按其工作介质是油还是水（乳化液），液压机可分为油压机和水压机两种，本节介绍一种以油为介质的 YA32-200 型四柱万能液压机。该液压机主缸最大压制力为 2000kN。液压机要求液压系统完成的主要动作是：主缸滑块的快速下行、慢速加压、保压、卸压、快速回程及在任意点停止；顶出缸活塞的顶出、退回等。在作薄板拉伸时，有时还需要利用顶出缸将坯料压紧。这时顶出缸下腔须保持一定压力并随主缸一起下行，在一个工作循环内，系统中的压力和流量变化很大，因此要特别注意功率的合理利用。

8.3.2 YA32-200 型四柱万能液压机液压系统工作原理

图 8-3 所示为该机液压系统原理图。系统中有两个泵：主泵 1 是一个高压、大流量恒功率（压力补偿）变量泵，最高工作压力为 32MPa，由远程调压阀 4 调定；辅助泵 2 是一个低压小流量的定量泵，主要用以供给电-液换向阀的控制油液，其压力由溢流阀 5 调整。

1. 主缸运动

（1）快速下行　按下起动按钮，电磁铁 1YA、5YA 得电吸合。低压控制油使电液换向阀 6 切换至右位，同时经电磁阀 14 使液控单向阀 13 打开。主泵 1 供油经电液换向阀 6 右位、单向阀 8 至主缸 10 上腔，而主缸下腔油液经液控单向阀 13、电液换向阀 6 右位、电液换向阀 15 中位回油箱。此时主缸滑块在自重作用下快速下降，主泵 1 虽为最大流量，但还不足以补充主缸上腔空出的容积，因而上腔形成局部真空，置于液压机顶部的充液箱中的油液在大气压及油位作用下，经液控单向阀 13（充液阀）进入主缸上腔。

（2）慢速接近工件、加压　当主缸滑块上的挡铁压下行程开关 2st 时，电磁铁 5YA 失电，电磁阀 14 处于常态位，液控单向阀 13 关闭。主缸回油经背压（平衡）阀 12、电液换向阀 6 右位、电液换向阀 15 中位至油箱。由于回油路上有背压力，滑块单靠自重就不能下降，由主泵 1 供给的压力油使之下行，速度减慢。这时主缸上腔压力升高。带卸荷阀芯的充液阀 9 关闭。来自主泵 1 的压力油推动活塞使滑块慢速接近工件，当主缸活塞的滑块抵住工件后，阻力急剧增加，上腔油压进一步提高，主泵 1 的排油量自动减小，主缸活塞以极慢的速度对工件加压。

downwards by itself due to against pressure in the returned oil line. The pressurized oil supplied by pump 1 makes it slowly move downwards. The pressure in the upper port of the main cylinder is increased and closes fill oil valve 9. The pressurized oil from pump 1 makes the sliding block slowly close to the workpiece. Once the piston in the main cylinder presses the workpiece, the resistance against the piston increases dramatically and upper port pressure further increases. The delivery of variable pump 1 reduces automatically and the piston pressurizes the workpiece at a slow speed.

(3) Pressure-holding When the oil pressure in the upper port of the main cylinder reaches the setting value, pressure switch 7 signals to de-energize solenoid 1YA. Valve 2 returns to its neutral position, closing the upper and down ports of the main cylinder. In addition, pump 1 is under an unloaded condition via valves 6 and 15. It is the check valve 8 that ensures the precision sealing for the upper port of the main cylinder, accomplishing a pressure-holding for the upper port of the main cylinder. The time for pressure-holding can be adjusted by the time-relay which is controlled by pressure switch 7.

(4) Pressre-venting, retracting rapidly At the end of the pressure-holding process, the time-relay signals to actuate 2YA and the main cylinder is ready to retract. Note that because of the high pressure oil in the bigger diameter with long stroke in the main cylinder, the liquid saved a good number of energy due to the oil compressed in the pressure adding process. If the upper port is connected to the return oil line, the energy saved in the oil will release suddenly, resulting in hydraulic impulsion, severe vibration to the machine and pipelines and thus noise. It is necessary to emphasize that pressure-venting is necessary when pressre-holding is ended and the cylinder will be retracted.

When valve 6 is shifted to its left position, the pressure in the upper port of the main cylinder is still very high without pressure-venting action. Relief valve 11 (with damping orifice) is opened, which allows the oil flow from main pump 1 to reservoir via valves 6 (left position) and 11. Main pump 1 is in operation under a low pressure, which is not high enough to open the main valve core of check valve 9, but high enough to open its small discharged valve core. The high pressure oil in the upper port of the main cylinder is connected to fill reservoir through the orifice of the small discharged valve core. The pressure in the upper port is reduced gradually. This process lasts until the pressure in the upper port reduces to a relative low point. Then relief valve 11 is closed and the pressure supplied by pump 1increases to open the main valve core of check valve 9. Then the pressurized oil enters the downward port of the main cylinder via valve 6 (left position) and check valve 13. The oil in the upper port returns back to fill oil reservoir through valve 9 and accomplishes a rapid returning for the main cylinder.

(5) Stop When distance switch 1st is pressed by block on the sliding block of the main cylinder, 2YA is released. The piston of the main cylinder is sealed by valve 6 with M type neutral function, which indicates the finished retraction. The oil from pump 1 opens to the reservoir via valves 6 and 15 and pump 1 is under unloaded condition. Note that the main cylinder can be stopped at any position in practice.

2. Cylinder for extending

Cylinder 16 can only operate when the main cylinder is at stop state. Pressurized oil must pass through valve 6 before it reaches valve 15, which is used to control the movement of cylinder 16.

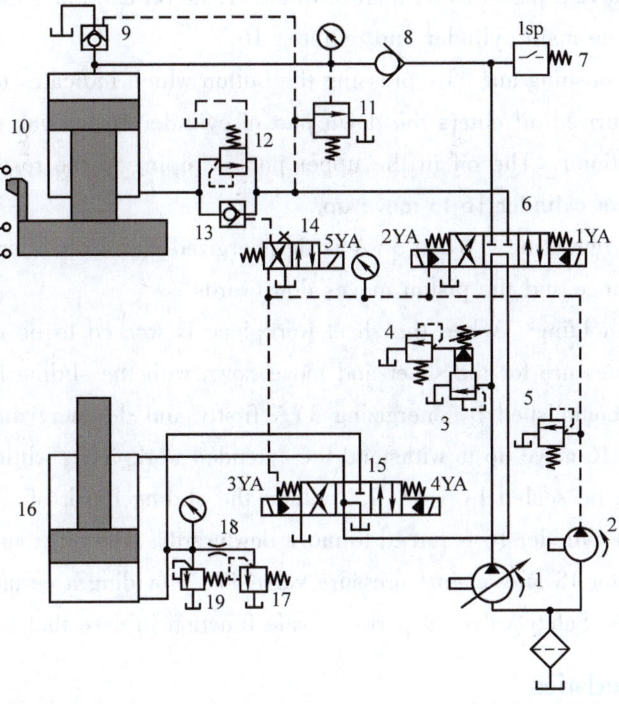

Fig. 8-3　Hydraulic system of YA32-200 type four-column universal press machine

（YA32-200 型四柱万能液压机液压系统原理图）

1—Main pump（主泵）　2—Auxiliary pump（辅助泵）　3，5—Relief valve（溢流阀）
4—Remote pressure regulating valve（远程调压阀）　6，15—Electro-hydraulic directional valve（电液换向阀）
7—Pressure switch（压力继电器）　8—Check valve（单向阀）
9—Filling oil valve with a small discharged core（带卸荷阀芯的充液阀）　10—Main cylinder（主缸）
11—Relief valve with damping orifice（带阻尼孔的卸荷阀）　12，17—Against pressure valve（背压阀）
13—Hydraulic operated check valve（液控单向阀）　14—Solenoid valve（电磁阀）
16—Cylinder for extending（顶出缸）　18—Restrictor（节流器）　19—Safety valve（安全阀）

（3）保压　当主缸上腔的油压达到预定值时，压力继电器 7 发出信号，使电磁铁 1YA 失电，电液换向阀 6 恢复中位，将主缸上、下油腔封闭。同时主泵 1 经电液换向阀 6、电液换向阀 15 的中位卸荷。单向阀 8 保证了主缸上腔良好的密封性，主缸上腔保持高压。保压时间可由压力继电器 7 控制的时间继电器调整。

（4）卸压、快速回程　保压过程结束，时间继电器发出信号，使电磁铁 2YA 得电，主缸处于回程状态。但由于液压机的油压高，且主缸的直径大，行程长，缸内液体在加压过程中受到压缩而贮存相当大的能量。如果此时上腔立即与回油相通，缸内液体积蓄的能量突然释放出来，产生液压冲击，造成机器和管路的剧烈振动，发出很大的噪声。为此，保压后必须先卸压然后再回程。

当电液换向阀 6 切换至左位后，主缸上腔还未卸压，压力很高，带阻尼孔的卸荷阀 11 呈开启状态，主泵 1 的油经电液换向阀 6 左位、带阻尼孔的卸荷阀 11 回油箱。这时主泵 1 在低压下运转，此压力不足以打开液控单向阀 13 的主阀阀芯，但能打开带卸荷阀芯的充液阀 9 中的卸荷小阀芯，主缸上腔的高压油经此卸荷小阀芯的开口而泄回充液箱，压力逐渐降低。这一过程持续到主缸上腔压力降至较低值时，带阻尼孔的卸荷阀 11 关闭，主泵 1 的供

i. e. , only when valve 6 performs its neutral function, the oil can enter cylinder 16 to accomplish the interlock between the main cylinder and cylinder 16.

(1) Cylinder pushing out By pressing the button which indicates for pushing out, 3YA is energized. The pressurized oil enters the down port of cylinder 16 via valve 6 (medium position) and valve 15 (left position). The oil in the upper port is piping to the reservoir through valve 15 and causing the piston of cylinder 16 to move up.

(2) Cylinder retracting When 3YA is de-energized and 4YA is energized, the inlet line and returned line exchange and the piston moves downwards.

(3) Floating-holding When the sheet-workpiece is worked to be extended, cylinder 16 must provide a certain pressure for the sheet and move down with the sliding block of the main cylinder. This function is accomplished by energizing 3YA firstly and de-energizing it afterwards. Energizing makes the cylinder 16 move up to withstand the extended workpiece, while de-energizing allows oil in the down port of to be sealed by valve 15. When the sliding block of the main cylinder is pressed down, the piston of cylinder 16 is forced to move downwards. The oil from the down port is piped into reservoir via restrictor 18 and against pressure valve 17, providing a certain floating-holding force for the sheet-workpiece. Safety valve 19 performs safe function in case that restrictor 18 were blocked.

8.3.3 Characteristics

1) The employment of a constant-power variable pump with high pressure and large flow rate not only meets the technological requirements but also saves energy.

2) The piston moves down rapidly by deadweight and a fill valve is used for the main cylinder oil compensating. This type of quick movement circuit is simple in structure and employs fewer components.

3) Check valve 8 is used for pressure-holding. A pressure-venting circuit composed of relief valve 11 and check valve with small discharged core 9 is designed for the hydraulic impulsion abatement from pressure-holding process to quick returning.

4) Interlock function accomplished by the cylinder for push-out and the main cylinder. Only when valve 6 performs its neutral function and the main cylinder stops, can the pressurized oil enter into valve 15 to allow cylinder 16 to move. This is one method for safety precautions.

8.4 Hydraulic System of Manipulators

8.4.1 Introduction

Hydraulic manipulators are candidates for imitating actions of human hands to automation accomplish snatching, conveying and operator by a given setting of sequence. They can be used for replacing human working and operated in harsh environments such as high temperature, high pressure, more dusty, flammability, explosive, radiation, ponderosity, humdrum and repeated operation. The main advantages are their large payload with respect to volume and mass, their reliability and their robustness. Hydraulic manipulators have been applied widely in industry.

油压力升高，推开带卸荷阀芯的充液阀9的主阀芯。此时主泵1的压力油经电液换向阀6左位、液控单向阀13进入主缸下腔；而主缸上腔油液经带卸荷阀芯的充液阀9回油至充液箱，实现主缸快速回程。

（5）停止　当主缸滑块上的挡铁压下行程开关1st时，电磁铁2YA失电，主缸活塞被中位为M机能的电液换向阀6锁紧而停止运动，回程结束。此时主泵1油液经电液换向阀6、15回油箱，主泵处于卸荷状态。实际使用中，主缸随时都可处于停止状态。

2. 顶出缸运动

顶出缸16只是在主缸停止运动时才能动作。由于压力油先经过电液换向阀6后才进入控制顶出缸运动的电液换向阀15，亦即电液换向阀6处于中位时，才有油通向顶出缸，保证主缸和顶出缸的运动互锁。

（1）顶出　按下顶出按钮，3YA得电，压力油由主泵1经电液换向阀6中位、电液换向阀15左位进入顶出缸16下腔，上腔油液则经电液换向阀15回油，活塞上升。

（2）退回　3YA失电，4YA得电，油路换向，顶出缸的活塞下降。

（3）浮动压边　作薄板拉伸压边时，要求顶出缸既保持一定压力，又能随主缸滑块的下压而下降。这时3YA得电，使顶出缸上升到顶住被拉伸的工件，然后3YA失电，顶出缸下腔的油液被阀15封住。主缸滑块下压时，顶出缸活塞被迫随之下行，顶出缸下腔回油经节流器18和背压阀17流回油箱，使缸下腔保持所需的压边力。图中安全阀19在节流器18阻塞时起安全保护作用。

8.3.3　YA32-200型四柱万能液压机液压系统特点

1）系统采用高压大流量恒功率变量泵供油可节省能量。

2）利用活塞滑块自重的作用实现快速下行，并用充液阀对主缸充液。这种快速运动回路结构简单，使用元件少。

3）系统采用单向阀8进行保压。为了减小由保压转换为快速回程时的液压冲击，采用了带阻尼孔的卸荷阀11和带卸荷阀芯的充液阀9组成的卸压回路。

4）顶出缸16与主缸10运动互锁。只有电液换向阀6处于中位，主缸不运动时，压力油才能进入电液换向阀15，使顶出缸运动。这是一种安全措施。

8.4　机械手液压系统

8.4.1　概述

机械手是模仿人的手部动作，按给定程序、轨迹和要求实现自动抓取、搬运和操作的自动装置。它特别是在高温、高压、多粉尘、易燃，易爆、放射性等恶劣环境中，以及笨重、单调、频繁的操作中能够取代操作人员的作业，因此获得日益广泛的应用。

机械手一般由执行机构、驱动系统、控制系统及检测装置三大部分组成，智能机械手还具有感觉系统和智能系统，驱动系统多数采用电液（气）机联合传动。

本节介绍的JS01工业机械手属于圆柱坐标式、全液压驱动机械手，具有手臂升降、伸缩、回转和手腕回转四个自由度。执行机构相应由手臂、手腕、手臂伸缩机构、手臂升降机构、手臂回转机构和回转定位装置等组成，每一部分均由液压缸驱动与控制。它完成的动作循环为：

Generally manipulator consists of three parts actuators, transmission, control and detection units. Intelligent manipulator equips with sense and brainpower units. The transmission unit is usually done by the electro-hydraulic (or pneumatic) machine.

In this section JS01 industry manipulator will be introduced. It is a cylindrical coordinate and whole hydraulic with the features of 4 free-degree, i. e. , arm rising and going-down, extending and retracting, turn around and wrist gyration action. Thus the actuators are corresponding constituted by arm, wrist, arm extending and withdrawing machine, rising and descending machine, arm turn-around machine and gyration-orientation machine. They are driven and controlled by hydraulic system. They should accomplish sequences below:

The peg is inserted for fixing → arm is extended onward → finger is stretched and opened → finger clamped and snatched at goods → arm goes up → arm withdraws → wrist turns 180°→ the peg is drawn → arm turns 95°→ the peg is inserted for fixing again →arm extends onward again → arm stops (at this time the clamp in the main machine goes down to hold goods) → loose finger (at this time the clamp in the main machine rises with the goods held) →finger is closed→arm withdraws → arm goes down → wrist returns to original position → draw the peg → arm turns to original position → system waits for the next task with pump unloaded.

8.4.2 Operating principle and characteristics of hydraulic system for JS01 industry manipulator

The principle sketch of hydraulic system is shown in the Fig. 8-4. The action procedures of solenoids are in listed Tab. 8-3.

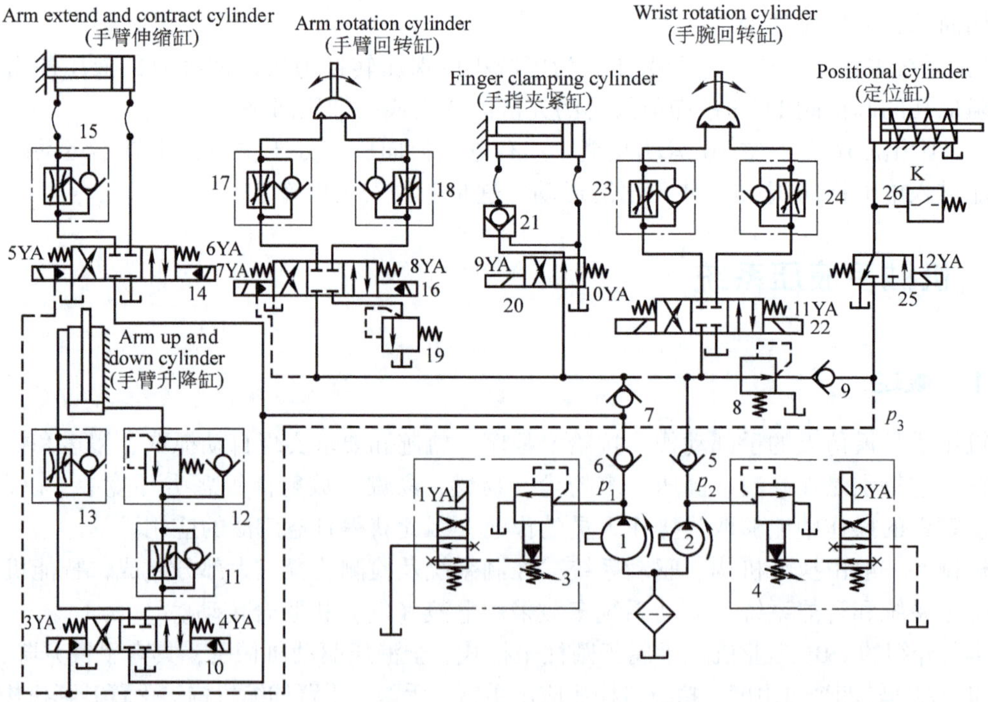

Fig. 8-4 Hydraulic system for JS01industry manipulator (JS01 工业机械手液压系统图)

插定位销→手臂前伸→手指张开→手指夹紧并抓料→手臂上升→手臂缩回→手腕回转 180°→拔定位销→手臂回转 95°→插定位销→手臂前伸→手臂中停（此时主机的夹头下降夹料）→手指松开（此时主机夹头夹着料上升）→手指闭合→手臂缩回→手臂下降→手腕回转复位→拔定位销→手臂回转复位→待料，泵卸荷。

8.4.2 JS01 工业机械手液压系统工作原理及特点

JS01 工业机械手液压系统图如图 8-4 所示。其电磁铁与压力继电器的动作顺序见表 8-3。

Tab. 8-3 Action procedures of solenoid and pressure relay for JS01 industry manipulator
（JS01 工业机械手液压系统电磁铁与压力继电器的动作顺序表）

Action scheduled（动作顺序）	1Y	2Y	3Y	4Y	5Y	6Y	7Y	8Y	9Y	10Y	11Y	12Y	K26
Peg inserted（插定位销）	+											+	+
Arm extending onward（手臂前伸）					+							+	+
Finger open（手指张开）	+								+			+	+
Finger holding goods（手指抓料）	+											+	+
Arm goes up（手臂上升）			+									+	+
Arm withdrew（手臂缩回）						+						+	+
wrist turns around（手腕回转）	+									+		+	+
Peg drew（拔定位销）	+												
Arm turned（手臂回转）	+						+						
Peg inserted（插定位销）	+											+	+
Arm extends onward（手臂前伸）					+							+	+
Arm stopped（手臂中停）												+	+
Finger stretched（手指张开）	+								+			+	+
Arm closed（手指闭合）	+											+	+
Arm withdrew（手臂缩回）						+						+	+
Arm goes down（手臂下降）				+								+	+
Wrist turned reversely（手腕反转）	+										+	+	+
Peg drew（拔定位销）	+												
Arm turned reverse（手臂反转）	+							+					
Waiting for next task, pump unloaded（待料卸荷）	+	+											

The principle of operating for hydraulic system can be analyzed by Fig. 8-4 and Tab. 8-3. It has some features below:

1) It utilizes two pumps to supply oil to system and the rating pressure is 6.3MPa. When arm goes up or goes down, or extends or withdraws, two pumps are needed to supply the oil with the flow rate (35 + 18) L/min. When arm or wrist turns around, or fingers loose and fixed cylinder works, the lower flow rate pump 2 (with flow rate 18L/min) is needed to supply oil to system only, and the big flow rate pump 1 (with flow rate 35L/min) is unloaded automatically. Since the fixed cylinder and control oil line need low pressure oil, thus pressure-reducing valve 8 is connected in series to obtain stable pressure of 1.5-1.8MPa.

2) A cylinder with one rod and double-action is used for driving the arm to extend and withdraw, go up and go down. And the speeds are controlled by check speed-regulating valve 15, 13 and 11. Arm and wrist's turn around is driven by swing hydraulic cylinder and the speeds are controlled by check speed-regulating valves 17, 18, 23 and 24.

3) The actuator's orientation and cushions are the key problems for manipulator to get a stable operation. The precision of arm extending and wrist turning is ensured by taking a fixed baffle. When it is moving near the end of the cylinder, the oil line is off to let it move slowly and get cushion. Before arm withdraws and goes to upper, we cut the oil line to let it move slowly and fix it by an electronic signal. Besides, to turn two operation states beween arm extending or withdrawing cylinder and going-up or going-down cylinder is accomplished by an electro-hydraulic directional valve.

4) A hydraulic actuated check valve 21 is used to ensure the precision of holding pressure during clamping workpiece and it is not affected by pressure fluctuating in system.

5) The cylinder used for arm's going up and going down is installed vertically so that the counterbalance circuit (check-sequence valve 12) is used to support and balance the deadweight parts during arm motion.

各执行元件动作的油路请读者根据液压系统图和电磁铁动作顺序表自行分析。该液压系统的特点归纳如下：

1）系统采用了双联泵供油，额定压力为 6.3MPa。手臂升降及伸缩时由两个泵同时供油，流量为（35 +18）L/min。手臂及手腕回转、手指松紧及定位缸工作时，只由小流量泵 2 供油，大流量泵 1 自动卸荷。由于定位缸和控制油路所需压力较低，在定位缸支路上串联有减压阀 8，使之获得稳定的 1.5 ~1.8MPa 压力。

2）手臂的伸缩和升降采用单杆双作用液压缸驱动，手臂的伸出和升降速度分别由单向调速阀 15、13 和 11 实现回油节流调速；手臂及手腕的回转由摆动液压马达驱动，其正反向运动也采用单向调速阀 17 和 18、23 和 24 回油节流调速。

3）执行机构的定位和缓冲是机械手工作平稳可靠的关键。该机械手手臂伸出、手腕回转由挡铁定位保证精度，到达端点前发信号切断油路，滑行缓冲；手臂缩回和手臂上升由行程开关适时发信号，提前切断油路滑行缓冲并定位。此外，手臂伸缩缸和升降缸采用了电液换向阀换向。

4）为使手指夹紧缸夹紧工件后不受系统压力波动的影响，保证牢固地夹紧工件，系统采用液控单向阀 21 实现锁紧。

5）手臂升降缸为立式液压缸，为支承平衡手臂运动部件的自重，采用了单向顺序阀 12 的平衡回路。

Chapter 9 Design of Hydraulic Transmission System

The design of hydraulic transmission system is an important part of the whole mechanical design and also a summarizing and comprehensive application of knowledge in previous chapters. The aim of this chapter is to show through an example how hydrostatic transmission circuits can be designed to satisfy a given set of requirements.

9.1 The Approach for Design of Hydraulic System

There exists a close connection between the design of a hydraulic transmission system and that of a mainframe. They are usually designed at the same time to correspond with each other. Before a hydraulic transmission system is designed, the requirements of the mainframe to a transmission system in the aspects of actions, performances, working environment and so on must be determined. For an actuator, the requirements include: mode of motion, ranges of distance and speed regulation, load conditions, smooth and precise position motion, work cycle, working environments, space for installation, simplification in design of configuration, reliability and safety in working, high efficiency, prolonged life expectancy, cost saving, easy to use and maintain, etc.

A hydraulic system can be designed according to the process shown in Fig. 9-1. All the steps are related to the system performance except for the last step (8). These steps affect and interact with each other. A system designed according to Fig. 9-1 must be modified repeatedly until the design requirements are satisfied. The steps and methods are introduced as follows.

9.1.1 Determination of design requirements

Before a hydraulic system is designed, the working conditions of the mainframe in the mechanical equipment must be detailly analyzed. Then the mainframe requirements to the system can be clearly determined. Concrete contents that must be cleared incorporate:

1) The purpose of the mainframe, the overall arrangement, the limitations to the position of the hydraulic devices and their sizes.

2) The mainframe work cycle, the system actions, their sequence, interlock function, automatization.

3) The load on actuators and the ranges of speed, motion stability, position accuracy, conversion accuracy, etc.

4) Working environment and working conditions.

5) Working efficiency, safety, reliability, cost, etc.

第 9 章　液压系统的设计

液压系统设计是液压主机设计的重要组成部分，也是对前面各章内容的概括总结和综合应用。本章主要阐述液压系统设计的一般步骤、内容和计算方法，并通过实例来说明液压系统的设计过程。

9.1　液压系统的设计步骤

液压系统设计与主机的设计是紧密联系的，两者往往同时进行，互相协调。设计液压系统时应首先明确主机对液压系统在动作、性能、工作环境等方面的要求，如执行元件的运动方式、行程、调速范围、负载条件、运动平稳性和精度、工作循环及周期、工作环境、安装空间大小、结构简单、工作安全可靠、效率高、使用寿命长、经济性好、使用维修方便等设计原则。

液压系统设计步骤大体上可按图 9-1 所示的内容和流程进行。这里除了最后一项（8）外，均属性能设计范围。这些步骤是相互关联、相互影响的，必须经反复修改才能完成。设计步骤及方法介绍如下。

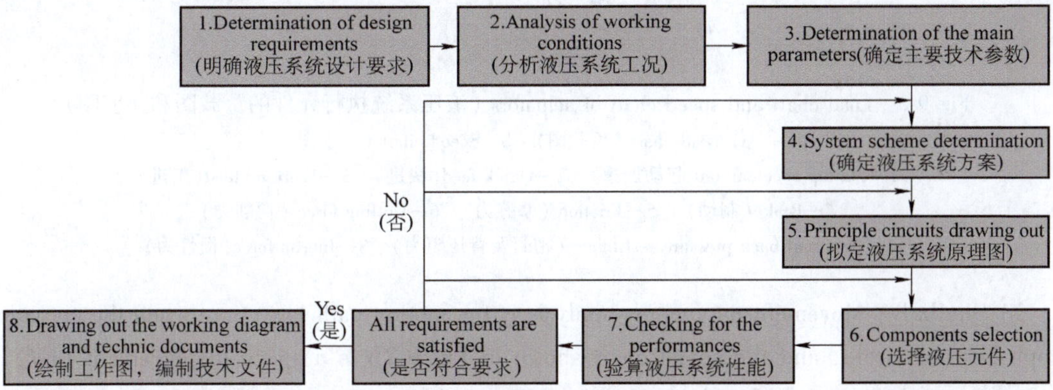

Fig. 9-1　General design process for a hydraulic transmission system（液压传动系统的一般设计流程）

9.1.1　明确系统的设计要求

设计液压系统时，首先要对液压主机的工况进行分析，明确主机对液压系统的要求，具体包括：

1）主机的用途、总体布局、对液压装置的位置和空间尺寸的限制。

2）主机的工作循环、液压系统应完成的动作、动作顺序或互锁要求，以及自动化程度的要求。

3）液压执行元件的负载和运动速度的大小及其变化范围，运动平稳性、定位精度及转

9.1.2 Analysis of working conditions and determination of main parameters

1. Analysis of working conditions

To analyze working conditions is to analyze the moving speed of each actuator in operation and the changing pattern of the load characteristics. After analyzing working conditions, the hydraulic system scheme can be drawn out and the gist to select or design of hydraulic components can be determined. The analysis of working conditions includes two parts of dynamic parameters and movement parameters analysis:

1) Dynamic parameters analysis is accomplished by calculating the load on the hydraulic actuators, estimating its acting direction and analyzing the impact, vibrations and over-load that may occur to each actuator in operation. A load diagram can be drawn out based on the determined extrinsic loads, see Fig. 9-2a, which is load chart. The load chart is actually a load-displacement (F-l) curve, which can express the load overcome by each actuator in each phase.

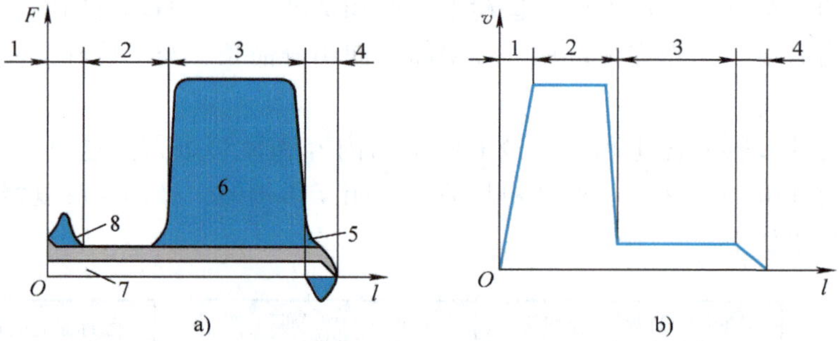

Fig. 9-2 Load chart and speed chart of actuators (液压系统执行元件的负载图和速度图)

a) Load chart (负载图) b) Speed chart (速度图)

1—Startup acceleration(起动加速) 2—Quick feed(快进) 3—Working feed(工进)

4—Brake(制动) 5—Friction (摩擦力) 6—Cutting force (切削力)

7—Seal and back pressure resistance (密封及背压阻力) 8—Inertia force(惯性力)

2) Similarly, Movement parameters analysis is the motion path law curve when the actuator accomplishes one circle, and the speed chart shown in Fig. 9-2b is a speed-displacement (v-l) curve drawn to express the speed of each actuator in different phases. For a simple hydraulic system, these two charts can be neglected.

When the actuator works in line reciprocating, the extrinsic load will be

$$F = F_L + F_f + F_a \tag{9-1}$$

1) The working load F_L relates to the working conditions of equipment, for instance, the cutting force is the working load for a tool machine, and the heavy motion is the working load for an elevator or hydraulic jack. The working load can be a fixed or variable value, and it can also be a positive or negative value.

2) The friction resistance force, F_f, relates to the status of motion parts and the shape of the supporting surface, placement, lubrication and motion conditions. It is

换精度等要求。

4）液压系统的工作环境和工作条件。

5）工作效率、安全性、可靠性以及经济性等要求。

9.1.2 分析系统工况，确定主要参数

1. 工况分析

工况分析，就是分析主机在工作过程中各执行元件的运动速度和负载的变化规律。它是拟定液压系统方案，选择或设计液压元件的依据。工况分析包括动力参数分析和运动参数分析两个部分。

1）动力参数分析就是通过计算液压执行元件的载荷大小和方向，并分析各执行元件在工作过程中可能产生的冲击、振动及过载等。对于动作较复杂的机械设备，根据工艺要求，将各执行元件在各阶段所需克服的负载用图 9-2a 所示的负载-位移（F-l）曲线表示，称为负载图。

2）运动参数分析是指液压执行元件在完成一个工作循环时的运动规律，其在各阶段的速度用图 9-2b 所示的速度-位移（v-l）曲线表示，称为速度图。设计简单的液压系统时，这两种图可省略不画。

液压缸驱动执行机构进行直线往复运动时，所受到的外负载为

$$F = F_L + F_f + F_a \tag{9-1}$$

（1）工作负载 F_L　工作负载与设备的工作状况有关，如在机床上切削力是工作负载，而对于提升机、千斤顶等来说所移动的物体的重量就是工作负载。工作负载可以是定量或变量，可以是正值或负值。

（2）摩擦阻力负载 F_f　摩擦阻力负载与运动部件的支承面的形状、放置情况、润滑条件以及运动状态有关。即有

$$F_f = fF_N \tag{9-2}$$

式中，F_N 为运动部件及外负载对支承面的正压力；f 为摩擦因数，分为静摩擦因数（$f_s \leqslant 0.2 \sim 0.3$）和动摩擦因数（$f_d \leqslant 0.05 \sim 0.1$）。

（3）惯性负载 F_a　惯性负载是指当运动部件的速度变化时，由其惯性而产生的负载，由牛顿第二定律得

$$F_a = ma = \frac{G}{g} \cdot \frac{\Delta v}{\Delta t} \tag{9-3}$$

式中，m 为运动部件的质量（kg）；a 为运动部件的加速度（m/s^2）；G 为运动部件的重力（N）；g 为重力加速度（m/s^2）；Δv 为速度的变化量（m/s）；Δt 为速度变化所需的时间（s）。

除以上三种负载力外，液压缸的受力还有粘性阻力、密封阻力和背压阻力等。

如果执行机构为液压马达，其负载力矩的计算方法与液压缸相类似。

2. 确定主要参数

这里是指确定液压执行元件的工作压力和最大流量。

执行元件的工作压力，可以根据负载图中的最大负载来选择（表 9-1），也可以根据主机类型来选择（表 9-2）；而最大流量则由执行元件速度图中的最大速度计算出来。这两者与执行元件的结构参数（指液压缸的有效工作面积 A 或液压马达的排量 V_M）有关。一般的

$$F_f = fF_N \tag{9-2}$$

Where F_N is the force that motion heavy and extrinsic load act on the supporting surface; f is the friction coefficient and divided into static friction coefficient($f_s \leqslant 0.2\text{-}0.3$) and dynamic friction coefficient ($f_d \leqslant 0.05\text{-}0.1$).

3) The inertia load F_a is a load due to the speed change of motion part, which can be described by the Newtonian's second law

$$F_a = ma = \frac{G}{g} \cdot \frac{\Delta v}{\Delta t} \tag{9-3}$$

Where m is the mass of motion parts, kg; a is acceleration of the motion parts, m/s^2; G is the gravity, N; g is acceleration of gravity, m/s^2; Δv is a change value of speed, m/s; Δt is required time of the speed change, s.

Besides three loads mentioned above, an actuator also stands viscosity resistance force, sealed resistance force and back pressure resistance, etc.

If we choose a hydraulic motor as driving actuator, the extrinsic load calculation of it is similar to those of the hydraulic cylinder.

2. Determination of main parameters

The main parameters here include the working pressure of the actuators and the largest flow rate.

The working pressure for the actuators can be selected according to the largest load in the load chart (Tab. 9-1) or the mainframe type (Tab. 9-2). However the largest flow rate must be calculated by the highest speed shown in the speed chart. These two parameters are related to the structure parameters of the actuators (the effective working area A of the cylinder or the delivery V_M of the motor). Generally, the working pressure p is first selected and the effective working area A and the delivery V_M are calculated according to the largest load and the evaluated mechanical efficiency of the actuators. The structure parameters calculated must be rounded to a standard value in a handbook. Finally, the maximum flow rate q_{max} is calculated.

The lowest working speed v_{min} or ω_{min} ($\omega = 2\pi n_{min}/60$) must satisfy the following requirements:

For cylinders

$$\frac{q_{min}}{A} \leqslant v_{min}$$

For motors

$$\frac{q_{min}}{V_M} \leqslant \omega_{min}$$

Where q_{min} is the lowest stable flow rate of throttle valves, speed-regulating valves or the variable delivery pump, q_{min} can be obtained from the product performance table.

For some occasions, the stability of the piston rod in the cylinder must be checked, which is usually accomplished during the period of parameters determination.

A or V_M should be modified until the check results meet the given requirements. Finally, the structure parameters of the actuators must be rounded to a standard value (see national standard GB/T 2347—1980 and GB/T 2348—1993).

做法是，先选定工作压力 p，再按最大负载和预估的执行元件机械效率求出 A 或 V_M，经过各种必要的验算、修正和圆整后定下这些结构参数，最后再算出最大流量 q_{max}。

Tab. 9-1 Working pressure selected for an actuator according to load
（按负载选择执行元件工作压力）

Load（负载）F/N	<5 000	5 000-10 000	10 000-20 000	20 000-30 000	30 000-50 000	>50 000
Working pressure（工作压力）p/MPa	<0.8-1	1.5-2	2.5-3	3-4	4-5	>5-7

Tab. 9-2 Working pressure selection for an actuator according to mainframe type
（按主机类型选择执行元件工作压力）

Mainframe type（主机类型）	Machine tools（机床）				Agricultural machinery, small-sized engineering machinery, auxiliary devices for engineering machinery（农业机械、小型工程机械、工程机械辅助机构）	Hydraulic press, large-sized excavator, heavy machinery, lifting and conveying machinery（液压机中、大型挖掘机、重型机械、起重运输机械）
	Grinder（磨床）	Modular machine tools（组合机床）	Planer（龙门刨床）	Drawing bench（拉床）		
Working pressure（工作压力）p/MPa	≤2	3-5	≤8	8-10	10-16	20-32

在本步骤的验算中，必须使执行元件的最低工作速度 v_{min} 或 ω_{min}（$\omega = 2\pi n_{min}/60$）符合下述要求：

液压缸 $$\frac{q_{min}}{A} \leqslant v_{min}$$

液压马达 $$\frac{q_{min}}{V_M} \leqslant \omega_{min}$$

式中，q_{min} 为节流阀或调速阀、变量泵的最小稳定流量，由产品性能表查出。

此外，有时还需对液压缸的活塞杆进行稳定性验算，验算工作常和这里的参数确定工作交叉进行。

以上的一些验算结果如不能满足有关的规定要求时，A 或 V_M 的量值就必须进行修改。这些执行元件的结构参数最后还必须圆整成标准值（见国标 GB/T 2347—1980 和 GB/T 2348—1993）。

3. 绘制液压执行元件工况图

在执行元件结构参数确定之后，根据设计任务要求，算出工作循环各阶段中的实际工作压力、流量和功率。这时就可绘制出执行元件在一个工作循环中的压力、流量和功率对时间或位移的变化曲线图，如图 9-3 所示。工况图显示液压系统在实现整个工作循环时三个参数（压力、流量和功率）随时间的变化情况。当系统中包含多个执行元件时，其工况图是各个执行元件工况图的综合。

3. Drawing of working condition chart of the actuators

After the structure parameters of the actuators are determined, the practical working pressure, the flow rate and the power can be calculated according to the design task and its requirements. Then pressure (flow rate, or power)-time (or displacement) curves can be drawn out, as shown in Fig. 9-3. This working condition chart indicates the changing trends of these three parameters (pressure, flow rate and power) during a whole work cycle of the hydraulic system. As to the system with multiple actuators, the working condition chart is the synthesis of each actuator working condition chart.

The working chart of the actuator is not only the basis of other hydraulic components and hydraulic basic circuit selections, but also the basis of system scheme determination.

9.1.3 Determination of hydraulic system diagram

This is the most important step of a whole system design. It decides the system performance, rationality and whether the design scheme is economical or not.

1. Analysis and selection of the hydraulic scheme

It includes the determination of oil source (pump) and actuators, speed regulating method (throttle, displacement, or their combination), oil line circulation mode (open or close), etc.

2. Analysis and selection of hydraulic circuits

The main circuits which have decisive effects on the main performances of the mainframe must be firstly determined according to its working features and performances. This includes the hydraulic system of machine tool, speed-regulating and speed exchanging circuits, because the speed ranges, the speed characteristics, the speed exchanging stability and position accuracy are usually the key issues of the hydraulic system of machine tools. Then the general circuits can be designed after the main circuits are determined. Usually a general hydraulic system incorporates pressure-regulating circuit, directional control circuit, relief circuit, etc. In addition, balancing circuit must be considered for vertically moving parts, braking circuit for the systems with larger load inertia to prevent hydraulic impulsion and sequence action circuit or the circuit with interlock function for the systems with multiple actuators. Sometimes there are many schemes for a designed system. In this case the designers have to carefully analyze the schemes, compare one with another, and then design the given system by referring to and absorbing the similar better circuits, which have been practically tested.

3. Determination of the hydraulic system diagram

To design the whole hydraulic system, the designers have to combine and arrange the above circuits and add some components or auxiliary circuits if necessary. In this process, it must be paid attention to the aspects of simplicity, safety, stability, energy saving, heat and impulsion reducing, easy to maintain, etc.

9.1.4 Calculation and selection of hydraulic components

1. Selection of pumps

The maximum working pressure provided by the pump should be equal to or surpass the sum of

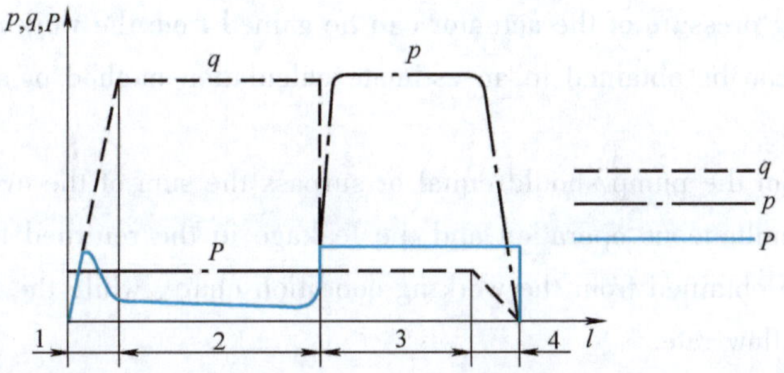

Fig. 9-3　Working condition chart of actuators（执行元件的工况图）

1—Startup acceleration（起动加速）　2—Quick feed（快进）　3—Working feed（工进）　4—Brake（制动）

液压执行元件的工况图是选择系统中其他液压元件和液压基本回路的依据，也是拟订液压系统方案的依据。

9.1.3　拟定液压系统图

拟定液压系统图是整个设计工作中最主要的步骤，它对系统的性能以及设计方案的合理性、经济性具有决定性的影响。其内容包括：

1. 液压系统类型的分析与选择

其主要内容包括：油源（液压泵）和液压执行元件类型的分析与选择，调速方式（节流调速、容积调速和容积节流调速）的分析与选择，以及油路循环方式（开式油路和闭式油路）的分析与选择等。

2. 液压回路的分析与选择

根据液压主机的工作特点和性能要求，首先确定对主机主要性能起决定性影响的主要回路。例如机床液压系统，调速和速度换接回路是主要回路。因为调速范围、速度-负载特性、速度换接的平稳性及位置精度是机床液压系统的核心问题。而后是一般回路，一般液压系统都必须设置调压回路、换向回路、卸荷回路等。对垂直运动部件的系统要考虑平衡回路；对负载惯性较大的系统，为防止产生液压冲击，要设置制动系统；对多个执行元件的系统，应根据需要设置顺序动作回路、动作互锁回路等。当选择液压回路出现多种可能方案时，要反复进行分析、对比外，应多参考和吸收同类型液压系统中使用的并被实践证明是比较好的回路。

3. 液压系统图的拟订

将挑选出来的各个回路加以综合、归并、整理，再增加一些必要的元件或辅助回路，使之成为一个完整的液压系统。进行回路综合，拟订液压系统时，在满足液压主机所要求的各项功能的前提下，力求系统简单、工作安全可靠、动作平稳、节省能源、减少发热、减少冲击和调整维护方便等。

9.1.4　液压元件计算与选择

1. 液压泵的选择

液压泵的最大工作压力必须等于或超过液压执行元件最大工作压力与进油路上总压力损

the maximum working pressure for an actuator and the overall pressure loss along the inlet oil line. The maximum working pressure of the actuator can be gained from the working state chart, while the overall pressure loss can be obtained in an estimate calculation method or according to experiential datum (Tab. 9-3).

The flow rate from the pump should equal or surpass the sum of the overall maximum flow rate for the actuator in simultaneous operation and the leakage in the returned lines. The overall maximum flow rate can be obtained from the working condition chart, while the leakage is 10%-30% of the overall maximum flow rate.

According to maximum working pressure and provided flow rate calculated above, a proper pump can be selected from product samples. For the sake of safety, the chosen pump must be with a rated pressure 20%-60% higher than the maximum working pressure (described above) for pressure reserve , while with a rated flow rate just meeting the maximum flow rate to avoid an exceeding loss of power.

When pump is operating under a rated pressure and a rated flow rate, the power needed to drive the motor can be obtained from the product samples. The power of motor can also be calculated from the concrete working condition (see related hydraulic engineering handbooks for the needed expressions and datum).

2. Selection of valves

The specifications for the valve components can be determined according to the maximum pressure of the hydraulic system and the practical flow rate through these valves. The rated pressure and flow rate of the selected valves must be higher than the system maximum working pressure and the flow rate through the valves. Note that the flow rate through the valves can be increased in necessary, but not higher than 20% of the rated flow rate to avoid exceeding pressure loss, heat generation and noise. To choose the throttle valves and the speed-regulating valves, the minimum stable flow rate must be ascertained to meet the design requirements. While choosing the directional control valves, not only the pressure and flow rate, but also their neutral functions and manipulating methods must be paid attention to.

3. Selection of valve arrangements

For hydraulic equipments, the power source and the valves are usually set in the hydraulic station outside the mainframe. This can make it easy to install, maintain and reduce vibration caused by power source and the effect on the mainframe working accuracy due to oil temperature change. There are three types of arrangements, namely, modular type, manifold plate type and integrated block.

(1) Modular type Whose structure, features and so on have been introduced in the chapter of hydraulic control valves, neglected here.

(2) Manifold plate type The manifold plate is actually a thicker installing plate for hydraulic components. The plate type valve components are mounted on this plate by bolt and the oil line between components are actually the machined holes in this plate. There are pipe fittings mounted on the side of the plate.

(3) Integrated block This block is a universal hexahedron. The designers can mount pipe fittings on one of its sides and plate type valves on the other three. Like manifold plate type, the oil

失两者之和。液压执行元件的最大工作压力可以从工况图中找到；进油路上的总压力损失可以通过估算求得，也可以按经验资料估计（表 9-3）。

Tab. 9-3　Experiential values of the pressure loss in the inlet oil line（进油路压力损失经验值）

System configuration（系统结构情况）	Overall pressure loss（总压力损失）Δp/MPa
Simple system with general throttle speed-regulating components and pipelines（一般节流调速及管路简单的系统）	0. 2-0. 5
Complex system with speed-regulating valves and pipelines（进油路有调速阀及管路复杂的系统）	0. 5-1. 5

液压泵的流量必须等于或超过几个同时工作的液压执行元件总流量的最大值与回路中泄漏量两者之和。液压执行元件总流量的最大值可以从工况图中找到（当系统中备有蓄能器时此值应为一个工作循环中液压执行元件的平均流量）；而回路中泄漏量则可按总流量最大值的10% ~30%估算。

根据上面计算得到的液压泵最大工作压力和最大供油流量，即可从产品样本中选取合适的型号和规格。为了工作安全可靠，通常泵的额定压力应选得比上述最大工作压力高20% ~60%，以便留有压力储备；额定流量则只需选得能满足上述最大流量需要即可，以免造成过大的功率损失。

液压泵在额定压力和额定流量下工作时，其驱动电动机的功率一般可以直接从产品样本上查到。电动机功率也可以根据具体工况计算出来，有关的计算式和数据见液压工程手册。

2. 阀类元件的选择

阀类元件的选择是根据液压系统的最大工作压力和实际通过该阀的最大流量来选取其规格。所选用的阀类元件的额定压力和额定流量要大于系统的最大工作压力和实际通过该阀的最大流量。必要时，可允许增大通过阀的流量，但不得超过额定流量的20%，以免引起过大的压力损失、发热和噪声。选择节流阀和调速阀时，还要考虑它的最小稳定流量是否符合设计要求。选择换向阀时除考虑压力、流量外，还应考虑其中位机能及操纵方式。

3. 液压阀配置形式的选择

常将液压系统的动力源、阀类元件等几种液压设备安装在主机外的液压站上，使安装、维修方便，并消除动力源的振动与油温变化对主机工作精度的影响。阀类元件在液压站上的配置形式主要有三种，即：

（1）叠加阀式　其结构、特点等在液压控制阀一章中已作介绍，不再重复。

（2）油路板式　油路板是一块较厚的液压元件安装板，将板式阀类元件用螺钉安装在板的正面，元件之间的连接油路由板内加工的孔道形成。管接头安装在板的侧面。

（3）集成块式　集成块是一块通用化的六面体，四周除一面安装通向执行元件的管接头外，其余三面均可安装板式液压阀。块内由钻孔形成元件之间的油路。一个液压系统往往由几块集成块组成。块的上、下两面作为块与块之间的结合面，各集成块与顶盖、底板一起用长螺栓叠装起来，即组成整个液压系统。总进油口与回油口开在底板上，通过集成块的公共孔直通顶盖。

lines are formed by bored holes on the block. Usually a hydraulic system consists of several integrated blocks. The up and down surfaces are used as the joint faces between blocks. The whole hydraulic system is then designed when each integrated block, their coping and soleplate are fixed together as a nappe by long bolts. The inlet and outlet ports are set on the soleplate and the oil reaches the coping via the public holes on the integrated block.

9.1.5 Performance check for a hydraulic system designed

After the hydraulic system has been initially determined, it is necessary to check its main technic performance parameters. By then the selected hydraulic components and the hydraulic system parameters can be further adjusted. The events needed to check are different for different systems, but generally the pressure loss in the system and the temperature rise due to heat generation must be covered.

1. Checking of the system pressure loss

The pressure loss includes frictional loss along the pipelines, local loss in the pipelines and to a less produced by the valve components. The pressure loss of along or local can be calculated from the involved expressions shown in Chapter 2, while the local loss of the valve components can be obtained from the product samples. When the practical flow rate q of a valve component doesn't equal the nominal flow rate q_n, its practical pressure loss Δp and the rated pressure loss Δp_n meet the following expression

$$\Delta p = \Delta p_n \left(\frac{q}{q_n}\right)^2 \tag{9-4}$$

The returned line pressure loss must be calculated independently for different working phases. The pressure loss in the returned line is usually converted to that in the inlet oil line. Note that the design should be modified if the calculated overall pressure loss is far from the assumed pressure loss for the hydraulic components.

2. Checking of the temperature rise caused by heat generation

This item is checked through estimating the temperature rise in the oil according to heat balance principle. The heat Q entering into the hydraulic system per unit time is the difference of the input power P_1 from the pump and the effective power P_0 of the actuators. Suppose all this heat emits out from the reservoir, regardless of other effect of heat transmission in the system, then the expressions for temperature rise in oil estimation can be found from involved handbooks according to different conditions. For example, when the dimension ratios of three sides of the reservoir are between 1:1:1 and 1:2:3, the oil lever is 80% of the reservoir height, and the reservoir is in good ventilation condition, then the expression for the oil temperature rise ΔT can be approximatively described by the input heat $Q(W)$ per unit time and the effective volume $V_2(\mathrm{m}^3)$ of the reservoir

$$\Delta T = \frac{Q}{\sqrt[3]{V^2}} \times 10^{-2}\,(℃) \tag{9-5}$$

Once the checked oil temperature rise surpasses the allowable value, a proper cooler must be set in the system. The allowable oil temperature rise $[\Delta T]$ in the reservoir is different for different mainframes, usually $[\Delta T]$ is 25-30℃ for general machine tools, 35-40℃ for engineering machinery, etc.

9.1.5　液压系统的性能验算

液压系统初步确定之后，需要对它的主要技术性能进行必要的验算，以便对所选液压元件和液压系统参数作进一步调整。根据液压系统的不同，需要验算的项目也有所不同，但一般都要进行系统压力损失验算和发热温升验算。

1. 液压系统压力损失的验算

压力损失包括管道内的沿程损失和局部损失以及阀类元件的局部损失三项。管道内的这两种损失可用第 2 章中的有关公式计算；阀类元件的局部损失则需从产品样本中查出。当通过阀类元件的实际流量 q 不是其公称流量 q_n 时，它的实际压力损失 Δp 与其额定压力损失 Δp_n 间将成如下的近似关系

$$\Delta p = \Delta p_n \left(\frac{q}{q_n}\right)^2 \tag{9-4}$$

计算液压系统压力损失时，不同的工作阶段要分开来计算。其压力损失一般都需折算到进油路上。计算时所得的总压力损失如果与计算液压元件时假定的压力损失相差太大，则应对设计进行必要的修改。

2. 液压系统发热温升验算

这项验算是用热平衡原理来对油液的温升值进行估计的。单位时间内进入液压系统的热量 Q（以 W 计）是液压泵输入功率 P_1 和液压执行元件有效功率 P_0 之差。假如这些热量全部由油箱散发出去，不考虑系统其他部分的散热效能，则油液温升的估算公式可以根据不同的条件分别从有关的手册中找出来。例如，当油箱三个边的尺寸比例在 1:1:1 到 1:2:3 之间、油面高度是油箱高度的 80% 且油箱通风情况良好时，油液温升 ΔT（℃）的计算式可以用单位时间内输入热量 Q（W）和油箱有效容积 V_2（m^3）近似地表示成

$$\Delta T = \frac{Q}{\sqrt[3]{V^2}} \times 10^{-2} \tag{9-5}$$

当验算出来的油液温升值超过允许数值时，系统中必须考虑设置适当的冷却器。油箱中油液允许的温升［ΔT］随主机的不同而异；一般机床为 25～30℃，工程机械为 35～40℃。

9.1.6　绘制工作图，编制技术文件

工作图一般包括液压系统图、液压系统装配图、液压缸等非标准元件装配图及零件图。

液压系统图应附有液压元件明细表，表中表明各液压元件的规格、型号和压力、流量调整值。对于复杂的系统还应绘出各元件的工作循环图和电磁铁动作顺序表。

液压系统装配图是液压系统的安装施工图，包括液压泵装置图、集成或叠加阀油路装配图和管路安装图等。

编制技术文件一般包括设计计算说明书，液压系统使用及维修技术说明书，零部件目录表，标准件、通用件及外购件汇总表等。

9.2　液压系统的设计计算实例

本节以一台卧式单面多轴钻孔组合机床液压系统设计实例来说明组合机床液压系统的设计过程。已知：机床工作时轴向切削力 $F_L = F_t = 30\ 000\text{N}$；移动部件总重力 $G = 9500\text{N}$；快

9.1.6 Drawing of working diagrams and technical documents

The working diagrams usually incorporate hydraulic system figure, assembly drawing and the assembly and part drawings of non-standard components such as hydraulic cylinder.

There must exist a list of components attached to the hydraulic system figure, with the clearly listed specifications, type, settings of pressure and flow rate. For more complex system, the work cycle chart and the table of sequence actions electromagnetisms must also be drawn.

The system assembly figure is the construction and erection drawing for the hydraulic system, which indicates the pump's configuration, the pipeline assembly of integrated or modular valves, the pipeline arrangement, etc.

Generally, the technical documents include the explanations of design calculation, manual to use and maintain, content of parts, standard parts, universal parts, overall table of outsourcing parts, etc.

9.2 A Design Example of Hydraulic System

This section is to study the hydraulic system of a single side multi-spindle drill machine tool. The purpose is to introduce the design process of hydraulic system in any machine tool. Given the axial cutting force $F_L = F_t = 30\ 000\text{N}$; the overall gross weight $G = 9500\text{N}$; the distances covered at rapid speed $l_1 = 100\text{mm}$ and at work speed $l_2 = 50\text{mm}$; the rapid feed and retract velocity $v_1 = v_3 = 5\text{m/min}$, and the work feed velocity $v_2 = 0.05\text{m/min}$; the time for acceleration and deceleration $\Delta t = 0.2\text{s}$; the friction coefficient of rest $f_s = 0.2$, and friction of motion $f_d = 0.1$; the mechanical efficiency of the cylinder $\eta_m = 0.9$. The dynamic slipway is required to achieve such a work cycle: rapid feed→work feed→block dwell→rapid retract→stop at the original position.

9.2.1 Analysis of working conditions

The axial cutting force $F_L = F_t = 30\ 000\text{N}$, which is also the only force whose direction is horizontal (for the working parts moves in the horizontal direction). The other forces must be discomposed along the horizontal direction. Then the inertia force F_a, the friction of rest F_{fs} and the friction of motion F_{fd} against the guide way can be calculated as follows

$$F_a = m\frac{\Delta v}{\Delta t} = \frac{G}{g}\frac{\Delta v}{\Delta t} = \frac{9500 \times 5}{9.8 \times 0.2 \times 60}\text{N} \approx 404\text{N}$$

$$F_{fs} = f_s F_N = 0.2 \times 9500\text{N} = 1900\text{N}$$

$$F_{fd} = f_d F_N = 0.1 \times 9500\text{N} = 950\text{N}$$

The loads on the cylinder in each work step are listed in Tab. 9-4, where the back pressure in the oil returned chamber is neglected. The load and speed diagrams are indicated in Fig. 9-4a and Fig. 9-4b.

9.2.2 Determination of main parameters of the actuator

From Tab. 9-1 and Tab. 9-2, we know that p_1 is 4-5MPa when the maximum load is 30 000-

进行程长 $l_1 = 100\text{mm}$，工进行程长 $l_2 = 50\text{mm}$；快进速度为 v_1、快退速度为 v_3，$v_1 = v_3 = 5\text{m/min}$，工进速度 $v_2 = 0.05\text{m/min}$；加速、减速时间 $\Delta t = 0.2\text{s}$；静摩擦因数 $f_s = 0.2$，动摩擦因数 $f_d = 0.1$；液压缸的机械效率 $\eta_m = 0.9$。设计要求滑台实现"快进→工进→死挡铁停留→快退→原位停止"的工作循环。

液压系统的设计过程如下。

9.2.1 工况分析

因为工作部件是卧式装置，除轴向切削力 $F_L = F_t = 30\,000\text{N}$，为水平方向（卧式）外，其他的力均应转化到水平方向，则惯性力 F_a、导轨的静摩擦力 F_{fs} 和动摩擦力 F_{fd} 分别为

$$F_a = m\frac{\Delta v}{\Delta t} = \frac{G}{g}\frac{\Delta v}{\Delta t} = \frac{9500 \times 5}{9.8 \times 0.2 \times 60}\text{N} \approx 404\text{N}$$

$$F_{fs} = f_s F_N = 0.2 \times 9500\text{N} = 1900\text{N}$$

$$F_{fd} = f_d F_N = 0.1 \times 9500\text{N} = 950\text{N}$$

这里暂不考虑回油腔的背压力。因此液压缸在各个工作阶段的负载见表 9-4。负载图、速度图分别如图 9-4a、b 所示。

Tab. 9-4 Loads on the cylinder in each work step（液压缸在各工作阶段的负载值）

Working condition（工况）	Load combination（负载组成）	Load（负载值）F/N	Thrust（推力）$\frac{F}{\eta_m}$/N
Startup（起动）	$F = F_{fs}$	1900	2111
Acceleration（加速）	$F = F_{fd} + F_a$	1354	1504
Deceleration（快进）	$F = F_{fd}$	950	1056
Work feed（工进）	$F = F_{fd} + F_t$	30 950	34 389
Rapid retract（快退）	$F = F_{fd}$	950	1056

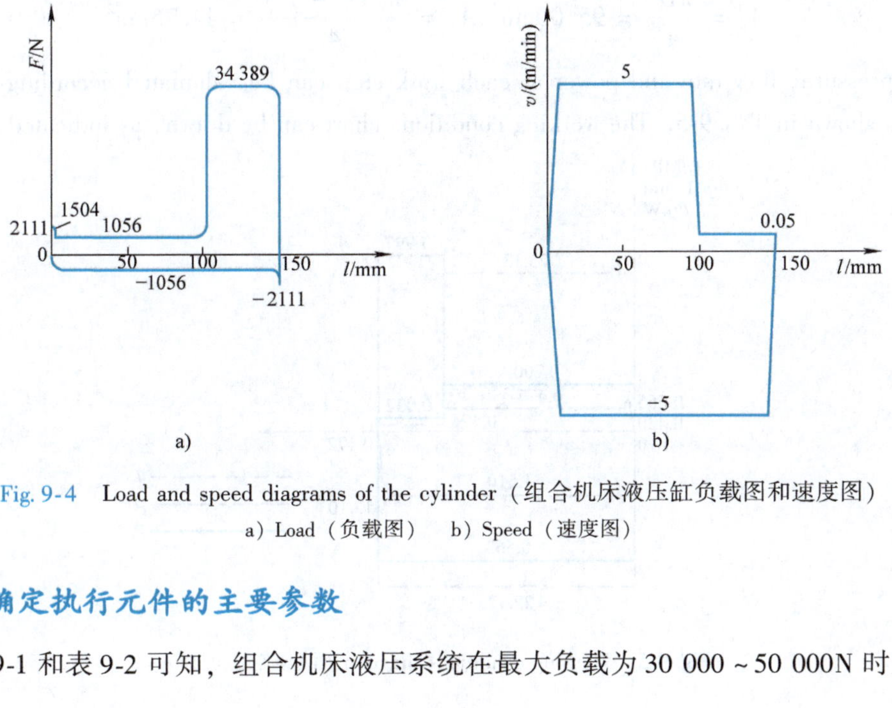

Fig. 9-4 Load and speed diagrams of the cylinder（组合机床液压缸负载图和速度图）
a）Load（负载图） b）Speed（速度图）

9.2.2 确定执行元件的主要参数

由表 9-1 和表 9-2 可知，组合机床液压系统在最大负载为 30 000 ~ 50 000N 时宜取工作

50 000N, here a 4MPa of p_1 is adopted. The cylinder is a single piston rod type and in differential connection for rapid feed. The working area of the non-rod end chamber (A_1) is double that of the rod end chamber (A_2), i. e., $d = 0.707D$, where d and D are the diameters of the rod and the cylinder barrel respectively.

Note that when a hole is being machined, there must exist a certain back pressure p_2 (here a 0.8MPa of p_2 is adopted) on the oil return line to prevent the slipway from suddenly rushing forward when the hole is finished. At rapid feed period, the pressure in the rod chamber must be a little higher than that in the non-rod chamber for there still exists pressure loss in the pipeline even the cylinder is in differential connection. Considering that the pressure loss in the pipeline is relatively small, we can set the pressure in the rod chamber 0.3MPa higher than that in the rod chamber. At rapid retract period, p_2 is considered as 0.6MPa due to the existence of the back pressure in the return oil chamber.

The cylinder area can be calculated from the thrust at work feed period, i. e.

$$\frac{F}{\eta_m} = A_1 p_1 - A_2 p_2 = A_1 p_1 - \frac{A_1}{2} p_2$$

Then $$A_1 = \frac{F}{\eta_m \left(p_1 - \frac{p_2}{2}\right)} = \frac{34\ 389}{\left(4 - \frac{0.8}{2}\right) \times 10^6} \mathrm{m}^2 \approx 0.0096\mathrm{m}^2 = 96\mathrm{cm}^2$$

$$D = \sqrt{\frac{4A_1}{\pi}} = 11.06\mathrm{cm}, d = 0.707D = 7.82\mathrm{cm}$$

D and d must be corrected to standard values according to GB/T 2348—1993: $D = 11\mathrm{cm}$, $d = 8\mathrm{cm}$, then the practical effective areas of the two chambers of the cylinder are

$$A_1 = \frac{\pi D^2}{4} = 95.00\mathrm{cm}^2, A_2 = \frac{\pi(D^2 - d^2)}{4} = 44.75\mathrm{cm}^2$$

The pressure, flow rate and power at each work step can be calculated according to D and d values, as shown in Tab. 9-5. The working conditions chart can be drawn, as indicated in Fig. 9-5.

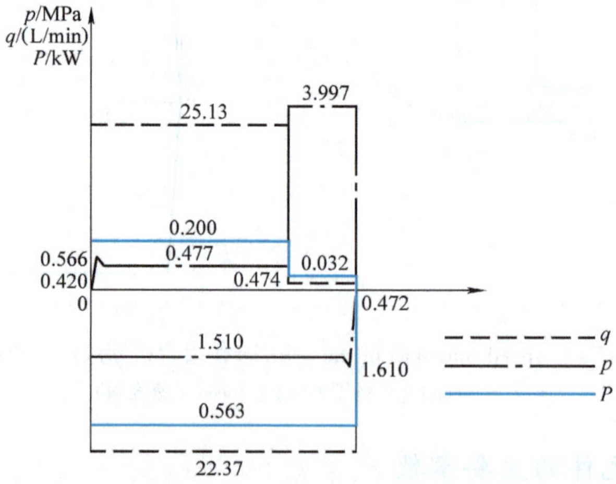

Fig. 9-5 Working conditions diagram (液压缸工况图)

压力$p_1=4\sim5\text{MPa}$，本设计取 $p_1=4\text{MPa}$。液压缸采用单活塞杆式，并在快进时作差动连接。此时液压缸无杆腔的工作面积 A_1 应为有杆腔工作面积 A_2 的两倍，即活塞杆直径 d 与缸筒直径 D 的关系为 $d=0.707D$。

在钻孔加工时，液压缸回油路上必须具有背压 p_2，以防孔被钻通时滑台突然前冲，可取 $p_2=0.8\text{MPa}$。快进时液压缸虽作差动连接，但由于油管中有压力损失，有杆腔的压力应略大于无杆腔，但其值较小，可先按 0.3MPa 考虑。快退时回油腔中是有背压的，这时 p_2 可按 0.6MPa 估算。

由工进时的推力计算液压缸面积，即

$$\frac{F}{\eta_m}=A_1p_1-A_2p_2=A_1p_1-\frac{A_1}{2}p_2$$

故

$$A_1=\frac{F}{\eta_m\left(p_1-\dfrac{p_2}{2}\right)}=\frac{34\ 389}{\left(4-\dfrac{0.8}{2}\right)\times10^6}\text{m}^2\approx0.0096\text{m}^2=96\text{cm}^2$$

$$D=\sqrt{\frac{4A_1}{\pi}}=11.06\text{cm},d=0.707D=7.82\text{cm}$$

将这些直径按 GB/T 2348—1993 圆整成就近标准值得：$D=11\text{cm}$，$d=8\text{cm}$，由此求得液压缸两腔的实际有效面积分别为

$$A_1=\frac{\pi D^2}{4}=95.00\text{cm}^2,A_2=\frac{\pi(D^2-d^2)}{4}=44.75\text{cm}^2$$

根据上述 D 与 d 值，可估算液压缸在各个工作阶段中的压力、流量和功率值，见表9-5。并据此绘制出工况图如图 9-5 所示。

Tab. 9-5　The pressure, flow rate and power at each step
（液压缸在不同阶段的压力、流量和功率值）

Working condition（工况）		Load（负载）F/N	Pressure in the oil return chamber（回油腔压力）p_2/MPa	The pressure in the inlet chamber（进油腔压力）p_1/MPa	Input flow rate（输入流量）(L/min)	Input power（输入功率）P/kW	Expressions（计算式）
Rapid feed (differential) 快进（差动）	Startup（起动）	2111	$p_2=0$ （$\Delta p=0$）	0.420	—	—	$p_1=\dfrac{F+A_2\Delta p}{A_1-A_2}$ $q=(A_1-A_2)\ v_1$ $P=p_1q$
	Acceleration（加速）	1504	$p_2=p_1+\Delta p$ （$\Delta p=0.3\text{MPa}$）	0.566	—	—	
	Constant（恒速）	1056		0.477	25.13	0.200	
Work feed（工进）		34 389	0.8	3.997	0.47	0.032	$p_1=\dfrac{F+p_2A_2}{A_1}$ $q=A_1v_2$ $P=p_1q$
Rapid retract（快退）	Start up（起动）	2111	0	0.472	—	—	$p_1=\dfrac{F+p_2A_1}{A_2}$ $q=A_2v_3$ $P=p_1q$
	Acceleration（加速）	1504	0.6	1.610	—	—	
	Constant（恒速）	1056		1.510	22.37	0.563	

9.2.3 Design of hydraulic system

1. Selection of hydraulic circuits

From the working conditions diagram (Fig. 9-5), we can clearly see the features for the designed hydraulic system: low power, low speed of the slipway, small change of working load. So an entrance throttle governing circuit in open cycle can be adopted for speed control circuit. Oil supply and pressure regulating methods can also be determined after the speed control circuit is designed. A back pressure valve is set on the oil return line.

The cylinder is required to provide large flow rate with low pressure and low flow rate with high pressure alternately, as indicated in Fig. 9-5. The ratio of the maximum flow rate to the minimum flow rate is about 53, while the time covered by the rapid feed and rapid retract periods is shorter than that by the work feed period. For the sake of high efficiency and energy saving, it is recommended to adopt a double pumps hydraulic system or a displacement throttle speed-regulating system composed by a pressure-limited variable pump and speed-regulating valves, rather than a single pump of constant delivery.

Considering that the rapid feed and the rapid retract speeds are relatively high, the cylinder is set in differential connection for rapid feed to ensure a reliable direction changing. In this design, five-position, three-way Y type electro-hydraulic directional valves are adopted to achieve direction changing and directional connection.

A pressure-reducing valve and a check valve are set inline on the branch to ensure a reliable clamping. Both valves can be regulated independently. To achieve the sequence actions of 'positioning→clamping', a sequence check valve is set on the oil line that connects to the clamping cylinder. Only when the pressure in the positioning cylinder equals or surpasses the sequence valve setting can the clamping cylinder begin to act. A pressure switch is set at the inlet of the clamping cylinder to prevent the feed cylinder from acting before the workpiece is firmly clamped. Only when the clamping force surpasses the pressure switch setting can the switch signal to actuate the three-position, five-way directional valve on the oil line that connects to the feed cylinder. The feed cylinder begins to feed rapidly after this actuation.

2. Design of hydraulic system

From the above analysis and the designed scheme, the principle chart of hydraulic system can be determined by reasonably combined each chosen circuit, as shown in Fig. 9-6. The serial number and name of components in the Fig. 9-6 relate to the Tab. 9-6.

3. Selection of hydraulic valves

Here we adopt a modular type. Fig. 9-7 shows the system figure in which modular valves are adopted. The serial number and name of components in the Fig. 9-7 relate to the Tab. 9-7.

9.2.4 Selection of hydraulic components

1. Pump

The maximum working pressure of the cylinder during the whole work cycle is 3.997MPa. If the pressure loss along the inlet oil line is assumed 0.8MPa, the pressure switch setting must be

9.2.3 液压系统图的拟订

1. 液压回路的选择

首先选择调速回路。由工况图（图 9-5）可知，这台机床液压系统的功率小，滑台运动速度低，工作负载变化小，可采用进口节流的调速方式和开式循环。为了解决进口节流调速回路在孔钻通时的滑台突然前冲现象，回油路上要设置背压阀。

从工况图可以清楚地看到，在这个液压系统的工作循环内，液压缸交替地要求油液提供低压大流量和高压小流量。

最大流量和最小流量之比约为 53，而快进、快退时所需时间比工进时所需时间小得多，因此从提高系统效率、节省能量的角度来看，采用单个定量泵作为油源显然是不合适的。宜采用双泵供油系统，或采用限压式变量泵与调速阀组成的容积节流调速系统，本方案采用前者。

在调速方案确定以后，供油方式与调压方式也随之确定。

本机床快进、快退速度较大，为保证换向平稳，且液压缸在快进时为差动连接，故采用三位五通 Y 型电-液换向阀来实现运动换向，并实现差动连接。

2. 拟订液压系统图

综合上述分析和所拟定的方案，将各种回路合理地组合成为该机床的液压系统原理图，如图 9-6 所示。图 9-6 中各元件的序号和名称与表 9-6 中的相对应。

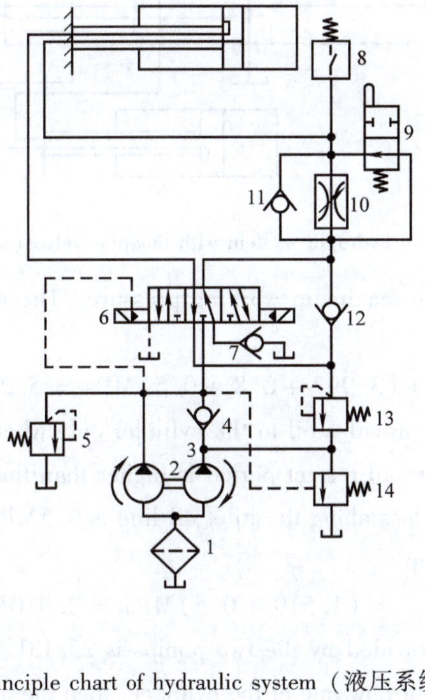

Fig. 9-6 Principle chart of hydraulic system（液压系统原理图）

3. 液压阀配置形式的选择

本设计采用叠加阀式配置形式，图 9-7 所示为叠加阀式液压系统原理图。图 9-7 中各元件的序号和名称与表 9-7 中的相对应。

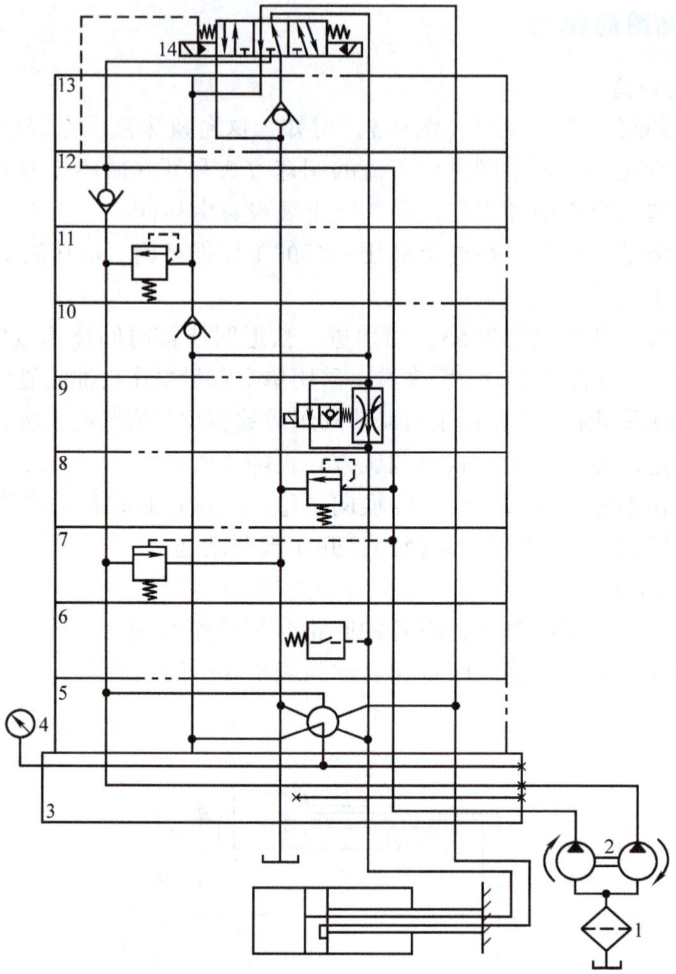

Fig. 9-7 Principle chart of hydraulic system with modular valves(叠加阀式液压系统原理图)

0.5MPa higher than the system maximum working pressure. The maximum working pressure of the low flow rate pump is

$$p_{p1} = (3.997 + 0.8 + 0.5)\text{MPa} = 5.297\text{MPa}$$

The large flow rate pump provides oil to the cylinder only when the cylinder needs to move rapidly. The working pressure at rapid retract period is higher than that at rapid feed period, as indicated in Fig. 9-5. If the pressure loss along the inlet oil line is 0.5MPa, then the highest working pressure of the large flow rate pump is

$$p_{p2} = (1.510 + 0.5)\text{MPa} = 2.010\text{MPa}$$

The maximum flow rate provided by the two pumps is 25.13L/min. If the leakage in the return line is supposed 10% of the input flow rate of the cylinder, then the overall flow rate of the two pumps is

$$q_p = 1.1 \times 25.13\text{L/min} \approx 27.64\text{L/min}$$

Because the minimum steady flow rate of the pressure relief valve is 3L/min, and the input flow rate to the cylinder for work feed is 0.47L/min, the flow rate provided by the low flow rate pump must be at least 3.47L/min.

9.2.4 液压元件的选择

1. 液压泵

液压缸在整个工作循环中的最大工作压力为3.997MPa，如取进油路上的压力损失为0.8MPa，压力继电器调整压力应比系统最大工作压力高出0.5MPa，则小流量泵的最大工作压力应为

$$p_{p1} = (3.997 + 0.8 + 0.5)\text{MPa} = 5.297\text{MPa}$$

大流量泵是在快速运动时才向液压缸输油的，由图9-5可知，快退时液压缸中的工作压力比快进时大，如取进油路的压力损失为0.5MPa，则大流量泵的最高工作压力为

$$p_{p2} = (1.510 + 0.5)\text{MPa} = 2.010\text{MPa}$$

两个液压泵应向液压缸提供的最大流量为25.13L/min，若回路中的泄漏按液压缸输入流量的10%估计，则两个泵的总流量为

$$q_p = 1.1 \times 25.13\text{L/min} \approx 27.64\text{L/min}$$

由于溢流阀的最小稳定溢流量为3L/min，工进时输入液压缸的流量为0.47L/min，所以小流量泵的流量规格最少应为3.47L/min。

根据以上压力和流量的数值查阅产品目录，最后确定选取PV2R12-6/26型双联叶片泵，其小流量泵和大流量泵的排量分别为6mL/r和26mL/r，当液压泵的转速$n_p = 940\text{r/min}$时该液压泵的理论流量为30.08L/min，若取液压泵的容积效率为$\eta_V = 0.9$，则液压泵的实际输出流量为

$$q_p = [(6 + 26) \times 940 \times 0.9/1000]\text{L/min} = (5.1 + 22)\text{L/min} = 27.1\text{L/min}$$

由于液压缸在快退时输入功率最大，这相当于液压泵输出压力为2MPa、流量为27.1L/min时的情况。如取双联叶片泵的总效率为$\eta_p = 0.75$，则液压泵驱动电动机所需的功率为

$$P = p_p q_p / \eta_p = [2 \times 10^6 \times 27.1/(60 \times 10^3)]/0.75\text{W} \approx 1204\text{W} = 1.204\text{kW}$$

根据此数值查阅电机产品目录，选取Y100L-6型电动机，其额定功率$P_n = 1.54\text{kW}$，额定转速$n_n = 940\text{r/min}$。

2. 阀类元件及辅助元件

根据液压系统的工作压力和通过各个阀类元件和辅助元件的实际流量，可选出这些元件的型号及规格，见表9-6。采用叠加阀方案，其型号及规格见表9-7。

3. 油管

各元件间连接管道的规格按液压元件接口处的尺寸决定，液压缸进、出油管则按输入、排出的最大流量计算。由于液压泵选定之后液压缸在各个工作阶段的进、出流量已与原定数值不同，所以要重新计算，见表9-8。

由表9-8可以看出，液压缸在各个工作阶段的时间运动速度符合设计要求。若油液在压油管中的流速$v = 3\text{m/s}$，则液压缸无杆腔与有杆腔相连的油管内径分别为

$$d = 2 \times \sqrt{q/(\pi v)} = 2 \times \sqrt{(51.23 \times 10^6/60)/(\pi \times 3 \times 10^3)}\text{mm} \approx 19.04\text{ mm}$$

$$d' = 2 \times \sqrt{(27.1 \times 10^6/60)/(\pi \times 3 \times 10^3)}\text{mm} \approx 13.85\text{ mm}$$

这两根油管都按YB231—1970选用内径ϕ15mm、外径ϕ18.2mm的15号冷拔无缝钢管。

From the pressure and flow rate calculated above, the PV2R12-6/26 type pair vane pump can be adopted according to involved product contents. The deliveries of the small pump and the large pump are 6mL/r and 26mL/r respectively. The theoretical flow rate of this pump is 30.08L/min at a rotate speed of 940r/min. If the volumetric efficiency of this pump (η_V) is 0.9, then the practical output flow rate of this pump is

$$q_p = [(6+26) \times 940 \times 0.9/1000]\text{L/min} = (5.1+22)\text{L/min} = 27.1\text{L/min}$$

The output power reaches maximum when the cylinder is at rapid retract period, which happens when the pump provides a pressure of 2MPa and a flow rate of 27.1L/min. If the overall efficiency of the pair vane pump is 0.75, then the power required for the pump to drive the motor is

$$P = p_p q_q/\eta_p = [2 \times 10^6 \times 27.1/(60 \times 10^3)]/0.75\text{W} \approx 1204\text{W} = 1.204\text{kW}$$

According to involved motor product contents, we can select the Y100L-6 type motor which has a rated power of 1.54kW and a rated rotate speed of 940r/min.

2. Valves and auxiliary elements

The valves and auxiliary elements can be selected according to the system working pressure and the practical flow rate through them, see Tab. 9-6. Tab. 9-7 shows the valves and auxiliary elements selected for the scheme with modular valves.

Tab. 9-6 Types and specification of the components(元件的型号及规格)

Number (序号)	Name (元件名称)	Flow rate estimated (估计通过流量)/($L \cdot min^{-1}$)	Type and specification (型号及规格)
1	Filter(过滤器)	30	XU-J63X80
2	Pair vane pump (双联叶片泵)	5.1 + 22	PV2R-6/26
3	Pressure gauge switch (压力表开关)	—	KF3-E3B
4	Check valve(单向阀)	25	AF3-Ea10B
5	Pressure relief valve (溢流阀)	5	YF3-E10B
6	Five-position, three-way solenoid valve (三位五通电磁阀)	60	35DY-E10B
7	Check valve(单向阀)	60	AF3-Ea10B
8	Pressure relay (压力继电器)	—	PF-B8L
9	Stroke valve (行程阀)	50	
10	Speed control valve (调速阀)	0.5	AXQ-E10B
11	Check valve(单向阀)	60	
12	Check valve(单向阀)	25	AF3-Ea10B
13	Back pressure valve (背压阀)	0.5	YF3-E10B
14	Hydraulic-controlled sequence valve(液控顺序阀)	25	XF3-E10B

Tab. 9-7　Types and specification of the components (modular valves)

元件的型号及规格（叠加阀系列）

Number （序号）	Name （元件名称）	Flow rate estimated （估计通过流量）/L · min^{-1}	Type and specification （型号及规格）
1	Filter（过滤器）	30	XU-J63X80
2	Pair vane pump （双联叶片泵）	5. 1 + 22	PV2R-6/26
3	Soleplate block（底板块）	27. 1	EDKA-10
4	Pressure gauge（压力表）		Y-100T
5	Pressure gauge switch （压力表开关）	—	4K-F10D3
6	Pressure switch （压力继电器）	—	PD-F10D3-A
7	Hydraulic-controlled sequence valve （液控顺序阀）	25	X-F10D3-P/T（P_1）
8	Pressure relief valve （溢流阀）	5	Y1-F10D3-P/T
9	Electromagnetic speed control check valve （电磁单向调速阀）	—	Q_1AD-F6/10D3-B
10	Check valve（单向阀）	25	A-F10D3-T
11	Back pressure valve （背压阀）	0. 5	PA-Fa6/10D3-B（A）
12	Check valve（单向阀）	25	A-F10D3-T
13	Check valve（单向阀）	60	A-F10D3-T
14	Five-position, three-way solenoid valve （三位五通电磁阀）	60	35DY-E10B

3. Pipeline

The pipeline specification for each component is decided according to their port size, while the inlet and outlet pipelines of the cylinder must be decided according to the input and output maximum flow rate. The input and output flow rate from the cylinder in each working step might be different from the original calculated values after the exact pump is selected, see Tab. 9-8.

As can be seen from the above table, the cylinder speed meets the design requirements in each work step. If the velocity of flow in the pressurized oil pipeline is 3m/s, then the inner diameters of the oil pipelines that connect the non-rod chamber and the rod chamber of the cylinder are

$$d = 2 \times \sqrt{q/(\pi v)} = 2 \times \sqrt{(51.23 \times 10^6/60)/(\pi \times 3 \times 10^3)}\,\text{mm} \approx 19.04\text{mm}$$

$$d' = 2 \times \sqrt{(27.1 \times 10^6/60)/(\pi \times 3 \times 10^3)}\,\text{mm} \approx 13.85\text{mm}$$

According to YB231—1970, we can choose the 15-sized cold-drawn seamless steel tube which has an inner diameter of 15mm and outer diameter of 18.2mm.

4. Reservoir

The volume of the reservoir can be estimated from $V = \xi q_p$, where ξ is 7 by experience, then

$$V = \xi q_p = 7 \times 27.1\text{L} = 189.7\text{L}$$

V can be corrected to a standard value of 250L according to JB/T 7938—1995.

9.2.5 Performance check for hydraulic system

1. Pressure loss

Because the concrete pipeline arrangement hasn't been determined, the overall pressure loss cannot be estimated. We can only calculate the pressure loss caused by the valves. When the whole pipeline arrangement is finished, we can calculate the frictional loss and the local loss and plus them to the pressure loss caused by the valves. The pressure loss must be checked according to different steps in a work cycle, neglected here.

2. Oil temperature rise

The time for the work feed period takes up 95% of the whole work cycle, so the system heat generation and the oil temperature rise can be calculated according to the work feed condition.

The effective power of the cylinder at work feed can be calculated as follows

$$P_0 = p_2 q_2 = F v_2 = \frac{34\,389 \times 0.05}{60}\text{W} \approx 28.7\text{W}$$

During work feed period, the large flow rate pump is unloaded by sequence valve 7 with an unloading pressure of 0.037MPa, while the small flow rate pump provides oil under a high pressure of 4.92MPa. The total output power from these two pumps is

$$P_1 = \frac{p_{p1} q_{p1} + p_{p2} q_{p2}}{\eta} = \frac{0.037 \times 10^6 \times \frac{22}{60} \times 10^{-3} + 4.92 \times 10^6 \times \frac{5.1}{60} \times 10^{-3}}{0.75}\text{W} \approx 575.7\text{W}$$

Then power of heat generation of the hydraulic system is

$$Q = P_1 - P_0 = (575.7 - 28.7)\text{W} = 547\text{W}$$

The heat-transfer area of the reservoir is

Tab. 9-8 **Input and output flow rate of the cylinder**（液压缸的进、出流量）

	Input flow rate（输入流量）/（L/min）	Output flow rate（排出流量）/（L/min）	Speed（运动速度）/（m/min）
Rapid feed（快进）	$q_1 = A_1 q_p / (A_1 - A_2)$ $= 95.00 \times 27.1 / (95.00 - 44.75)$ $= 51.23$	$q_2 = A_2 q_1 / A_1$ $= 44.75 \times 51.23 / 95.00$ $= 24.13$	$v_1 = q_p (A_1 - A_2)$ $= 27.1 \times 10 / (95.00 - 44.75)$ $= 5.4$
Work feed（工进）	$q_1 = 0.47$	$q_2 = A_2 q_1 / A_1$ $= 44.75 \times 0.474 / 95.00$ $= 0.24$	$v_2 = q_1 / A_1$ $= 0.47 \times 10 / 95.00$ $= 0.049$
Rapid retract（快退）	$q_1 = q_p = 27.1$	$q_2 = A_1 q_1 / A_2$ $= 95.00 \times 27.1 / 44.75$ $= 57.53$	$v_3 = q_1 / A_2$ $= 27.1 \times 10 / 44.75$ $= 6.06$

4. 油箱

油箱容积按 $V = \xi q_p$ 估算，取经验数据 $\xi = 7$，故其容积为

$$V = \xi q_p = 7 \times 27.1\text{L} = 189.7\ \text{L}$$

按 JB/T 7938—1995 规定，取最靠近的标准值 $V = 250$ L。

9.2.5 液压系统的性能验算

1. 系统压力损失验算

由于系统的具体管路布置尚未确定，整个回路的压力损失无法估算，故只能估算阀类元件的压力损失，待设计好管路布局图后，加上管路的沿程损失和局部损失即可。压力损失的验算应按一个工作循环中不同阶段分别进行，这里估算从略。

2. 油液温升验算

工进在整个工作循环中所占的时间比例达 95%，所以系统发热和油液温升可用工进时的情况来计算。

工进时液压缸的有效功率为

$$P_0 = p_2 q_2 = F v_2 = \frac{34\,389 \times 0.05}{60}\text{W} \approx 28.7\text{W}$$

这时大流量泵通过顺序阀 7 卸荷（卸荷压力为 0.037MPa），通过计算小流量泵在高压（4.92MPa）下供油，所以这两个泵的总输出功率为

$$P_1 = \frac{p_{p1} q_{p1} + p_{p2} q_{p2}}{\eta} = \frac{0.037 \times 10^6 \times \frac{22}{60} \times 10^{-3} + 4.92 \times 10^6 \times \frac{5.1}{60} \times 10^{-3}}{0.75}\text{W} \approx 575.7\text{W}$$

由此得出液压系统的发热功率为

$$Q = P_1 - P_0 = (575.7 - 28.7)\text{W} = 547\text{W}$$

油箱的散热面积为

$$A = 6.5\sqrt[3]{V^2} = 6.5\sqrt[3]{(250 \times 10^{-3})^2}\text{m}^2 \approx 2.58\text{m}^2$$

$$A = 6.5\sqrt[3]{V^2} = 6.5\sqrt[3]{(250 \times 10^{-3})^2}\text{m}^2 \approx 2.58\text{m}^2$$

The heat transfer coefficient K is $9\text{W}/(\text{m}^2 \cdot ℃)$, then the oil temperature rise is

$$\Delta t = \frac{Q}{KA} = \frac{575.7}{9 \times 2.58}℃ \approx 24.8℃$$

Δt doesn't surpass the allowable range, so cooler isn't needed for this hydraulic system.

查得油箱的散热系数 $K=9\mathrm{W}/(\mathrm{m}^2\cdot℃)$，则油液温升为

$$\Delta t = \frac{Q}{KA} = \frac{575.7}{9\times 2.58}℃ \approx 24.8℃$$

此温升没有超出允许范围，故液压系统不必设置冷却器。

Part Two

Pneumatic Transmission

第2篇

气压传动

Chapter 10 Air Supply Devices and Pneumatic Components

Similar to hydraulic transmission, pneumatic transmission performs the function of energy transfer and control by the medium of flow. They have a great deal in common in the aspects of working principle, components' structure, system composition and graphic symbol. The readers should notice their differences.

A pneumatic system incorporates the following components and devices:

(1) Air supply devices they include pressure generation devices, air regeneration and storing devices, F. R. L (air filter, pressure reducing valve and oil mist lubricator) combination, etc. They provide qualified compressed air for pneumatic systems.

(2) Actuators they convert the compressed air pressure into mechanical energy, such as air cylinders and air motors.

(3) Control elements they are used to control the pressure, flow rate and moving direction for the components, such as valves.

(4) Pneumatic attachments they incorporate the components which are used to cool, silence, purify, lubricate compressed air and those to joint each component.

10.1 Air Supply Devices

Air supply devices are used to provide compressed air for pneumatic systems. An air supply device is an important part of a pneumatic system, which supplies qualified compressed air (with a certain lever of pressure, flow rate and purification) for the pneumatic system. Generally, for the delivery surpasses or equals 6-12m^3/min, a compressed air station is needed. But for the delivery less than 6-12m^3/min, the air compressor can be mounted besides the pneumatic equipment directly.

An air supply device incorporates four parts: pressure generation devices, air regeneration and storing devices, air lines and F. R. L. combination.

Normally the first two parts are located in compressed air station as a center to supply gas in industry or workshop, as shown in Fig. 10-1.

The functions of each component:

Air compressor 1: pressurize atmospheric air;

Cooler 2: cool compressed air; condense water and hydrocarbon vapours;

Oil and water separator 3: remove gross solid and liquid particulates;

Receivers 4, 7: air storage; settling chamber for particulates; cool air;

Air dryer 5: further reduce water and oil vapour content;

Air filter 6: further remove desiccant dust and fines.

第 10 章　气源装置与气动元件

气压传动与液压传动一样，都是利用流体作为介质进行能量传递和控制的传动形式，其工作原理、元件结构、系统组成与图形符号等有很多相似之处，因此在学习气压传动部分时，主要应注意两者不同之处。

气动系统由以下元件及装置组成：

（1）气源装置　它为气动系统提供符合质量要求的压缩空气。包括气压发生装置、净化与贮存压缩空气的装置及气源处理装置等。

（2）气动执行元件　将压缩空气的压力能转变为机械能的元件，如气缸、气马达。

（3）气动控制元件　控制气体的压力、流量及运动方向的元件，如各种阀类。

（4）气动辅件　使压缩空气冷却、消声、净化、润滑以及实现元件间连接所需的元件。

10.1　气源装置

向气压传动系统提供压缩空气的装置称为气源装置，它是气动系统的重要组成部分，为气动系统提供满足一定质量要求的压缩空气，即具有一定压力和流量，以及一定的净化程度的压缩气体。一般规定：排气量大于或等于 6 ~ 12m^3/min 时，就应独立设置压缩空气站；若小于 6 ~ 12m^3/min 时，可将空压机直接安装在气动设备旁。

气源装置一般由四个部分组成：①气压发生装置；②净化、贮存压缩空气的装置和设备；③传输压缩空气的管道系统；④气源处理装置。

往往将前两部分设备布置在压缩空气站内，作为工厂或车间统一的气源，如图 10-1 所示。图 10-1 中，空气压缩机 1 用以产生压缩空气。其吸气口装有空气过滤器，以减少进入空气压缩机内气体的杂质量。冷却器 2 用以降温冷却压缩空气，使汽化的水、油凝结出来。油水分离器 3 用以分离并排出降温冷却凝结的水滴、油滴、杂质等。气罐 4 和 7 用以贮存压缩空气，稳定压缩空气的压力，并除去部分油分和水分。干燥器 5 用以进一步吸收或排除压缩空气中的水分及油分，使之变成干燥空气。过滤器 6 用以进一步过滤压缩空气中的灰尘、杂质颗粒。气罐 4 输出的压缩空气可用于一般要求的气压传动系统，气罐 7 输出的压缩空气可用于要求较高的气动系统（如气动仪表及射流元件组成的控制回路等）。

10.1.1　气压发生装置

1. 空压机的类型

空压机是一种气压发生装置，是将机械能转换为气体压力能的转换装置。其种类可按工作原理、结构形式和性能参数进行分类。

1）按工作原理分为容积型空压机和速度型空压机。

2）按结构形式分类，如图 10-2 所示。

3）按输出压力大小可分为：低压空压机（0.2 ~ 1.0MPa）、中压空压机（1.0 ~

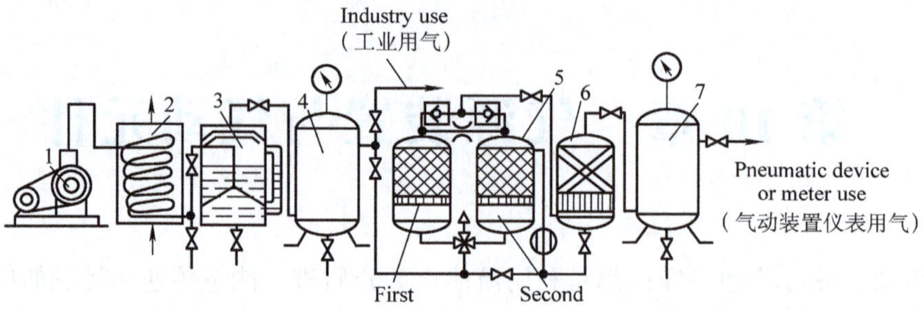

Fig. 10-1 Schematic view of air supply system(气源系统组成示意图)

1—Air compressor(空气压缩机) 2—Cooler(冷却器) 3—Oil and water separator(油水分离器)
4,7—Receiver(气罐) 5—Air dryer(干燥器) 6—Air filter(过滤器)

In a pneumatic system, the compressed air flowing out from receiver 4 is for general use and that from receiver 7 is for special use (such as the control circuits composed by pneumatic meters or fluidic elements).

10.1.1 Pressure generation devices

1. Classification of compressors

Compressor is one type of pressure generation devices, which converts mechanical energy into pressure energy. Compressors can be classified according to their working principles, structures or performance properties.

1) According to working principles: displacement and dynamic.

2) According to structures: as shown in Fig. 10-2.

3) According to output pressures: low pressure (0.2-1.0MPa), medium pressure (1.0-10MPa), high pressure (10-100MPa) and superhigh pressure (>100MPa).

4) According to output flow rates (delivery): mini (<1m^3/min), small (1-10 m^3/min), medium (10-100 m^3/min) and large (>100 m^3/min).

2. Working principles of compressors

Take a piston compressor for example: In a piston compressor, the piston moves reciprocatingly for air suction and compression (to increase the air pressure) through a crank and connecting rod mechanism. Fig. 10-3 shows its operation, which is similar to that of the single piston pump operation (see Fig. 3-1 in Chapter 3). The crank slider mechanism is driven by the motor to develop the reciprocating motion of piston 3, i.e., the rotation of crank 7 is converted into the reciprocating motion of slider 5 and piston 3. Most air compressors are the combination of multi-cylinder and multi-piston.

Piston compressors have the disadvantages of serious vibration, noise and output pressure fluctuation, which can be overcome by setting a receiver. On the contrary, vane and screw compressors do not suffer these problems and receivers are not needed. Compressors of these two types are similar to vane and screw pumps in both structures and operation.

10.1.2 Purification and storing devices and equipments for compressed air

1. Demands for better quality of compressed air

The compressed air flowing out from the compressor includes many contaminants, such as solid

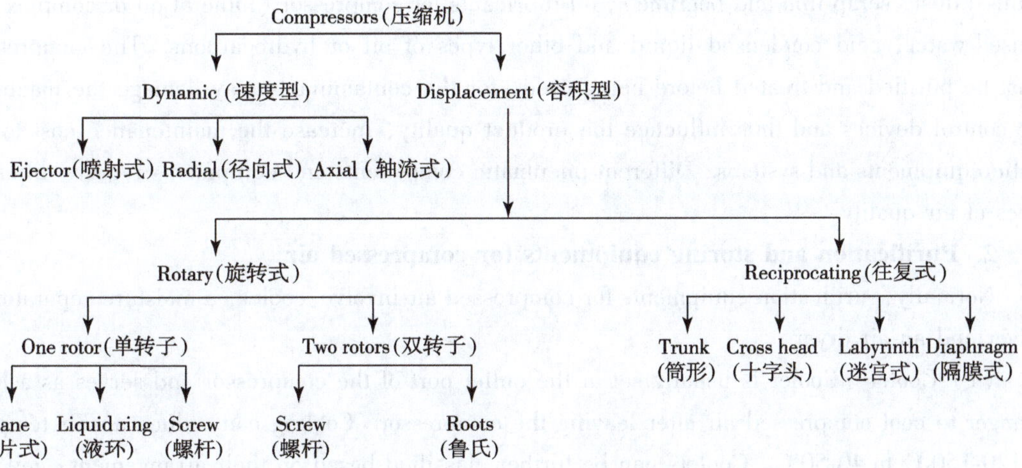

Fig. 10-2　Basic compressor types（空气压缩机基本类型）

10MPa）、高压空压机（10～100MPa）及超高压空压机（＞100MPa）。

4）按输出流量（排量）大小可分为：微型（＜$1m^3$/min）、小型（1～$10m^3$/min）、中型（10～$100m^3$/min）及大型（＞$100m^3$/min）。

2. 空压机的工作原理

以活塞式压缩机为例介绍空压机的工作原理。图 10-3 所示为活塞式空压机工作原理图，它通过曲柄连杆机构使活塞作往复运动而实现吸、排气，并达到提高气体压力的目的。其工作原理和第 3 章图 3-1 所示的单柱塞式液压泵工作原理相似，其中活塞 3 的往复运动由电动机带动曲柄滑块机构实现，即由曲柄 7 的旋转运动转换为滑块 5 和活塞 3 的往复运动。大多数空气压缩机是多缸和多活塞的组合。

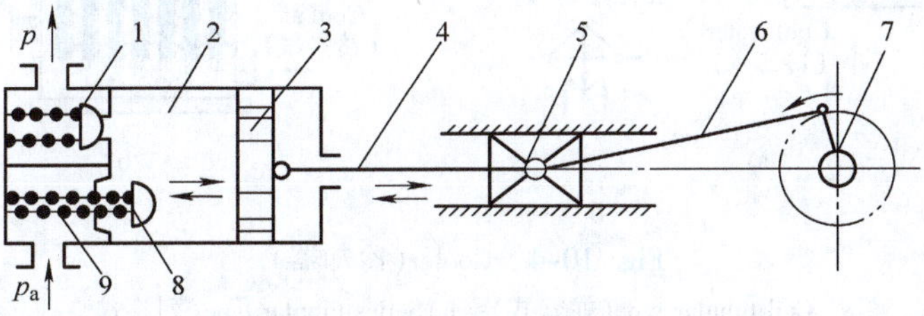

Fig. 10-3　Operation of piston type compressor（活塞式空压机的工作原理图）

1—Exhaust valve（排气阀）　2—Cylinder（气缸）　3—Piston（活塞）　4—Piston rod（活塞杆）
5—Sliding block（滑块）　6—Connecting rod（连杆）　7—Crank（曲柄）
8—Exhale valve（吸气阀）　9—Spring（弹簧）

活塞式空压机的缺点是振动大、噪声大，输出压力脉动大，需设置气罐。而叶片式和螺杆式空压机则噪声小，输出压力脉动小，无须设置气罐。这两种空压机的结构、工作原理和叶片泵、螺杆泵类似，这里不再赘述。

10.1.2　压缩空气净化和贮存的装置和设备

1. 气动系统对压缩空气质量的要求

气压传动所使用的压缩空气必须经过净化处理后才能使用。因为从空压机输出的压缩空

grains (dust, scrap iron and begrime), oil lubricants for compressor (mineral oil or complex), condensed water, acid condensed liquid and other types of oil or hydrocarbons. The compressed air must be purified and treated before being used, for the contaminants may damage the machine and the control devices and thus influence the product quality, increase the maintenance cost for pneumatic equipments and systems. Different pneumatic components and equipments require different degrees of air quality.

2. Purification and storing equipments for compressed air

Normally, purification equipments for compressed air involve: cooler, a moisture separator, a receiver and an air dryer.

(1) Cooler Cooler is usually set at the outlet port of the compressor and serves as a heat exchanger to cool compressed air after leaving the compressor. Cooling can reduce an air temperature of 120-150℃ to 40-50℃. Coolers can be further classified based on their arrangement: coil tubular type (Fig. 10-4a), series tubular type (Fig. 10-4b), thermosyphon system and tube drive type. There are two types of cooling: water cooling and air cooling.

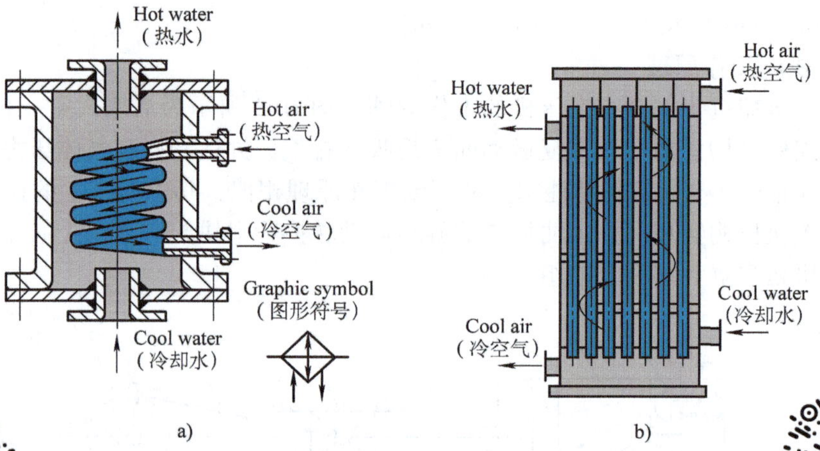

Fig. 10-4 Cooler(冷却器)
a) Coil tubular type(蛇管式) b) Series tubular type(列管式)

(2) Moisture separator It is used to separate the impurities such as condensed water and oil stain from the compressed air to gain an initial purified compressed air. When compressed air enters the separator, its flow rate and velocity will change sharply and the impurities with a higher density such as the water droplets and oil will be separated by inertia effect. By then the compressed air is initially purified. The configurations of separators incorporate: ring revolving type, strike and return type, centrifugal rotary type, water-bath type, etc. Fig. 10-5 shows a strike and return type moisture separator. When the compressed air enters the separator, it strikes against the baffle and turns sharply before falling down. When air falls down, it again strikes the case bottom and then rises up. The water droplets and oil are then separated from the compressed air by the centrifugal force produced by the sharp turning action.

(3) Receiver It is used to store a certain amount of compressed air, reduce its pressure fluctu-

气总会有不少污染物，例如灰尘、铁屑和积垢等固态颗粒，压缩机润滑油（矿物油或合成物）、冷凝水和酸性冷凝液以及其他油类和碳氢化合物等。如果不除去这些污染物，将导致机器和控制装置故障，损害产品质量，增加气动设备和系统的维护成本。不同的气动元件、设备，对空气质量的要求不同。

2. 压缩空气净化贮存设备

压缩空气净化设备一般包括冷却器、油水分离器、气罐、干燥器。

（1）冷却器　冷却器一般安装在空压机的出口管路上，将温度高达 120～150 ℃的空压机排出的气体冷却到 40～50℃。冷却器的结构形式有：蛇管式（图 10-4a）、列管式（图 10-4b）、散热片式、套管式等，其冷却方式有水冷式和风冷式两种。

（2）油水分离器　其作用是将经冷却器降温析出的冷凝水滴、油滴等杂质从压缩空气中分离出来，使它得到初步净化。其工作原理是：使压缩空气进入油水分离器后产生流量和速度的急剧变化，再依靠惯性作用，将密度比压缩空气大的水滴、油滴等杂质分离出来，使压缩空气得到初步净化。其结构形式有：环形回转式、撞击折回式、离心旋转式和水浴式等。图 10-5 所示为撞击折回式油水分离器。压缩空气进入油水分离器后撞击挡板急剧转折下降，又撞击容器底部而上升，依靠急剧转折时的离心力作用析出水滴和油滴。

（3）气罐　其作用是贮存一定数量的压缩空气，减小压力波动，保证供气的连续性和稳定性。有立式和卧式两种，为节省占地面积，多采用立式结构，如图 10-6 所示。气罐的容积 V 一般是以空压机每分钟的排气量 q 为依据选择的。当 $q<6\text{m}^3/\text{min}$ 时，取 $V=1.2\text{m}^3$；$q=6\sim30\text{m}^3/\text{min}$ 时，取 $V=1.2\sim4.5\text{m}^3$；当 $q>30\text{m}^3/\text{min}$ 时，取 $V=4.5\text{m}^3$。

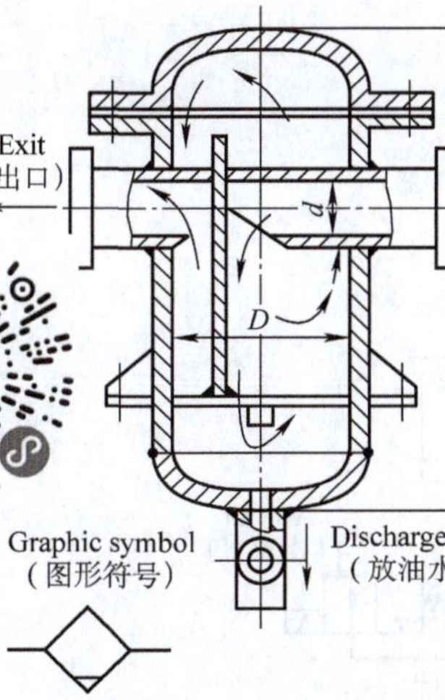

Fig. 10-5　Strike and return with ring revolving type moisture separator
（撞击折回式油水分离器）

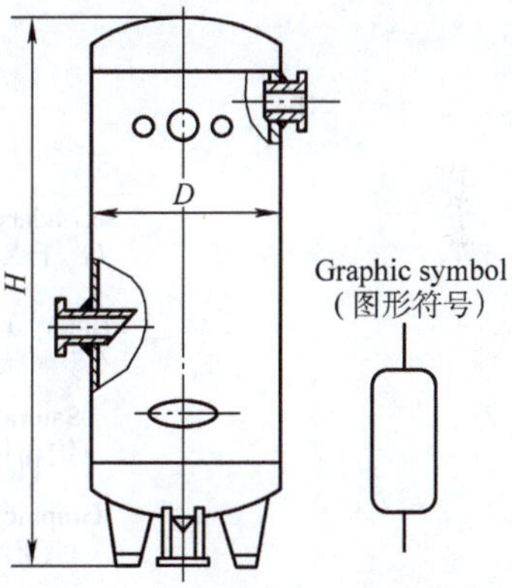

Fig. 10-6　Receiver（气罐）

目前，冷却器、分水排水器和气罐三者一体的结构形式已得到广泛应用。

（4）空气干燥器　经过冷却器、油水分离器和气罐净化处理的压缩空气已能满足一般气动系统的使用要求。但对一些精密机械、仪表等装置还不能满足要求，还需经过进一步的净化处理。干燥器的作用就是进一步除去压缩空气的水分、油分和微颗粒杂质。其主要方法有机械法、离心法、冷冻法和吸附法等。图 10-7 所示为一种不加热再生式干燥器，其中 1、2 为填满干燥剂的容器，工作时压缩空气从容器 1 的底部流到顶部，水分被干燥剂吸收而输

ation and provide gas continuously and stably. There are two types of receivers: horizontal and vertical. Vertical type receivers are more often adopted to save room, as shown in Fig. 10-6. The volume required for a selected receiver is determined by the delivery q. For $q < 6\text{m}^3/\text{min}$, $V = 1.2\ \text{m}^3$; for $q = 6\text{-}30\text{m}^3/\text{min}$, $V = 1.2\text{-}4.5\ \text{m}^3$; for $q > 30\text{m}^3/\text{min}$, $V = 4.5\ \text{m}^3$.

Nowadays, the arrangement with the combination of cooler, moisture separator and receiver has found many applications.

(4) Air dryer The compressed air after having been purified and treated by the cooler, moisture separator and receiver can meet the requirements of general pneumatic systems. But for other precise devices such as precise machines and meters, the compressed air must be further purified or treated. Air dryers have the function of further removing water vapour, oil mist and other particle impurities with mini diameter. Methods of air drying incorporate mechanical, centrifugal force, refrigeration, adsorption, etc. Fig. 10-7 shows a heatless regeneration dryer. Components 1 and 2 are containers filled with dessicant, which absorbs the water vapor when compressed air flows from the bottom of container 1 to its top. Part of the desiccant compressed air flows from the top of container 2 to its bottom and exhausts out via solenoid valves 3 and 4 to the atmosphere. This action also removes the water vapour in the dessicant. By controlling solenoid valve 3, the two containers operates in turn and alternately to accomplish a continuous dry air output. To prevent dessicant powder from sneaking into the dried compressed air, a high-grade filter can be set at the outlet port. Two-way, two-position valve 4 is used to pre-fill pressure and prevent flow fluctuation when the regeneration container turns to work by absorption.

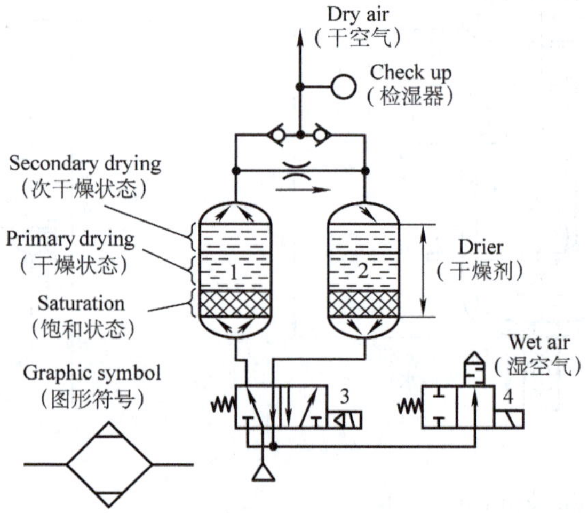

Fig. 10-7 Heatless regeneration dryer (不加热再生式干燥器)

10. 1. 3 F. R. L. combination

F. R. L. combination is short for the combination of an air filter, a pressure reducing valve and an oil mist lubricator. Their installation order is shown in Fig 10-8, i. e., first comes the air filter (secondary filter), followed by the pressure reducing valve and the oil mist lubricator is set behind (seen from the air inlet direction). F. R. L. combination supplies better quality compressed air for

出干燥的压缩空气。其中的一部分干燥后的压缩空气从容器 2 的顶部流到底部，把干燥剂中的水分带走，经电磁阀 3 和 4 排向大气。控制电磁阀 3 可实现两容器轮流干燥和再生交替工作，可得到连续输出的干燥压缩空气。为防止干燥剂的粉末混入干燥后的压缩空气，在其输出端应安装精过滤器。电磁阀 4 的作用是使再生容器在转为吸附干燥前预先充压，防止交替时输出流量的波动。

10.1.3　气源处理装置

空气过滤器、减压阀、油雾器一起称为气源处理装置，其安装次序根据进气方向依次为空气过滤器（二次过滤器）、减压阀和油雾器，如图 10-8 所示。压缩空气经过气源处理装置的最后处理，将进入各气动元件及气动系统。因此，气源处理装置是气动元件及气动系统使用压缩空气质量的最后保证。

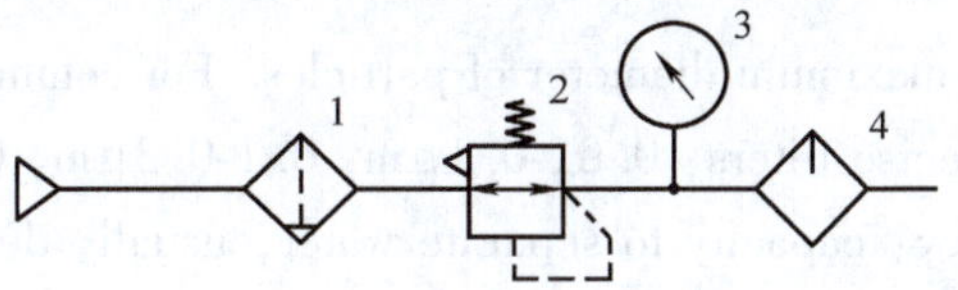

Fig. 10-8　F. R. L.　combination（气源处理装置）

1—Air filter（空气过滤器）　2—Pressure reducing valve（减压阀）

3—Pressure gauge（压力表）　4—Oil mist lubricator（油雾器）

1. 空气过滤器

（1）工作原理　空气过滤器的作用是滤去压缩空气中的灰尘、杂质、油污并将水分分离出来。如图 10-9 所示，当压缩空气从输入口进入后被引入旋风叶片 1，旋风叶片上有许多成一定角度的缺口，使气流沿切线方向强烈旋转，于是其中较大的灰尘、杂质、油污和水滴等因质量较大受离心力作用被高速甩到存水杯内壁产生碰撞而从空气中分离出来，并流到杯底；而微粒灰尘和雾状水汽则由滤芯 2 滤除后从输出口输出。挡水板 4 是为了防止杯中污水卷起而破坏空气过滤器的过滤作用。污水由排水阀 5 放掉。

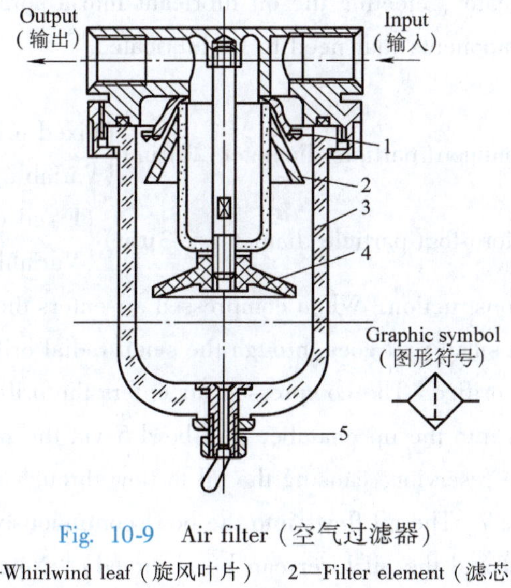

Fig. 10-9　Air filter（空气过滤器）

1—Whirlwind leaf（旋风叶片）　2—Filter element（滤芯）

3—Bowl（水杯）　4—Baffle（挡水板）　5—Drain valve（排水阀）

pneumatic components or systems.

1. Air filter

(1) Operation An air filter is used to remove dust, impurities, oil stain and separate water vapour from the compressed air. Fig. 10-9 shows its operation, as compressed air passes the inlet port and enters the filter section, it passes through the directional louvers forcing it into a whirling flow pattern. Liquid particles and heavy solid are thrown against the inside wall of the bowl 3 by centrifugal force. The liquid then runs to the bottom of the bowl where it is either removed by the automatic drain assembly or by the manual drain. The baffle 4 maintains a "quite zone" in the lower portion of the bowl to prevent air turbulence from picking up the liquid and returning it to the air stream. The air then passes through the filter element 2 to remove solid contaminants. The automatic drain assembly dumps the liquid as it collects.

(2) Performance

1) Filter rating (μm): maximum diameter of particles. For common filters: 5-10μm, 10-20μm, 25-40μm, 50-75μm; for precise filters: 0.01-0.1μm, 0.1-0.3μm, 0.3-3μm, 3-5μm.

2) Water trap efficiency: capacity to separate water, usually denoted by symbol η, $\eta \geqslant 80\%$. Then

$$\eta = \frac{\varphi_1 - \varphi_2}{\varphi_1} \tag{10-1}$$

Where φ_1 and φ_2 are relative humidities before and after the air filter respectively.

3) Dust removal efficiency: the ratio of separated dust weight to the total dust weight in the air filter. For secondary filters, the dust removal efficiency surpasses 80%; for higher efficiency filters (for special use), the dust removal efficiency reaches up to 99%.

4) Flow rate characteristics: the difference of input and output pressures must be less than 5% at a rating flow rate.

2. Oil mist lubricator

(1) Operation and construction An oil mist lubricator is one type of oiling devices. The compressed air enters the lubricator, ejecting the oil lubricant into a spraying fog which flows with the compressed air into the components that need to be lubricated.

Classification:

Oilmist lubricators
- Common (particle diameter: 20μm)
  - Fixed orifice
  - Variable orifice
- Micro-fog (particle diameter: 2-3μm)
  - Fixed orifice
  - Variable orifice

Fig. 10-10 shows its construction. When compressed air enters the inlet port, most of it puts out from the outlet port, only a small part goes through the small radial orifice *a* on gas-guide pulverizing tube 2 and then the center orifice. The compressed air enters the orifice to prop up the steel ball of the check valve and passes into the up chamber *d* of bowl 5 via the radial orificec on valve seat 4. The air then pressurizes the reservior, causing the oil to flow through siphon tube 6 and prop up the steel ball of the check valve 7. The oil flows into the bowl continuously through the orifice of throttle valve 8 (lid 9 is transparent and the oil lever can be observed). Note that there is a square orifice

（2）性能指标

1）过滤度。是指能允许通过的杂质颗粒的最大直径。常用的规格有：5～10μm、10～20μm、25～40μm 及 50～75μm 四种；精过滤的规格有：0.01～0.1μm、0.1～0.3μm、0.3～3μm 及 3～5μm 四种。

2）水分离率。是指分离水分的能力，用符号 η 表示，则有

$$\eta = \frac{\varphi_1 - \varphi_2}{\varphi_1} \tag{10-1}$$

式中，φ_1 为空气过滤器前空气的相对湿度；φ_2 为空气过滤器后空气的相对湿度。

一般要求空气过滤器的水分离率大于80%。

3）滤灰效率。指空气过滤器分离灰尘的质量和进入空气过滤器的灰尘质量之比。对二次过滤器其滤灰效率为80%以上；在某些特殊要求的场合，可采用高效过滤器，其滤灰效率可达99%。

4）流量特性。额定流量下，输入压力与输出压力之差不超过输入压力的5%。

2. 油雾器

（1）工作原理及结构　油雾器是一种注油装置。当压缩空气流过时，它将润滑油喷射成雾状，随压缩空气一起流进需要润滑的部件，达到润滑的目的。

油雾器的分类如下：

- 油雾器
  - 普通型油雾器（油雾粒径为20μm）
    - 固定节流式
    - 变节流式
  - 微雾型油雾器（油雾粒径为2～3μm）
    - 固定节流式
    - 变节流式

图 10-10 所示为普通油雾器的结构图。压缩空气由输入口进入后，大部分从出口输出，其中一小部分通过导气雾化管 2 上的径向小孔 a 的中心孔后，顶开单向阀钢球，经阀座 4 上

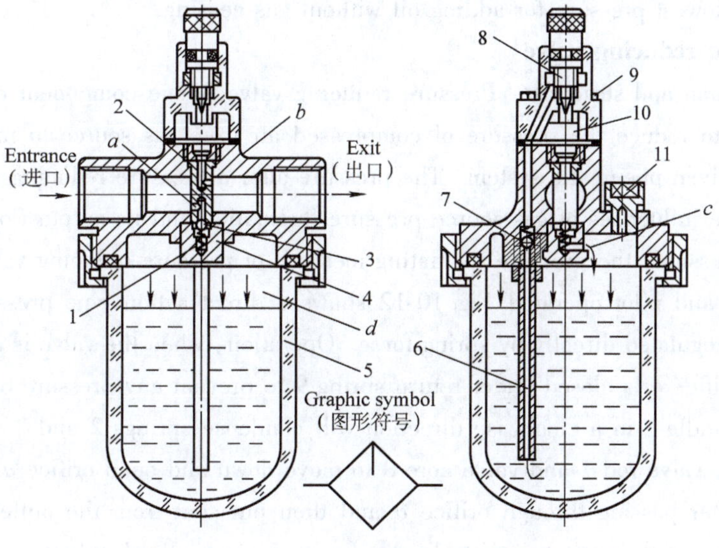

Fig. 10-10　Oil mist lubricator（油雾器）

1—Spring（弹簧）　2—Gas-guide pulverizing tube（导气雾化管）　3—Steel ball（钢球）　4—Valve seat（阀座）　5—Bowl（贮油杯）　6—Syphon tube（吸油管）　7—Check valve（单向阀）　8—Throttle valve（节流阀）　9—Lid（盖）　10—Seal gasket（密封垫）　11—Oil plug（油塞）

(whose side length is smaller than the diameter of steel ball of check valve 7) in pipeline upside the steel ball to prevent it from closing its upper pipeline. The oil passing into the bowl then goes through the center orifice and radial orifice *b* on gas-guide pulverizing tube 2 and is jeted out by the gas flow in the main pipeline via small orifice *b* (pressure difference principle). The oil then converts into constant density fog and continues downstream to the point of application. The drip rate can be regulated by adjusting throttle valve 8.

Oil can be added without air ceasing for this type of oil mist lubricator. It is the special check valve (key component, composed by valve seat 4, steel ball 3 and spring 1) that makes it possible. Fig. 10-11 shows the three states of the oil mist lubricator: at rest, at work, adding oil. When the lubricator is at the state of adding oil, oil plug 11 is loosened off to reduce the pressure in chamber *d* as it connects with atmosphere. Steel ball 3 is pressed on valve seat 4 by compressed air (Fig. 10-11c), cutting off the compressed air channel to chamber *d*. In addition, the compressed air won't flow back into the bowl for siphon tube 6 is closed by steal ball 7.

Fig. 10-11 is a one-stage oil mist lubricator, or common lubricator. Two-stage oil mist lubricator performs two injections giving a fine fog (with a particle diameter of 5μm); with this system a central lubricant can feed several stations.

(2) Main Performance and indices

1) Flow characteristics: describe how input and output pressure drop change as air flow under given output pressure.

2) Fogging oil flow rate: when the oil level is at normal state in the bowl and the throttle valve is fully open, the inlet pressure must meet the specification. Minimum fogging oil flow rate is considered 40% that of the rated flow rate.

Other performances of oil mist lubricator incorporate: oil drip regulation, fog size, fluctuation characteristics, lowest pressure for adding oil without gas ceasing.

3. Pressure reducing valve

(1) Operation and structure Pressure reducing valve is one component of F. R. L. combination. It is used to reduce the pressure of compressed air from gas source to meet the pressure requirement of a given pneumatic system. The pressure after a pressure reducing valve is kept stable, avoiding both the influences of gas-source pressure fluctuation and its output flow rate change. Like hydraulic transmission, the pressure regulating methods of pressure reducing valves incorporate: direct action type and pilot operated. Fig. 10-12 shows a direct action type pressure reducing valve. The pressure is regulated directly by spring force. Operation: when the valve is at the position shown in the figure, orifice *a* is closed under return spring 9 to prevent any pressure output from the outlet port. Turning handle 1 in a clockwise direction will condense springs 2 and 3, pushing spring seat 4, diaphragm 5, valve rod 7 and valve core 8 to move down and open orifice *a*. The pressure of the air falls down after passing through orifice *a* and then puts out from the outlet port. At the same time, a small part of the output air acts on diaphragm 5 via feedback tube 6 and produces a thrust, which then balances out with the spring force. In this way, the pressure reducing valve puts out air with stable pressure. When the output pressure surpasses the setting, the diaphragm leaves the balancing position and convexes up. Valve core 8 moves up under return spring 9, causing orifice *a* to

的径向孔 c 进入贮油杯5的上腔 d，油面受压，使油经吸油管6将单向阀7的钢球顶起。钢球上部管口为一个边长小于钢球直径的四方孔，所以钢球不可能将上部管口封死，油能不断经节流阀8的节流口滴入杯内（盖9为透明的，可观察滴油情况），经导气雾化管2上部的中心孔和径向小孔 b，被主管道中的气流从小孔 b 引射出来（压差原理），雾化后从输出口输出而进入需要润滑的部位。调节节流阀8可以调节滴油量。

油雾器可以在不停气状态下加油。实现不停气加油的关键部件是由阀座4、钢球3和弹簧1组成的特殊单向阀，如图10-10所示，其工作情况如图10-11所示。其中图10-11a所示为油雾器不工作时的状态，图10-11b所示为工作时的状态，图10-11c所示为加油时的状态。加油时松开油塞11使 d 腔与大气相通而压力下降，于是钢球3被压缩空气压在阀座4上（图10-11c），切断了压缩空气进入 d 腔的通道；又因单向阀7钢球的作用，封住了吸油管，压缩空气也不会从吸油管6倒流入贮油杯中，所以可在不停气状态下从油塞口中加油。

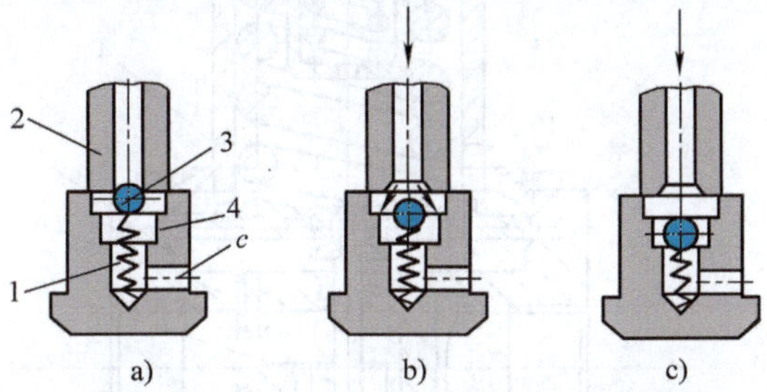

Fig. 10-11　Three operation states of the special check valve（特殊单向阀的三种工作状态）

Note：The components of 1，2，3，4 are the same as that in Fig. 10-10

（1、2、3、4所表示的零件与图10-10相同）

a）At rest（不工作时）　b）At working（工作时）　c）Adding oil（加油时）

图10-10所示的油雾器为一次油雾器，也称普通油雾器。二次油雾器能使油滴在油雾器内进行两次雾化，使油雾粒度更小，更均匀，输送距离更远。二次油雾器粒径可达5μm。

（2）主要性能和指标

1）流量特性。指油雾器在给定进口压力下，其通过流量变化时，进、出口压降的变化情况。

2）起雾流量。贮油杯中油位处于正常工作油位，节流阀全开，油雾器进口压力为规定值，起雾时的最小空气流量规定为额定流量的40%。

油雾器的其他性能指标还有滴油量调节、油雾粒度、脉冲特性、最低不停气加油压力等。

3. 减压阀

（1）工作原理及结构　气源处理装置所用的减压阀，其作用是把气源输出的压缩空气的压力减小到气压系统所要求的压力，并保持其压力稳定，不受气源压力波动和减压阀输出流量变化的影响。和液压传动一样，减压阀调压方式有直动式和先导式两种。图10-12所示减压阀为直动式，它是依靠改变弹簧力来直接调整压力的。其工作原理是：当阀处于图示位置，阀芯8在复位弹簧9的作用下，阀口 a 处于关闭状态，输出口无气压输出。调压时，若顺时针方向调节手柄1，调压弹簧2、3被压缩，推动弹簧座4、膜片5和阀杆7，阀芯8下移，使阀口 a 打开，气流通过阀口 a 后压力降低，从输出口输出。与此同时，有一小部分输

reduce and the output pressure falls down to the setting. However, when the output pressure falls down, the diaphragm concaves down. Orifice a enlarges and the output pressure returns to the setting. So the output pressure can be kept stable. The output pressure can be adjusted simply by controlling the pressure reducing valve orifice, which can be accomplished by adjusting handle 1.

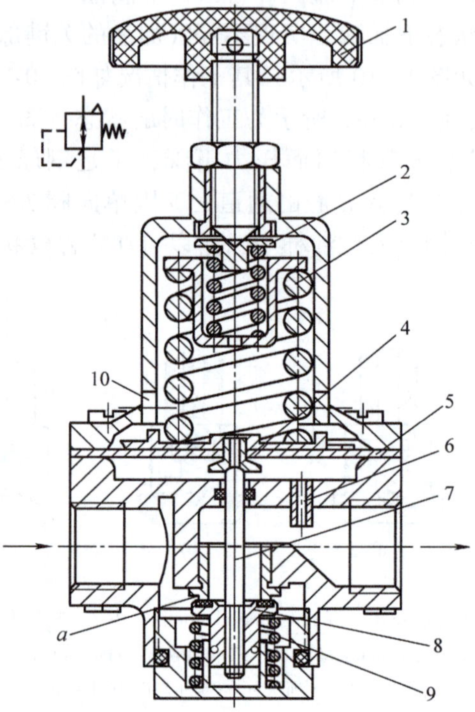

Fig. 10-12 Pressure reducing valve (减压阀)

1—Handle (手柄) 2,3—Regulating spring (调压弹簧) 4—Spring seat (弹簧座) 5—Diaphragm(膜片) 6—Feedback tube (反馈管) 7—Valve rod(阀杆) 8—Valve core(阀芯) 9—Return spring(复位弹簧)

(2) Main performance

1) Pressure characteristics: describes the fluctuation of output pressure p_2 with the input pressure p_1 under a given output flow rate q. Fig. 10-13a shows its characteristic curves. When p_2 is lower than p_1 with a certain level, p_2 won't change with p_1, as indicated in the figure.

2) Flow rate characteristics: describes the fluctuation of the output pressure p_2 with output flow rate q under a given input pressure p_1. Fig. 10-13b shows its characteristic curves. The lower the p_2, the slighter influence it suffers from the flow rate; the fluctuation of p_2 is larger when the output flow rate is lower, as can be seen from Fig 10-13b.

10.2 Pneumatic Attachments

10.2.1 Mufflers

Unlike hydraulic circuits, there are no return pipelines in pneumatic circuits. The compressed air exhausts to the atmousphere directly after having been used. The air will sharply expand due to

出气流经反馈管6作用在膜片5上的推力与弹簧力相平衡，减压阀便有稳定的压力输出。其过程是：若输出压力超过调定值时，膜片离开平衡位置向上变形，阀芯8在复位弹簧9的作用下而上移，阀口a关小，减压作用增强，使输出压力降低到调定值；反之，若输出压力下降，膜片向下变形，阀口a开大，减压作用减弱，使输出压力回升到调定值，保证输出压力稳定。手柄1用以控制减压阀阀口的大小，即可调节一定的输出压力值。

（2）主要性能

1）压力特性。压力特性是指输出流量q一定时，输入压力p_1的波动引起输出压力p_2波动的特性，可用特性曲线表示，如图10-13a所示。由图可看出，输出压力p_2必须低于输入压力p_1一定值后才基本上不随输入压力的变化而变化。

2）流量特性。流量特性是指输入压力p_1一定时，输出流量q的变化引起输出压力p_2变化的特性，可用特性曲线表示，如图10-13b所示。由图可看出，输出压力p_2越低，受流量的影响越小，但在减压阀输出流量较小时，输出压力p_2波动较大。

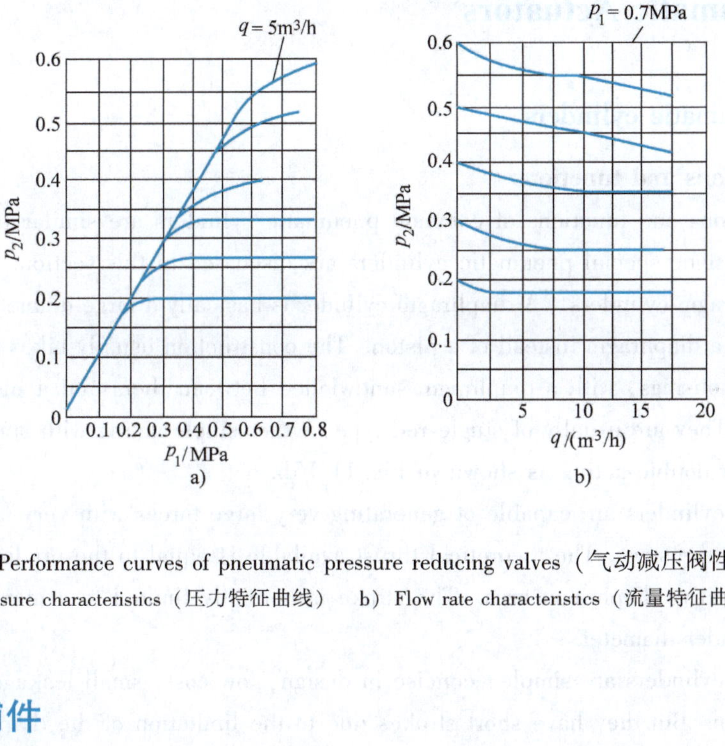

Fig. 10-13 Performance curves of pneumatic pressure reducing valves（气动减压阀性能曲线）

a）Pressure characteristics（压力特征曲线） b）Flow rate characteristics（流量特征曲线）

10.2 气动辅件

10.2.1 消声器

气动回路与液压回路不同，它没有回气管道，压缩空气使用后直接排入大气，较高的压差使气体体积急剧膨胀，产生强烈的排气噪声。为了降低噪声，一般在排气口装设消声器。消声器就是通过阻尼或增加排气面积来降低排气的速度和功率，从而降低噪声的。

气动元件上使用的消声器的类型一般有三种：吸收型消声器、膨胀干涉型消声器、膨胀干涉吸收型消声器。图10-14所示吸收型消声器是目前使用最广泛的一种，其消声套采用了由铜或聚苯乙烯颗粒烧结而成的多孔的吸声材料。

high pressure difference and produce serious exhausting noise. Usually a muffler is set at the discharge port to reduce the velocity and power of the jet noise by means of damp or increasing exhausting area.

Three types of muffler are available: absorptive mufflers, expansive mufflers, expansive and absorptive mufflers. Fig. 10-14 shows an absorptive muffler structure. The muffler body is made of multi-hole absorptive material, sintered by copper or polystyrene bean. Today absorptive mufflers are also most wildly used.

10. 2. 2 Pipeline Connectors

Pipeline connectors incorporate tubes and pipe fittings. There are rigid tubes and hose tubes. Pipe couplings include bite type tube fittings, flared type tube fittings, corpling of borsing and quick release fittings.

10. 3 Pneumatic Actuators

10. 3. 1 Pneumatic cylinders

1. Operations and functions

The operations and functions of common pneumatic cylinders are similar to those of hydraulic cylinders. Only some special pneumatic cylinders are discussed in this section.

(1) Diaphragm cylinders A diaphragm cylinder is basically a large diameter, short stroke cylinder fitted with a diaphragm instead of a piston. The construction usually takes the form of a pair of shallow convex housings, with a diaphragm sandwiched between them and a piston rod attached to the diaphragm. They are usually of single-rod type, either single-acting with spring return as shown in Fig. 10-15a or double-acting as shown in Fig. 11-15b.

Diaphragm cylinders are capable of generating very large forces with very short strokes and are commonly called thrusters. The theoretical thrust available is equal to the product of the applied air pressure and effective diaphragm area. The stroke is normally limited to a maximum of about one-third of the cylinder diameter.

Diaphragm cylinders are simple, concise in design, low cost, small leakage, easy to manufacture and maintain. But they have short strokes due to the limitation of the diaphragm deformation. Diaphragm cylinders are mainly used in pneumatic clamping devices and the working occasions with short stroke.

(2) Air-hydraulic cylinders When outside load is large enough, there may occur poor stability to the movement of a unique pneumatic cylinder. This can be solved by an air-hydraulic cylinder. Hydro-pneumatic working is achieved by mechanically coupling a pneumatic cylinder and a low-pressure hydraulic cylinder so that they have a common movement normally initiated by compressed air applied to the air cylinder. Liquids have slight compressibility and air flow is easy to control, which leads to a stable movement and an adjustable speed.

The normal configuration is back-to-back mounting with a common piston rod, see Fig. 10-16.

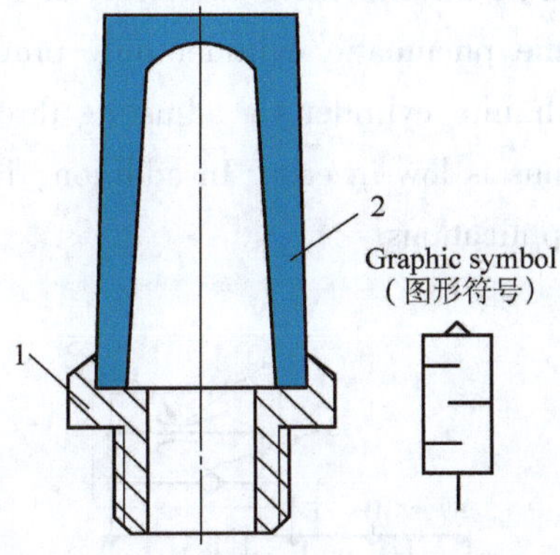

Fig. 10-14 Absorptive muffler（吸收型消声器）

1—Connecting screw（联接螺纹） 2—Muffler body（消声罩）

10.2.2 管道连接件

管道连接件包括管子和各种管接头。管子可分为硬管及软管两种。气动系统中使用的管接头分为卡套式、扩口螺纹式、卡箍式、插入快换式等。

10.3 气动执行元件

10.3.1 气缸

1. 气缸的工作原理及功用

普通气缸的工作原理及功用与液压缸类似，此处不再赘述，下面仅讨论特殊气缸。

（1）薄膜式气缸　薄膜式气缸是以膜片取代活塞带动活塞杆运动的。其结构常为盘状。也有单作用与双作用式之分。单作用气缸带有复位弹簧，如图 10-15a 所示；双作用气缸如图 10-15b 所示。依靠膜片在气压作用下的变形来使活塞杆运动。

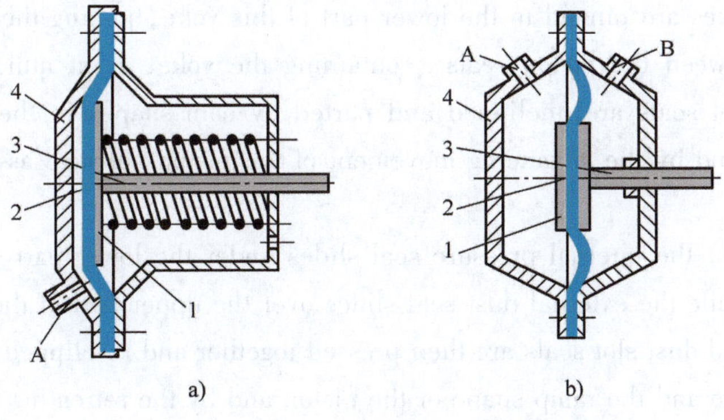

Fig. 10-15 Diaphragm cylinder（薄膜式气缸）

a）Single-acting（单作用） b）Double-acting（双作用）

1—Diaphragm plate（膜盘） 2—Diaphragm（膜片） 3—Piston rod（活塞杆） 4—Cylinder body（缸体）

A，B—Inlet/Outlet port（进/出气口）

The piston speed can be adjusted by throttle valve 4. Reservior 5 and check valve 3 are used for oil compensation. In this circuit, the pneumatic cylinder only provides driving force and the stable movement depends upon the hydraulic cylinder by adjusting throttle valve 4. This circuit does not need any hydraulic source and thus is low in cost. In addition, it makes full use of pneumatic and hydraulic and thus finds many applications.

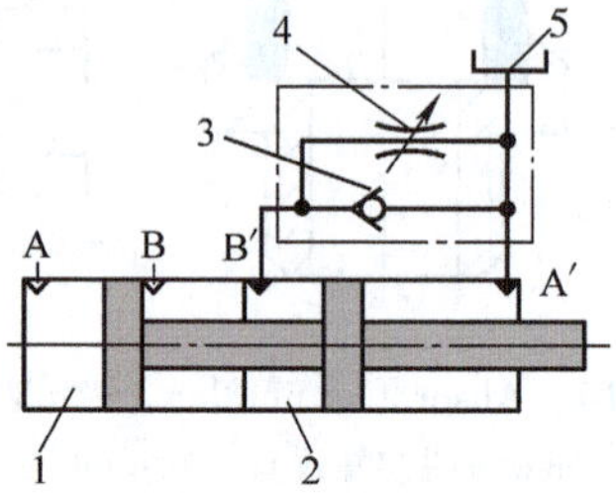

Fig. 10-16 Air-hydraulic cylinders(气-液阻尼缸)

1—Pneumatic cylinder(气缸) 2—Hydraulic cylinder(液压缸)

3—Check valve(单向阀) 4—Throttle valve(节流阀) 5—Reservior(油箱)

(3) Rodless cylinders Recently-developed rodless cylinders provide an ideal solution for long stroke applications. In one rodless design the cylinder barrel has a slit along the barrel and the piston in the barrel is connected directly through the slit to a mounting on the outside of the cylinder.

Thus when the piston is moved by applying air on either side, the load connected to the external mounting is moved. The problem of providing a continuous and moving seal between the piston and the external mounting has now been achieved with a high degree of reliability. Not only can this design handle long strokes effectively but can also accept lateral loading at the mounting point.

An illustration of the construction of one type of rodless cylinder is shown in Fig. 10-17. The drive from the piston to the carriage is via a substantial drive tongue which passes through the barrel slot. This drive tongue forms the center part of a yoke which is incorporated in the carriage extrusion. The piston halves are pinned to the lower part of this yoke, joining the piston to the carriage.

The volume between the piston seals, containing the yoke, is at atmospheric pressure. The pressure and dust slot seals are unclipped and parted by cam shapes in the yoke within this non-pressurized section and by the advancing movement of the piston carriage assembly along the cylinder.

After unclipping, the internal pressure seal slides under the lower part of the yoke adjacent to the piston halves, while the external dust seal slides over the upper part of the yoke within the carriage. The pressure and dust slot seals are then pressed together and re-clipped by the spring roller assembly in the carriage and the ramp shape of the piston and by the retreating movement of the piston carriage assembly along the cylinder body. Thus the drive is taken up from the piston to the carriage through the slot as this assembly traverses the cylinder.

(4) Impact cylinders An impact cylinder is one arranged so that the speed developed is high enough for impact work, such as forging, piercing, etc. It converts the pressure energy of the com-

薄膜式气缸能在很小的行程内产生很大的力，又称推进器。其最大行程约为气缸直径的1/3，理论推力为空气压力与薄膜有效面积的乘积。这种气缸结构简单、紧凑，制造容易、成本低、泄漏小，维修方便。但因膜片变形量有限，行程小，仅适用于气动夹具及行程短的工作场合。

（2）气-液阻尼缸　气缸因气体有较大的可压缩性，在负载较大时，运动平稳性较差，若要提高其运动平稳性，可采用气-液阻尼缸。气-液阻尼缸是由气缸和液压缸组合而成的，它以压缩空气为动力源，利用油液的可压缩性小和流量容易控制的特点，可达到获得运动平稳和速度可调的目的。

图 10-16 所示为串联式气-液阻尼缸。气缸活塞的右行速度可由节流阀 4 来调节，油箱 5 和单向阀 3 起补油作用。在这里，气缸只作驱动力，靠液压缸的阻尼调节作用获得平稳的运动。它不需要液压源，经济性好，同时具有气压和液压的优点，因而获得了广泛的应用。

（3）无杆气缸　无杆气缸是最近几年才出现的，它适用于行程较长的场合。

图 10-17 所示为无杆气缸。在气缸筒内沿轴向方向开有一条槽，为防止内部压缩空气泄漏和外界杂物侵入，槽由内、外防尘密封件 4 和 1 密封且两密封件互相夹持（图 10-17b），其动密封性能良好。无杆活塞 5 通过销与缸筒内的传动舌片 3 下部相嵌接，传动舌片又与导架相连。活塞两端分别进、排气时，活塞将在缸筒内往复运动。该运动通过传动舌片带动与负载相连的导架一起移动。此时，传动舌片将防尘密封件 1 和 4 组成的密封带挤开，但它们在缸筒的两端仍然是互相夹持的，因此传动舌片与活塞导架组件在气缸上移动时无空气泄漏。值得一提的是，无杆气缸不仅行程长，其导架还能承受轴向载荷。

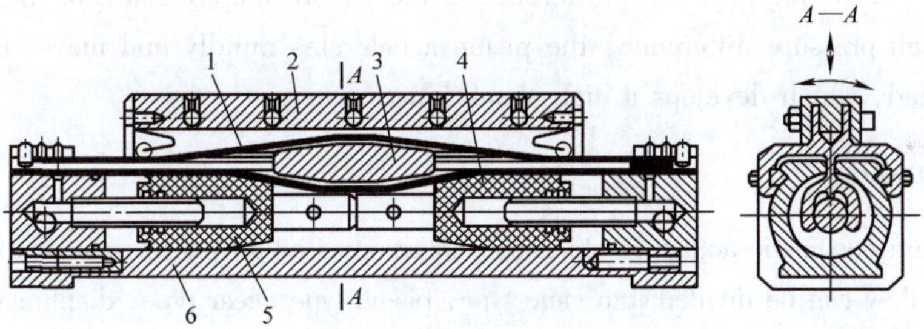

a)

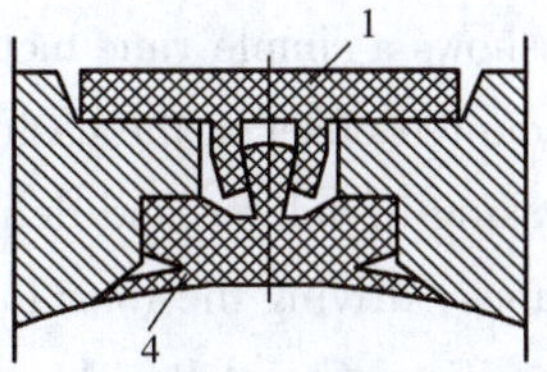

b)

Fig. 10-17 Rodless cylinder（无杆气缸）

a）Rodless cylinder construction（无杆气缸结构图） b）Cylinder tube grooved seal arrangement（缸筒槽密封布置）

1，4—Dust slot seals（防尘密封带） 2—Carriage（导架） 3—Drive tongue（传动舌片）

5—Rodless piston（无杆活塞） 6—Cylinder barrel（缸筒）

（4）冲击气缸　冲击气缸是将压缩空气的压力能瞬间转化为活塞高速运动能量的一种气缸，活塞速度可达每秒十几米，以适应冲击性工作场合，如锻造、冲孔、下料、铆接和破碎等多种作业。

pressed air into piston's high speed (more than ten meters per second) motion energy in a very short time.

Fig. 10-18 shows an impact cylinder design. Compared with common cylinders, there is an extra middle lid 3 (fitted with the cylinder body) with nozzle *D* and exhausting small orifice *E*. The cylinder is divided into three chambers by this middle lid and piston 2. For the sake of analysis, the working process can be simply divided into three steps.

1) Reset. As is shown in Fig. 10-18a, the piston rod chamber A is charged with air and the air in energy storing chamber *C* exhausts out. The piston moves up until the seal gasket on the piston seals nozzle *D* on middle lid 3. Piston chamber *B* connects with the atmospheric via exhausting port E. The pressure in chamber *A* builds up to air source pressure, while the pressure in chamber *C* falls down to atmospheric pressure.

2) Energy storage. As is shown in Fig. 10-18b, when air enters chamber *C*, it passes through the nozzle and acts on the piston. The area of the nozzle is so small that the air pressure cannot overcome the resultant of the up thrust (produced by air exhausting in the piston rod chamber) and the friction against the cylinder. The nozzle is still closed and the pressure in chamber *C* builds up gradually.

3) Impacting. As is shown in Fig. 10-18c, when the ratio of the pressure in chamber *C* to that in the piston rod chamber surpasses the ratio of the acting area in the piston rod chamber to the nozzle area, the piston moves down to open the nozzle. Immediately this happens, the compressed air accumulated in chamber *C* acts on the whole of the piston via the nozzle, causing the piston to move down rapidly due to the high pressure difference of the piston rod. Because of the large thrust produced by this high pressure difference, the piston accelerates rapidly and moves down with an extremely high speed, which develops a high shock kinetic energy.

10.3.2 Air motors

Air motors are pneumatic actuators which convert compressed air energy into rotating motion. Based on arrangement, they can be divided into vane type, piston type, gear type, diaphragm type, etc.

1. Operation and characteristics of air motors

(1) Vane motors Fig. 10-19a shows a simple vane motor operation. The rotor and the vanes are the only moving parts. The rotor is mounted offset in the casing, as can be seen. Air is admitted when the crescent shaped chamber is increasing in volume and delivered when decreasing. Torque is developed by pressure difference on the vanes, driving the rotor to turn in a counter clockwise direction for this motor. The air exhausts through orifices C and B. This direction of the rotor can be changed by merely changing the inlet port, i. e., air is admitted from orifice B for this motor (Fig. 10-19a). Fig. 10-20 shows the soft characteristics of a vane motor under a certain working pressure.

Small size motors are used for hand tools such as drills, grinders and screwdrivers. Larger sizes are used for winches, pump drivers, drive motors on pneumatic drill rigs and general industrial applications.

(2) Radial piston motors Piston motors operate at much lower speeds than vane motors because of the greater inertia of the reciprocating parts.

The typical radial motor has all connecting rods mounted on a common crank (crank and con-

冲击气缸的工作原理如图 10-18 所示，它是在普通气缸中间增加一个带有喷嘴口 D 和排气小孔 E 的中盖 3，中盖与缸体固接在一起，中盖和活塞把气缸分成三个腔室，即：活塞杆腔 A、无杆腔 B 和蓄能腔 C。其工作过程可简单地分为三个阶段：

1）复位段　如图 10-18a 所示，活塞杆腔 A 进气时，蓄能腔 C 排气，活塞 2 上移，直至活塞上的密封垫封住中盖 3 上的喷嘴口 D。活塞腔 B 经泄气口 E 与大气相通，使活塞杆腔压力升至气源压力，蓄能腔压力减至大气压力。

2）贮能段　如图 10-18b 所示，压缩空气进入蓄能腔 C，其压力只能通过喷嘴口的小面积作用在活塞上，不能克服活塞杆腔的排气压力所产生的向上推力及活塞与缸体间的摩擦力，喷嘴仍处于关闭状态，蓄能腔的压力将逐渐升高。

3）冲击段　如图 10-18c 所示，当蓄能腔的压力与活塞杆腔压力的比值大于活塞杆腔作用面积与喷嘴面积之比时，活塞下移，使喷嘴口开启，聚集在蓄能腔中的压缩空气通过喷嘴口突然作用于活塞的全面积上。此时，活塞一侧的压力可达活塞杆一侧压力的几倍乃至几十倍，使活塞上作用着很大的向下推力。活塞在此推力作用下迅速加速，在很短的时间内以极高的速度向下冲击，从而获得很大的动能。

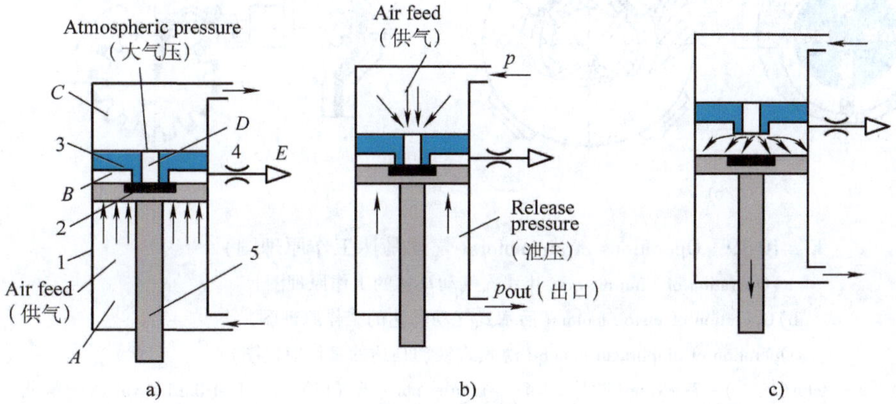

Fig. 10-18　Operation of impact cylinder（冲击气缸工作原理）

1—Cylinder tube（缸筒）　2—Piston（活塞）　3—Middle cover（中盖）

4—Control valve（控制阀）　5—Piston rod（活塞杆）

A—Piston rod chamber（活塞杆腔）　B—Non-rod chamber（无杆腔）　C—Energy storing chamber（蓄能腔）

D—Nozzle（喷嘴口）　E—Exhausting port（泄气口）

10.3.2　气动马达

气动马达是利用压缩空气的能量实现旋转运动的气动执行元件。按结构形式可分为叶片式、活塞式、齿轮式和膜片式等。

1. 气动马达工作原理及特性

（1）叶片式气动马达　如图 10-19a 所示。压缩空气由 A 口输入后分为两路，一路经定子两端的密封盖上的槽进入叶片底部（图中未示出），将叶片推出顶在定子内壁；另一路进入由叶片、定子、转子及两端密封盖构成的月牙形密闭空间。由于定子与转子偏心布置，两叶片伸出长度不同，使气压力的作用面积不等而产生转矩差，这个转矩差通过叶片带动转子逆时针方向旋转做功后的空气经 C 口和 B 口排出。若由 B 口输入压力气体则转子作顺时针方向

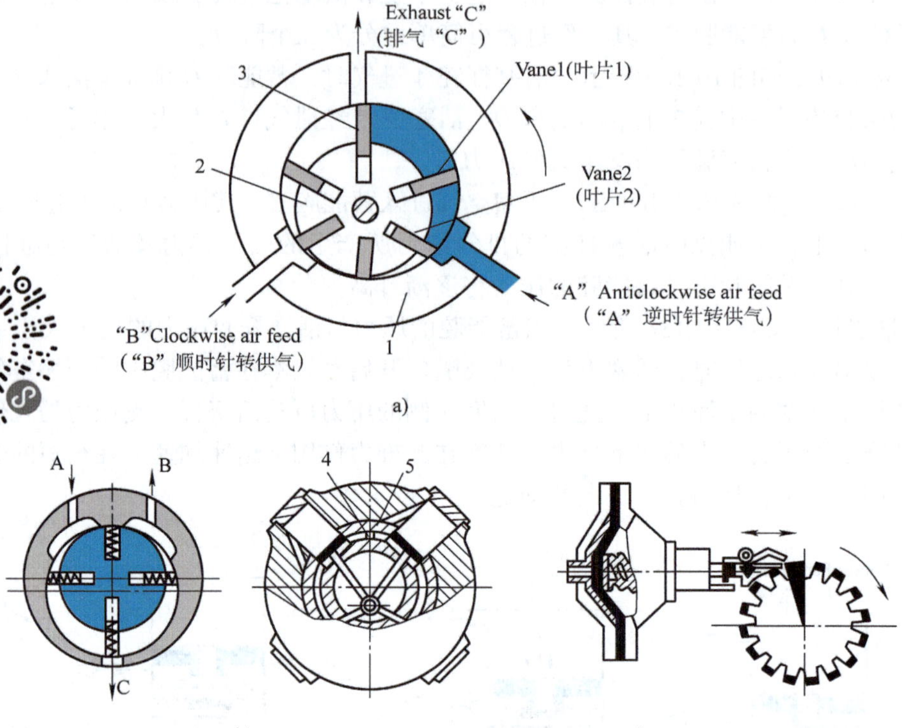

Fig. 10-19 Operations of air motors(气动马达工作原理图)

a) Operation of vane motors(叶片式气动马达的工作原理图)

b) Operation of piston motors(活塞式气动马达的工作原理图)

c) Operation of diaphragm motors(薄膜式气动马达的工作原理图)

1—Stator(定子) 2—Rotor(转子) 3—Vane(叶片) 4—Air inlet port(进气口) 5—Distribution valve(分配阀)

necting rod mechanism, see Fig. 10-19b. The compressed air passes through the distribution valve from the inlet port and then enters the cylinder, driving part of the pistons to move reciprocatingly. The pistons' reciprocating motion converts into crankshaft rotation through the crank and connecting rod mechanism. The crankshaft then drives the distribution valve which is fixed on it to move simultaneously, causing the compressed air to enter different cylinders according to the angle change of distribution valve. The pistons move in turn to develop a continuous crankshaft rotation. At the same time, the cylinder opposite to the inlet cylinders is connected to the atmosphere and the high pressure air is released. The soft feature of the power-speed and the torque-speed characteristic curves of piston motors under different working pressures are also very evident, as can be seen in Fig. 10-21.

(3) Diaphragm motors Diaphragm motors are basically an application of diaphragm cylinders. As is shown in Fig. 10-19c, the cylinder's reciprocating motion converts into the ratchet wheel's intermittent rotary output motion via the pawl at the end of handspike. There also exists soft characteristic due to the gas compressibility, i. e. , the output torque falls off with the increase of intermittent transmission speed under the constant working pressure.

旋转。图 10-20 所示为叶片式气动马达在某工作压力下的特性曲线，可以明显地看出其特性。

小型叶片式气马达用于手动工具，比如钻床、磨床，以及气动旋具。大型叶片式气马达的应用有绞盘、泵驱动、气动钻探设备的马达驱动等。

（2）径向活塞式气动马达　活塞式气动马达由于运动部件惯量大，其运行速度与叶片式气马达相比要小得多。

如图 10-19b 所示，径向活塞式气动马达所有的连杆都安装在曲柄上，构成了曲柄连杆机构。压缩空气经进气口进入分配阀（又称配气阀）后再进入气缸，经曲柄连杆机构将活塞往复运动变为曲轴转动而输出曲轴转动的同时，带动固定在轴上的分配阀同步运动，使压缩空气随着分配阀角度位置的改变而进入不同气缸内完成配气，依次推动各个活塞运动，经曲柄连杆使曲轴连续回转，与进气缸处于相对位置的气缸则同时排气。图 10-21 所示为活塞式气动马达在不同工作压力下的功率-转速和转矩-转速曲线，软特性十分明显。

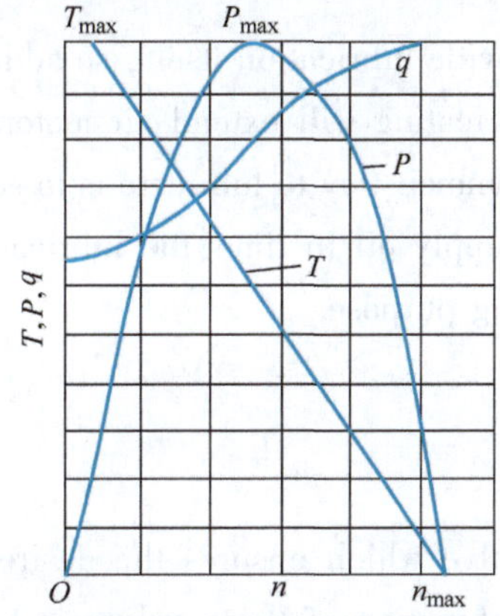

Fig. 10-20　Characteristics of a vane motor
（叶片式气动马达的特性曲线）

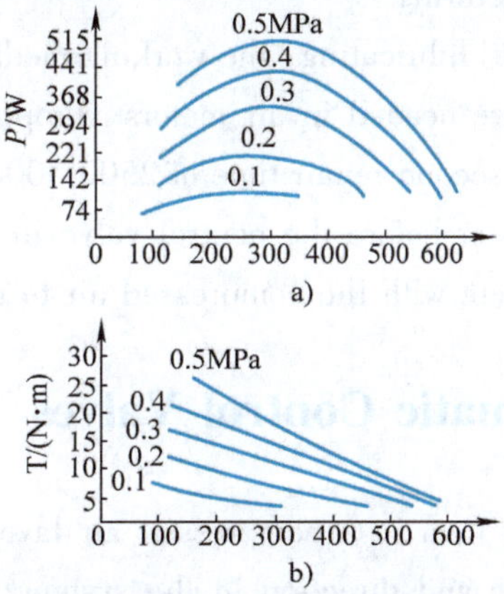

Fig. 10-21　Work characteristic curve of piston pneumatic motor
（活塞式气动马达工作特性曲线）
a）Power-speed curve（功率-转速曲线）
b）Torque-speed curve（转矩-速度曲线）

（3）薄膜式气动马达　如图 10-19c 所示，它实际上是薄膜式气缸的具体应用，将气缸的往复运动经推杆端部的棘爪使棘轮作间歇性单向转动，由于气体的可压缩性，其仍具有软特性，即也是当工作压力不变时随间歇传动速度的增加而输出转矩下降。

2. 气动马达的选择、应用及润滑

1）气动马达的选择。主要是从负载状态考虑，在变负载的场合使用时，主要考虑的因素是速度范围及满足工作情况所需的转矩；在恒负载时，工作速度是重要因素。在满足上述性能因素要求的前提下，还要考虑其他因素，如工作压力、耗气量和安装方式等。

2）气动马达的应用。气动马达应用随着生产力的发展越来越广，涉及各行各业。它应用在需要安全、无级变速、起动换向频繁、防爆及负载起动且易过载的场合，也适用于潮湿、高温、振动及不便于工人直接操作的恶劣环境中，如矿山、汽车制造等。

3）气动马达的润滑。因为气动马达采用压缩空气作为工作介质，空气无润滑能力，所以气动马达必须润滑，如果能正确使用，并且使其得到正确良好的润滑，可以延长气动马达的使用寿命，二次检修时间可达 2500～3000h 或更长。常用的方法是在气动马达控制阀前加油雾器，按期补油，随压缩空气进入系统进行润滑。

2. Selection and application of air motors

1) Air motors' selection: the selection of air motors is mainly based on its load conditions. In variable load occasions, the speed range and the torque needed to meet working conditions are principally considered. Working speed is also an important factor for constant load occasions. Other factors must be considered except for the above performance-related factors, such as working pressure, gas consumption, installation type, etc.

2) Air motors' application: air motors find more and more applications as the developing of the productivity. They can be used in all walks of life, particularly applicable for the occasions which require safety-needed, stepless speed changing, frequent direction changing for startup, resistant exploding, load startup and that easy to be overloaded. They are equally suitable to the harsh environment with moisture, high temperature, vibration or that inconvenient to operate directly, such as mine, car manufacturing.

3) Air motors' lubricating: the working medium of air cannot provide lubrication itself, so additional lubricants are needed in air motors. Proper used and nice lubricating will extend air motors' use life, with the second repair time of 2500-3000h or longer. The common way to lubricate is to set an oil mist lubricator before the control valve in the air motor and supply oil in time; the lubricant will enter the system with the compressed air to accomplish lubricating purpose.

10.4 Pneumatic Control Valves

An important item in a compressed air layout is a regulating valve which ensures the desired pressure, flow rate and direction in the system. The first important function of these valves is for safety; the second function is to provide for a part of the system to operate at a lower pressure than the main supply. There are various types of regulating valves; generally they can be divided into three main categories defined by the functions they perform: pressure control, flow control and directional control. Tab. 10-1 shows the three types of regulating valves, with their characteristics.

The pneumatic regulating valves listed in Tab. 10-1 are fixed value switching regulating valves. Similar to hydraulic regulating valves, there are also pneumatic proportional regulating valves and pneumatic servo-control valves (see related reference books). The recently-developed valve terminal has the advantages of concise construction, easy to install and maintain, high degree of modularization, and can achieve centralized control in medium or small pneumatic systems.

Tab. 10-1 Pneumatic control valves(压力控制阀)

Type (类型)	Name (名称)	Graphic symbol (图形符号)	Characteristics (特点)
Pressure control valves (压力控制阀)	Pressure reducing valves (减压阀)	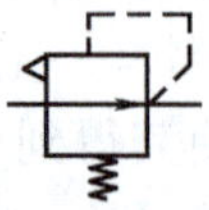	A valve having its outlet port connected to the system so that with varying inlet pressure or outlet flow, the outlet pressure remains substantially constant. Inlet pressure must remain substantially higher than the selected outlet pressure (出口与系统相连以调整或控制气压的变化,保持压缩空气减压后稳定在需要值。进口压力应大于出口压力)

（续）

Type （类型）	Name （名称）	Graphic symbol （图形符号）	Characteristics （特点）
Pressure control valves（压力控制阀）	Pressure relief valves (safety valves)[溢流阀(安全阀)]	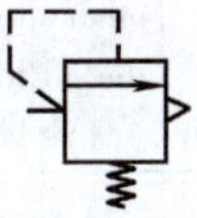	A valve having its inlet port connected to the system which limits the maximum pressure by exhausting fluid when the system pressure exceeds the preset pressure of the valve. Under normal operating conditions the valve remains closed, only operating when unusual system conditions prevail （进口与系统相连，为保证气动回路或气罐的安全，当压力超过某一调定值时，实现自动向外排气，使压力回到某一调定值范围内，起过压保护作用。也称为安全阀。在正常工作状态下阀处于关闭状态）
	Sequence valves（顺序阀）	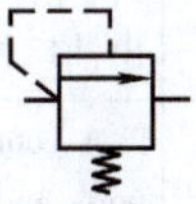	A valve with its inlet port connected to one part of the system and its outlet port connected to another part of the system and set so that until the inlet pressure rises above the preset opening pressure the fluid medium is prevented from flowing to the outlet port （进出口均与系统相连，当进口压力大于调定压力时阀打开通流）
Flow control valves（流量控制阀）	Throttle valves（节流阀）	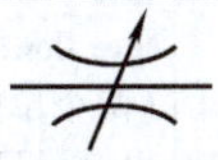	A valve where an orifice size is set to control the flow rate through the valve （通过改变阀的流通面积来实现流量调节）
	Exhaust muffler throttle valves（排气消声节流阀）	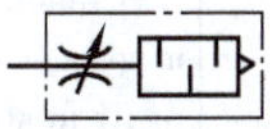	A valve situated at exhaust port and set to control exhausting flow rate, moving speed of actuator and thus reduce exhaust noise （装在执行元件主控阀的排气口处，调节排入大气中气体的流量。用于调整执行元件的运动速度并降低排气噪声）
Directional control valves（方向控制阀）	Shift directional valves（换向型控制阀） Pneumatic pressures（气压控制换向阀）	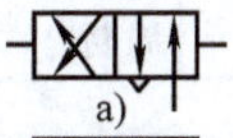 a) 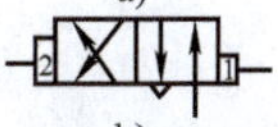 b)	A shift directional valve driven by pneumatic pressure. Safe and reliable in operation, available for poor conditions a) Controlled by pressure or pressure-relief b) Controlled by differential pressure （以气压为动力切换主阀，使气流改变流向操作安全可靠。适用于易燃、易爆、潮湿和粉尘多的场合） a)通过压力控制 b)通过压差控制
	Electrically（电磁控制换向阀）	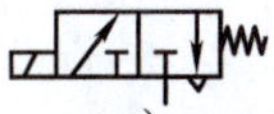 a) 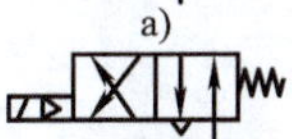 b) 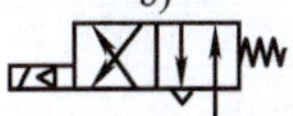 c)	A shift directional valve driven by solenoid, including spring loaded type and pilot operated type a) Direct air-operated type b) Pilot operated type controlled by pneumatic pressure c) Pilot operated type controlled by pneumatic pressure-relief （用电磁力的作用来实现阀的切换以控制气流的流动方向。分为直动式和先导式两种） a)直控式控制 b)气压加压控制先导式控制 c)气压卸压控制先导式电磁阀
	Mechanically（机械控制换向阀）	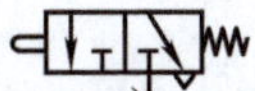 a) 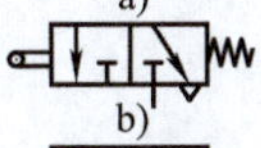 b) 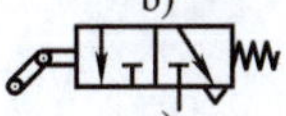 c)	A shift directional valve driven by cam, block or other mechanism, commonly used as signal valve in stroke control system (stroke valve) a) Direct air-operated type b) Roller-type c) Passable type （依靠凸轮、撞块或其他机械外力推动阀芯使其换向，多用于行程程序控制系统，作为信号阀使用，也称为行程阀） a)直动式机控阀 b)滚轮式机控阀 c)可通过式机控阀
	Manually（人力控制换向阀）	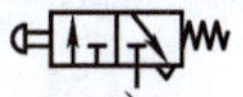 a) 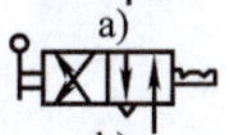 b) 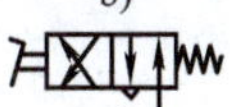 c)	By hand or pedal etc a) Button b) Lever c) Pedal （分为手动和脚踏两种操作方式） a)按钮式 b)手柄式 c)脚踏式

(续)

Type (类型)	Name (名称)		Graphic symbol (图形符号)	Characteristics (特点)
Directional control valves (方向控制阀)	One-way control valves (单向型控制阀)	Check valves(单向阀)		A valve allows flow of the air in one direction only, the other direction through the valve being at all times blocked to the airflow (气流只能一个方向流动而不能反向流动)
		shuttle valves(梭阀)	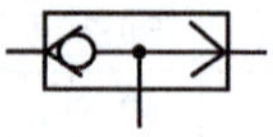	A combination of two one-way valves, giving an outlet from either one of two pressurized inlets (whichever is the greater value). (两个单向阀的组合。两个进油口中压力较大的一个与出油口相连)
		Dual pressure valves(双压阀)	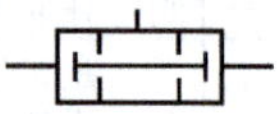	A combination of two one-way valves, which has two inlet ports and one outlet port. Only when both the inlet ports are simultaneously pressurized can the outlet port put out pressure flow. (两个单向阀的组合结构形式,有两个输入口和一个输出口。只有当两个输入口同时有输入时,才有输出)。
		Quick exhaust valves(快速排气阀)	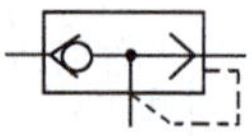	A quick exhaust valve is designed to increase the speed of movement of a cylinder by allowing the exhaust air to vent directly to atmosphere (常装在换向阀与气缸之间,它使气缸不通过换向阀而快速排出气体,从而加快气缸的往复运动速度,缩短工作周期)

10.4 气动控制阀

在气动系统中，气动控制阀用来调节压缩空气的压力、流量和流动方向，以确保系统安全并为部分系统提供低于气源压力的压缩空气。气动控制阀有很多种，按功能不同可分为压力控制阀、流量控制阀和方向控制阀三大类。表 10-1 列出了三大类气动控制阀及其特点。

表 10-1 列出的是定值开关式气动控制阀，与液压控制阀一样，也有气动比例控制阀和气动伺服控制阀，详见有关参考书。近年来出现的阀岛，具有结构紧凑、便于安装、易于维护、模块化程度高等特点，可实现中小型气动系统的集中控制。

Chapter 11 Basic Pneumatic Circuits and Their Application

All pneumatic systems are composed of some basic circuits with different function; familiarizing and mastering these basic circuits is a necessary foundation for analyzing and designing pneumatic systems.

The basic circuits may be classified based on function: pressure and force control (Tab. 11-1), direction control (Tab. 11-2), speed control (Tab. 11-3), etc.

第 11 章　气动基本回路及其应用

气动系统都是由各种不同功能的基本回路组成的，熟悉和掌握基本回路是分析和设计气动系统的必要基础。

气动基本回路按其功能不同分为压力和力控制回路（表 11-1）、换向控制回路（表 11-2）、速度控制回路（表 11-3）等。

Tab. 11-1　Pressure and force control circuits（压力和力控制回路）

Type（类型）	Basic circuits（基本回路）	Name, operation and application characteristics（回路名称、工作原理及其应用特点）
	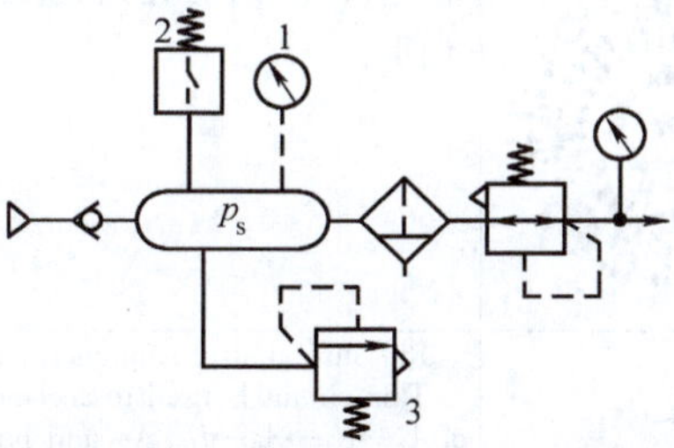	One-stage pressure control circuit This circuit is used to control the gas pressure out put from the receiver in case it surpassed the setting of pressure. A gauge 1 (or a pressure relay 2) is used to control the compressor to rotate or stop. Component 3 is a safety valve 一次压力控制回路 用于控制气罐输出气体压力 p_s，使之不超过规定的压力值。压力表 1（或压力继电器 2）用于控制空气压缩机的转、停。阀 3 是安全阀
Pressure control（压力控制回路）	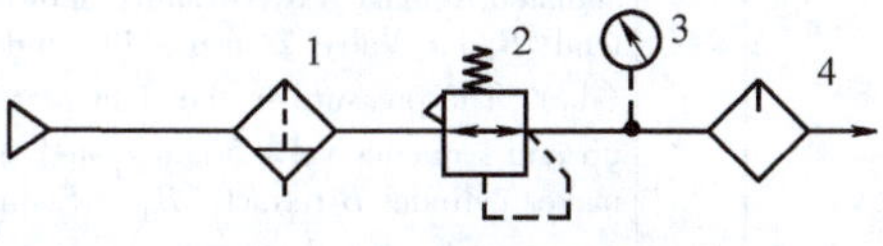a) 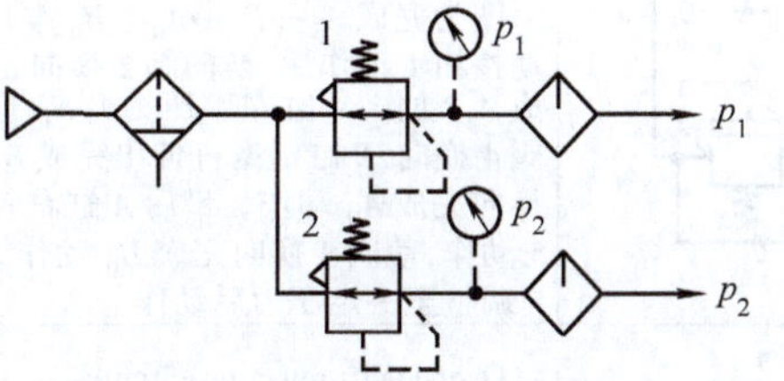b)	Two-stage pressure control circuit The circuit in Fig. a incorporates F. R. L combination, which is mounted at the inlet port of the pneumatic equipment. The needed pressure can be obtained by adjusting valve 2. The circuit shown in Fig. b is used to supply multiple pressure grades 二次压力控制回路 图 a 所示回路由空气过滤器、减压阀、油雾器组成，安装在气动设备的气源入口处。调节液压阀 2 能得到气动设备所需的工作压力。若系统需要多种不同的压力，可采用图 b 所示回路
	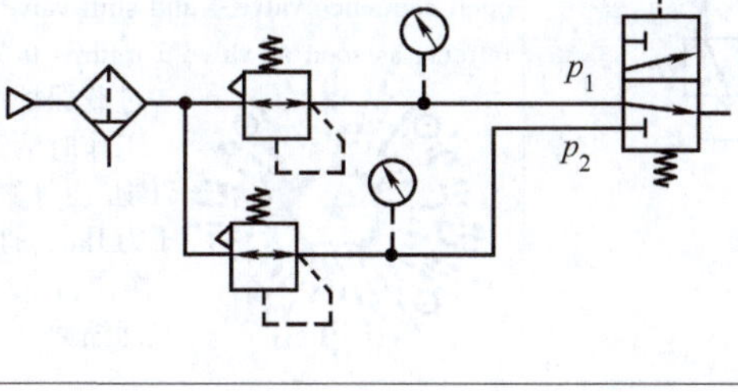	High and low pressures changeover circuit Directional control valve and pressure reducing valve are used to accomplish high (p_1) and low (p_2) pressures changeover 高低压转换回路 利用换向阀和减压阀实现高（p_1）、低（p_2）压力切换输出

(续)

Type (类型)	Basic circuits (基本回路)	Name, operation and application characteristics (回路名称、工作原理及其应用特点)
	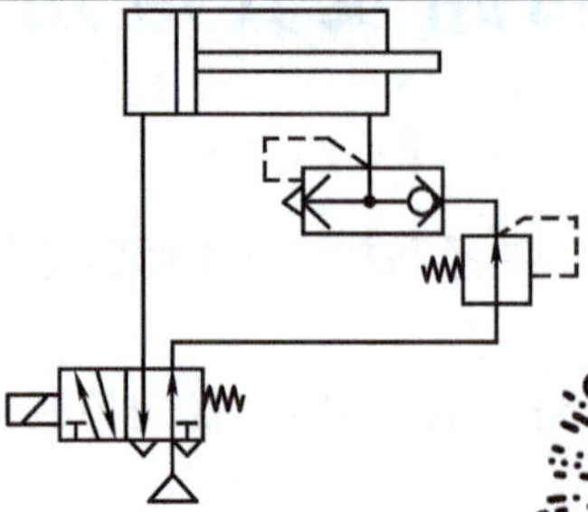 a) 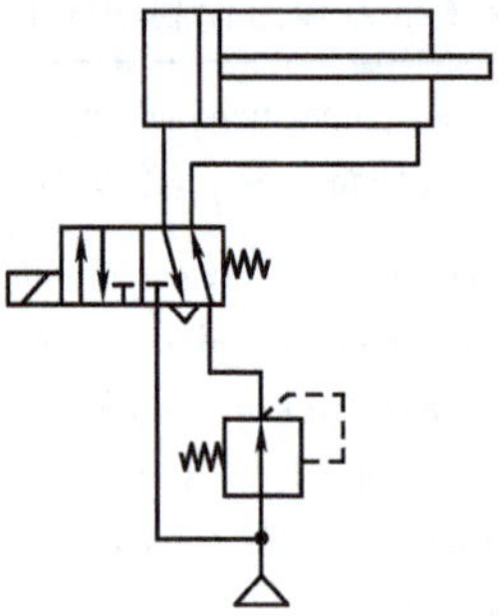 b)	Pressure difference circuit This circuit can minimize air consumption and reduce impact. Fig. a and b show a pressure difference circuit composed by a double-acting cylinder. The inlet pressures are different when the piston extends (working stroke) and retracts (idle stroke) 差压回路 采用差压回路，可以减少压力空气消耗量，并减小冲击。图a与b所示为双作用气缸的差压回路。活塞杆伸出(工作行程)和返回(空程)时的进气压力不同
Pressure control (压力控制回路)	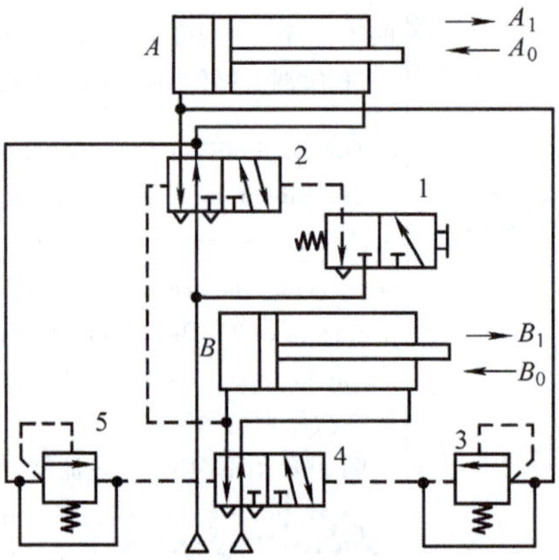	Pressure control sequence circuit This circuit is used to accomplish the sequence actions of $A_1 \to B_1 \to A_0 \to B_0$. As start button 1 is pressed, valve 2 shifts and cylinder A extends (A_1). The pressure in the left port of cylinder A increases. After a time sequence 3 is actuated to make valve 4 shift, allowing cylinder B to extend (B_1). Valve 2 also shifts and cylinder A retracts (A_0). The pressure in the right port of cylinder A builds up and sequence valve 5 is operated to shift valve 4, which makes cylinder B retract (B_0). Sequence valves 3 and 5 operate only after their settings are reached 压力控制顺序回路 图为完成$A_1 \to B_1 \to A_0 \to B_0$顺序动作的回路。起动按钮1动作后，换向阀2换向，$A$缸活塞杆伸出完成$A_1$动作；$A$缸左腔压力升高，顺序阀3动作推动阀4换向，$B$缸活塞杆伸出完成$B_1$动作，同时使阀2换向完成$A_0$动作。最后$A$缸右腔压力增高，顺序阀5动作，使阀4换向完成$B_0$动作。此处顺序阀3及5调整至一定压力后动作
	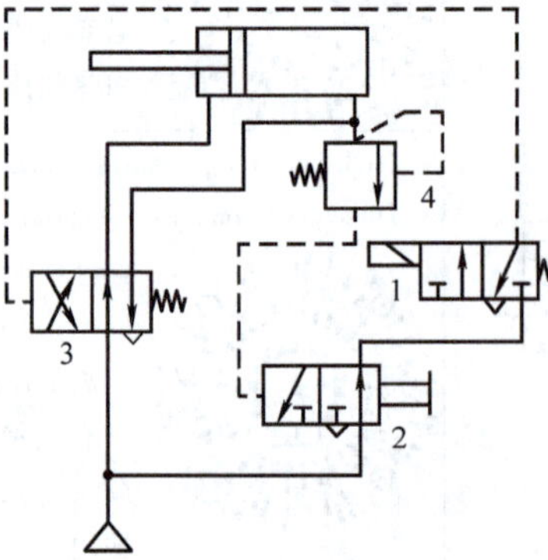	Over-load protection circuit The piston extends out at work. Once the piston barges up against a barrier and thus the cylinder is overloaded, the pressure in the non-rod chamber increases to open sequence valve 4 and shift valve 2. The piston rod retracts as soon as valve 3 returns to its start position 过载保护回路 工作时活塞伸出，当遇到障碍物，使气缸过载时，无杆腔压力升高，打开顺序阀4使阀2换向，阀3复位，使活塞杆立即缩回

（续）

Type （类型）	Basic circuits （基本回路）	Name，operation and application characteristics （回路名称、工作原理及其应用特点）
	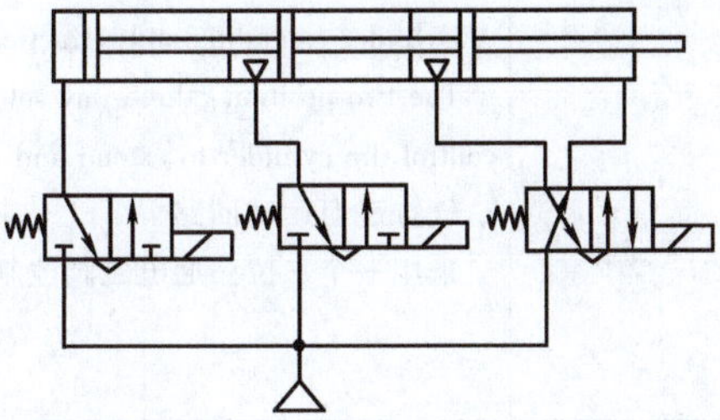	Force increased circuit by multiple (three here) in-line cylinders The cylinder rod output depends upon the number of energized solenoid valves. The more the in-line cylinders, the larger the output 三段式活塞缸串联增力回路 通过控制电磁阀的得电个数，实现对活塞杆输出推力的控制。活塞缸串联段数越多，输出的推力越大
Force control （压力控制回路）	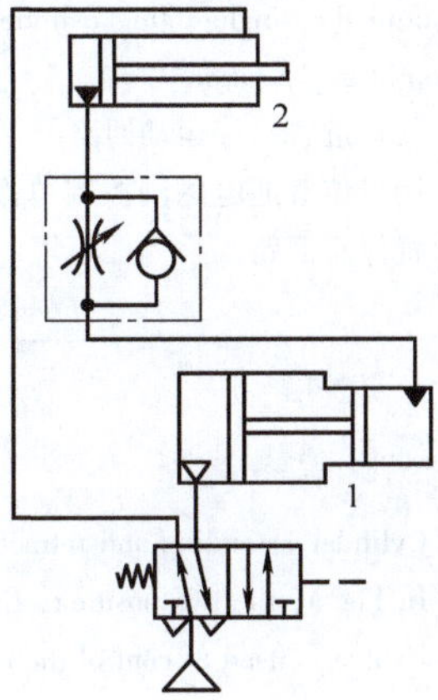	Force increased circuit by an air-hydraulic intensifier Gas with lower pressure becomes higher after air-hydraulic intensifier 1，increasing the output force of air-hydraulic intensifier 2 气-液增压器增力回路 利用气-液增压缸 1 把较低的气体压力变为较高的液压力，以提高气-液缸 2 的输出力
	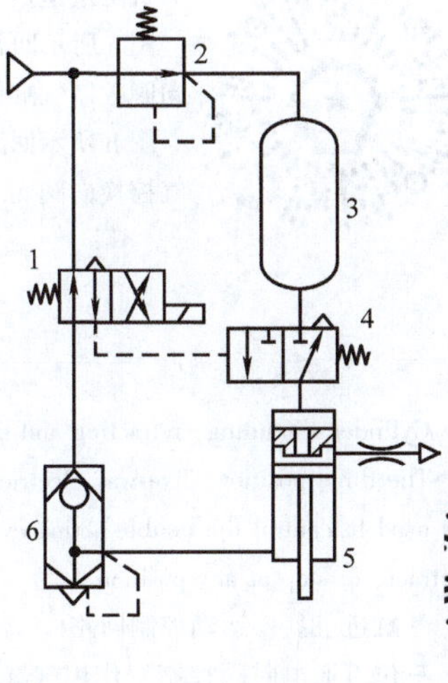	Impact cylinder circuit Solenoid directional control valve 1 is energized to make the down port of impact cylinder 5 open to atmosphere via quick exhaust valve 6. Valve 4 shifts under air pressure and the compressed air in receiver 3 enters the impact cylinder directly，causing the piston to move down rapidly. The kinetic energy of the piston translates into high shock load，which can be adjusted by pressure reducing valve 2 冲击气缸回路 换向阀 1 电磁铁得电，冲击气缸 5 的下腔由快速排气阀 6 通大气，换向阀 4 在气压作用下换向，气罐 3 内的压缩空气直接进入冲击气缸，使活塞以极高的速度运动，该活塞所具有的动能转换成很大的冲击力输出。减压阀 2 可以用来调节冲击力的大小

Tab. 11-2 Directional control circuits(换向控制回路)

Type (类型)	Basic circuits (基本回路)	Name, operations and application characteristics (回路名称、工作原理及其应用特点)
Single-acting cylinder directional control circuits (单作用气缸换向回路)	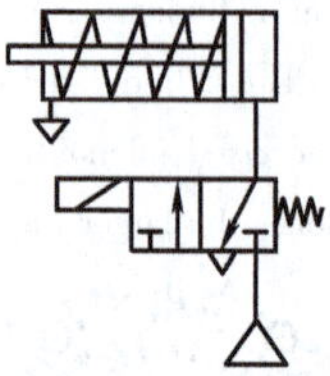	Cylinder extending and retracting circuit The two-position, three-way solenoid valve is used to control the cylinder to extend and retract 气缸进退运动回路 采用一个二位三通电磁阀控制单作用气缸伸、缩
	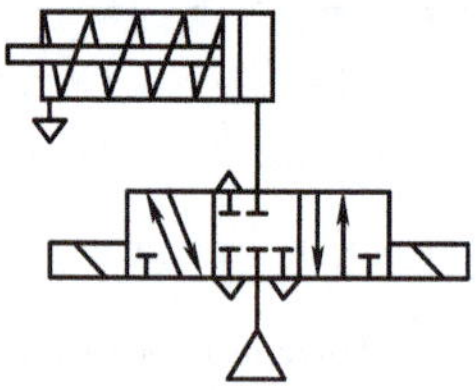	Cylinder extending, retracting and stop circuit The three-position, five-way solenoid valve is used to control the single-acting cylinder to extend, retract or stop at any position 气缸进、退、停运动回路 用三位五通电磁阀控制单作用气缸伸、缩及任意位置停止回路
Double-acting cylinder directional control circuits (双作用气缸换向回路)	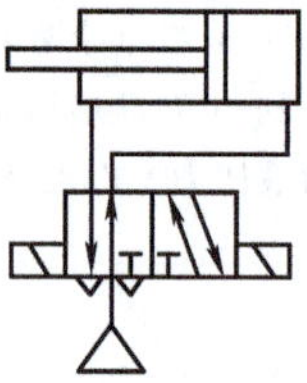 a) 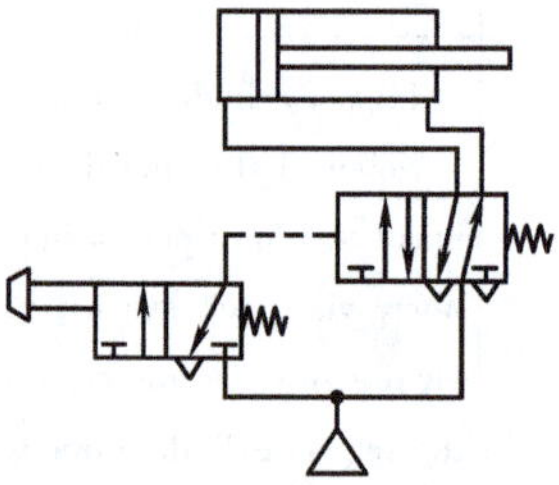 b)	Cylinder extending and retracting circuit In Fig. a, the two-position, five-way directional control valve is used to control the cylinder to extend or retract In Fig. b, the stroke valve is used to control the pneumatic directional valve to extend or retract 气缸进退运动控制回路 图 a 所示回路用二位五通换向阀控制气缸伸、缩 图 b 所示回路用行程阀控制气控换向阀实现气缸伸、缩
	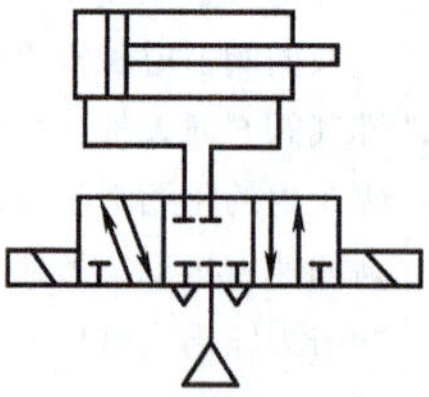	Cylinder extending, retracting and stop circuit The three-position, five-way electrically control valve is used to control the double-acting cylinder to extend, retract, or stop at any position 气缸进、退、停运动控制回路 三位五通电磁阀控制双作用气缸伸、缩及任意位置停止

Tab. 11-3　Speed control circuits（速度控制回路）

Type （类型）	Basic circuits （基本回路）	Name, operations and application characteristics （回路名称、工作原理及其应用特点）
Single-acting cylinder speed control circuits （单作用气缸速度控制回路）	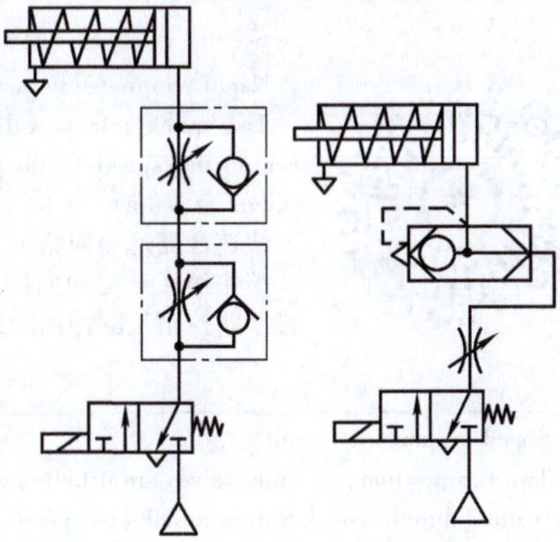 a)　b)	Single-acting cylinder one-direction speed control circuit In Fig. a, two check throttle valves are used to control the extend and retract speeds of the cylinder. In Fig. b, the throttle valve and the quick release valve are mounted in series to control the extending speed of the cylinder rod and make it retract rapidly 单作用气缸单向调速回路 图 a 所示回路用两个单向节流阀控制活塞杆的伸出和退回速度。 图 b 所示回路用一个节流阀和一个快速排气阀串联来控制活塞杆的伸出速度和快速退回
Double-acting cylinder speed control circuits （双作用气缸速度控制回路）	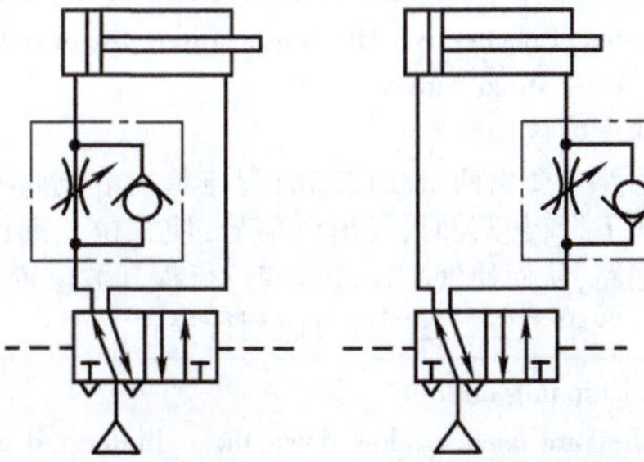 a)　b)	1）Inlet one-direction throttling speed control circuit In Fig. a, a throttle check valve is set in the inlet port. Because there is no damping for the exhausting air, there may occur "runaway" operation when the directions of the load and the piston are the same and "crawling" when opposite. This leads to an unstable movement. 2）Exhaust one-direction throttling speed control circuit In Fig. b, a throttle check valve is set at the exhaust port, which serves as a damper for the exhaust air. The piston speed changes little with the load and a stable movement can be obtained. In this circuit, the piston can bear reverse load (the direction of the load is different from that of the piston movement) 1）进气单向节流调速回路 在气缸进气口接一单向节流阀，如图 a 所示。这种回路因气缸的排气无阻尼，当负载方向与活塞方向相同时，因排气无阻尼，而产生所谓“跑空”现象；当负载方向与活塞方向相反时，而产生所谓的“爬行”现象，运动不很平稳 2）排气单向节流调速回路 在气缸排气口接一单向节流阀，如图 b 所示。这种回路缸的排气须经节流阀，起阻尼作用。活塞速度随负载变化很小，运动较平稳，能承受反向负载（负载方向与活塞运动方向相同）
	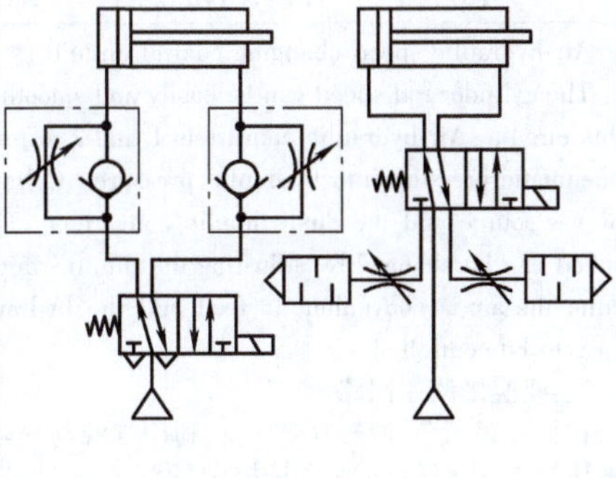 a)　b)	1）Double-direction throttling speed control circuit. In Fig. a, two throttle check valves are set at the inlet and outlet ports. They are used to control the extending and retracting speeds of the piston rod. This circuit can achieve a stable movement. 2）Exhaust throttling speed control circuit. In Fig. b, two two-position, five-way valves with two mufflers set at the exhaust ports are used to control the extending and retracting speeds of the cylinder. This design can achieve a stable movement and low noise level 1）双向节流调速回路 在气缸进、排气口各接一单向节流阀，如图 a 所示。实现活塞杆双向调速度，运动较平稳 2）排气节流调速回路 在二位五通换向阀的排气口各接一带有消声器的排气节流阀，实现活塞杆双向调速，运动较平稳、噪声较小

(续)

Type (类型)	Basic circuits (基本回路)	Name, operations and application characteristics (回路名称、工作原理及其应用特点)
Rapid reciprocating motion circuits (快速往复运动回路)	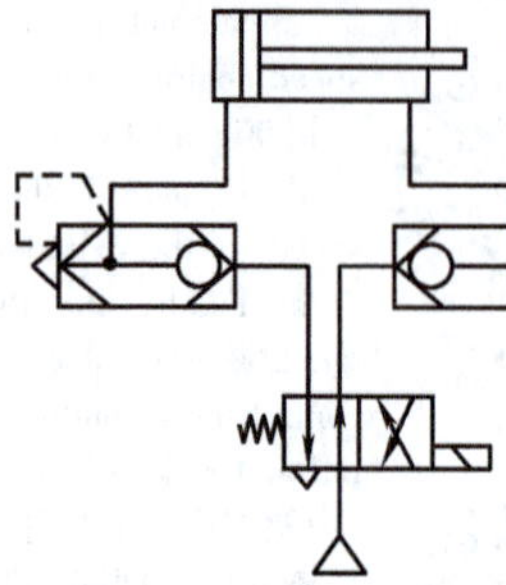	Rapid reciprocating movement circuit Two quick release valves are used to control the speed of the cylinder rod to extend or retract 快速往复运动回路 该快速往复运动回路通过两个快速排气阀控制气缸伸缩的速度
Speed exchanging circuits (速度换接回路)	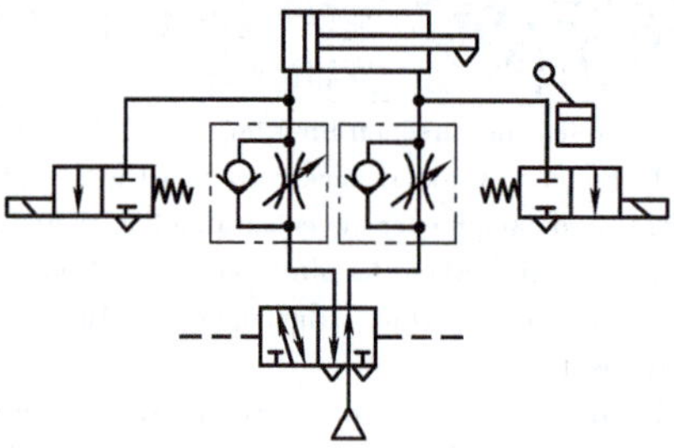	Speed exchanging circuit Two two-position, two-way valves are usually mounted in parallel with a throttle check valves to achieve speed exchanging control. When the stroke switch is pressed by the damp block, the two-position, two-way valves shift to change the exhaust path and achieve speed changing control. The stroke switch position can be adjusted if necessary. The two-position, two-way valves can be replaced by stroke valves 速度换接回路 回路主要由两个二位二通阀与单向节流阀并联组成,当挡块压下行程开关时,发出电信号,使二位二通阀换向,改变排气通路,从而使气缸速度改变。行程开关位置,可根据需要调整。图中二位二通阀也可改用行程阀
Cushioning circuits (缓冲回路)	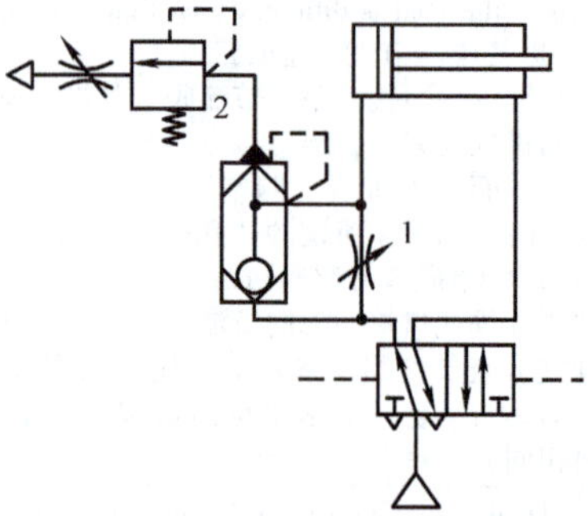	Cushioning circuits They are used to slow down the cylinder rod at the end of its stroke, particularly for the long stroke, high speed and large inertia occasions. As is shown in the left figure, when the cylinder rod returns to its stroke end, the pressure in the left is not high enough to open sequence valve 2. The air can only exhaust via throttle valve 1 and achieve a deceleration control 缓冲回路 缓冲回路应用于气缸行程末端要求获得缓冲的场合,如行程长、速度快、惯性大的情况。左图中当活塞返回到行程末端时,其左腔压力已降低到打不开顺序阀 2 的程度,余气只能经节流阀 1 派出,因此活塞得到缓冲
Air-hydraulic speed control circuits (气-液换接速度控制回路)	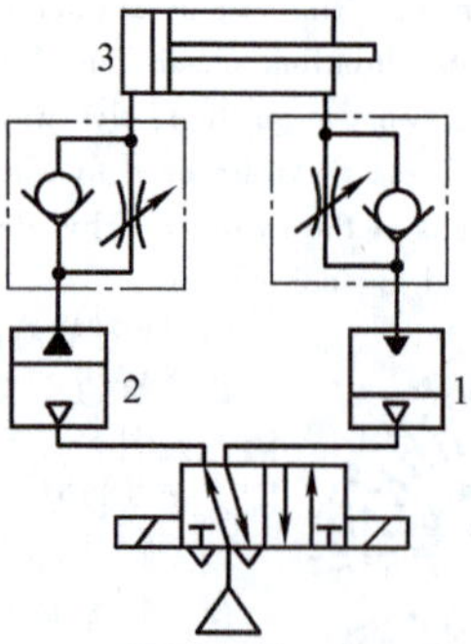	Air-hydraulic speed changing control circuit The cylinder rod speed can be easily and smoothly controlled by this circuit. Air-hydraulic converters 1 and 2 are used to translate pneumatic pressure into hydraulic pressure. Cylinder 3 is driven by gas source and the air-hydraulic converters. The cylinder rod speed can be changed by adjusting the throttle valve. In such circuit, the air is convenient to feed and the hydraulic pressure is easy to be controlled 气-液换接调速回路 在回路设置气-液转换器 1、2,利用气源与气-液转换器驱动液压缸 3,从而得到平稳易控制的活塞运动速度,调节节流阀,就可以改变活塞的运动速度。这种回路充分发挥了气动供气方便和液压速度容易控制的特点

（续）

Type （类型）	Basic circuits （基本回路）	Name, operations and application characteristics （回路名称、工作原理及其应用特点）
Speed control for air-hydraulic damping cylinder （气-液阻尼缸的速度控制回路）	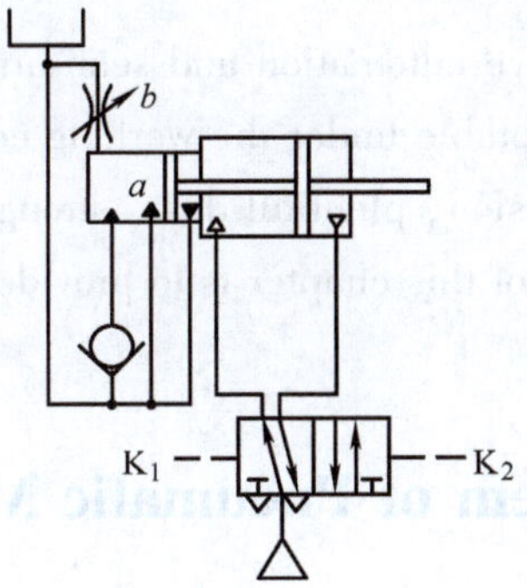	Speed changing circuit Such circuit can achieve the working cycle of machine tools, i. e. , quick feed→work speed→quick retract. When signal K_2 occurs, the five-way, two-position valve is actuated to make the cylinder rod move in the left. Oil in the cap end of the cylinder enters the rod end of the cylinder via orifice *a* and the cylinder rod moves in the left rapidly. When orifice *a* is closed by the piston, the oil in the cap end is forced to enter the rod end through orifice *b*. The piston then moves at work speed. When signal K_1 occurs, the five-way valve shifts to make the piston return back rapidly in the right 气缸变速回路 实现机床工作循环中常用的快进→工进→快退的动作。当 K_2 有信号时，二位五通阀换向，活塞向左运动，液压缸无杆腔中的油液通过 *a* 口进入有杆腔，气缸快速向左前进；当活塞将 *a* 口关闭时，液压缸无杆腔中的油液被迫从 *b* 口经节流阀进入有杆腔，活塞工作进给；当 K_1 输入信号时，五通阀换向，活塞向右快速返回

Chapter 12 Examples of Pneumatic Transmission System

The pneumatic technology is a method to achieve automation and semi-automation in industrial production. Pneumatic transmission systems are adaptable under the working conditions of high temperature, oscillation, corrosion, flammability, explosion, plentiful dust, strong magnetism, radiate, etc. , which lead to its increasing use. The purpose of this chapter is to provide a variety of practical applications in mechanical field.

12. 1 Pneumatic Transmission System of Pneumatic Manipulators

Manipulators are one of the important devices in automatic production equipment and production line. They can simulate real human hands to accomplish some actions according to the working needs of each automatic equipment. In other words, they can accomplish the actions of picking, conveying, workpiece placing and removal, and even automatic tool changing according to the preset program, track and technical requirements. Manipulators are wildly used in machining, pressing, forging, founding, mounting, heat treatment and so on to relieve labour intensity. As one type of manipulators, they have the advantages of light in weight, simple in design, rapid, stable and reliable act, energy saving, no pollution to environment, etc.

Fig. 12-1 shows a pneumatic manipulator structure. In this design, cylinders *A*, *B*, *C* and *D* are used for fingers to accomplish holding, arm to extend and retract, column to rise, fall, and rotate. Cylinder *A* is used for the gripper to clamp and loosen workpieces; cylinder *B* for the long arm to extend and retract; cylinder *C* for the column to rise and fall; cylinder *D* for the column to rotate. There is a gear-rack mechanism in cylinder *D*. Two pistons are mounted on the two ends of the piston rod, along which the rack is built. The rack's reciprocating motion drives the gear on the column to rotate and thus brings the column and arm into turn operation. Fig. 12-2 shows a universal manipulator operation. The finger of this manipulator is actually a vacuum chuck, i. e. , cylinder *A* is replaced by a vacuum chuck. The work cycle for this system is: column rises→arm extends→column rotates clockwise→vacuum chuck snatches at workpiece→

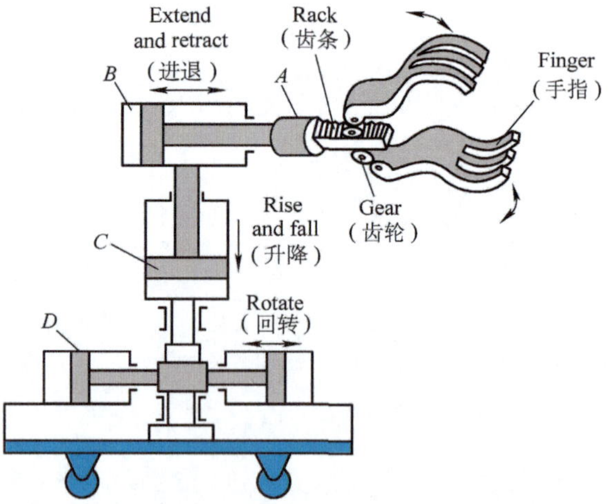

Fig. 12-1 Pneumatic manipulator structure
(气动机械手的结构示意图)

第 12 章　气压传动系统实例

气压传动是实现工业自动化和半自动化的方式之一。由于气压传动系统使用安全、可靠，可以在高温、震动、腐蚀、易燃、易爆、多尘埃、强磁和辐射等恶劣环境下工作，所以应用日益广泛。本章介绍气压传动在机械行业中的几个应用实例。

12.1 气动机械手气压传动系统

机械手是自动化生产设备和生产线上的重要装置之一，它可以根据各种自动化设备的工作需要，模拟人手的部分动作，按着预定的控制程序、轨迹和工艺要求实现自动抓取、搬运，完成工件的上料、卸料和自动换刀。因此，在机械加工、冲压、锻造、铸造、装配和热处理等生产过程中被广泛应用，以减轻工人的劳动强度。气动机械手是机械手的一种，它具有结构简单，重量轻，动作迅速、平稳、可靠，节能和不污染环境等优点。

图 12-1 所示为气动机械手的结构示意图。该系统由 *A*、*B*、*C*、*D* 四个气缸组成，能实现手指夹持、手臂伸缩、立柱升降和立柱回转四个动作。其中，缸 *A* 为抓取工件的松紧缸；缸 *B* 为长臂伸缩缸，可实现手臂的伸出与缩回动作；缸 *C* 为立柱升降缸；缸 *D* 为立柱回转缸，该气缸为齿轮齿条缸，它有两个活塞，分别装在带齿条的活塞杆两端，齿条的往复运动带动立柱上的齿轮旋转，从而实现立柱及手臂的回转。图 12-2 所示为一种通用机械手的气动系统。此机械手手指部分为真空吸头，即无松紧缸。要求其完成的工作循环为：立柱上升→伸臂→立柱顺时针方向转动→真空吸头吸取工件→立柱逆时针方向转动→缩臂→立柱下降。

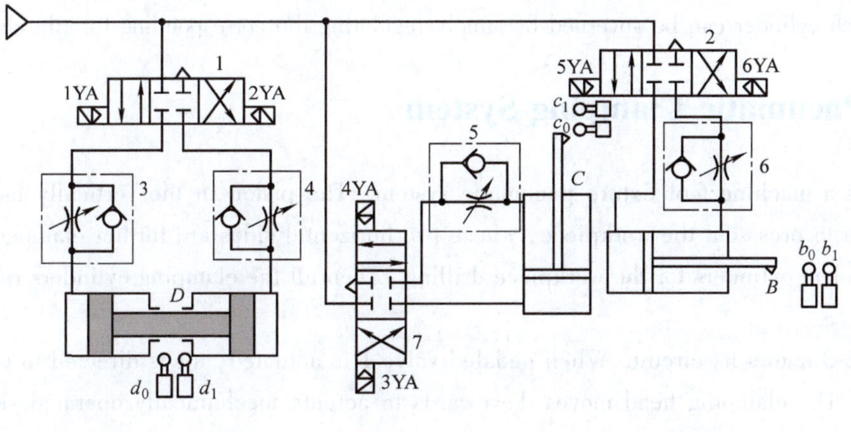

Fig. 12-2　Pneumatic transmission system of universal manipulators（通用机械手气动系统）

三个气缸分别与三个三位四通双电控换向阀 1、2、7 和单向节流阀 3、4、5、6 组成换向、调速回路。各气缸的行程位置均由电气行程开关进行控制。该机械手的电磁铁动作顺序

column rotates counterclockwise→arm retracts→column falls down.

The three cylinders are separately combined with three four-way, three-position double electric control directional valves 1,2,7 and throttle check valves 3,4,5,6 to accomplish direction and speed control. The stroke positions of each cylinder are controlled by electrical position switches. Tab. 12-1 shows the sequence actions of electromagnets.

Analysis of the work cycle for the pneumatic manipulator:

When the start button is depressed, 4YA is energized and valve 7 shifts to its upper position. Compressed air enters the down chamber of vertically located cylinder C, allowing the piston rod (column) to rise.

When the block on the piston rod in cylinder C touches the electrical position switch c_1, 4YA is released and 5YA is energized, which makes valve 2 shift to its left position. The piston rod (arm) in horizontally located cylinder B extends, which drives the vacuum chuck into operation and to grip workpieces.

When the block on the piston rod in cylinder B touches the electrical position switch b_1, 5YA is released and 1YA is energized, which makes valve 1 shift to its left position. Cylinder D(column) rotates in a clockwise direction, allowing the vacuum chuck to remove workpieces.

When the electrical position switch d_1 is pressed by the block on the piston rod in cylinder D, 1YA is released and 2YA is actuated, which makes valve 1 shift to its right position and cylinder D returns to its original position. Then the block on cylinder D touches the electrical position switch d_0. By this 6YA is energized and 2YA is released, which makes valve 2 shift to its right position and the piston rod (arm) in cylinder B retract.

As the piston rod (arm) in cylinder B retracts, the block touches the position switch b_0 after a certain time. Then 6YA is released and 3YA is energized to make valve 7 shift to its down position, allowing the piston rod(column) in cylinder C to move down. As soon as the piston rod returns to its start position, it touches the position switch c_0 and 3YA is released. By then a work cycle is accomplished. The next work cycle can be started by simply pressing the start button.

The positions of the electrical position switches can be adjusted if necessary, while the stroke and speed of each cylinder can be governed by simply regulating the corresponding throttle check valves.

12.2 Pneumatic Clamping System

This is a machine tool fixture pneumatic system. The piston in the vertically located cylinder moves down to press out the workpiece, whose two horizontal sides are further clamped by the independent tandem cylinders for the workpiece drilling. Then all the clamping cylinders retract to loosen the workpiece.

Fig. 12-3 shows its circuit. When pedaled valve 1 is actuated, air is directed to the cam end of cylinder A. The clamping head moves downwards to actuate mechanically operated stroke valve 2. Air flows into the two-position, three-way directional pneumatic valve 4 through check adjustable restrictor 6. Valve 4 shifts (this action can be delayed by adjusting the orifice of restrictor 6) to allow the compressed air to enter the cap ends of cylinders B and C via main valve 3. The air in the piston rod chamber exhausts out via valve 3 and the piston rods in cylinders B and C then extend out for

见表12-1。

Tab. 12-1 Sequence actions of electromagnets（电磁铁动作顺序表）

Electromagnet（电磁铁） / Action（动作）	1YA	2YA	3YA	4YA	5YA	6YA
Column rises（立柱上升）				+		
Arm extends（手臂伸出）				−	+	
Column rotates（立柱转位）	+				−	
Column resets（立柱复位）	−	+				
Arm retracts（手臂缩回）		−				+
Column falls down（立柱下降）			+			−

气动机械手工作循环分析：

按下起动按钮，4YA通电，阀7处于上位，压缩空气进入垂直缸C下腔，活塞杆（立柱）上升。

当垂直缸C活塞杆上的挡块碰到电气行程开关c_1时，4YA断电，5YA通电，阀2处于左位，水平缸B活塞杆（手臂）伸出，带动真空吸头进入工作点并吸取工件。

当水平缸B活塞上的挡块碰到电气行程开关b_1时，5YA断电，1YA通电，阀1处于左位，回转缸D（立柱）顺时针方向回转，使真空吸头进入卸料点卸料。

当回转缸D活塞杆上的挡块压下电气行程开关d_1时，1YA断电，2YA通电，阀1处于右位，回转缸D复位。回转缸复位时，其上的挡块碰到电气行程开关d_0时，6YA通电，2YA断电，阀2处于右位，水平缸B活塞杆（手臂）缩回。

水平缸B活塞杆（手臂）缩回时，挡块碰到电气行程开关b_0，6YA断电，3YA通电，阀7处于下位，垂直缸C活塞杆（立柱）下降，到达原位时，碰到电气行程开关c_0，使3YA断电，至此完成一个工作循环。如再给起动信号，可进行同样的工作循环。

根据需要只要改变电气行程开关的位置，调节单向节流阀的开度，即可改变各气缸的行程和运动速度。

12.2 气动夹紧系统

图12-3所示机床夹具的气动夹紧系统，其动作循环是：垂直缸活塞下降将工件压紧，两侧夹紧，然后进行钻削加工，最后各夹紧缸退回，松开工件。其工作原理如下所述。

用脚踏下脚踏阀1，空气进入缸A的无杆腔，夹紧头下降与机动行程阀2接触后发出信号，压缩空气经单向节流阀6进入二位三通气控换向阀4（调节节流阀开度可以控制阀4的延时接通时间）。因此，压缩空气通过主阀3进入两侧气缸B和C的无杆腔，活塞杆腔气体经主阀3排向大气，使活塞杆前进，钻头开始钻孔。

与此同时，流过主阀3的一部分压缩空气经过单向节流阀5进入主阀3右端，经过一段时间（由节流阀控制）后主阀3右位接通，两侧气缸后退到原来位置。同时，一部分空气作为信号进入脚踏阀1的右端，使脚踏阀1右位接通，压缩空气进入缸A的下腔。夹紧头上升的同时使机动行程阀2复位，使空气换向阀4也复位（此时主阀3右位接通）。气缸B、C的无杆腔通过主阀3和二位三通气控换向阀4排气，主阀3自动复位到左端接入工作状态，

hole drilling.

Partly compressed air flowing through main valve 3 is also directed to its right end via check restrictor 5. After a time (controlled by restrictor 5) main valve 3 is actuated by this flow path, allowing the tandem cylinders to retract to their original positions. In addition, part of the air is also directed to the right end of valve 1 for actuation. The compressed air flows into the down port of cylinder *A*, allowing the clamping head to move up. By this action, valve 2 and thus valve 4 are reset. The cap ends of cylinders *B* and *C* exhaust via valves 3 and 4, causing main valve 3 to return to its original condition by its spring after a time. *A* working cycle is accomplished by this point.

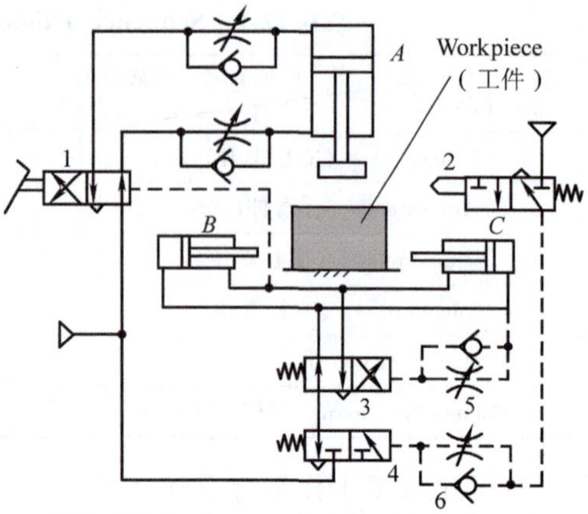

Fig. 12-3 Pneumatic clamping system (气动夹紧系统)

12.3 Auto-close Pneumatic System of Sliding Doors

In this system, the linear motion of the piston rod is converted into the open-and-close action of the sliding door through a linkage mechanism. Any actuation (by passers-by) on the pedal is detected by super-low pressure pneumatic valves. Two pedals 11 and 12 are placed in and out of the sliding door respectively. Some fully closed rubber hoses with one side connected to the pilot ports of super-low pressure pneumatic valves 10 and 13 are located under the pedals. Once the pedals are pressed (passers-by stand on the pedals), the pressure in the rubber hoses increases to actuate the super-low pressure stroke valves.

Fig. 12-4 shows its pneumatic circuit. When manual valve 2 is actuated, air is directed to the cap end of cylinder 5 via pneumatic directional valve 3 and check restrictor 4. The piston rod in cylinder 5 extends out (the door is closed). When a passer-by stands on pedal 11, valve 10 is actuated. Air passes through shuttle valve 9, check restrictor 8 and receiver 7 to actuate valve 3. Compressed air is directed to the rod end of cylinder 5 and the piston rod retracts (the door is opened).

When a passer-by stands on pedal 12, valve 13 is actuated to close the upper port of shuttle valve 9 and open its down port (by then the passer-by has left pedal 11 and valve 10 has been reset). The air in receiver 7 exhausts through check restrictor 8, shuttle valve 9 and valve 13 (valve 13 is reset after the passer-by leaves pedal 12). After a time (regulated by restrictor 8) valve 3 returns to its original condition. Air is directed to the cap end of cylinder 5 and the piston rod extends to close the sliding door. In addition, this circuit allows passers-by to go through the sliding door from either side. If the manual valve is reset, this auto-close sliding door becomes a hand-operated door.

12.4 Pneumatic Transmission System of Air-Hydraulic Power-Slipway

Fig. 12-5 shows its operation. A working cycle with working feed in two opposite directions can

完成一个工作循环。

12.3　拉门自动启闭气动系统

该装置通过连杆机构将气缸活塞杆的直线运动转换成拉门的启闭运动，利用超低压气动阀来检测行人的踏板动作。在拉门内、外装踏板11和12，踏板下方装有完全封闭的橡胶管，管的一端与超低压气动换向阀10和13的控制口连接。当人站在踏板上时，橡胶管里压力上升，超低压气动阀动作。其气动系统如图12-4所示。

首先使手动阀2上位接入工作状态，压缩空气通过气动换向阀3与单向节流阀4进入气缸5的无杆腔，将活塞杆推出（门关闭）。当人站在踏板11上时，气动换向阀10动作，压缩空气通过梭阀9、单向节流阀8和气罐7使气动换向阀3换向，压缩空气进入气缸5的有杆腔，活塞杆退回（门打开）。

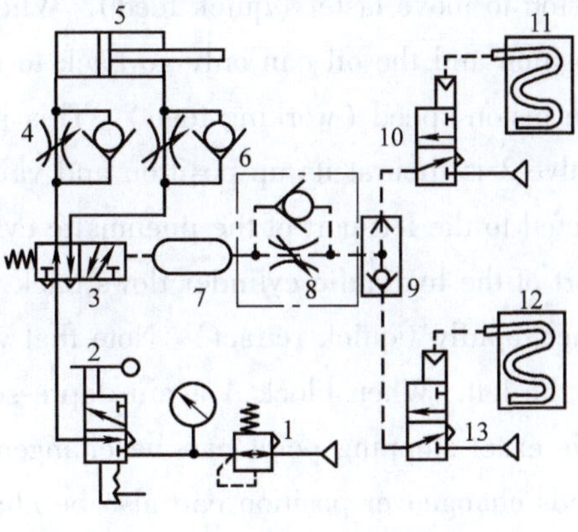

Fig. 12-4　Auto-close pneumatic system of sliding door（拉门自动启闭气动系统）

当行人经过门后踏上踏板12时，气动换向阀13动作，使梭阀9上面的通口关闭，下面的通口接通（此时由于人已离开踏板11，气动换向阀10已复位），气罐7中的压缩空气经单向节流阀8、梭阀9和气动换向阀13放气（人离开踏板12后，气动换向阀13已复位），经过延时（由节流阀8控制）后气动换向阀3复位，气缸5的无杆腔进气，活塞杆伸出（关闭拉门）。行人无论从门的哪一边进出均可，十分便利。如将手动阀复位，则可变为手动门。

12.4　气液动力滑台气压传动系统

图12-5所示为气液动力滑台的气动系统原理图。该滑台以气-液阻尼缸作为执行元件，能完成正、反向均能工进的工作循环。

1. 快进→工进→快退→停止

如图12-5所示，当将换向阀3切换到下位时，压缩空气经换向阀1、3进入气缸右腔，推动活塞向左运动，液压缸左腔中的油液经行程阀6、单向阀7回液压缸右腔，实现了快进；当快进到挡铁*B*压下行程阀6，阀处下位时，油液只能经节流阀5回液压缸右腔，开始工进；当工进到挡铁*C*压下行程阀2，阀处上位时，使接通换向阀3控制通路，换向阀3复位。此时压缩空气进入气缸左腔，使活塞右行。液压缸右腔油液经

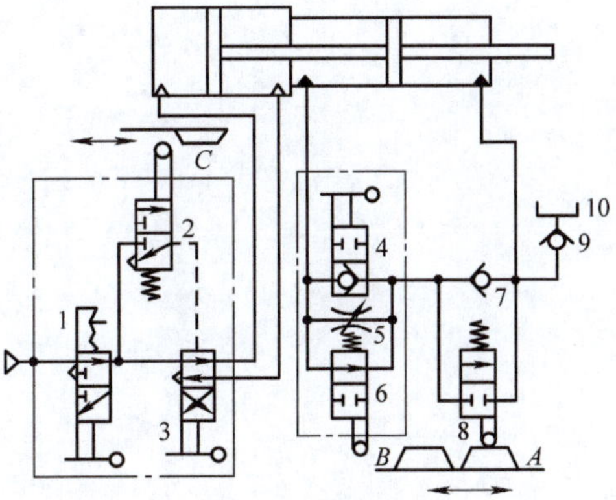

Fig. 12-5　Pneumatic transmission system of air-hydraulic power-slipway（气液动力滑台气压传动系统）

be realized by an air-hydraulic damping cylinder.

1. Quick feed→working feed→quick retract→stop

As is shown in Fig. 12-5, when manual valve 3 is operated, compressed air is directed to the right port of the pneumatic cylinder via valves 1 and 3, causing the piston to move in the left. The oil in the up port of the hydraulic cylinder flows into the right port via valves 6 and 7, allowing the piston to move faster (quick feed). When valve 6 is actuated by block *B*, valve 6 shifts to its down position and the oil can only go back to the right port of the cylinder through restrictor 5, reducing the piston speed (working feed). This process lasts until stroke valve 2 is depressed by block *C*. Valve 2 is then at its up position and valve 3 returns to its original condition. Compressed air is directed to the left port of the pneumatic cylinder and the piston moves in the right. The oil in the right port of the hydraulic cylinder flows back to the left port via valves 8 and 4, causing the piston to retract rapidly (quick retract). Note that valve 8 has been released by block *A* when the piston moves in the left. When block *A* again depresses valve 8, the return path is cut off and the piston stops. The exact stopping point can be changed by adjusting the position of block *A*. The quick-working feeds changeover position can also be changed simply by changing the position of block *B*.

2. Quick feed→working feed→slow retract (inversely working feed)**→stop**

This working cycle can be realized simply by the actuation of manual 4. The quick feed, working feed operations are the same as the previous working cycle. When valve 2 is actuated by block *C*, valve 3 is shifted to its up position and the piston in the pneumatic cylinder moves in the right. The oil in the right port of the hydraulic cylinder flows back to the left port via valve 8 and restrictor 5. It is the restrictor that makes the piston retract slowly (inversely working feed). When valve 6 is released by block *B*, the returned oil flows through the upper port of valve 6, allowing for a quick retraction. The piston will not stop until valve 8 is actuated by block *A*.

Note that a reservoir 10, usually replaced by a blow, is located in the high position as shown in the figure. Its purpose is to make up the oil leakage. Valves 1,2,3 and valves 4,5,6 are two integrated valve blocks.

行程阀 8 上位（当活塞向左运动时，挡铁 A 已将行程阀 8 释放）和换向阀 4 中的单向阀回液压缸左腔，实现了快退；当挡铁 A 再次压住行程阀 8 时，回油路被切断，活塞停止运动。改变挡铁 A 的位置，就改变了停的位置；改变挡铁 B 的位置，就改变了快进和工进速度换接的位置。

2. 快进→工进→慢退（反向工进）→停止

把换向阀 4 关闭，即使其处于上位时，就实现了这一双向进给程序。其快进、工进的动作原理与上述相同。当工进至挡铁 C 切换阀 2 时，输出信号使阀 3 切换到上位，气缸活塞右行，这时液压缸右腔油液经行程阀 8 和节流阀 5 回到左腔，实现了反向进给。当挡铁 B 离开行程阀 6 后，回油可经过行程阀 6 上位，于是开始了快退，到挡铁 A 切换行程阀 8 时活塞停止运动。

图 12-5 中高位油箱 10 是为了补充液压部分的漏油而设的。阀 1、2、3 及阀 4、5、6 分别为两个组合阀块。

Appendix A　Exercises

课后练习

Chapter 1

1-1　What are the differences and similarities among hydraulic pressure transmission, pneumatic pressure transmission and electrical towage?

1-2　How many parts does a hydraulic or pneumatic pressure transmission system consist of? What are they and what are their functions?

1-3　There are two significant features and two basic parameters in hydraulic or pneumatic pressure transmission operation. What are they?

Chapter 2

2-1　A hydraulic cylinder has a volume of 50L at the atmospheric pressure, as the pressure increases the volume decreases to 49.9L, how much is the rising value of the pressure? The hydraulic oil bulk elastic module K is 7000×10^5Pa.

2-2　A hydraulic oil (200mL) having a mass density of 850 kg/m^3 is flowing through an orifice for 155s tested by Engler viscometer. The same volume of water flows through it for 51s at 20℃. How much are the $^\circ E$ of this hydraulic oil, the dynamic viscosity μ(Pa·s), and the kinematic viscosity v(m^2/s)?

2-3　How much is the pressure on the ear drums of a scuba diver when he is 18m below the surface of the ocean? Salt water density is 960kg/m^3.

2-4　A hydraulic oil having a mass density of 1 000kg/m^3 is flowing through an orifice 6.35×10^{-4}m in diameter at a rate of 21.3L/min. The orifice coefficient c is 0.80. How much is the pressure differential across the orifice?

2-5　A hydraulic system reservoir contains an oil with a volume of 170L. The initial oil level is 0.9m above a 2.54cm diameter drain hole. If the drain were opened, and if the oil were assumed frictionless, how quickly would the tank drain?

2-6　A reservoir is 0.6m in diameter. If an oil is flowing into the reservoir at a rate of 13.25L/min and drawn out at a rate of 7.57L/min, how rapidly is the oil level rising?

2-7　As shown in Fig. A-1, an oil having a mass density of 900kg/m^3 is flowing around a $\theta = 45°$ bend in a 1.9cm tube having 0.09cm walls at a rate of $q = 56.78$L/min. How much are the forces F on the tube?

2-8　As shown in Fig. A-2, a container with a certain vacuum degree is located inversely in a groove (open to the atmosphere) through a tube, with an oil rising lever of 0.5m in the tube. If the density of the oil is 1000kg/m^3, how much is the vacuum degree in the container?

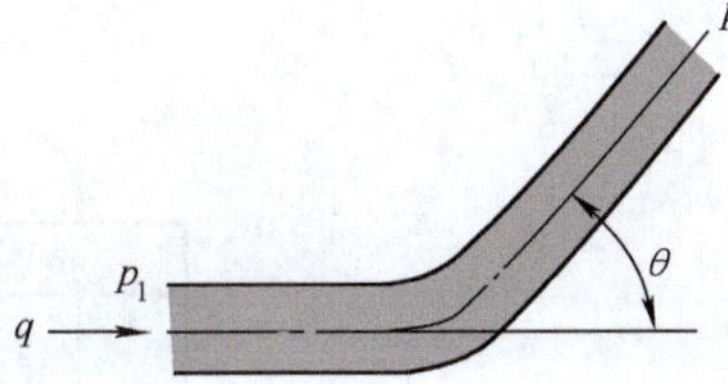

Fig. A-1

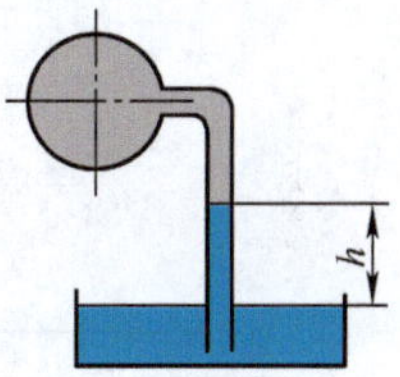

Fig. A-2

2-9 A hydraulic pump at a flow rate of 16L/min is located under the oil surface. The oil kinematic viscosity $\upsilon = 0.11\text{cm}^2/\text{s}$, the oil density $\rho = 880\text{kg/m}^3$, the coefficient of local resistance in the bend position $\varepsilon = 0.2$, and the other dimensions are shown in the Fig. A-3. How much is the absolute pressure at the inlet port of the hydraulic pump?

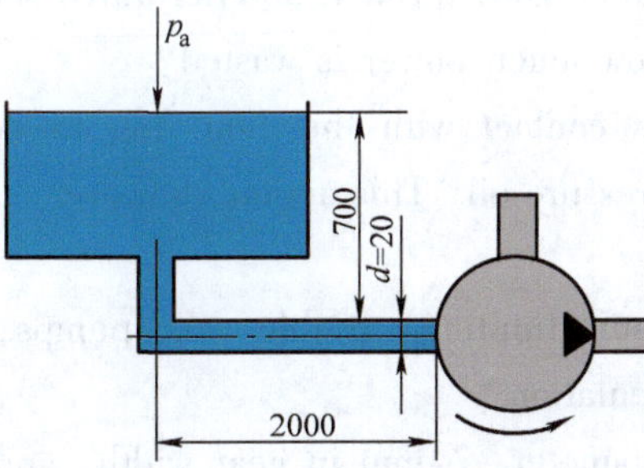

Fig. A-3

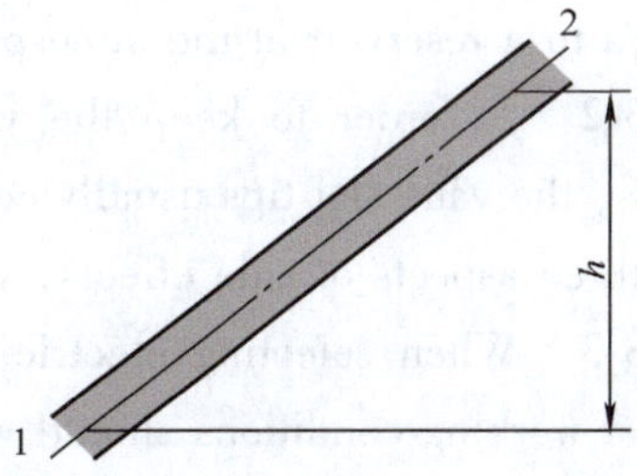

Fig. A-4

2-10 A type of liquid with a density of 900kg/m^3 is flowing through a tube. Here h is 15m (see the Fig. A-4), the pressures at 1 and 2 are 4.5×10^5Pa and 4×10^5Pa respectively. What is the flowing direction of the liquid in the tube?

2-11 As shown in Fig. A-5, a flat is inserted into the free jet of water and vertical to the jet axes, cutting off part of the jet flow q_1 and causing the rest jet flow to swerve an angle of α. If the jet speed $v = 30\text{m/s}$, the total flow rate $q = 30\text{L/s}$, and $q_1 = 12\text{L/s}$, how much is the angle of α and the applying force F on the flat?

2-12 As shown in Fig. A-6, a piston with 20mm in diameter is moving downwards under a force of 150N, which forces the oil in the hydraulic cylinder out to the atmosphere through a narrow gap ($\delta = 0.05\text{mm}$). Assume that the piston and the cylinder tube are set inline and the gap length is 70mm, and the oil dynamic viscosity $\mu = 50 \times 10^{-3}\text{Pa} \cdot \text{s}$. How long would it experience for the piston to drop down by 0.1m?

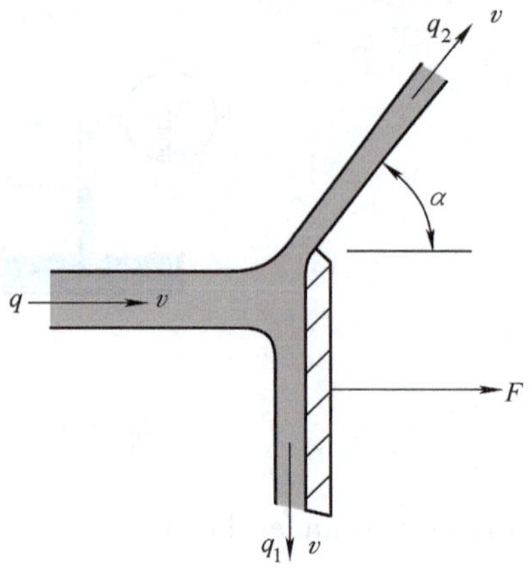

Fig. A-5

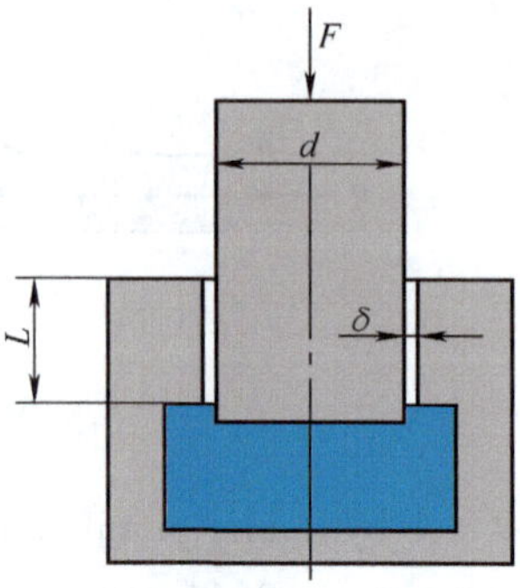

Fig. A-6

Chapter 3

3-1 A fixed delivery pump operating at 1750 rpm delivers 9. 8mL/r. The load requires an average flow rate of 13. 25L/min. The balance of the flow goes through a relief valve set at 1.03×10^4kPa to a reservoir at the atmospheric pressure. How much power is wasted?

3-2 In order to keep the vane in continuous contact with the cam ring in double-acting pumps, the vane slot tips usually connect with high pressure oil. This action, however, may bring about three aspects of side effects, what are they?

3-3 When selecting electric machine for pressure-limiting variable vane pumps, how much kind of working conditions should we base on for calculation?

3-4 A gear pump is 48mm in the top circle diameter, 24mm in gear width, and has a gear number of 14. If the highest working pressure is 10MPa, and the motor speed is 980r/min, what is the power of the motor? (Volumetric efficiency of the pump $\eta_{pV} = 0.90$, the total efficiency $\eta_p = 0.8$.)

3-5 An equal clearance 0. 04mm exists between the two side surfaces and the gears ($s_1 = s_2 = 0.04$mm) in a gear pump. The rotating speed is 1000r/min, the output flow rate is 20L/min under a working pressure of 2. 5MPa and the volumetric efficiency is 0. 90. After working for a certain period, s_1 expands to 0. 042mm and s_2 to 0. 048mm due to friction (the other clearances were assumed to be not changed). If the working pressure and the rotating speed of the pump are constant, how much is the volumetric efficiency? (Reminder: the surface clearance leakage takes up 85% of the total leakage when $s_1 = s_2 = 0.04$mm.)

Chapter 4

4-1 A speed reducer requires the hydraulic motor to put out a torque of 52. 5N · m and a rotating speed of 30r/min. The hydraulic motor delivery is 12. 5cm^3/r, and both the volumetric and mechanical efficiencies of the hydraulic motor are 0. 9. How much are the flow rate and the pressure needed?

4-2 A hydraulic motor has a delivery of 70mL/r, providing an oil pressure of 10MPa and an input flow rate of 100L/min. If its volumetric and mechanical efficiencies are 0. 92 and 0. 94 respectively, and the backpressure is 0. 2MPa in the return oil chamber of the hydraulic motor, try to calculate:

(1) the output torque of the hydraulic motor;

(2) the rotating speed of the hydraulic pump.

4-3 A hydraulic motor has a delivery of 40mL/r, and the practical input flow rate is 63L/min under a pressure of 6. 3MPa and a rotating speed of 1450r/min. If the practical output torque is 37. 5N m, what are the volumetric, mechanical and total efficiencies of the hydraulic motor?

4-4 As shown in Fig. A-7, the effective areas of two hydraulic cylinders are $50cm^2$ and $20cm^2$ respectively. The loads they are undertaking are 5000N and 4000N respectively, and the flow rate of the hydraulic pump is 3L/min. This system is assumed lossless, how much are the working pressures of the two cylinders and how quickly the two pistons would move?

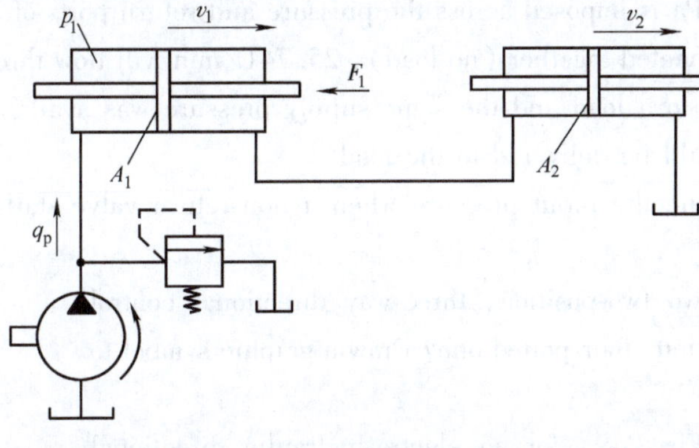

Fig. A-7

4-5 For a differential hydraulic cylinder, if the feed speed is three times the retract speed, how much is the ratio of the cross areas of the two piston rods?

4-6 A hydraulic cylinder has a piston 100mm in diameter and a piston rod 70mm in diameter. The oil flow rate is 25L/min and the pressure is 20×10^5Pa. The backpressure of the return oil is 2×10^5Pa. Try to calculate the largest thrusts, the moving speeds and their directions at the situations of a, b, c(see the Fig. A-8).

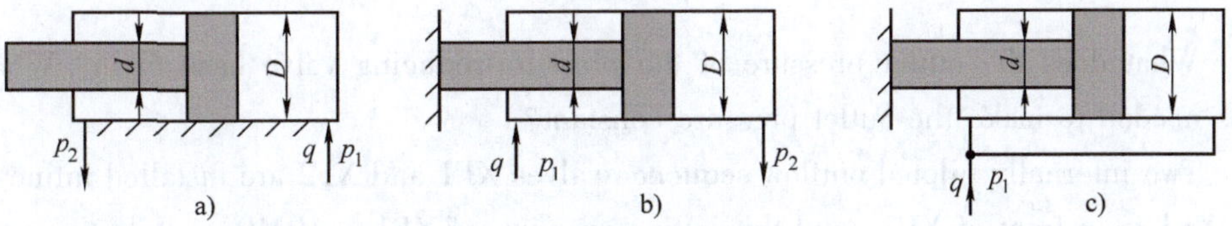

Fig. A-8

4-7 A single-vane swing hydraulic cylinder is 40mm in vane shaft radius, 100mm in cylinder body radius and 10mm in vane width. If the load torque is 600N · m, how much is the pressure of the input oil?

4-8 A single-rod hydraulic cylinder is in differential connection when moves forwards. But pressurized oil is put into the rod chamber when retracts. If the cylinder tube and the piston rod diameters are 100mm and 70mm respectively, the piston rod is under pressure when moves at a low speed, the load is 25000N, the input flow rate is 25L/min, the backpressure of the return oil is 2×10^5Pa, then:

(1) How much is the moving speed of the piston?

(2) How thick is the cylinder tube wall when the allowable stress of the cylinder material is 50MPa?

Chapter 5

5-1 The supply pressure at a valve inlet port is 1.8×10^4kPa. The pressure drop from the return port to the reservoir is 1.7×10^3kPa. The reservoir is pressurized at 413.7Pa. The valve is sized to provide maximum power to the load when the load flow is 22.71L/min. How much is the maximum power?

5-2 When 2×10^4kPa is imposed across the pressure and return ports of a valve and when the two cylinder ports are connected together (no load), 25.74L/min will flow through the valve. If the same valve was connected to a load and the same supply pressure was available, how much is the maximum power which could be delivered to the load?

5-3 How to determine the input pressure when a non-return valve starts to work? The open pressure is 0.04MPa.

5-4 Can we use two two-position, three-way directional control valve instead of a two-position, four-ported one? Draw a graphic symbol to illustrate it.

5-5 As shown in Fig. A-9, for an electro-hydraulic directional control valve, the cylinder is motionless even when solenoid is actuated, analyze the reasons and propose some improvement measures.

5-6 For a pilot operated relief valve, in which the damping hole on the main valve is blocked up by dirties, what possible barriers may occur? Why?

5-7 When the pre-compression of pressure-adjusting spring in a relief valve is constant, why will the inlet pressure fluctuate with the flow rate?

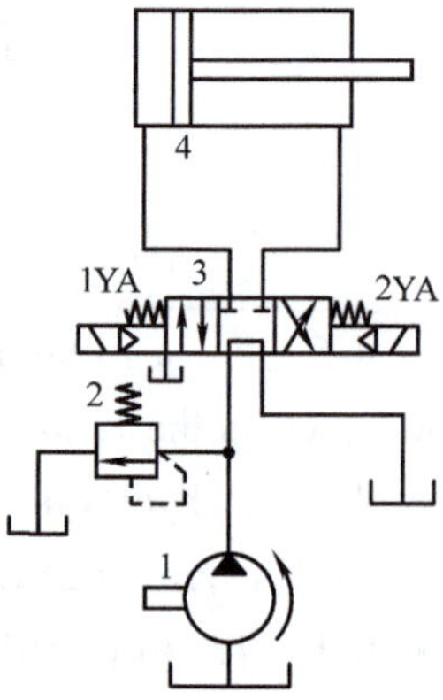

Fig. A-9

5-8 What does the outlet pressure of the pressure-reducing valve base upon? What are the conditions needed to make the outlet pressure constant?

5-9 Two internally piloted outflow sequence valves XF1 and XF2 are installed inline in a same oil path. XF1 is in front of XF2, and the setting pressure of XF1 is 10MPa, while the setting pressure of XF2 ranges from 5MPa to 15MPa. How would the inlet pressure of XF1 change with the pressure of XF2? At which state the valve would be?

5-10 Compare the relief valves, pressure-reducing valves and the sequence valves (internally piloted outflow type) and describe their differences and similarities.

5-11 If the flow rate through the throttle valve is constant, change the openings of the throttle valve, which parameters would change and how?

5-12 If the inlet and outlet ports of a speed regulation valve are installed inversely, will the speed regulation valve still work normally? Why?

5-13 When the speed regulation valve is installed in the oil return path or the bypass oil path, why can we use common fixed displacement pressure-reducing valve or throttle valve?

5-14 Why cannot a bypass speed regulation valve regulate the oil speed for the oil return path?

Chapter 6

6-1 An accumulator is used as the power source, whose volume is 4L and the pressure of filled air is 3.2MPa. The highest and lowest working pressures of this system are 8MPa and 5MPa respectively, how much is the discharge volume of the accumulator? (The working state of the accumulator can be viewed as an isothermal process.)

6-2 The pressure of filled air in an accumulator is 9MPa, and an oil is provided by a pump at a flow rate of 5L/min. The accumulator exhausts the oil to the system quickly when the pressure raises to 20MPa. The discharge volume is 5L when the pressure drops to 10MPa. How much is the volume of the accumulator?

6-3 Why don't we install a pure filter in the inlet port of the hydraulic pump?

6-4 In a hydraulic system, the working pressure is 7MPa, and the flow rate is 60L/min.

(1) How much is the pressure of the feed pipe?

(2) How to select the oil pipe material and the wall depth?

Chapter 7

7-1 Try to make a three stages pressure-regulating and unloading circuit and describe briefly its working principle, using a pilot operated relief valve, two distant pressure regulation valves and two solenoid slide valves with two-position and two-way.

7-2 As is shown in the Fig. A-10, the effective working areas of the hydraulic cylinders $A_{11} = A_{21} = 100\text{cm}^2$, $A_{12} = A_{22} = 50\text{cm}^2$. The working loads of two cylinders: $F_{L1} = 35\ 000\text{N}$, $F_{L2} = 25\ 000\text{N}$, the setting pressures of the relief valve, sequence valve and pressure-reducing valve are 5MPa, 4MPa and 3MPa respectively. If friction resistance, inertial force, the pressure loss in pipelines or directional control valve are neglected, how much are the pressures p_1, p_2 and p_3 under the working conditions as follows?

(1) The hydraulic pump starts and the two directional control valves are at middle positions.

(2) Solenoid 2YA is actuated and cylinder Ⅱ reaches the end stop when moving forwards at working feed speed.

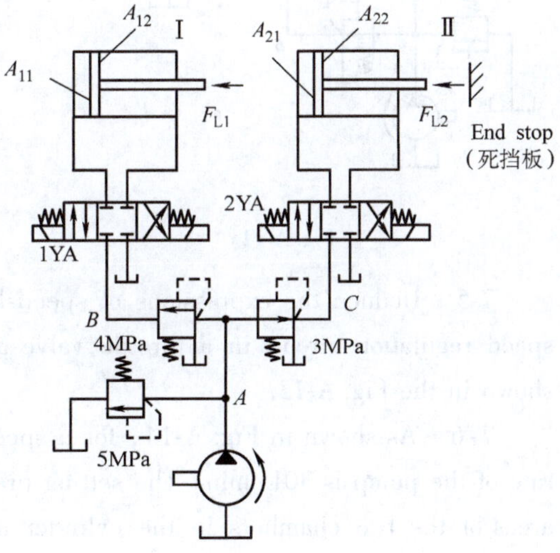

Fig. A-10

(3) Solenoid 2YA is released and 1YA is actuated, cylinder Ⅰ loses load suddenly when reaching the end.

7-3 As shown in Fig. A-11, for a clamping circuit, if the setting pressures of the relief valve and the pressure-reducing valve are 5MPa and 2.5MPa respectively, analyze:

(1) Before clamping a workpiece, the clamping cylinder would move without load, then how much are the pressures of points A, B and C?

(2) After clamping a workpiece, the outlet pressure of the pump is 5MPa, how much are the pressures of points A and C?

(3) After clamping a workpiece, the outlet pressure would reduce to 1.5MPa due to other actuators' quick feed actions, how much are the pressures of points A and C?

7-4 A hydraulic system is shown in the Fig. A-12; the total weight of the vertical hydraulic piston and the moving parts is G. The areas of the two chambers are A_1 and A_2 respectively, and the largest working pressures of pump 1 and pump 2 are p_1 and p_2 respectively. If the pressure in the pipelines were assumed lossless, then:

(1) What types of valve do valves 4, 5, 6, 9 belong to? And what are their functions in the system?

(2) How to adjust the pressures of valves 4, 5, 6, 9?

(3) Which basic circuits does this system consist of ?

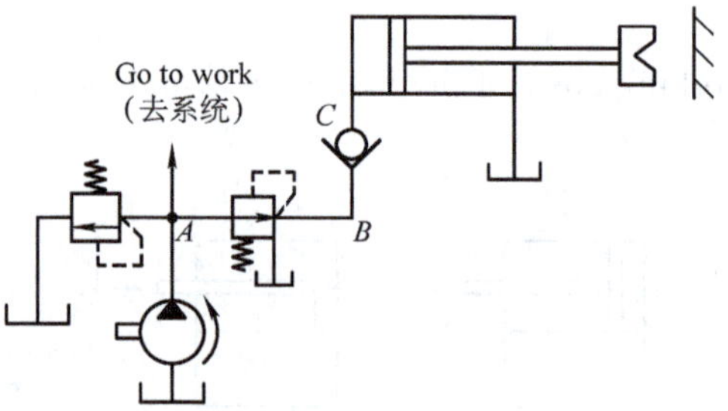

Fig. A-11

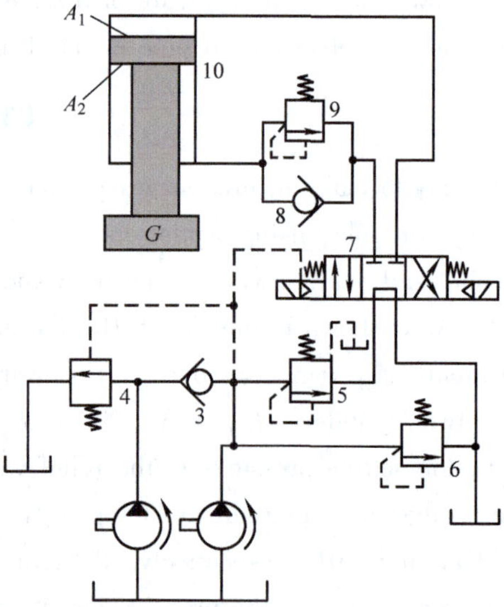

Fig. A-12

7-5 Deduce the expressions of speed-load characteristic, speed stiffness and efficiency of a speed regulation circuit in a throttle valve-used oil return path. The other known conditions are shown in the Fig. A-13.

7-6 As shown in Fig. A-14, for a special milling machine hydraulic system, the output flow rate of the pump is 30L/min. The setting pressure of the relief valve is 2.4MPa, and the operation areas of the two chambers in the cylinder are 50cm^2 and 25cm^2 respectively. The cutting load is 9000N and the friction load is 1000N. The flow rate through the adjusting valve when cutting is

1.2L/min. If the leakage of the components and the pressure loss are neglected, then:

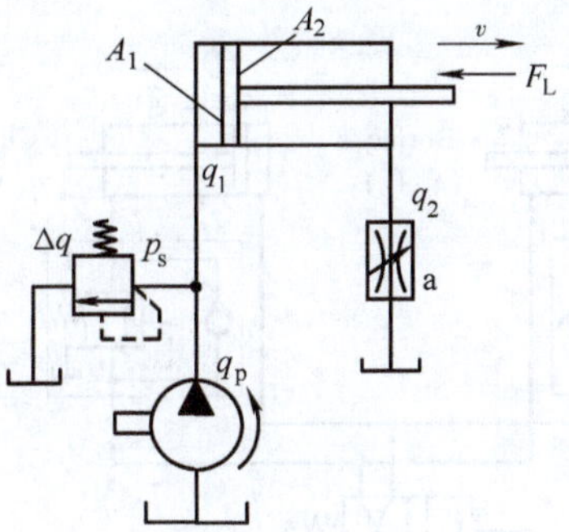

Fig. A-13

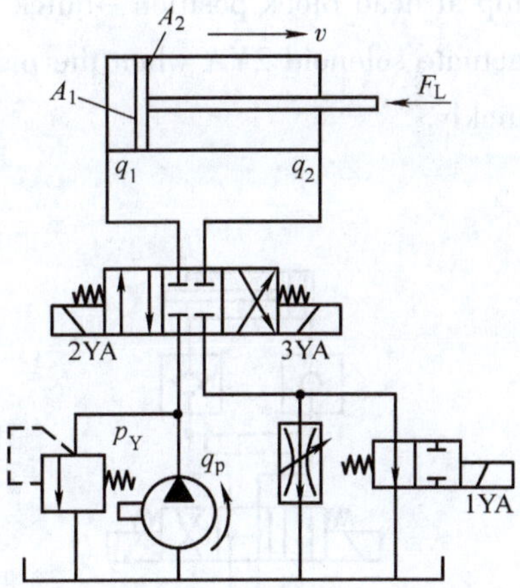

Fig. A-14

(1) How much is the feed speed v_1 of the piston and the circuit efficiency η_1 when the piston is closing to the workpiece at a high speed?

(2) How much is the working feed speed v_2 and the circuit efficiency η_2 during cutting at feed speed?

7-7 In a variable displacement pump-fixed displacement motor circuit, the rotating speed and the delivery of the variable displacement pump are 1500r/min and 8mL/r respectively; while the delivery of the fixed displacement motor is 10mL/r; the setting pressure of the safety valve is 40×10^5Pa; and the volumetric and mechanical efficiencies of the pump and motor are all 0.95, i.e., $\eta_{pV} = \eta_{pm} = \eta_{MV} = \eta_{Mm} = 0.95$.

(1) How much is the delivery of the pump when the rotating speed of the motor is 1000r/min?

(2) How much is the rotating speed of the motor when the load torque of the motor is 8N · m?

(3) How much is the largest output power of the pump?

7-8 Draw the operating principle chart of a displacement speed regulation circuit composed by a throttle valve mounted in the oil return path and a pressure differential type variable displacement pump.

7-9 In a two stages speed conversion circuit achieved by two speed regulation valves shown in the Fig. A-15, why is the open area of the front speed regulation valve larger than that of the later speed regulation valve? If three stages feed speed conversion is to be achieved without adding any speed regulation valve, what improvement can be made based upon the present circuit?

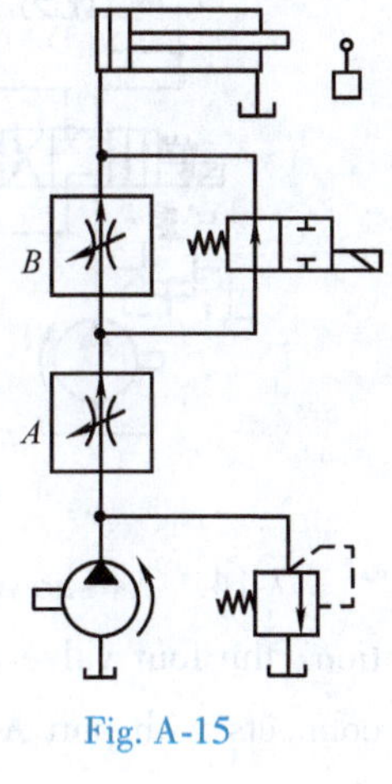

Fig. A-15

7-10 As shown in Fig. A-16, in a lock circuit shown in the figure, why should we use a directional control valve with H or Y type neutral configuration? If we use a valve with M type neutral configuration, what problems may occur?

7-11 For the sequence control circuit shown in the Fig. A-17, can the sequence control valve 3 be replaced by an outside control type? Illustrate it with a chart.

7-12 As shown in Fig. A-18, the working cycle of a hydraulic system is: quick feed→working feed→stop at dead block position→quick retract→stop at original position. The pressure switch signals to actuate solenoid 2YA when the piston is stopped by the dead block, and then the piston retracts quickly.

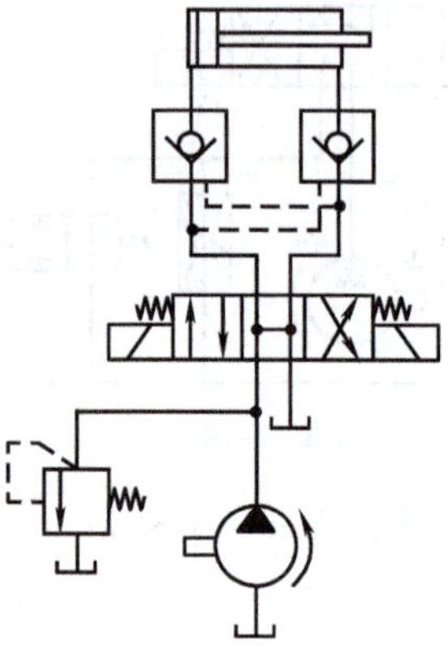

Fig. A-16

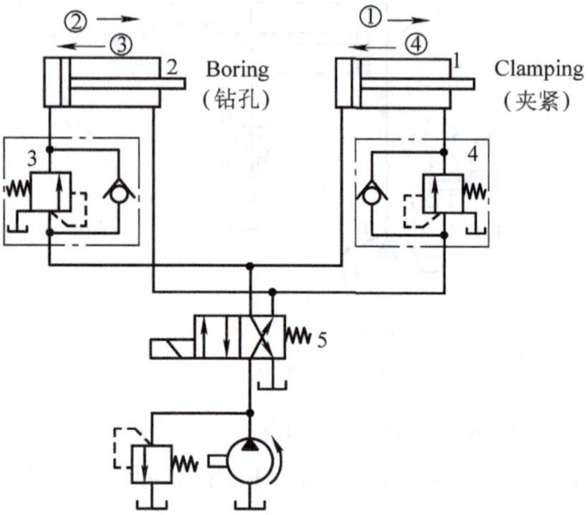

Fig. A-17

(1) How to determine the acting pressure of the pressure switch?

(2) If the circuit is changed into a speed regulation circuit in the oil return path with a throttle valve, how to install the pressure switch? Illustrate its operating principle.

7-13 Which basic circuits are included in the circuit shown in the Fig. A-19? What are the functions of cartridge valves 1, 2, 3, 4, and 5 in the circuit?

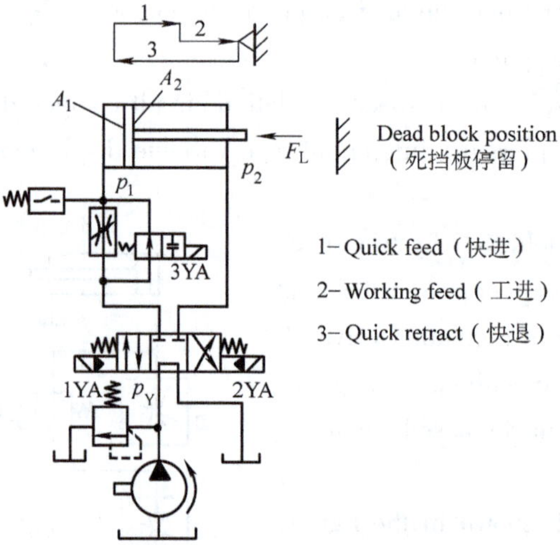

Fig. A-18

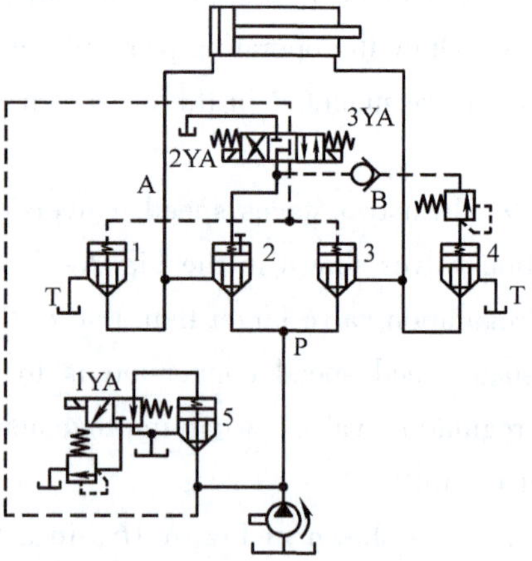

Fig. A-19

7-14 As shown in Fig. A-20, in a hand-operated slide valve with H type neutral configuration, the four valve openings are equal, i. e., $x_1 = x_2 = x_3 = x_4$. At the left position of the valve: P connects with port A, and port B connects with port T; at the right position of the valve: P connects with B, A connects with T. How do the openings of the four valves change when the slide valve shifts from the neutral position to the left position or the right position?

In addition, in engineering mechanical occasions, valves of this type are usually used to regulate speed, what is the operating principle and what kind of speed regulation circuit does it belong to?

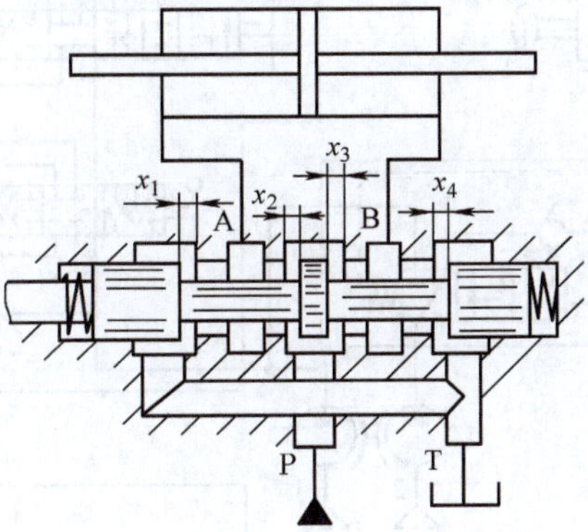

Fig. A-20

Chapter 8

8-1 The Fig. A-21 shows the operating principle chart of a ZL50 loader hydraulic system. The actuators include turning cylinders, bucket cylinders and boom cylinders. The power source is provided by a twin pumps and a single pump driven by a diesel engine, whose working speed ranges from 600r/min to 2200r/min. The working characteristics of the loader are: 1) flow switching valve 4 ensures a constant flow rate q_z to make the turning parts operate at a speed ranging from 600r/min to 1500r/min. q_z increases with n when $n > 1500$r/min. Steering valve 6 is a hand-operated servo valve. Operating the turning handwheel can open the valve which locates at the upstream of the steering cylinder, the valve returns to its neutral position by mechanical feedback when the turning handwheel is released. 2) Boom cylinder and bucket cylinder can operate either individually or simultaneously, without the influence of the turning actions.

(1) Which speed control type does the steering cylinder belong to? What is the function of valve 5 here?

(2) Which connection form does the multiple directional control valve 11 belong to? How does it meet the requirements of the bucket cylinder and the boom cylinder?

(3) How to regulate the speed of the bucket cylinder and the boom cylinder?

(4) What is the function of valve 10 in this system?

8-2 The principle chart of a shearing machine hydraulic system is shown in the Fig. A-22. The scissors move up and down driven by the main cylinder. The working circle is: start without load→idle stroke→cutting→down stroke with cushion→return stroke at a high speed. The main cylinder can stop and retract any time during the down stroke. In order to locate the scissors, the main cylinder must have the capacity of locating line under light pressure, then the force applying on the

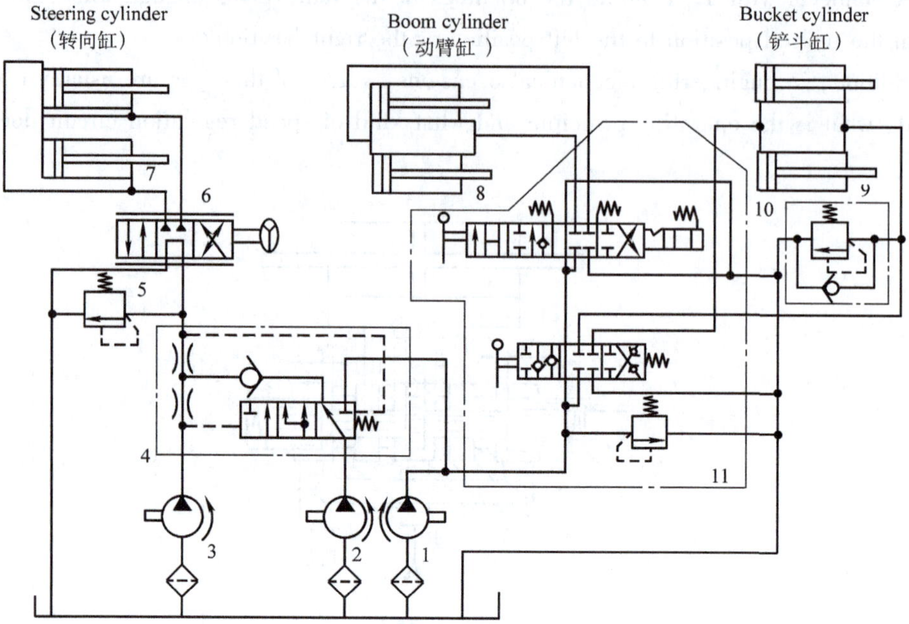

Fig. A-21

1—Operating pump(工作泵) 2—Auxiliary pump(辅助泵) 3—Steering pump(转向泵)
4—Flow swiching valve(流量转换阀) 5—Safety valve(安全阀) 6—Steering valve(转向阀)
7—Steering cylinder(转向缸) 8—Boom cylinder(动臂缸) 9—Buchet cylinder(铲斗缸)
10—Double acting safety valve(双作用安全阀) 11—Multiple directional control valve(多路换向阀)

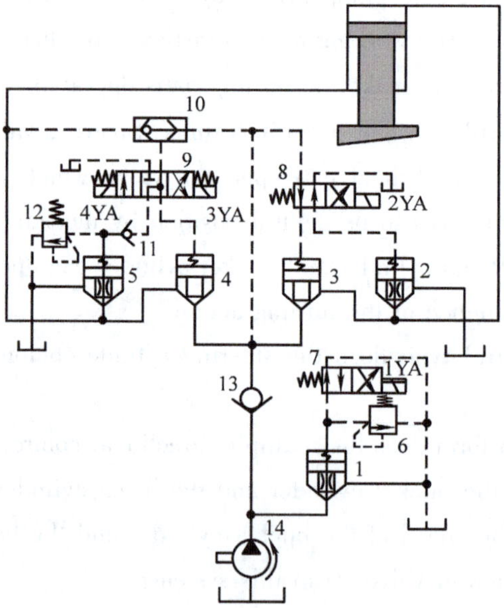

Fig. A-22

scissors in their return stroke would be quite small so as not to damage the plate material.

(1) What is the operating principle of this shearing machine hydraulic system? Mark the oil flow directions of each action and make a list of the solenoid action sequence.

(2) What are the functions of valves 5, 2, 1, and 10 in this system?

Chapter 9

9-1 In some rotation transmission devices with larger quality such as centrifugal machines and roller machines, the load of motor is virtually a fly wheel. The iron casting fly wheel is 1m in outer diameter, 200mm in width, 100mm in journal radius and under a heavy burden of 12.5kN. The transmission ratio of gear supercharge mechanism $i = n_1/n_2 = 0.2$ (see Fig. A-23), the steady speed of the fly wheel is 200r/min. Accelerating or decelerating time is set as 2s. The mechanical efficiencies of the hydraulic motor and the gear supercharge mechanism are 0.95 and 0.90 respectively, and the static and dynamic friction factors of the journal are 0.2 and 0.08 respectively. Draw the load circle chart of the hydraulic motor and calculate the largest output torque.

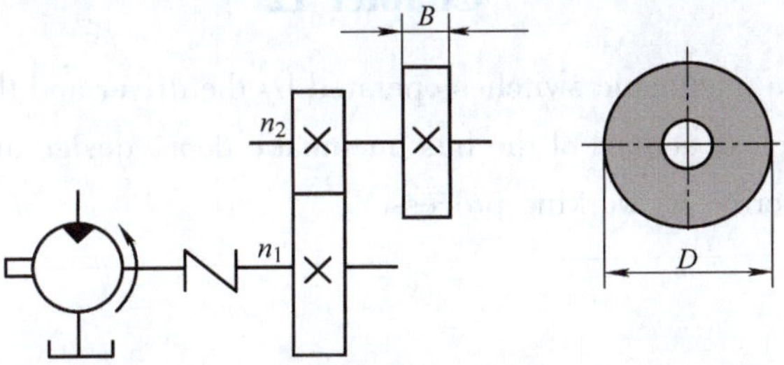

Fig. A-23

9-2 For a horizontal drilling modular machine tool with single-surface and multiple shafts, the working circle of the dynamic slipway is: quick feed→working feed→quick return→end. Requirements for the main performance of this hydraulic system are: axial cutting force $F_t = 24\ 000$N; total weight of the slipway moving parts is 5 000N; the accelerate or decelerate time is 0.2s; the static and dynamic friction factors are 0.2 and 0.1 when flat guide rail is employed; the quick feed stroke and working feed stroke are 200mm and 100mm respectively; both the quick feed speed and quick retract speed are equal to 3.5m/min, the working feed speed ranges from 30mm/min to 50mm/min. Try to design such a dynamic slipway hydraulic system with the capacities of moving steadily when working and being stopped any time.

Chapter 10

10-1 To what degree of quality does the compressed air require for the pneumatic system? How can we ensure such quality of compressed air? Name the equipments needed and illustrate briefly their working principle.

10-2 Describe the working process and principle of an impact cylinder.

10-3 What is the connection order of the three principal components? Why?

Chapter 11

11-1 Design a self-lock circuit with four cylinders.

11-2 A two-hand operation circuit (to protect the operator's hands) shown in the Fig. A-24 would not work if a spring in an operating valve is broken. Try to design another two-hand operation circuit to protect the operator's hands and can still work normally even a spring is broken.

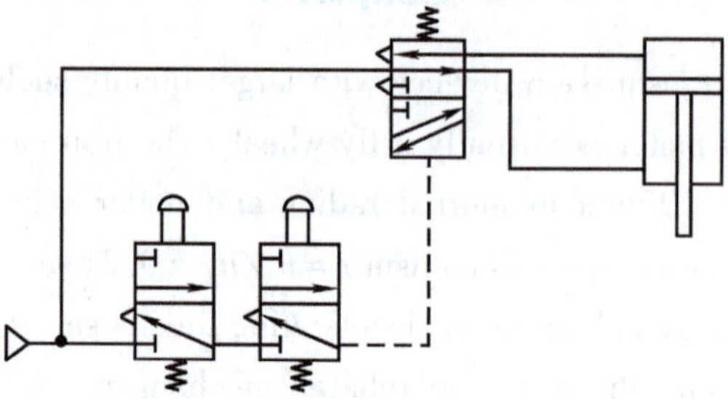
Fig. A-24

Chapter 12

12-1 There are two pneumatic switches operated by the driver and the ticket seller respectively to control the open and close actions of the bus pneumatic door, design an pneumatic control circuit for the bus door and describe its working process.

Appendix B　Graph Symbols of Common Hydraulic and Pneumatic Components

常用液压与气动元件图形符号

（GB/T 786.1—2009）

Table 1　Basic symbols, pipelines and connections

（基本符号、管路及连接件）

Name（名称）	Symbol（符号）	Name（名称）	Symbol（符号）
Working pipeline（工作管路）		Pipe end connects to the reservoir bottom（管端连接于油箱底部）	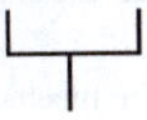
Control pipeline（控制管路）		Closed reservoir（密闭式油箱）	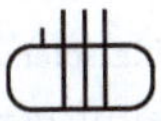
Connection pipe（连接管路）	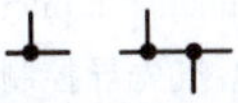	Exhaust directly（直接排气）	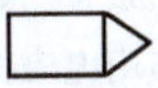
Crossing pipeline（交叉管路）	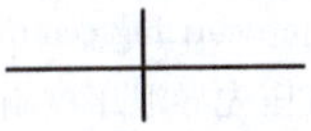	Exhaust indirectly（带连接排气）	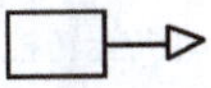
Flexiable pipeline（柔性管路）		Quick change coupler with check valve（带单向阀快换接头）	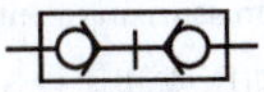
Composite component line（组合元件）		Quick change coupler without check valve（不带单向阀快换接头）	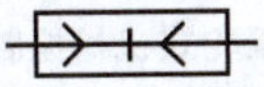
Reservoir above the oil surface of pipe mouth（管口在液面以上油箱）	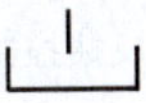	Single path rotary joint（单通路旋转接头）	
Reservoir below the oil surface of pipe mouth（管口在液面以下的油箱）	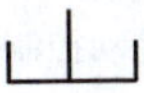	Three-way rotary joint（三通路旋转接头）	

Table 2 Control mechanisms and methods
(控制机构和控制方法)

Name(名称)	Symbol(符号)	Name(名称)	Symbol(符号)
Button control by manpower (按钮式人力控制)	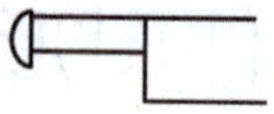	Hydraulic pilot control (液压先导控制)	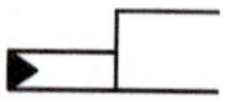
Handle control by manpower (手柄式人力控制)	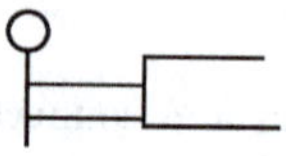	Hydraulic two-stage pilot control (液压二级先导控制)	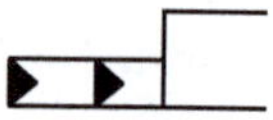
Spring control (弹簧控制)	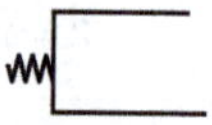	Air-hydraulic pilot control (气-液先导控制)	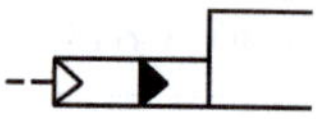
One-way roller mechanical control (单向滚轮式机械控制)	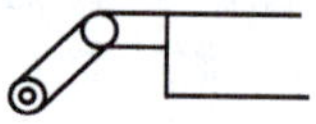	Inner pressure control (内部压力控制)	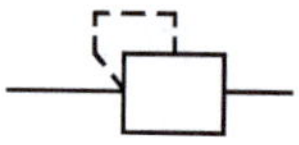
Single-acting solenoid control 单作用电磁控制	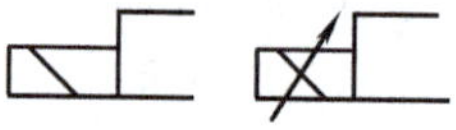	Electro-hydraulic pilot control (电-液先导控制)	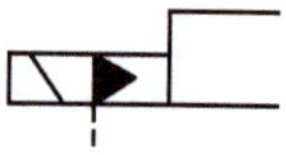
double-acting solenoid control (双作用电磁控制)	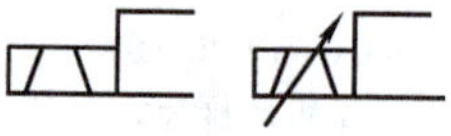	Electro-pneumatic pilot control (电-气先导控制)	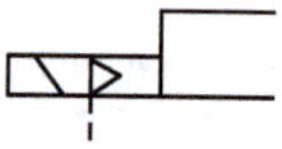
Motor rotary control (电动机旋转控制)	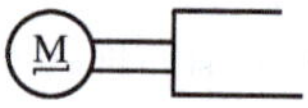	Hydraulic pilot depressurizing control (液压先导泄压控制)	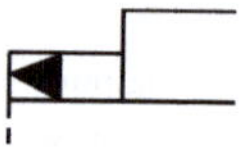
Pressurizing or depressurizing control (加压或卸压控制)	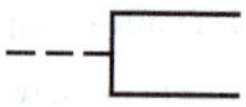	Outer pressure control (外部压力控制)	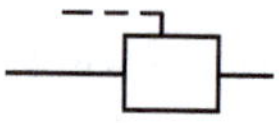
Roller mechanical control (滚轮式机械控制)	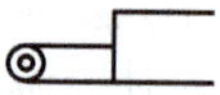	Pneumatic pressure pilot control (气压先导控制)	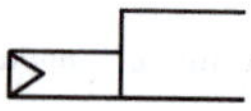
Pedal control by manpower (踏板式人力控制)	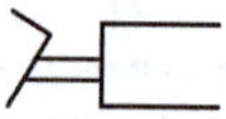	Electronic feedback control (电反馈控制)	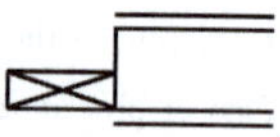
Eject-rod mechanical control (顶杆式机械控制)	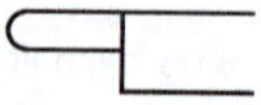	Differential control (差动控制)	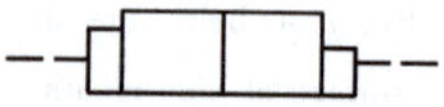

Table 3 Pumps, motors and cylinders
（泵、马达和缸）

Name(名称)	Symbol(符号)	Name(名称)	Symbol(符号)
One-way fixed displacement pump (单向定量液压泵)	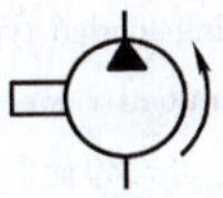	Double-way fixed displacement motor (双向定量马达)	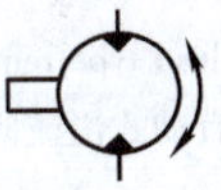
Double-way fixed displacement pump (双向定量液压泵)	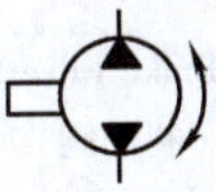	Fixed displacement pump/motor (定量液压泵/马达)	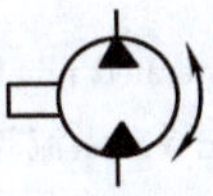
One-way variable displacement pump (单向变量液压泵)	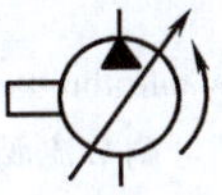	Variable displacement pump/motor (变量液压泵/马达)	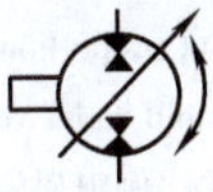
Double-way variable displacement pump (双向变量液压泵)	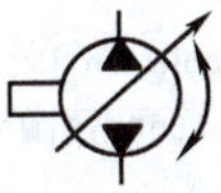	Double-acting single piston rod cylinder (双作用单活塞杆缸)	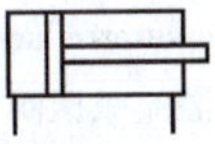
One-way fixed displacement motor (单向定量马达)	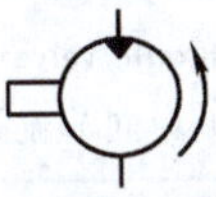	Double-acting double piston rod cylinder (双作用双活塞杆缸)	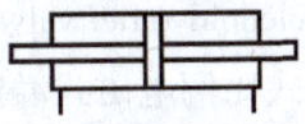
Hydraulic integer transmission device (液压整体式传动装置)	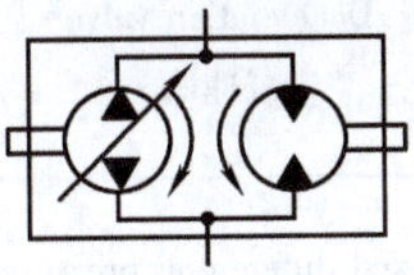	Double-way variable displacement motor 双向变量马达	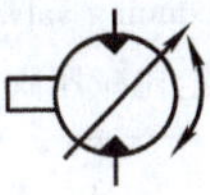
Swing motor (摆动马达)	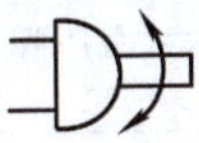	One-way buffer cylinder (单向缓冲缸)	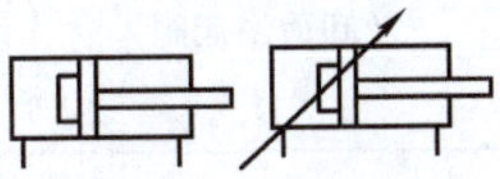
Single-acting cylinder returned by spring (单作用弹簧复位缸)	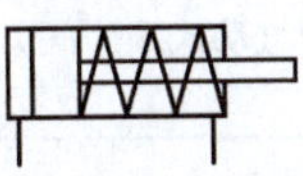	Double-way buffer cylinder (双向缓冲缸)	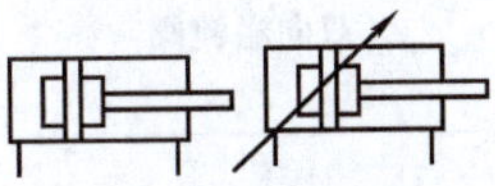
Single-acting telescopic cylinder (单作用伸缩缸)	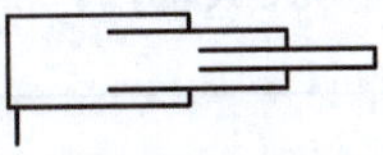	Double-acting telescopic cylinder (双作用伸缩缸)	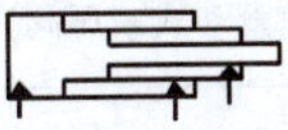
One-way variable displacement motor (单向变量马达)	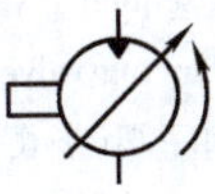	Supercharger (增压器)	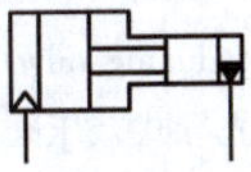

Table 4 Control elements

(控制元件)

Name(名称)	Symbol(符号)	Name(名称)	Symbol(符号)
Spring load type relief valve (直动式溢流阀)	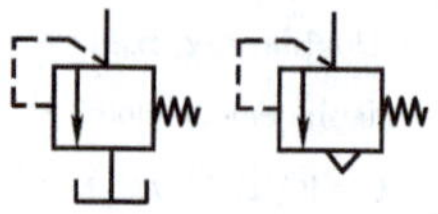	Spring loaded type unload valve (直动式卸荷阀)	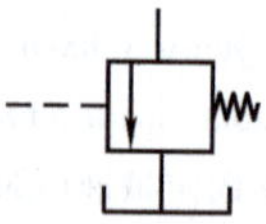
Pilot operated relief valve (先导式溢流阀)	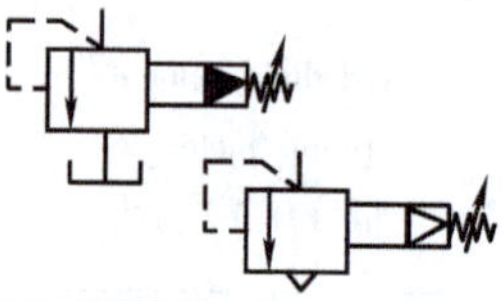	Brake valve (制动阀)	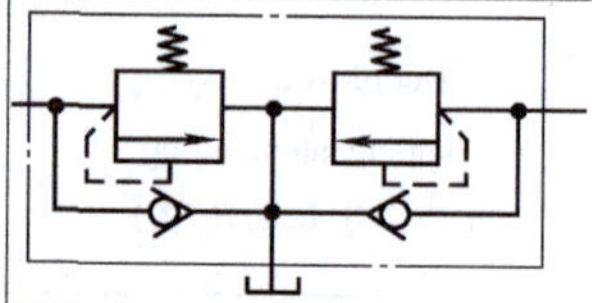
Pilot proportional solenoid relief valve (先导式比例电磁溢流阀)	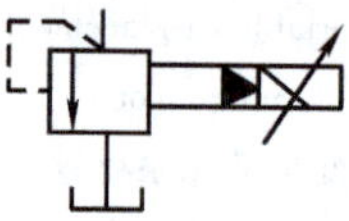	Unadjustable throttle valve (不可调节流阀)	
Relief pressure reducing valve (溢流减压阀)	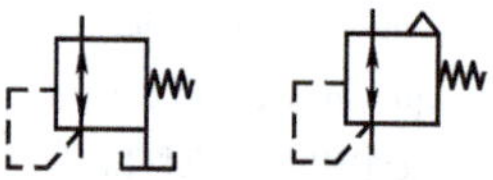	Adjustable throttle valve (可调节流阀)	
Pilot proportional solenoid relief valve (先导式比例电磁式溢流阀)	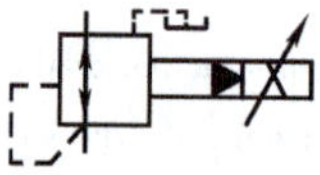	Adjustable check throttle valve (可调单向节流阀)	
Proportioning pressure reducing valve (定比减压阀)	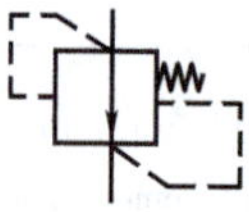	Deceleration valve (减速阀)	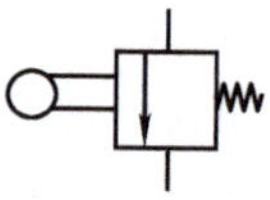
Unloading relief valve (卸荷溢流阀)	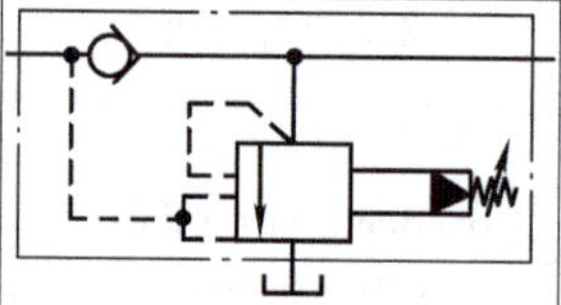	Fixed differential pressure reducing valve (定差减压阀)	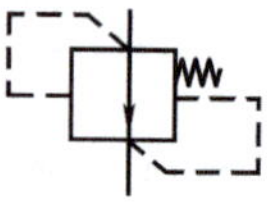
Double-way relief valve (双向溢流阀)	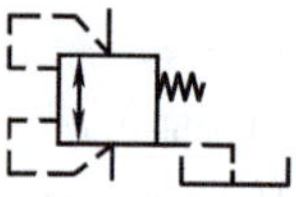	Spring load type sequence valve (直动式顺序阀)	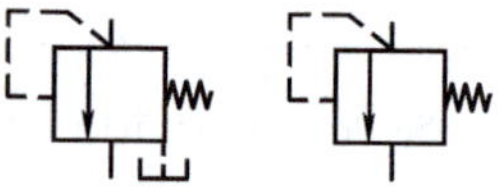
Sping loaded type pressure reducing (直动式减压阀)	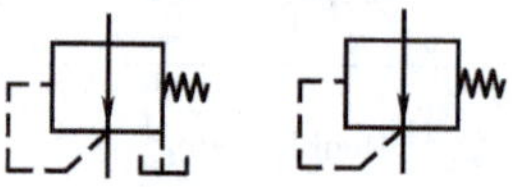	Pilot operated sequence valve (先导式顺序阀)	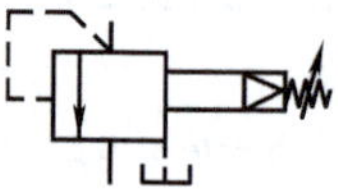
Pilot operated pressure reducing valve (先导式减压阀)	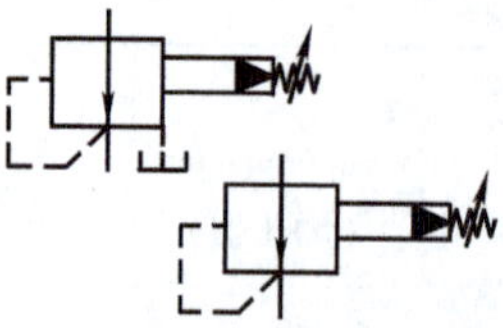	Check sequence valve (balancing valve) [单向顺序阀(平衡阀)]	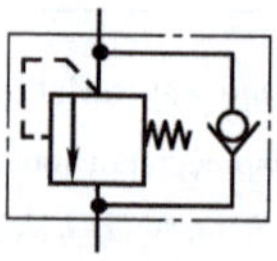

（续）

Name(名称)	Symbol(符号)	Name(名称)	Symbol(符号)
Flow-combining valve (集流阀)	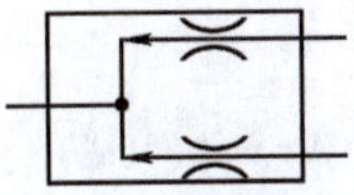	Diverting valve (分流阀)	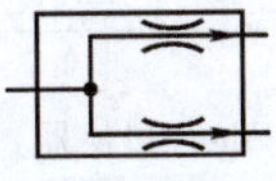
Flow dividing/combining valve (分流集流阀)	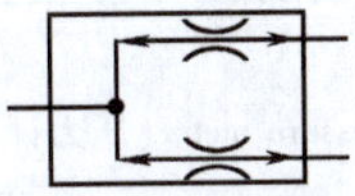	Three-position, four-way directional control valve (三位四通换向阀)	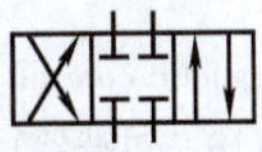
Check valve (单向阀)		Three-position, five-way directional control valve (三位五通换向阀)	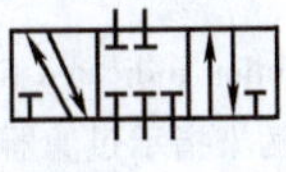
Pilot-controlled check valve (液控单向阀)		AND' shuttle valve (双压阀)	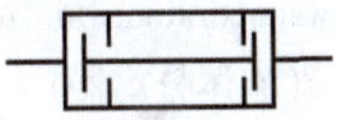
Hydraulic lock (液压锁)	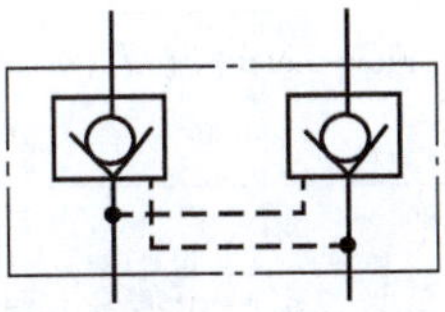	Quick exhaust valve (快速排气阀)	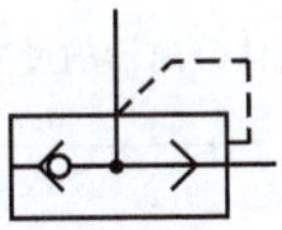
OR' shuttle valve (梭阀)	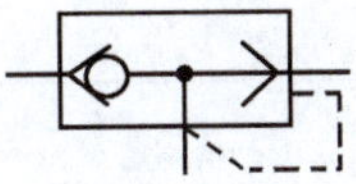	Two-position, two-way directional control valve (二位二通换向阀)	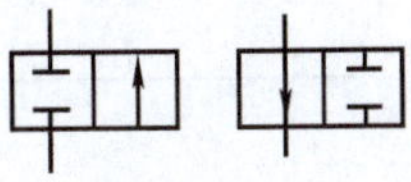
Throttle valve with silencer (带消声器的节流阀)	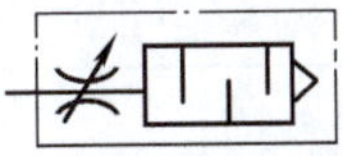	Two-position, three-way directional control valve (二位三通换向阀)	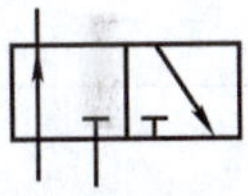
Speed regulating valve (调速阀)	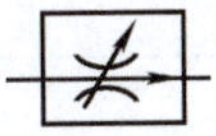	Two-position, four-way directional control valve (二位四通换向阀)	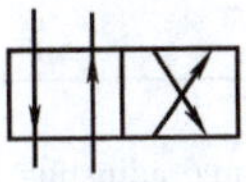
Temperature compensation speed regulating valve (温度补偿调速阀)	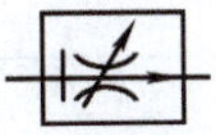	Two-position, five-way directional control valve (二位五通换向阀)	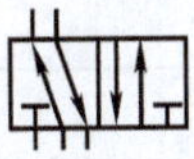
Bypass speed regulating valve (旁通型调速阀)		Four-way electro-hydraulic servo valve (四通电-液伺服阀)	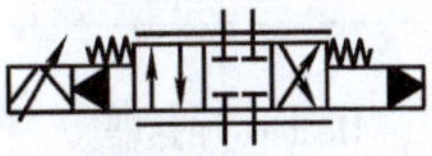
One-way speed regulating valve (单向调速阀)	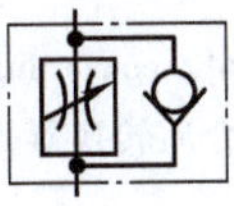		

Table 5 Assistant elements

(辅助元件)

Name(名称)	Symbol(符号)	Name(名称)	Symbol(符号)
Filter(过滤器)		Air tank(气罐)	
Magnetic core filter (磁芯过滤器)		Pressure meter(压力计)	
Pollution indicated filter (污染指示过滤器)	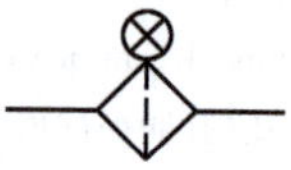	Liquid level gauge(液位计)	
Dewater drainage device (流体分离器)	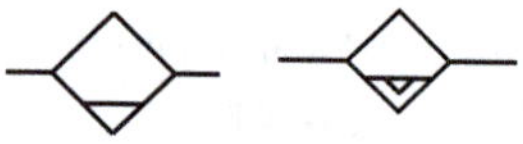	Temperature gauge(温度计)	
Air filter (带流体分离器的过滤器)	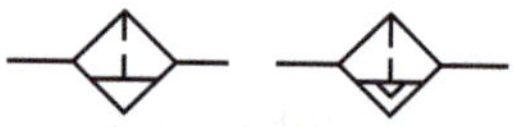	Flow meter(流量计)	
Oil filter(油雾分离器)	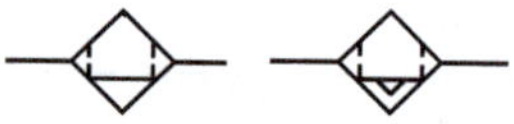	Pressure switch(压力继电器)	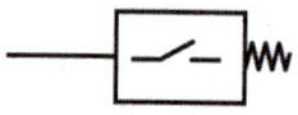
Air dryer(空气干燥器)	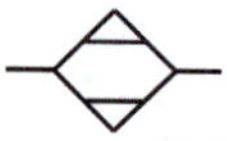	Silencer(消声器)	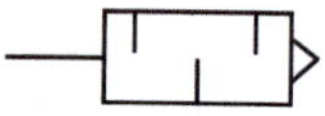
Oil mist lubricator(油雾器)	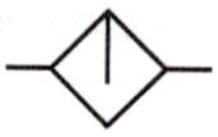	Hydraulic source(液压源)	
Air source adjusting device (气源处理装置)		Air pressure source(气压源)	
Cooler(冷却器)	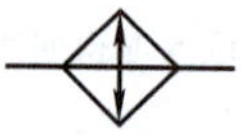	Motor(电动机)	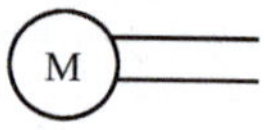
Heater(加热器)		Drive motor(原动机)	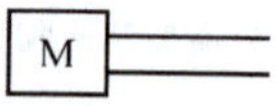
Receiver(蓄能器)		Air-hydraulic converting device (介质转换器)	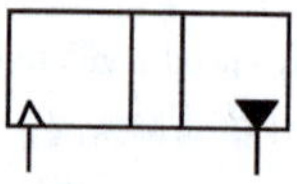

Appendix C Commonly Used Terminology in Hydraulic and Pneumatic Systems
液压与气动系统中常用的专业术语

Chapter 1 Introduction to Hydraulic Pneumatic Transmission 液压与气压概论

hydraulic cylinder	液压缸
Pascal's law	帕斯卡定律
reservoir	油箱
hydraulic jack	液压千斤顶
hydraulic pump	液压泵
flow rate	流量
oil pipe	油管
piston	活塞
heat exchanger	热交换器
outside load	外负载
pipe line	管道
energy portion	能源装置
actuator portion	执行元件
control element	控制元件
assistant element	辅助元件

Chapter 2 Fundament Hydraulic Fluid Mechanics 液压流体力学基础

bulk modulus of elasticity	体积弹性模量
compressibility	可压缩性
kinematic viscosity	运动粘度
relative viscosity	相对(条件)粘度
dynamic(absolute) viscosity	动力(绝对)粘度
°*E* viscometer	恩氏粘度计
viscometer	粘度计
hydrostatics	液体静力学
pressure	压力
vane pump	叶片泵
gear pump	齿轮泵
radial piston pump	径向柱塞泵
axial piston pump	轴向柱塞泵
screw pump	螺杆泵
absolute pressure	绝对压力
relative pressure	相对(表)压力
the principle of Pascal application	帕斯卡(静压传递)原理
equation of Bernoulli	伯努利方程
conservation of energy	能量守恒定律
equation of continuity	流量连续性方程
steady flow	恒定流动
equation of momentum	动量方程
conservation of momentum	动量守恒
conservation of mass	质量守恒
Venturi meter	文丘利流量计
mass flux	质量流量
cavitation	气穴
kinetic correction factor	动能修正系数
Reynolds number	雷诺数
critical Reynolds number	临界雷诺数
laminar flow	层流
turbulence flow	湍流
wetted perimeter	湿周
loss along circle parallel pipe	沿程压力损失
minor loss in pipe system	局部压力损失
cylinder spool orifice	圆柱滑阀阀口
orifice shape of poppet valve	锥阀阀口
Hydraulic lock	液压卡紧现象
hydraulic shock	液压冲击

Chapter 3 Hydraulic Pumps 液压泵

non-return valve	单向阀
plunger	柱塞
eccentric wheel	偏心轮
offset	偏心距
tight chamber	密闭容积
suction/discharge	吸油/压油
delivery	排量
flow-deploying	配流机构

self-priming suction	自吸能力
volume loss	容积损失
clearance	间隙
leakage	泄漏
suction pressure	吸入压力
working pressure	工作压力
rated pressure	额定压力
volumetric/mechanical/overall efficiency	容积/机械/总效率
power	功率
stroke	行程
theoretical/practical flow rate	理论/实际流量
sound level	噪声指标
fixed / variable delivery	定量/变量
transmission shaft	传动轴
thrust bearing	推力轴承
cylinder block	缸体
reaction ring	压环
sliding shoe	滑履
stator	定子
rotor	转子
controlling piston	控制活塞
load-sensitive variable displacement	负载敏感变量
swashplate	斜盘
bolzen	轴销
adjusting piston	变量活塞
hand wheel	手轮
variable displacement mechanism	变量机构
damping groove (or hole)	减振槽(孔)
bent axis axial piston pump	斜轴式无铰轴向柱塞泵
connecting rod	连杆
constant power variable displacement	恒功率变量
pressure-limiting variable displacement	限压式变量
pin	拨销
spring plate	弹簧座
dustproof ring	防尘圈
inlet port	吸油窗口
outlet port	压油窗口
spring ring	弹簧挡圈
port end cover	压盖
needle roller bearing	滚针轴承
key	键
seal support	密封座
seal ring	密封环
oil leakage path	泄油通道
decompression gap	卸荷沟
parallel pin	圆柱销
flow pulsation	流量脉动
oil film	油膜
phenomenon of surrounded oil	困油现象
gear modulus	齿轮模数
base pitch	基节
stationary crescent	月牙板

Chapter 4 Hydraulic Actuators
液压执行元件

crank shaft	曲轴
retaining ring	卡环
roller bearing	滚柱轴承
rolling group	滚轮组
rack-piston cylinder	齿条活塞液压缸
supercharge cylinder	增压缸
flange	法兰
glyd ring seal	格来圈密封
guide ring	导向环
cushion collar	缓冲套
cylinder tube	缸筒
O-Ring seal	O 形密封圈
buffer throttle valve	缓冲节流阀
step seal	斯特圈密封
retaining ring	护环
plastic composite type seal	橡塑组合密封
lip seal	唇形密封
cylinder block assembly	缸体组件
piston assembly	活塞组件
sealing device	密封装置
cushion device	缓冲装置
venting device	排气装置
bolted joint	螺栓联接
screw coupling	螺纹联接
inner semi-ring joint	内半环连接
thread coupling	螺纹联接
welding connection	焊接连接
roundwire snap ring connection	钢丝挡圈连接
nut	螺母
split pin	开口销
ring set	套环
tolerance	公差
venting plug	排气塞
creep	爬行
expand with heat and contract with cold	热胀冷缩
swing motor	摆动液压马达

Chapter 5 Hydraulic Control Valves 液压控制阀

solenoid operated valve	电磁控制阀
hydraulic operated valve	液控阀
gag bit	衔铁
pilot operated valve	先导阀
energized	（电磁铁）得电
de-energized	（电磁铁）失电
functions of neutral position	中位机能
three-position, four-way, solenoid operated valve	三位四通电磁换向阀
Eletro-hydraulic operated directional control valve	电-液换向阀
relief valve	溢流阀
reducing pressure valve	减压阀
sequence valve	顺序阀
servo valve	伺服阀
electro-hydraulic servo valve	电液伺服阀
electro-hydraulic proportional valve	电液比例阀
pressure control electro-hydraulic proportional servo valve	电液比例压力阀
flow control electro-hydraulic proportional servo valve	电液比例流量阀
flow sensor	流量传感器
electro-hydraulic proportional shift directional valve	电液比例换向阀
proportional solenoid	比例电磁铁
electro-hydraulic digital valve	电液数字阀
ball bearing screw	滚珠丝杠

Chapter 6 Auxiliary Components for Hydraulic Systems 液压辅件

position limited valve	限位阀
piston type	活塞/柱塞式蓄能器
bag type	囊式蓄能器
polytropic exponent	多变指数
adiabatic (isentropic) change	绝热状态
polytropic change	多变过程
adiabatic exponent	绝热指数
surface filter	表面型过滤器
depth filter	深度型过滤器
heat-exchanger	热交换器
suction oil pipe	吸油管
oil escape valve	放油塞
compound seals	组合密封圈
clip	卡套
lip style sealing	唇形密封圈
Y-style lip seal	Y 形密封圈
rubber pipe	胶管
pad	挡圈
V-style lip seal	V 形密封圈
rubber with proof against oil	耐油橡胶

Chapter 7 Basic Hydraulic Circuits 液压基本回路

pressure control circuit	压力控制回路
speed control circuit	速度控制回路
directional control circuit	方向控制回路
pressure regulated circuit	调压回路
pressure-venting circuit	卸载回路
pressure-reducing circuit	减压回路
pressure-increasing circuit	增压回路
pressure counter-balance circuit	平衡回路
pressure-holding circuit	保压回路
electric touch pressure gauge	电接触式压力表
displacement	排量
speed rigidity	速度刚性
outlet power	输出功率
torque	转矩
throttle speed-regulating circuit by fixed displacement pump	定量泵节流调速回路
speed-regulating circuit by variable displacement pump	变量泵容积调速回路
fast speed circuit	快速回路
shift oil flow direction circuit	换向回路
lock up circuit	锁紧回路
circuits for multi actuator control	多执行元件控制回路
sequence action circuit	顺序动作回路
synchronization circuit	同步回路
hands-off circuit	互不干扰回路

Chapter 8 Examples of Hydraulic Systems 液压系统实例

power-slipway	动力滑台
combined machine tools	组合机床
workbench	工作台
drilling	钻

enlarging	扩
reaming	铰
boring	镗
milling	铣
tapping	攻螺纹
clamping	夹紧
feeding	送料
plastic injection moulding machine	注塑机
cylinder for closing model-chamber	合模缸
nozzle	喷嘴
fixed /active-template	定/动模板
hydraulic motor for pre-plastic	预塑液压马达
cylinder for injection	注射缸
model-chamber	模腔
cylinder for push out product	顶出缸
four column universal press machine	四柱万能液压机
auxiliary pump	辅助泵
interlock function	互锁作用
cylinder for extending	液压缸伸出
cylinder for retracting	液压缸缩回
industry manipulator	工业机械手
peg	定位销
arm withdrew	手臂缩回
wrist turns around	手腕回转
finger stretch	手指张开

Chapter 9 Design of Hydraulic Transmission System 液压系统的设计

load chart	负载图
cycle view	循环图
start-up with acceleration	加速起动
fast moving onward	快进
working going onward	工进
brake	制动
fast return	快退
original position	原位
working situation	工况图
critical load	临界载荷

Chapter 10 Air Supply Devices and Pneumatic Components 气源装置与气动元件

air supply device	气源装置
air cylinder	气缸
air motor	气马达
pneumatic logical element	气动逻辑元件
pneumatic component	气动辅件
muffler	消声器
joint	接头
pneumatic sensor	气动传感器
signal processing device	信号处理装置
comparator	比值器
amplifier	放大器
pressure sensor	压力传感器
position sensor	位置传感器
air supply system	气源系统
air compressor	空气压缩机
aftercooler	后冷却器
oil and water separator	油水分离器
receiver	气罐
air dryer	干燥器
gas filter	过滤器
purification	净化
storage	贮存
diesel	柴油机
petrol engine	汽油机
gas filter for de-water	分水滤气器
oil mist lubricator	油雾器
filter rating	过滤度
water separator rating	水分离率
drainage valve	排水阀
pipeline connector	管道连接件
common cylinder	普通气缸
rodless cylinder	无杆气缸
diaphragm cylinder	膜片式气缸
impact cylinder	冲击气缸
proportional regulator	比例调节器
integral regulator	积分调节器
derivative regulator	微分调节器

Chapter 11 Basic Pneumatic Circuits and Their Application 气动基本回路及其应用

basic/common circuit	基本/常用回路
pressure control circuit	压力控制回路
high-pressure and low-pressure changeover circuit	高低压切换回路
over-load protection circuit	过载保护回路
force control circuit	力控制回路
force increased by multiple in-line cylinders	串联气缸增力回路

force increased by air-hydraulic intensifier 气-液增压器增力回路
impact cylinder circuit 冲击气缸回路
single/double-acting cylinder 单/双作用气缸
speed control circuit 速度控制回路
quick return circuit 快速返回回路
exhaust throttle valve 排气节流阀
buffer circuit 缓冲回路
air-hydraulic linkage speed control circuit 气-液联动速度控制回路
safety circuit 安全保护回路
two-hand operation circuit 双手操作回路
interlock circuit 互锁回路
counting circuit 计数回路

Chapter 12 Examples of Pneumatic Transmission Systems
气压传动系统实例

pneumatic manipulator 气控机械手
standard program 标准程序
obstruct 障碍
execute signal 执行信号
rack 齿条

References
参考文献

[1]James A Sullivan. Fluid Power: Theory and Applications[M]. 4th Ed. Columbus: Prentice Hall, 1998.
[2]Yeaple F. Fluid Power Design Handbook[M]. 2nd Ed. New York and Basel: Marcel Dekker Inc, 1990.
[3]Lambeck R P. Hydraulic pumps and motors: Selection and application for hydraulic power control system[M]. New York: Marcel Dekker Inc, 1983.
[4]雷天觉. 新编液压工程手册. 上、下册[M]. 北京:北京理工大学出版社,1998.
[5]许福玲,陈尧明. 液压与气压传动. 3 版[M]. 北京:机械工业出版社,2007.
[6]左健民. 液压与气压传动. 4 版[M]. 北京:机械工业出版社,2007.
[7]章宏甲. 液压与气压传动[M]. 北京:机械工业出版社,2005.
[8]袁承训. 液压与气压传动[M]. 2 版. 北京:机械工业出版社,2000.
[9]何存兴,张铁华. 液压与气压传动[M]. 2 版. 武汉:华中科技大学出版社,2000.

《液压与气压传动》(英汉双语)

陈淑梅　编著

读者信息反馈表

尊敬的老师：

您好！感谢您多年来对机械工业出版社的支持和厚爱！为了进一步提高我社教材的出版质量，更好地为我国高等教育发展服务，欢迎您对我社的教材多提宝贵意见和建议。另外，如果您在教学中选用了《液压与气压传动（英汉双语）》第2版（陈淑梅　编著），欢迎您对本书提出修改建议和意见。

授课教师如果需要多媒体电子教案，请填写信息反馈表，向主编发送电子邮件索取（smch2002@vip.sina.com）

一、基本信息

姓名：________　性别：________　职称：________　职务：________

邮编：______________________________　地址：______________________________

任教课程：________　电话________－________（H）________（O）

电子邮件：______________________________　手机：________________

二、您对本书的意见和建议

（欢迎您指出本书的疏误之处）

三、您对我们的其他意见和建议

请与我们联系：

100037·机械工业出版社·高等教育分社　余编辑　收

Tel：010-8837 9730

E-mail：yuhaohust@qq.com

http：//www.CmpEdu.com

http：www.CmpBook.com